Lecture Notes in Computer Science 15346

The series Lecture Notes in Computer Science (LNCS), including its subseries Lecture Notes in Artificial Intelligence (LNAI) and Lecture Notes in Bioinformatics (LNBI), has established itself as a medium for the publication of new developments in computer science and information technology research, teaching, and education.

LNCS enjoys close cooperation with the computer science R & D community, the series counts many renowned academics among its volume editors and paper authors, and collaborates with prestigious societies. Its mission is to serve this international community by providing an invaluable service, mainly focused on the publication of conference and workshop proceedings and postproceedings. LNCS commenced publication in 1973.

Vicente Julian · David Camacho · Hujun Yin ·
Juan M. Alberola · Vitor Beires Nogueira ·
Paulo Novais · Antonio Tallón-Ballesteros
Editors

Intelligent Data Engineering and Automated Learning – IDEAL 2024

25th International Conference
Valencia, Spain, November 20–22, 2024
Proceedings, Part I

Springer

Editors
Vicente Julian
Technical University of Valencia
Valencia, Valencia, Spain

David Camacho
Technical University of Madrid
Madrid, Spain

Hujun Yin
The University of Manchester
Manchester, UK

Juan M. Alberola
Universitat Politècnica de València
Valencia, Valencia, Spain

Vitor Beires Nogueira
University of Evora
Evora, Portugal

Paulo Novais
Universidade do Minho
Braga, Portugal

Antonio Tallón-Ballesteros
University of Huelva
Huelva, Spain

ISSN 0302-9743 ISSN 1611-3349 (electronic)
Lecture Notes in Computer Science
ISBN 978-3-031-77730-1 ISBN 978-3-031-77731-8 (eBook)
https://doi.org/10.1007/978-3-031-77731-8

This Springer imprint is published by the registered company Springer Nature Switzerland AG
The registered company address is: Gewerbestrasse 11, 6330 Cham, Switzerland

Preface

The International Conference on Intelligent Data Engineering and Automated Learning (IDEAL) is an annual international conference dedicated to emerging and challenging topics in intelligent data analytics and associated machine learning paradigms and systems.

The 25th edition, IDEAL 2024, was held in Valencia, Spain, on November 20–22. This edition represented an excellent opportunity for the scientific community to meet in person and to enhance the exchange of ideas and sharing of information. Moreover, Valencia, as the 2024 European Green Capital and a vibrant Mediterranean hub of innovation and culture, provided the perfect conditions to host a welcoming and successful event.

The IDEAL conference is designed as a global meeting point for researchers, fostering collaboration and the exchange of innovative ideas and insights. It provides an opportunity not only to present groundbreaking research but also to engage in meaningful discussions and share experiences that drive progress in these challenging scientific areas. The core themes of IDEAL 2024 included Big Data Analytics and Privacy, Machine Learning & Deep Learning for Real-World Applications, Data Mining and Pattern Recognition, Information Retrieval and Management, Bio- and Neuro-Informatics, and Hybrid Intelligent Systems and Agents.

In total 119 submissions were submitted. Through a single-blind peer review process, 57 papers were accepted for publication in the main track. The Programme Committee provided fundamental feedback and peer reviews to all the submissions.

In addition to the IDEAL 2024 main track, there were ten very interesting and relevant special sessions: Anomaly Detection with Machine Learning, Computational Intelligence for Imbalanced Classification, Computational Intelligence on Renewable Energies and Sustainable Automation, Data Selection in Machine Learning, Developing AI Curricula for Pre-University Education, Federated Learning, Intelligent Fusion and Applications (FLIFA), Example-Based Explainable Artificial Intelligence, Explainability and Fairness in Decision Support, Predictive and Prescriptive Models for Smart Cities Applications, and Quantum Computing for Machine Learning and Optimization (Q4ML-Opt). We would like to thank the Special Sessions Chairs for their invaluable efforts in organizing these sessions. We would also like to extend a special thank you to our esteemed keynote speakers – Pier Luigi Dragotti, Plamen Angelov, and Grzegorz Nalepa– for delivering their exceptional lectures.

We would like to thank all reviewers for their time and effort in reviewing the papers. The published papers have gone through a refinement process to accommodate discussions during the conference as well as comments from the reviewers who directed any necessary improvements. Also, we would like to thank all the members who participated and organizers in any way in the conference and especially the sponsors: Universitat Politècnica de València, School of Informatics, IEEE CIS UK and Ireland Chapter, and Springer LNCS.

We also extend our thanks to the organizers who managed the conference sessions, and all the authors who contributed to the scientific quality of the event. A final word to thank the local organising team at the Universitat Politècnica de València, who, with their work, made this event possible.

November 2024

Vicente Julián
David Camacho
Hujun Yin
Juan M. Alberola
Vitor Beires Nogueira
Paulo Novais
Antonio Tallón-Ballesteros

Organization

Conference Chairs

Vicente Julián	Universitat Politècnica de València, Spain
David Camacho	Universidad Politècnica de Madrid, Spain
Hujun Yin	University of Manchester, UK

Program Chairs

Juan M. Alberola	Universitat Politècnica de València, Spain
Vítor Beires Nogueira	University of Évora, Portugal
Paulo Novais	Universidade do Minho, Portugal
Antonio J. Tallón-Ballesteros	Universidad de Huelva, Spain

Local Chairs

Víctor Sánchez-Anguix	Universitat Politècnica de València, Spain
Jaume Jordan	Universitat Politècnica de València, Spain

Publications Chair

Pasqual Martí	Universitat Politècnica de València, Spain

Web Chair

Joan Ciprià Moreno	Universitat Politècnica de València, Spain

Steering Committee

Hujun Yin	University of Manchester, UK
Guilherme Barreto	Federal University of Ceará, Brazil
Jimmy Lee	Chinese University of Hong Kong, China
John Keane	University of Manchester, UK

Jose A. Costa	Federal University of Rio Grande do Norte, Brazil
Juan Manuel Corchado	University of Salamanca, Spain
Laiwan Chan	Chinese University of Hong Kong, China
Malik Magdon-Ismail	Rensselaer Polytechnic Institute, USA
Marc van Hulle	KU Leuven, Belgium
Ning Zhong	Maebashi Institute of Technology, Japan
Paulo Novais	University of Minho, Portugal
Peter Tino	University of Birmingham, UK
Samuel Kaski	Aalto University, Finland
Vic Rayward-Smith	University of East Anglia, UK
Yiu-ming Cheung	Hong Kong Baptist University, China
Zheng Rong Yang	University of Exeter, UK

Program Committee

Juan M. Alberola	Universitat Politècnica de València, Spain
Ricardo Aler	Universidad Carlos III de Madrid, Spain
Cesar Analide	University of Minho, Portugal
Jarvin Anton	Universidad de Sancti Spiritus Jose Marti, Cuba
Romis Attux	FEEC/UNICAMP, Brazil
Reyhan Aydoğan	Özyeğin Üniversitesi, Turkey
Clara Baltasar	Universidad Politécnica de Madrid, Spain
Bruno Baruque	Universidad de Burgos, Spain
Carmelo J. A. Bastos Filho	University of Pernambuco, Brazil
Riza Batista-Navarro	University of Manchester, UK
Vitor Beires Nogueira	University of Évora, Portugal
Fran Bellas	University of Coruna, Spain
Holger Billhardt	Universidad Rey Juan Carlos, Spain
Eduardo Borges	Universidade Federal do Rio Grande, Brazil
Lourdes Borrajo	University of Vigo, Spain
Juan A. Botia	University of Murcia, Spain
Vicent Botti	Universitat Politècnica de València, Spain
Jose Luis Calvo Rolle	University of A Coruña, Spain
David Camacho	Universidad Politécnica de Madrid, Spain
Hung Cao	University of New Brunswick, Canada
Roberto Carballedo	University of Deusto, Spain
Edianny Carballo	University of Ciego de Ávila, Cuba
Carlos Carrascosa	Universitat Politècnica de València, Spain
Luis Cavique	Universidade Aberta, Portugal
Songcan Chen	Nanjing University of Aeronautics and Astronautics, China

Ning Chen	Beijing Academy of Science and Technology, China
Stelvio Cimato	University of Milan, Italy
Manuel Cobo	University of Cadiz, Spain
Luís Conceição	Instituto Politécnico do Porto, Portugal
Roberto Confalonieri	Free University of Bozen-Bolzano, Italy
Juan M. Corchado	University of Salamanca, Spain
Paulo Cortez	University of Minho, Portugal
Raúl Cruz-Barbosa	Universidad Tecnológica de la Mixteca, Mexico
Sergio D'Antonio	Universidad Politécnica de Madrid, Spain
Andre de Carvalho	USP, Brazil
Fernando De la Prieta	University of Salamanca, Spain
Javier Díez-González	Universidad de León, Spain
Gracaliz Dimuro	Universidade Federal do Rio Grande, Brazil
Dinu Dragan	University of Novi Sad, Serbia
Nestor Duque Méndez	Universidad Nacional de Colombia, Colombia
Dalila Durães	Universidade do Minho, Portugal
Bapi Dutta	University of Jaén, Spain
Ziga Emersic	University of Ljubljana, Slovenia
Francisco Enguix	Universidad Politécnica de Valencia, Spain
Florentino Fdez-Riverola	University of Vigo, Spain
Alberto Fernandez	Universidad Rey Juan Carlos, Spain
Antonio Fernández-Caballero	Universidad de Castilla-La Mancha, Spain
Joao Carlos A. Ferreira	ISCTE-IUL, Portugal
Ruben Ferrero Guillen	Universidad de León, Spain
Cèsar Ferri	Universitat Politècnica de València, Spain
Felipe França	UFRJ, Brazil
Pedro Freitas	Universidade Católica Portuguesa, Portugal
Marcus Gallagher	University of Queensland, Australia
Diego Garces	Universidad Complutense de Madrid, Spain
Manuel Garcia-Piqueras	Universidad de Castilla-La Mancha, Spain
Isaías García-Rodríguez	Universidad de León, Spain
Diego Garcia-Zamora	University of Jaén, Spain
Javier Huertas	Universidad Politécnica de Madrid, Spain
María José Ginzo Villamayor	USC, Spain
Adrián Girón Jiménez	Universidad Politécnica de Madrid, Spain
Luis Gómez	Universidad de Las Palmas de Gran Canaria, Spain
Teresa Gonçalves	University of Évora, Portugal
Cesar Gonzalez	Universidad Politécnica de Valencia, Spain
Maciej Grzenda	Warsaw University of Technology, Poland
Mohamed Ali Hadj Taieb	University of Sfax, Tunisia

Eneko Osaba	TECNALIA, Spain
Cristina Padro-Ferragut	Universitat Politècnica de València, Spain
Javi Palanca	Universitat Politècnica de València, Spain
Jose Tomas Palma	University of Murcia, Spain
Peter Peer	University of Ljubljana, Slovenia
Marco Antonio Peña Cubillos	UNIA, Spain
Carlos Pereira	ISEC, Portugal
Yinela Pérez	Universidad de Holguín, Cuba
Iñigo Perez Delgado	Ayesa, Spain
Laura Po	University of Modena and Reggio Emilia, Italy
Alessandro Sebastian Podda	University of Cagliari, Italy
Radu-Emil Precup	Politehnica University of Timisoara, Romania
Abraham Prieto García	Universidade da Coruña, Spain
Akash Punhani	SRM Institute of Science and Technology Delhi, India
Paulo Quaresma	University of Évora, Portugal
Francisco Quesada Real	University of Jaén, Spain
Julio Quintana	Universidad de Ciego de Ávila, Cuba
Bruno Ramos Cruz	University of Jaén, Spain
Izabela Rejer	West Pomeranian University of Technology in Szczecin, Poland
Jaime Andres Rincon Arango	Universidad de Burgos, Spain
Elsa Rodrigues	Instituto Politécnico de Beja, Portugal
Sara Rodríguez González	University of Salamanca, Spain
Alejandro Romero	Universidade da Coruña, Spain
Ruben Ruiz-Gonzalez	Universidad de Burgos, Spain
Victor Sanchez-Anguix	Universitat Politècnica de València, Spain
María Inmaculada Santamaría	Universidad Politécnica de Madrid, Spain
Helida Santos	FURG, Brazil
Ricardo Santos	Escola Superior de Tecnologia e Gestão, Portugal
Matilde Santos Peñas	Universidad Complutense de Madrid, Spain
José Santos Reyes	University of A Coruña, Spain
Yuya Sasaki	Osaka University, Japan
Markus Schatten	University of Zagreb, Croatia
Marcin Szpyrka	AGH University of Science and Technology, Poland
Antonio J. Tallón-Ballesteros	University of Huelva, Spain
Joaquin Taverner	Universitat Politècnica de València, Spain
Nikolay Tcholtchev	Fraunhofer FOKUS, Germany
Jan Arne Telle	University of Bergen, Norway
Murat Testik	Hacettepe University, Turkey
Darian Tomašević	University of Ljubljana, Slovenia

Stefania Tomasiello	University of Salerno, Italy
Alexandros Tzanetos	Jönköping University, Sweden
Eiji Uchino	Yamaguchi University, Japan
Eva Vallada	Universitat Politècnica de València, Spain
Alfredo Vellido	Universitat Politècnica de Catalunya, Spain
Paula Verde	Universidad de León, Spain
María Fulgencia Villa	Universitat Politècnica de València, Spain
Matej Vitek	University of Ljubljana, Slovenia
Tzai-Der Wang	Cheng Shiu University, Taiwan
Agnieszka Wendland	Warsaw University of Technology, Poland
Michal Wozniak	University of Wrocław, Poland
Xin-She Yang	Middlesex University London, UK
Hua Yang	Zhongyuan University of Technology, China
Raciel Yera	University of Jaén, Spain
Hujun Yin	University of Manchester, UK

Special Session on Predictive and Prescriptive Models for Smart Cities Applications

Organizers

Vicent Botti	Universitat Politècnica de València, Spain
Jose Machado	Universidade do Minho, Portugal
Victor Sanchez-Anguix	Universitat Politècnica de València, Spain

Special Session on Computational Intelligence on Renewable Energies and Sustainable Automation

Organizers

J. Enrique Sierra García	University of Burgos, Spain
Matilde Santos Peñas	Complutense University of Madrid, Spain
Fares M'zoughi	University of the Basque Country, Spain

Special Session on Example-Based Explainable Artificial Intelligence

Organizers

Cesar Ferri	Universitat Politècnica de València, Spain
Jan Arne Telle	University of Bergen, Norway

Special Session on Explainability and Fairness in Decision Support

Organizers

Luis Martínez	University of Jaèn, Spain
Bapi Dutta	University of Jaèn, Spain
Raciel Yera	University of Jaèn, Spain

Special Session on Federated Learning, Intelligent Fusion and Applications (FLIFA)

Organizers

Cedric Marco-Detchart	Universitat Politècnica de València, Spain
Jaime A. Rincon Arango	Universidad de Burgos, Spain
Vicente Julián	Universitat Politècnica de València, Spain
Carlos Carrascosa Casamayor	Universitat Politècnica de València, Spain
Paulo Novais	Universidade do Minho, Portugal
Carlos Lopez-Molina	Universidad Pública de Navarra, Spain
Laura de Miguel Turullols	Universidad Pública de Navarra, Spain
Giancarlo Lucca	Universidade Católica de Pelotas, Brazil
Graçaliz P. Dimuro	Universidade Federal do Rio Grande, Brazil
Francisco Enguix Andrès	Universitat Politècnica de València, Spain

Special Session on Quantum Computing for Machine Learning and Optimization (Q4ML-Opt)

Organizers

Eneko Osaba	TECNALIA Research and Innovation, Spain
Esther Villar-Rodriguez	TECNALIA Research and Innovation, Spain
Izaskun Oregi	TECNALIA Research and Innovation, Spain

Special Session on Anomaly Detection with Machine Learning

Organizers

Jason J. Jung	Chung-Ang University, South Korea
Gen Li	Chengdu University, China
Longlong Yu	Chengdu University, China

Special Session on Developing AI Curricula for Pre-University Education

Organizers

Abraham Prieto García	Universidade da Coruña, Spain
Paulo Novais	Universidade do Minho, Portugal
Dalila Duraes	Universidade do Minho, Portugal
Peter Peer	University of Ljubljana, Slovenia
Žiga Emeršić	University of Ljubljana, Slovenia

Special Session on Data Selection in Machine Learning (6th Edition)

Organizers

Antonio J. Tallón-Ballesteros	University of Huelva, Spain
Ireneusz Czarnowski	Gdynia Maritime University, Poland

Special Session on Computational Intelligence for Imbalanced Classification

Organizers

Wenbin Pei	Dalian University of Technology, China
Bing Xue	Victoria University of Wellington, New Zealand
Antonio J. Tallón-Ballesteros	University of Huelva, Spain

Contents – Part I

Main

Contents – Part II

Main

Special Session on Computational Intelligence on Renewable Energies and Sustainable Automation

Special Session on Example-Based Explainable Artificial Intelligence

Special Session on Explainability and Fairness in Decision Support

Special Session on Federated Learning, Intelligent Fusion and Applications (FLIFA)

Special Session on Anomaly Detection with Machine Learning

Special Session on Developing AI Curricula for Pre-University Education

Special Session on Data Selection in Machine Learning (6th Edition)

Special Session on Computational Intelligence for Imbalanced Classification

Main

Quantitative Estimation of Reputation Risk

Peter Mitic[(✉)]

Department of Computer Science, UCL, Gower Street, London WC1E 6BT, UK
p.mitic@ucl.ac.uk
https://www.ucl.ac.uk/computer-science/ucl-computer-science

Abstract. We present a quantitative definition of reputation risk, formulated in terms of a reputation time series comprising daily sentiment measurements. Self Supported Learning is used to quantify reputation risk by progressively refining an initial proposal for a Minimum Acceptable Sentiment, calculated from descriptive statistics of the reputation data. The derived values are validated using a "sense test" based on a Loess quantile. The results show that the Minimum Acceptable Sentiment value is given approximately by a two standard deviation lower tail of the observed data.

Keywords: Reputation · Sentiment · Time Series · Descriptive Statistics · Value-at-Risk · Normal Mixture Distribution · CNN · Loess · Self Supervised Learning · SSL · NLP

1 Introduction and Motivation

Reputations, expressed as time series of real-valued sentiment measurements, have unique characteristics because sentiment measures are compiled starting at zero (neutral) every day. They have distinctive patterns of successive peaks and troughs, joined by steep upward and downward sentiment movements. Informally, they could be described as "choppy".

Given that low reputation is undesirable from a corporate viewpoint, a natural question to ask is "How low could reputation go?". That question should be understood in the context of a company risk manager who is faced with a sequence of falling sentiment measurements. He/She has two choices: either to do nothing and hope that sentiment recovers, or to take remedial action to boost sentiment. The analysis presented in this paper addresses the question of what sentiment level should be regarded as the limit at which to take action. Therefore we aim to define a definition for *Reputation Risk* (hitherto ill-defined), and a *Minimum Acceptable Sentiment* limit, such that action can be triggered if sentiment dips below that limit. It should be set "low, but not too low".

V. Julian et al. (Eds.): IDEAL 2024, LNCS 15346, pp. 3–14, 2025.
https://doi.org/10.1007/978-3-031-77731-8_1

2 Background and Related Work

2.1 Opinion, Sentiment and Reputation

Formal definitions for *opinion*, *sentiment* and *reputation* were presented in [11]. Briefly, *opinion* represents a view or judgement on a piece of text aimed at an entity T at time t, then mapped to a subset of the real numbers. *Sentiment*, $S_t(T)$, is a combination of opinions aimed at a common entity, T. The *Reputation* of T, $\{y_t(T)\}$, is a time series of *Sentiments* over an extended time period, and forms the basis of further analysis in this paper. It differs from previous qualitative definitions, which were formulated in terms of stakeholder perceptions and beliefs, and incorporate words such as "quality", "standards" and "trustworthiness". Dowling [5] notes fifty such definitions, none of which agree!

In Sect. 3.1, *Reputation Risk* (abbreviated to *RepRisk*) will be formulated in probabilistic terms using a proposed limiting value of $\{y_t(T)\}$, which we will call the *RepRisk Limit.* It represents a *Minimal Acceptable Sentiment.* A *Self Supported Learning* (*SSL*) procedure will be proposed wherein an initial estimation of *RepRisk* using descriptive statistics of the data is progressively refined in an iterative process. *SSL* obviates the need for partitioning the data into *training* and *test* sets by repeatedly using predictions on the entire data set in every *SSL* iteration.

2.2 Previous Work

Commercial organisations are reluctant to publish specific details of their methods, especially if they concern *Natural Language Processing* (*NLP*) techniques. This review therefore concentrates on academic studies of *reputation.*

In the context of reputational analysis, early objective approaches to sentiment analysis concentrated on using *NLP.* In that context, *reputation* and *sentiment* were used synonymously. Yang et al. [17] analysed online *Twitter (X)* comments on large corporates (the *RepLab 2012* project), with the *SentiWordNet* corpus. The same corpus was used by Li [8], whereas Colleoni et al. [4] used the *Neilson* corpus. In the context of corporate sentiment analysis, using social media is not ideal because input from the financial press is omitted. More recent and relevant sources of financial text data include *Reuters*, *Bloomberg* and *LSEG* (www.lseg.org).

Subjective user ratings (for example for consumer products) have been used frequently because they are easy to download from websites such as *Google* and *Amazon.* Alqwadri et al. [1] regarded *reputation* as synonymous with consumer ratings weighted by consumer characteristics such as activity frequency, reliability and experience. The weighting model used by Song, Lau and Xia [15] to investigate fraudulent e-markets used probabilities of fraudulent origin, calculated using characteristics of buyers and sellers in a logistic regression model. Rezvani and Rezvani [14] used weighted ratings in an iterative algorithm, calculating weights at each stage using a random sample of respondents.

In the discussion by Trcak et al. [16] the term *Reputation Risk* is used frequently, but is undefined. *Reputation* is formulated only for exchange-quoted corporates as *StockValue * SharesOutstanding - BookValue*. Effectively, *reputation* is then synonymous with *share price*. The approach is problematical because share price is often affected by other forces, and also because the approach is void for companies that have no share price (for example *IKEA* or *LIDL*).

A further common thread is manual sentiment labelling. Rantanen et al. [13] used manual classification of online opinions in terms of opinion holder reliability, with a *Convolutional Neural Network* (*CNN*) predictor. Klein et al. [7] used manual document labelling with *NLP* analysis of frequency counts of unigrams, and a *Support Vector Machine* predictor to derive numerical sentiment scores. A different approach is taken by Banguia et al. [3]. Based on a small data set of subjective views, a generative method is used to construct a much larger synthetic data set. A further deep learning model is then used for validation on actual data.

Overviews of *SSL* may be found in [6,9], or [12]. The "cookbook" approach of Balestriero et al. [2] clarifies many of the *SSL* subtleties.

3 *RepRisk* Estimation

In the absence of labels for instances of *Minimum Acceptable Sentiment* (such as *acceptable* and *unacceptable*), we adopt the iterative strategy of *Self-Supervised Learning* (*SSL*) to estimate *minimum acceptable sentiment*. An initial approximation using the distributional properties of the data acts as a *Pretext Task*, and yields *Pseudo-labels*. Intuitively, the initial estimate should be in the lower range of the historic sentiment data, but greater than the minimum value. The first approximation is progressively refined, using the previous refinement as the current objective. Finally, a way to replicate the results of that relatively complex process using simple distributional statistics of the data is found.

3.1 *RepRisk* Definition

Using the notation in Sect. 2.1 ($S_t(T)$ for sentiment and $\{y_t(T)\}$ for reputation), *Reputation Risk* is the probability that *sentiment* on any given day falls below the *RepRisk Limit* in a time period τ, $L(T,\tau)$. The precise specification of $L(T,\tau)$ will be deferred until a later stage (Sect. 4.5), following analysis of results. In Eq. 1, the symbol $\#(arg)$ denotes the cardinality of the argument.

$$R(T) = \Big\{P(y_t(T) < L(T,\tau))\Big\} = \Big\{\frac{\#(S_t(T) < L(T,\tau))}{\#(S_t(T))}\Big\} \quad t \in 1..\tau \tag{1}$$

The limit value $L(T,\tau)$ is intended to be the reputation equivalent of the well-known financial risk metric *Value-at-Risk* (*VaR*), but using the lower distribution tail instead of the upper.

3.2 *RepRisk* Measurement

In the context of reputation, there is no immediate objective with which to measure *RepRisk*. Therefore, *SSL* is used in conjunction with statistical features of reputation. The algorithm in Sect. 3.5 encapsulates that process. It incorporates an initial "label generation" stage based on 25% sentiment quantiles. Thereafter the *SSL* proceeds to "learn to represent" the *RepRisk Limit* by progressive iterations to refine the initial estimate. The iterative concept is similar to the rating weight calculation proposed by Rezvani and Rezvani [14].

Once completed, the *RepRisk Limit* values can be compared with easily calculated statistics of the data to derive a simple equivalent that is derived from the complex *SSL* calculations, and therefore has an objective basis. Ideally, for any given organisation, values of *RepRisk Limit* should be greater than minimum sentiment values.

3.3 Data and Pre-processing

Reputation time series for 130 corporate organisations, representing the principal world industrial and service sectors, were sourced from *Penta Group*[1]. Each ranged from July 2021 to June 2023 (730 days). Daily sentiments are recorded on a real-valued scale from −100 (the worst possible) to +100 (the best possible). Zero (or very near to zero) represents neutral sentiment.

For each data set, the following principal statistics were calculated and standardised to mean 0 and variance 1.

- *Mean, SD, Skewness, Kurtosis, Maximum, Minimum, Sum*
- *Sum and count of positive/negative sentiments*
- Quantiles *Q25* and *Q75*, and the median
- Gradient and sum of residuals for a linear fit against day
- Gradient and mean of cumulative sentiment residuals, fitted against day
- Mean up-move, mean down-move, turning point frequency and mean length of 'flat' period

3.4 Pre-requisites

The primary pre-requisite for the analysis that follows is an appropriate probability distribution for reputation time series. The probability distribution is used to extract a stable 25% quartile that is not directly extracted from empirical data. A *Normal* distribution is suitable for many, but a *Normal Mixture* distribution, with parameters $\{\mu_1, \sigma_1, \mu_2, \sigma_2, b\}$, is usually a better fit. Its density ϕ_M is shown in Eq. 2, in which $\phi(x)$ is the Normal density function, with argument x in the range $[-100, 100]$. Full details may be found in [11]. Goodness-of-fit was measured using the *TNA* test [10], which is robust with respect to data set size and outliers.

[1] https://pentagroup.co.

$$\phi_M(x, \mu_1, \sigma_1, \mu_2, \sigma_2, b) = b\phi(x, \mu_1, \sigma_1) + (1 - b)\phi(x, \mu_2, \sigma_2) \tag{2}$$

Each stage in the *SSL* iteration uses significant feature selection based on the current predicted *RepRisk* estimate. The most effective and direct way to quantify feature importance was *Successive Feature Subtraction*, in which successive features are removed, and a linear model is used to calculate a predictor and prediction error. Feature ranking is achieved by ordering features by prediction error. Features representing, proportionately, more than 75% of the sum of residuals are retained at each *SSL* iteration, with a minimum retention of 5.

A relaxation factor r_i is used to amend the risk estimate at each *SSL* iteration i to prevent divergence of the estimate. First, a ratio, λ, of the deviations of the maximum and minimum empirical data respectively from the minimum possible sentiment, -100, is calculated. Values of λ are therefore in the range [0, 2], and ensure that the final risk estimates do not differ markedly from the lower quartile boundary of the data. All relaxation factors use λ in an exponential factor which is applied only to the components of the current risk estimate that are greater than the minimum element of the current risk estimate R_i (Eq. 3 and step 3(d) in the Risk Estimation algorithm). The effect of the relaxation factor is to 'encourage' a downward trend for the current *RepRisk Limit*. The symbol $\mathbb{I}$ denotes an indicator function.

$$\begin{cases} \lambda = \frac{\max(D) - (-100)}{\min(D) - (-100)} \\ r_i = e^{-i/\lambda} \quad (i = 1, 2, ...) \\ R_i \rightarrow R_i - r_i \mathbb{I}(R_i > \min(R_{i-1})) \quad (i = 1, 2, ...) \end{cases} \tag{3}$$

3.5 Risk Estimation Algorithm

The algorithm generates a sequence of risk estimates $\{R_1, R_2, ...\}$ using relaxation factors r_i (Eq. 3), where i is the step number in the iterative process in the algorithm below. The sequence length is determined by a convergence criterion on successive error terms $|E_{i+1} - E_i| < 1$. The values of the E_i are typically in the range 10–20, and convergence is usually achieved after 10–15 iterations. The *CNN* is intended to replicate the characteristic patterns in reputational time series. The algorithm is stochastic, so it has to be applied multiple times, and predictions averaged, to obtain consistent results.

Algorithm 1

---

1. Initialise
   (a) For each data set D spanning times $1..\tau$:
      i. Fit a *Normal Mixture* distribution

   ii. Extract the 25% quantile
   (b) Select important features from all features
   (c) Use a *CNN* to estimate the initial risk $R_1 = CNN(D)$ at times $1..\mathcal{T}$
2. Set initial error indicator $E_1 = 0$ and iteration count: $i = 0$
3. Repeat until converged
   (a) Increment risk count: $i \to i + 1$
   (b) Reselect important features, min 5
   (c) Calculate *CNN* predictions: $R_{i+1} = CNN(R_{i+1})$ at times $1..\mathcal{T}$:
   (d) Calculate and apply a relaxation factor to R_{i+1} (Equation 3)
   (e) Calculate relative error $E_{i+1} = \frac{R_{i+1} - R_i}{R_i}$
   (f) Test convergence: is $|E_{i+1} - E_i| < 1$?
   (g) If converged, exit loop and record final step count, I
4. Return the final risk estimate using a linear predictor: $\hat{R} = LIN(R_I)$

3.6 Validation

Since there is no *a priori* objective function (i.e. the value of *RepRisk Limit*) in the algorithm outlined in Sect. 3.5, we seek a simple, but appropriate, way to validate the results of applying the algorithm. The result should "make sense" in the context of each reputation time series. The following strategy preserves the essential features of a reputation time series, is quick and easy to apply, and moderates outliers.

1. Smooth the reputation time series using *Loess* to minimise 'noise' and mitigate the effect of outlier 'spikes'.
2. Extract the minimum value on the smoothed time series.

Risk estimates can then be compared with their corresponding validation estimates on an organisation-wise basis using a t-test for difference of means. The null hypothesis for the theoretical mean difference μ is $\mu = 0$, and the alternative hypothesis is $\mu \neq 0$. For an observed mean m and an observed variance s^2 with sample size n, $\frac{(\mu - m)}{s/\sqrt{n}} \sim t(n-2)$. Acceptance of the null hypothesis signals agreements of the base risk estimates with the validations (i.e. the risk estimate method "makes sense").

4 Results

A general overview of *RepRisk* values is followed by an analysis of instances of *RepRisk* breaches, and then by validations.

4.1 Overall Reputation Risk

The numerical value of the *RepRisk Limit* depends heavily on the value of λ (step 3(d) of the algorithm in Sect. 3.5). The value of λ is data set-dependent, although the possibility exists to use a common value, derived using all data sets, throughout. The results in this section are for the data set-dependent $\lambda-$values. The algorithm is stochastic, and for consistency, the results shown are averages of 10 independent runs.

The distribution of *RepRisk Limits* is shown in Fig. 1, with summary statistics. It shows a concentration in the upper quartiles, with a significant number in the inter-quartile range. Notably, only 19.2% of *RepRisk Limits* are positive.

Figure 2 shows a comparison of the lower quartile *sentiment* and the *RepRisk Limit* for each organisation, calculated using Eq. 1. The risk estimates are shown in increasing order. The corresponding median and maximum *sentiments* in each case is also shown for comparison. The plot shows that for a majority of organisations (86.9%), the *RepRisk Limit* is greater than minimum *sentiment*. The *RepRisk Limit* lies below the lower quartile in all cases, showing that the effect of the Sect. 3.5 algorithm is to reduce the estimate relative to initial values, but not as far as the minimum value.

Summary statistics	
Minimum	-71.8
10% Quartile	-49.4
25% Quartile	-40.2
Median	-23.6
75% Quartile	-4.5
90% Quartile	7.3
Maximum	13.8
Mean	-22.5
SD	21.4

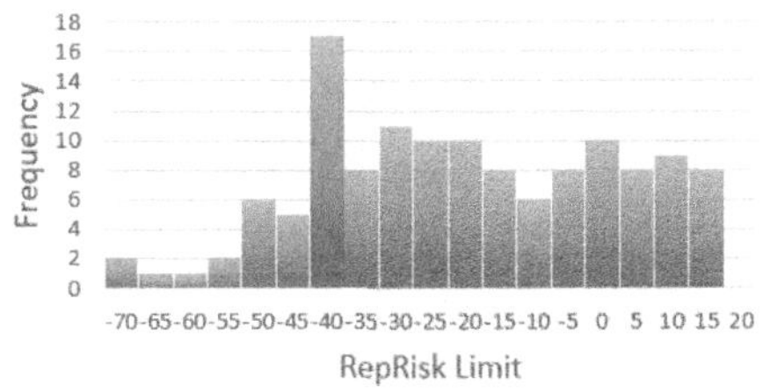

Fig. 1. *RepRisk Limit* distribution and summary statistics

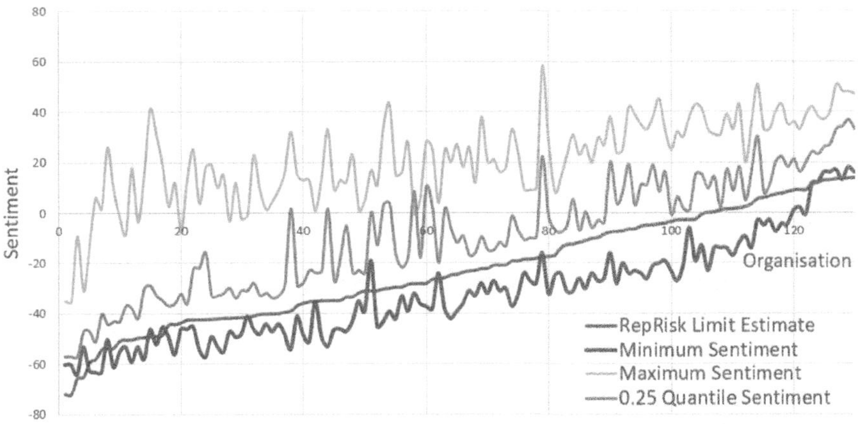

Fig. 2. *RepRisk Limit* estimates using Algorithm 1, ordered smallest to largest, with corresponding minimum, 25% quantile, and maximum sentiments per organisation.

4.2 Reputation Risk Breaches

Table 1 shows the percentage of organisations for which the calculated *RepRisk Limit* (calculated using the Sect. 3.5 algorithm) is greater than specified (low) boundary. Only a small percentage have a positive *RepRisk Limit*: a result that should be seen in relation to the mean sentiment for all data: −2.45. The lower quartile for all data was −20.00, and fewer than half of organisations had a *RepRisk Limit* greater than that value. The lower 5% 2-tail confidence bound for all data was −24.7 (which is lower tail *VaR* at 5%), exceeded by approximately only 50% of *RepRisk Limits.*

Table 1. Reputation Risk breaches: percentage of organisations for which *RepRisk Limit* is greater than specified boundary

Sentiment Boundary	% with *RepRisk* Limit > Boundary
10	0
5	13.1
0	19.2
−5	26.9
−10	33.1
−15	37.7
−20	43.8
−25	51.5

4.3 Feature Significance

The features introduced in Sect. 2.1 were selected on the basis that they might reasonably characterise individual reputation time series. Part 1(b) of the initialisation stage of the algorithm (Sect. 3.5) was to filter out non-significant features (i.e. those that have minimal affect the outcome of the *RepRisk* calculation). Applying the *Successive Feature Subtraction* method outlined in Sect. 3.4 resulted in retention of the features indicated in Table 2. The features listed represent 95% of the linear predictor variance. With each iteration of the algorithm in Sect. 3.5, the number of features diminishes (to a minimum 5).

Figure 3 shows correlations between the same features. The figure indicates that features with significant correlations are retained. It was found that removing highly correlated features triggered errors in the ensuing neural network analysis due to insufficient data.

Table 2. Most significant features (of those listed in Sect. 3.3)

Feature	Importance	
	Absolute	Cumulative
Median	14.936	0.357
Cumulative Gradient	5.155	0.48
Gradient	3.81	0.571
Standard Deviation	3.654	0.658
Turning Point Frequency	3.474	0.741
Maximum	3.388	0.822
Quantile Q25	2.733	0.887
Mean Up-Move	1.313	0.918
Mean Down-Move	0.944	0.941

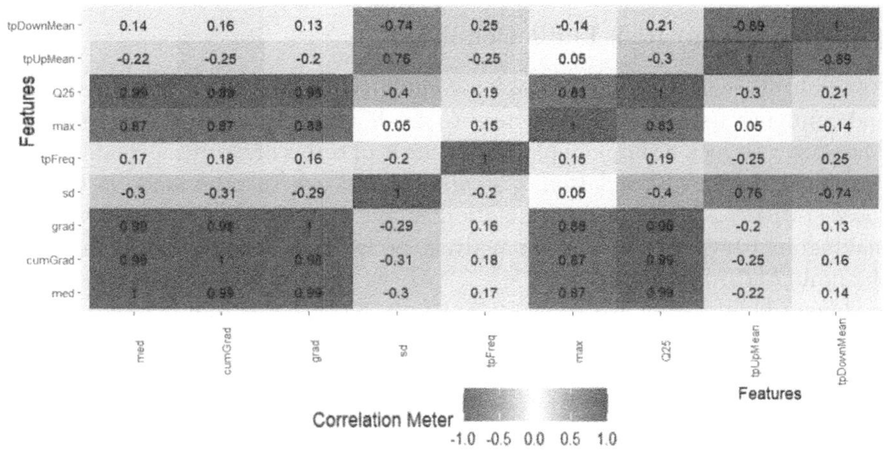

Fig. 3. Features Correlation matrix. The order of importance is from left to right, and from down to up.

4.4 Validation Results

The mean *Loess* (*minimum acceptable sentiment*) risk value was −23.17, which is similar to the mean *RepRisk Limit* value of −22.71, calculated using the Sect. 3.5 algorithm. Their standard deviations are similar: 22.8 for *Loess* and 20.9 for *RepRisk*. Figure 4 shows the *RepRisk* estimates with corresponding *Loess* validations.

The calculated value of t (Sect. 3.6) for the difference of means was 0.716 (with p-value 0.762, 128 degrees of freedom). The null hypothesis $\mu = 0$ is therefore not rejected. The conclusion is that the *Loess* validation method is in agreement with the proposed *RepRisk Limit* calculation.

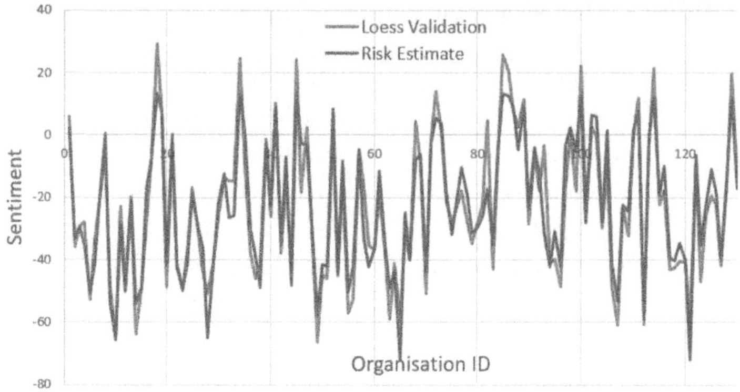

Fig. 4. Comparison of *RepRisk* estimates with corresponding *Loess* (*span* = 0.15) validations.

4.5 Reputation Risk Limit Definition

Applying the algorithm in Sect. 3.5 for any given data set is a lengthy process, and applying its output parameters to fresh data requires data preparation. Therefore, we seek a simple replacement for it in terms of easily calculable statistics of the data. A formulation of *RepRisk Limit* can then be made in terms of those statistics. Such a post-formulation is unusual, but it is consistent with not making an arbitrary decision pre-analysis. So we now complete the definition of the limit $L(T, \tau)$ in Eq. 1.

Quantile and standard error statistics were considered, and compared with the algorithm-derived *RepRisk Limits* using t-test for mean difference. After some investigation, the 0.04 quantile ($t = 0.975$; $p = 0.835$) proved to be optimal. A 95% two-tailed standard error bound ("Mean - 1.975*SD") was also acceptable ($t = 1.033$; $p = 0.848$). Extreme quantiles are prone to instability. For example, the 0.05 and the 0.025 quantiles both gave $t = 3.36$, whereas the 0.03 quantile gave $t = 0.971$. Therefore we prefer the more stable 95% two-tailed standard error metric, and define the *RepRisk Limit*, $L(T, \tau)$, for an organisation T with sentiment mean $m(\tau)$ and variance $s^2(\tau)$ measured in a time period τ in Eq. 4.

$$L(T, \tau) = m(\tau) - 1.975\, s(\tau) \tag{4}$$

So defined, $L(T, \tau)$ is the equivalent of 2-tail financial *VaR* at 95%, but using the lower tail.

5 Discussion

A general point to stress is that by monitoring reputation, risk managers can make better decisions [7], with a view to maintaining or boosting their reputation. To that end, our analysis, at its simplest, shows that only a minority of organisations have a positive *RepRisk Limit*: an undesirable state. Arguably, *Minimum Acceptable Sentiment should* be zero, on the basis that any negative sentiment is unacceptable. One could argue that it is not possible to completely eliminate negative sentiment. In that case, a negative *RepRisk Limit* should regarded as a 'loss', similar to a stock market loss, that could be recovered. A *Minimum Acceptable Sentiment* that is less than zero, makes sense in the context of organisations whose reputation is largely negative, but not across all organisations. The lower quartile from Sect. 4.2, −20, intuitively seems too low, but given the range of many reputations, becomes more realistic than zero.

6 Conclusions and Future Work

The final output of the *SSL* process in Sect. 3.5 is a very simple statistical formula, Eq. 4, to estimate a *RepRisk Limit*. It is easy to apply, and gives day-by-day feedback to risk managers, who are vary comfortable with the financial equivalent, *VaR* at 95%. The entire process could then be incorporated into a real-time risk measurement platform, in which remote clients would access a presentation layer linked to a data-mining and calculation engine.

There are possibilities for further work. First, the *SSL* approach as it stands does not take into account the effects of varying the value of τ, or, more generally, the period spanned by the reputation data. Variations using a moving window could serve to stabilise the result. Secondly, a link to predictive techniques (for example *ARIMA* as in [11]) could be investigated. As an alternative approach, preliminary trials using *Fourier* analysis to eliminate noise and detect periodicity have produced promising results. The idea would be to use a severely smoothed Fourier approximation to a reputation time series, detect local minima and calculate their mean value. Fine tuning would be required.

References

1. Alqwadri, A., Azzeh, M., Almasalha, F.: Application of machine learning for online reputation systems. Int. J. Autom. Comput. **18**, 492–502 (2021). https://doi.org/10.1007/s11633-020-1275-7
2. Balestriero, R., Ibrahim, M., et al.: A cookbook of self-supervised learning (2023). arXiv:2304.12210v2
3. Banguia, H., Ge, M., Buhnova, B.: Deep-learning based reputation model for indirect trust management. In: Proceedings of ANT 2023, Leuven (2023). Procedia Comput. Sci. **220**, 405–412. https://doi.org/10.1016/j.procs.2023.03.052
4. Colleoni, E., Arvidsson, A., Hansen, L.K., Marchesini, A.: Measuring corporate reputation using sentiment analysis. In: 15th International Conference on Corporate Reputation, New Orleans (2011). https://research-api.cbs.dk/ws/portalfiles/portal/58852755
5. Dowling, G.R.: Defining and measuring corporate reputations. Eur. Manag. Rev. **13**(3), 207–223 (2016). https://doi.org/10.1111/emre.12081
6. Jaiswal, A., Babu, A.R., Zadeh, M., Banerjee, D., Makedon, F.: A survey on contrastive self-supervised learning. Technologies **9**(1), 2 (2021). https://doi.org/10.3390/technologies9010002

7. Klein, A., Riekert, M., Dinev, V.: Accurate retrieval of corporate reputation from online media using machine learning. In: Proceedings of FedCSIS, Leipzig, vol. 18, pp. 43–46 (2019). https://doi.org/10.15439/2019F169. https://wi2.uni-hohenheim.de/analytics
8. Li, S.: Short or long review? - Text analytics and machine learning approaches to online reputation. IJBMR **9**(1), 28–40 (2021). https://doi.org/10.37391/IJBMR.090105
9. Liu, X., Zhang, F., Hou, Z., et al.: Self-supervised learning: generative or contrastive. IEEE Trans. Knowl. Data Eng. **35**(1), 857–876 (2023). https://doi.org/10.1109/TKDE.2021.3090866
10. Mitic, P.: Improved goodness-of-fit measures. J. Oper. Risk **10**(1), 77–126 (2015). https://doi.org/10.21314/JOP.2015.159
11. Mitic, P.: Reputation, sentiment, time series and prediction. In: Proceedings of DATA2024, Dijon, pp. 71–81 (2024). https://doi.org/10.5220/0012763100003756
12. Raina, R., Battle, B., Lee, H., Packer, B., Ng, Y.: Self-taught learning: transfer learning from unlabeled data. In: Proceedings of ICML 2007, pp. 759–766 (2007). https://doi.org/10.1145/1273496.1273592
13. Rantanen, A., Salminen, J., Ginter, F., Jansen, B.J.: Classifying online corporate reputation with machine learning: a study in the banking domain. Internet Res. **30**(1), 45–66 (2020). https://doi.org/10.1108/INTR-07-2018-0318
14. Rezvani, M., Rezvani, M.: A randomized reputation system in the presence of unfair ratings. ACM Trans. Manag. Inf. Syst. **11**(2), 1–16 (2020). https://doi.org/10.1145/3384472
15. Song, L., Lau, R., Xia, Y.-Q.: A novel fraudulent transaction detection model for enhanced reputation management at e-markets. In: International Conference on Machine Learning and Cybernetics, Xi'an, China, pp. 2013–2018 (2012). https://doi.org/10.1109/ICMLC.2012.6359685
16. Trcak, N., Itani, R., Mirza, N.: A framework of the financial reputational risk: a quantitative analysis to convert risk into opportunity. In: Salman, A., Tharwat, A. (eds.) AUEIRC 2020. ASTI, pp. 5–31. Springer, Cham (2024). https://doi.org/10.1007/978-3-031-49302-7_2
17. Yang, C., Bhattacharya, S., Srinivasan, P.: Lexical and machine learning approaches toward online reputation management. In: Proceedings of CEUR, Rome (1178) (2012). https://ceur-ws.org/Vol-1178/CLEF2012wn-RepLab-YangEt2012.pdf

Dissecting Data Practices in Android Apps: A Comparative Study of Data Collection and Sharing Behaviors

Triet M. Nguyen, Nghiem T. Pham(✉), Hieu M. Doan, Khoa D. Tran, Bao Q. Tran, Khiem G. Huynh, Nam B. Tran, and Khanh H. Vo

FPT University, Can Tho, Vietnam
trietnm3@fe.edu.vn, nghiemptce160353@fpt.edu.vn

Abstract. The exponential growth of mobile apps within the Android ecosystem has underscored the critical need for robust user privacy and data protection measures. Central to these concerns are the privacy policies that serve as the primary channel of communication between organizations and users, detailing data collection, utilization, and sharing practices. However, the efficacy of these policies is often undermined by their inaccessibility and the legalese that obfuscates their intent, presenting a barrier to informed user consent. This study addresses these challenges by harnessing the capabilities of Generative AI (GenAI) to perform a detailed analysis of data practices in Android apps. Our methodology extends beyond the traditional scope of AI-assisted analysis by not only identifying third-party entities but also by elucidating their data handling purposes. We introduce a classification system that distinguishes between 'Regular' and 'Irregular' app behaviors, offering a benchmark for app evaluation and compliance assessment. Our comparative analysis across various apps reveals patterns and anomalies in data management, providing actionable insights for developers, regulators, and users.

Keywords: Android Apps · Data Collection · Data Sharing · Privacy Policy · Data Usage Behavior · Mobile Privacy

1 Introduction

In today's digital age, mobile apps have become deeply ingrained in the fabric of our daily lives, encompassing a wide array of functionalities from basic communication to complex utility services. The Android ecosystem, in particular, hosts a myriad of such apps, propelling it to the forefront of concerns surrounding user privacy and data protection [16]. Privacy policies are the bedrock of user privacy on these platforms, serving as the official conduit for informing users about data collection, utilization, and sharing practices [8]. However, the efficacy of these policies in truly safeguarding user privacy has been called into question due to their barriers and the complexity of their language [18].

The journey to locate a privacy policy within an app can often be a "challenge" process, as illustrated in Fig. 1 which depicts the pathway to accessing

V. Julian et al. (Eds.): IDEAL 2024, LNCS 15346, pp. 15–26, 2025.
https://doi.org/10.1007/978-3-031-77731-8_2

Facebook's privacy policy in the Google Play Store. These critical documents, when located, present users with verbose and legalistic text, replete with jargon and intricate structures that can be daunting for the layperson [17]. The legalese employed in these policies not only impedes the average user's understanding but also veils their ability to make truly informed choices about their personal data. This gap between policy accessibility and user comprehension poses a significant threat to the autonomy of users over their personal information and highlights a pervasive issue in digital transparency [8].

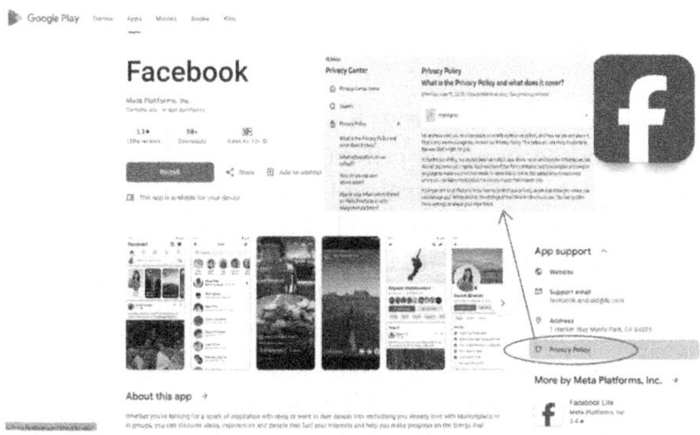

Fig. 1. The interface of the Facebook app's listing on the Google Play Store, highlighting the 'Privacy Policy' link that directs users to detailed information

The advent of AI-Assisted Privacy Policy Analysis has indeed enhanced our capabilities to deconstruct and interpret the complexities of privacy policies. Groundbreaking research by Story et al. [20], Yu et al. [24], and Liu et al. [9] has paved the way for the automation of privacy policy analysis, demonstrating the transformative potential of AI in mediating the disconnect between the legalese of privacy policies and the user's grasp of their content. Nevertheless, the persistent challenge is the effective communication of privacy practices in a manner that is both comprehensible and actionable for users.

The focus within the domain of Android App Privacy and Security has been on leveraging AI to detect post-development security vulnerabilities and align privacy policies with actual app behaviors, as seen in the works of Senanayake et al. [14]. These studies have made valuable strides, but they have not typically addressed the preemptive analysis of privacy practices nor the establishment of behavioral norms that could serve as benchmarks for privacy compliance. Moreover, the literature indicates a significant need for enhanced transparency and user empowerment, particularly in the decision-making processes related to data sharing. Manogaran et al. [10] and Di [4] have delved into this aspect, offering substantial insights into user consent mechanisms and behavior. However, their work has often not fully captured the dynamic and varied nature of app data practices that today's digital environment exhibits.

Our study introduces a pioneering use of Generative AI (GenAI) to conduct a granular analysis of data sharing activities within Android apps. This methodology does not merely identify third-party entities; it elucidates the intent behind their data interactions. Crucially, we delineate 'Regular' from 'Irregular' behaviors within app data practices, thereby providing a framework for evaluating apps against established norms. Our innovative approach facilitates a comprehensive comparative analysis, unearthing consistent and deviant patterns of data handling across a multitude of apps. The insights garnered are invaluable for developers aiming to adopt best practices, for regulatory bodies enforcing compliance, and for users navigating the complex choices regarding their data privacy. Our work thus expands the conversation, offering a detailed, GenAI-powered perspective on the nuances of Android apps' data practices.

2 Related Work

2.1 Privacy Violations and Compliance

The study of privacy violations and compliance in Android apps has seen significant contributions from various researchers. Nguyen et al. [12] tackled the challenging issue of privacy violations by developing an automated tool for detecting unexpected data access. In a related vein, Nguyen et al. [13] conducted a pioneering large-scale study on consent notices in Android apps, evaluating their compliance with the GDPR. They developed an automated approach to identify consent practices, revealing significant gaps in GDPR compliance.

Hashmi et al. [5] provided a longitudinal perspective on the compliance of Android apps with their privacy policies. By using machine learning techniques to compare the stated data collection practices with actual app behavior, they have uncovered a worrying trend of non-compliance over time. On the other hand, Shezan et al. [15] introduced NL2GDPR, an innovative tool that translates natural language descriptions into GDPR-compliant policies while ensuring the app's functionalities align with regulation. Lastly, Bui [2] provided a comprehensive assessment of privacy risks in mobile and web apps/services. Through a series of automated frameworks, the dissertation explored the alignment between privacy notices, actual data practices, and user preferences.

2.2 Privacy Policies and Transparency

Privacy policies and transparency in mobile apps have been the focus of significant research due to their implications on user data privacy. Zhao et al. [25] contributed to this field by addressing the opaque nature of privacy policies associated with third-party libraries (TPLs) in mobile apps. They developed ATPChecker, an automated system to scrutinize TPLs' compliance with privacy-related regulations, revealing that a significant portion of TPLs and host apps do not adequately disclose their data usage practices. Del et al. [3] performed a systematic mapping study to synthesize the diverse methods used to assess privacy in Android apps.

Their work offers an overview of techniques from 2016 to 2020, identifying critical gaps and suggesting future research directions. Moreover, Zimmeck et al. [26] introduced the Mobile App Privacy System (MAPS), a tool that combines code analysis and machine learning to conduct privacy evaluations on a large scale. Furthermore, Khandelwal et al. [7] discussed the introduction of privacy nutrition labels on app distribution platforms, which are intended to simplify the understanding of an app's data practices. Lastly, Zimmeck et al. [27] developed PrivacyFlash Pro, an innovative privacy policy generator for iOS apps that uses static analysis to ensure the policies reflect actual app practices.

2.3 User Perspective and Data Sharing

User perspective and data sharing in mobile apps are critical components in understanding the complex dynamics of user privacy. Bemmann et al. [1] explored the impact of transparency and control features on users' willingness to share data in adaptive mobile apps. In the same vein, Tang et al. [22] addressed the transparency of app analytics and the extent of user interaction data collection, a domain where privacy policies often fall short. By analyzing top analytic libraries and developing a standardized collection claim template, they provided a methodology to evaluate an app's data collection practices against its privacy policy claims. Nema et al. [11] presented a methodology to analyze user concerns about privacy in mobile app reviews at scale. Leveraging NLP techniques and a vast dataset of reviews, they extracted and categorized privacy concerns, providing empirical evidence of dominant privacy themes. Finally, Hatamian et al. [6] proposed a tool named Mobile App Reviews Summarization (MARS) to extract privacy-relevant information from user reviews on app markets. By employing machine learning, NLP, and sentiment analysis, MARS assists developers in understanding privacy aspects that concern users.

2.4 Technical Analysis and Frameworks

Technical analysis and frameworks are essential in enhancing our comprehension of privacy practices and behaviors in mobile apps. Wang et al. [23] illuminated the security risks associated with third-party libraries in mobile apps that surreptitiously harvest private user data. Sun et al. [21] responded to the privacy and security challenges heightened by the COVID-19 pandemic with the development of a comprehensive framework for assessing mobile apps' privacy practices. The framework stands out for its holistic approach, encompassing the establishment of a knowledge base and evaluation system.

For regular and irregular issues in the Android apps, Son et al. introduced a series of app and user behavior analyses that analyze the several dimensions of the Android apps. The first two papers [16,18], in particular, focused on the regular behavior metric definition based on static and dynamic analysis. They, on the one hand, defined the user installation behavior based on their previous installation app [17]. On the other hand, they exploited the app behavior for the specific medical app in the web-app services [19] and deployed them in the Blockchain platform

[8]. These studies collectively provide robust technical methodologies for analyzing privacy practices in the evolving landscape of mobile and IoT apps. They underscore the importance of a thorough and technical approach to understanding and mitigating privacy risks in the digital domain.

3 Approach

3.1 Analyzing Data Practices (i.e., Collection/Sharing) Architecture

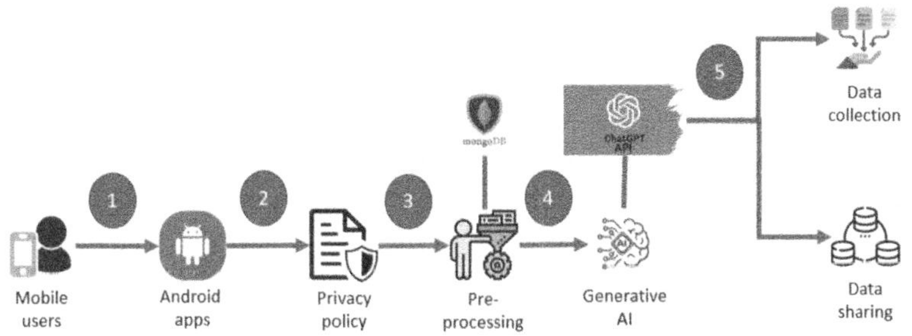

Fig. 2. Analyzing data practices architecture in Apps' privacy policy

The architecture presented in Fig. 2 outlines a novel methodology for examining the data practices of Android apps through a series of interconnected components. The process commences with mobile users engaging with various Android apps, which form the cornerstone of the study. The second component involves the extraction of these privacy policies, which serve as a written testament to the app's data handling protocols. The third stage leverages MongoDB, a dynamic NoSQL database, to store and manage the processed privacy policy data. At the heart of the architecture is the Generative AI component, powered by the ChatGPT API. This advanced AI model is employed to interpret the complex language of privacy policies and to identify specific instances of data collection and sharing practices.

The final component of the architecture focuses on the output of the AI analysis. It categorizes the findings into two primary streams: data collection and data sharing behaviors. Data collection refers to the types of user data that apps are designed to gather, such as location, personal identifiers, and usage habits. Data sharing, on the other hand, outlines the practices of distributing collected data, potentially to third parties, advertisers, or other entities.

3.2 Manual Clustering Process of Data Types

Generative AI tools are proficient in identifying and extracting terms related to data types from the text of privacy policies. However, due to the nuanced

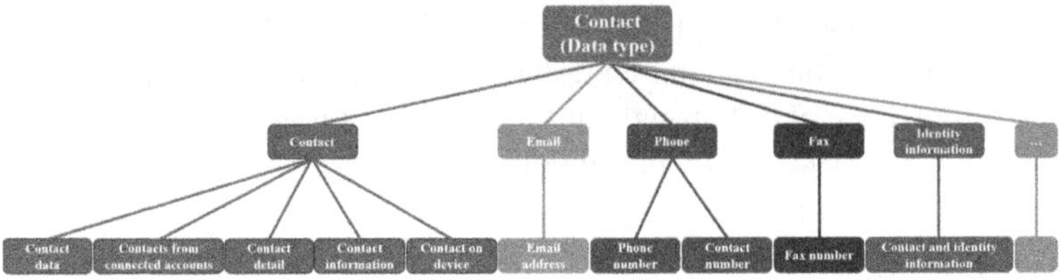

Fig. 3. Manual clustering sample for the contact data type

language used across different documents, there is a need for a standardized classification system. The manual clustering process involves categorizing these varied terms into coherent groups that reflect their underlying similarity in function or purpose.[1] The clustering begins with the establishment of primary data type categories, which are broad groups that represent the major types of data described in privacy policies. These categories are determined based on common privacy concerns and regulatory frameworks, such as contact information, financial information, or usage data.

Within these primary categories, we then identify sub-categories. These are more specific groups of data types that provide additional detail and granularity. For example, under the 'Contact' category (see Fig. 3), we may have sub-categories such as 'Email', 'Phone', and 'Fax', each representing a different mode of contact. Once the primary categories and sub-categories are established, we proceed with the manual clustering. Each term detected by GenAI is reviewed and assigned to the appropriate sub-category based on its meaning and use in the context of the privacy policy. This task requires careful consideration to ensure that each term is placed in the most fitting group, reflecting how the app may collect, use, or share that particular piece of data.

3.3 Categorization Process of Data Practices

Figure 4 exhibits the categorization process used to classify the data practices of Android apps into 'Regular' and 'Irregular' behaviors. This classification is essential to understanding the standard data handling norms within specific app categories and identifying outliers that may warrant further scrutiny. The architecture begins with a collection of Android apps that are grouped based on their category. This grouping is significant as it allows for comparative analysis within a homogenous set of apps that serve similar functions and are likely to have comparable data practices. By examining apps within the same category, the analysis can reveal patterns and standards that are specific to that niche. The central focus of this analysis is on two main aspects of data practices: data collection and data sharing. These are represented by two distinct pathways that

[1] In the practices, we collected 1,351 data types (e.g., Name, Email, Transaction, etc.) from the privacy policy of 303 randomly selected Android apps. We cluster them into 17 category data types - e.g., User Personal Information; Technical Data; Location.

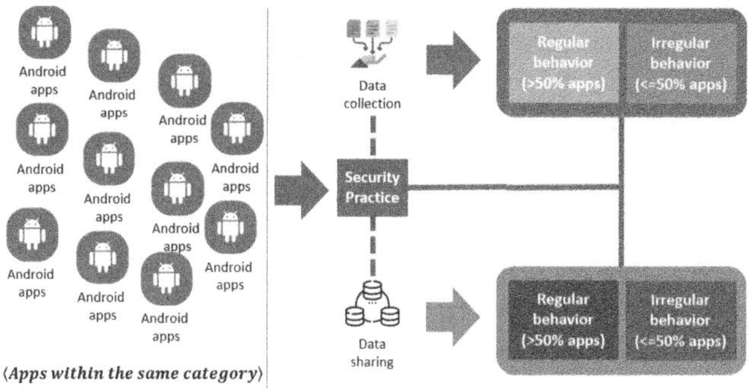

Fig. 4. Categorization of Data Practices (i.e., Regular and Irregular Behaviors) for Android Apps

emerge from the examination of the grouped apps' privacy policies. Data collection refers to the strategies and types of data that apps gather from their users. Conversely, the data sharing path investigates the policies governing how apps disseminate the collected data. This involves examining whether the data is shared with third parties, used for advertising purposes, or shared with affiliates and partners. Upon analyzing these aspects, the apps are divided into two categories: those that exhibit 'Regular behavior' and those that demonstrate 'Irregular behavior'. Regular behavior is characterized by data practices that align with the majority of apps in the category. If over 50% of the apps in the same category engage in a particular data practice, it is considered regular. This behavior is indicative of the industry standard and reflects a consensus among the apps on how user data is managed. Irregular behavior, on the other hand, is less common and is defined by data practices that are not as widespread, occurring in 50% or fewer of the apps within the same category. These behaviors deviate from the norm and can either represent more privacy-conscious practices or, conversely, more invasive approaches to data handling.

4 Evaluation and Discussion

4.1 Android App Collection

In the pursuit of a comprehensive understanding of data practices across a spectrum of Android apps, our study employed a methodologically sound approach to the random selection of apps. This process was meticulously designed to ensure a representative sample that encompasses a diverse array of categories and varying levels of popularity, as determined by the number of downloads (Fig. 5).

The selection began by delineating distinct categories within the Android ecosystem, ranging from productivity and communication tools to lifestyle and entertainment. Our aim was to capture the multifaceted nature of app usage and its implications for data privacy. To this end, we included 16 different categories,

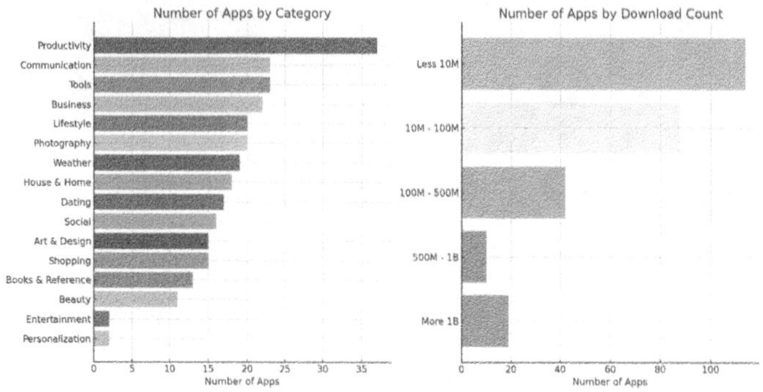

Fig. 5. The number of apps (i.e., 303 apps) by category and their download count

each representing a unique facet of user interaction and functionality. This categorization enabled us to draw parallels and contrasts between data practices across different domains. Within each category, apps were chosen through a randomization process that negated any potential for selection bias, thus providing an unbiased cross-section of the market. The random selection was stratified by the number of downloads, which served as a proxy for app popularity and user reach. The apps were categorized into five brackets based on the number of downloads: less than 10 million; 10 million to 100 million; 100 million to 500 million; 500 million to 1 billion; and more than 1 billion. The rationale behind this stratification was twofold: firstly, to assess whether data practices scale with the user base and, secondly, to understand if the magnitude of exposure influences the privacy framework adopted by the app developers. In total, 303 apps were selected, with the majority falling into the 'less than 10 million downloads'.

4.2 Data Collection and Data Sharing Behavior Analysis

In the analysis process, we found 273 apps on 303 apps mentioning at least one data type for data collection or data sharing. We consider the raw data type (i.e., before the manual clustering) for the data collection and data sharing with 1,040 and 415, respectively. After the manual clustering, we have 512 data types for the data collection belonging to 14 clusters. For the data sharing behavior, we have 62 data types belonging to 17 clusters (see the noted in Fig. 7).

Data Collection. Figure 6 is a stacked bar chart that categorizes Android applications according to the type of data they collect, as stated in their privacy policies. Categories like Productivity, Social, and Tools appear to have a higher overall data collection count, suggesting that apps in these categories tend to collect more types of data compared to others. Certain data types, such as Location and Personal Info, are prevalent across multiple app categories, highlighting the widespread collection of this type of data.

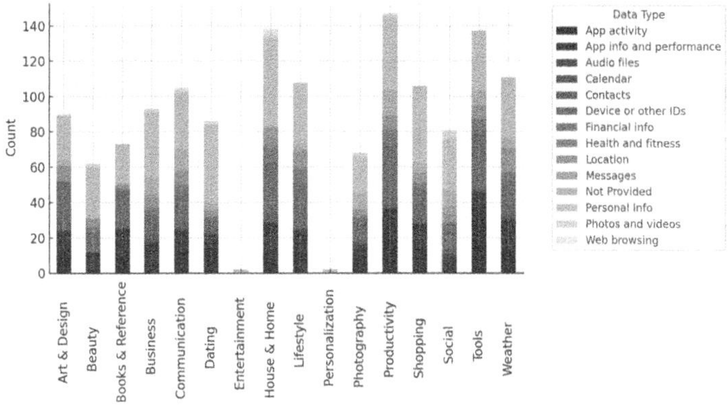

Fig. 6. The regular behavior of the data collection by app's category

Data Sharing. Since the number of apps mentioned the data sharing behavior in their privacy policy is few, we process the outcome in the total rather than the separated app's category. Figure 7 presents the view of the stacked bar chart showing the distribution of specific data types within each data type category, as shared by Android apps. Each bar, representing categories like 'User Personal Information' or 'Technical Data', comprises various colors that detail the contribution of individual data types. The disparity in the number of Android apps mentioning data sharing practices—65 over 303 apps—highlights several key aspects of privacy policy formulation and disclosure in the app ecosystem:

- **Varying Levels of Transparency**: A significant number of apps may not explicitly mention data sharing practices in their privacy policies. This lack of transparency could be due to various reasons, such as the absence of data sharing activities or an oversight in policy formulation.
- **Differing Data Handling Practices**: Not all apps engage in data sharing activities. Some apps might collect data solely for in-app functionalities and not share it with third parties.
- **Complexity of Privacy Policies**: The complexity and variability in how privacy policies are written can lead to inconsistencies in reporting data sharing practices.
- **Regulatory Compliance**: Apps that are more diligent about complying with privacy regulations like GDPR or CCPA may be more likely to disclose their data sharing practices clearly.
- **Nature of the App**: The purpose and functionality of an app significantly influence its data practices. For instance, social networking or e-commerce apps are more likely to share user data for advertising or analytics purposes than utility apps.
- **User Base and Privacy Expectations**: Apps with a user base more sensitive to privacy concerns may be more transparent and cautious about data sharing.

– **Evolution of Privacy Practices**: The landscape of digital privacy is continually evolving. Some apps may have older privacy policies that do not reflect current practices.

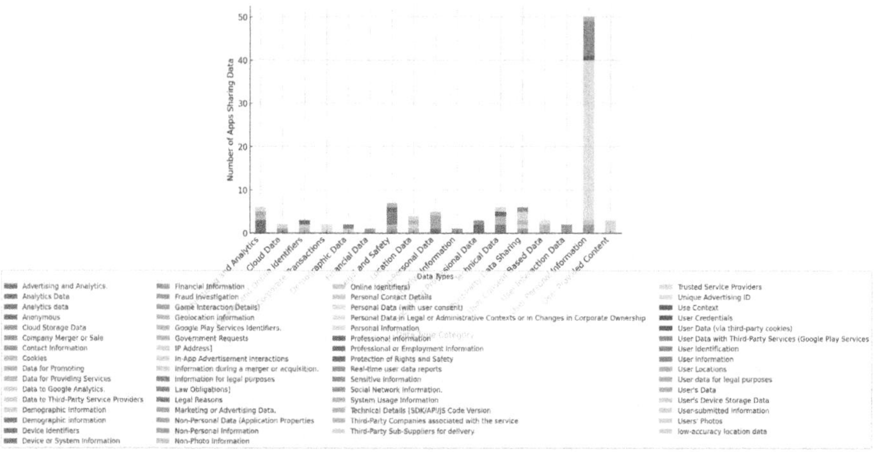

Fig. 7. The #apps (i.e., total - 65 apps) by the data type sharing

4.3 Regular and Irregular Data Practices Behavior

Data Collection. Our analysis reveals that certain types of data, such as 'App activity', 'App info and performance', 'Device or other IDs', and 'Personal info', are commonly collected across multiple app categories, suggesting these data types are integral to the core functionality and user experience offered by these apps. Notably, 'Personal info' stands out as a regularly collected data type, highlighting its critical role in personalization and user engagement within apps. Conversely, data types listed under 'irregular behavior' show a more varied pattern of collection across categories. These include more sensitive or less commonly required information such as 'Financial info', 'Health and fitness', 'Audio files', and 'Web browsing'. The presence of such data under 'irregular behavior' indicates that their collection is not as widespread and may point to specific functionalities or niche requirements within certain apps.

Data Sharing. Our empirical analysis revealed a notable deviation from what one might expect regarding data sharing behaviors. Among the 17 different data types we identified, only one—User Personal Information—was consistently shared by a majority of apps in our study, thereby qualifying as 'Regular behavior' for data sharing. This finding was unexpected given the breadth of data categories that apps typically handle. This singular instance of Regular behavior stands in stark contrast to the myriad of data types that apps are known to collect. It suggests a conservative approach to data sharing among Android apps, at least within the scope of our dataset. The implications of this are significant; it indicates that most apps either do not share data (i.e., locally processing data)beyond the essential User Personal Information or do so in a manner that is not transparently disclosed within their privacy policies.

5 Conclusion and Future Work

Our comprehensive study has embarked on an analytical journey through the landscape of Android apps, unveiling the intricacies of data practices with the aid of Generative AI. By dissecting privacy policies and app behaviors, we have illuminated the often opaque third-party data interactions, providing clarity on the intentions behind data collection and sharing. The classification of app behaviors into 'Regular' and 'Irregular' categories has furnished stakeholders with a yardstick for measuring privacy compliance and user transparency. The insights derived from our analysis bear significant implications. For developers, they offer a blueprint for best practices in data privacy. For regulatory bodies, they provide a robust framework for enforcing compliance. For users, they ensure a better understanding of privacy rights and greater control over personal information. Our GenAI-powered methodology stands as a vanguard in the quest for digital privacy, transcending traditional analysis and advocating for a user-centric approach to privacy policy design and implementation.

References

1. Bemmann, F., et al.: The influence of transparency and control on the willingness of data sharing in adaptive mobile apps. Proc. ACM Hum.-Comput. Interact. **6**(MHCI), 1–26 (2022)
2. Bui, D.: Assessment of privacy risks in mobile and web applications/services. Ph.D. thesis (2022)
3. Del Alamo, J.M., et al.: Privacy assessment in Android apps: a systematic mapping study. Electronics **10**(16), 1999 (2021)
4. Di, M., et al.: Influencing user's behavior concerning Android privacy policy: an overview. Mob. Inf. Syst. **2021**, 1–19 (2021)
5. Hashmi, S.S., et al.: Longitudinal compliance analysis of Android applications with privacy policies. In: Hara, T., Yamaguchi, H. (eds.) MobiQuitous 2021. LNICST, vol. 419, pp. 280–305. Springer, Cham (2021). https://doi.org/10.1007/978-3-030-94822-1_16
6. Hatamian, M., et al.: Revealing the unrevealed: mining smartphone users privacy perception on app markets. Comput. Secur. **83**, 332–353 (2019)
7. Khandelwal, R., et al.: The overview of privacy labels and their compatibility with privacy policies. arXiv preprint arXiv:2303.08213 (2023)
8. Khiem, H., et al.: Applying blockchain technology for privacy preservation in Android platforms. In: Zhang, Y., Zhang, L.J. (eds.) ICWS 2023. LNCS, vol. 14209, pp. 47–61. Springer, Cham (2023). https://doi.org/10.1007/978-3-031-44836-2_4
9. Liu, S., et al.: APPCorp: a corpus for android privacy policy document structure analysis. Front. Comput. Sci. **17**(3), 173320 (2023)

10. Manogaran, G., et al.: Guest editorial introduction for the special section on deep learning algorithms and systems for enhancing security in cloud services (2022)
11. Nema, P., et al.: Analyzing user perspectives on mobile app privacy at scale. In: Proceedings of the 44th International Conference on Software Engineering, pp. 112–124 (2022)
12. Nguyen, T.T.: Understanding and measuring privacy violations in Android apps (2023)
13. Nguyen, T.T., et al.: Freely given consent? Studying consent notice of third-party tracking and its violations of GDPR in Android apps. In: Proceedings of the 2022 ACM SIGSAC Conference on Computer and Communications Security, pp. 2369–2383 (2022)
14. Senanayake, J., et al.: Android code vulnerabilities early detection using AI-powered *ACVED* plugin. In: Atluri, V., Ferrara, A.L. (eds.) DBSec 2023. LNCS, vol. 13942, pp. 339–357. Springer, Cham (2023). https://doi.org/10.1007/978-3-031-37586-6_20
15. Shezan, F.H., et al.: NL2GDPR: automatically develop GDPR compliant Android application features from natural language. arXiv preprint arXiv:2208.13361 (2022)
16. Son, H.X., Carminati, B., Ferrari, E.: A risk assessment mechanism for Android apps. In: 2021 IEEE International Conference on Smart Internet of Things (SmartIoT), pp. 237–244. IEEE (2021)
17. Son, H.X., Carminati, B., Ferrari, E.: PriApp-Install: learning user privacy preferences on mobile apps' installation. In: Su, C., Gritzalis, D., Piuri, V. (eds.) ISPEC 2022. LNCS, vol. 13620, pp. 306–323. Springer, Cham (2022). https://doi.org/10.1007/978-3-031-21280-2_17
18. Son, H.X., Carminati, B., Ferrari, E.: A risk estimation mechanism for Android apps based on hybrid analysis. Data Sci. Eng. **7**(3), 242–252 (2022)
19. Son, H.X., et al.: In2P-Med: toward the individual privacy preferences identity in the medical web apps. In: Garrigós, I., Murillo Rodríguez, J.M., Wimmer, M. (eds.) ICWE 2023. LNCS, vol. 13893, pp. 126–140. Springer, Cham (2023). https://doi.org/10.1007/978-3-031-34444-2_10
20. Story, P., et al.: Natural language processing for mobile app privacy compliance. In: AAAI Spring Symposium on Privacy-Enhancing Artificial Intelligence and Language Technologies, vol. 2, p. 4 (2019)
21. Sun, R., et al.: Measuring privacy practices and application behaviours. In: Understanding and Measuring Privacy and Security Assertions of Mobile and VR Applications, p. 109 (2023)
22. Tang, F., Østvold, B.M.: Transparency in app analytics: analyzing the collection of user interaction data. arXiv preprint arXiv:2306.11447 (2023)
23. Wang, J., et al.: Understanding malicious cross-library data harvesting on Android. In: 30th USENIX Security Symposium (USENIX Security 2021), pp. 4133–4150 (2021)
24. Yu, L., et al.: PPChecker: towards accessing the trustworthiness of Android apps' privacy policies. IEEE Trans. Softw. Eng. **47**(2), 221–242 (2018)
25. Zhao, K., et al.: Demystifying privacy policy of third-party libraries in mobile apps. In: 2023 IEEE/ACM 45th International Conference on Software Engineering (ICSE), pp. 1583–1595. IEEE (2023)
26. Zimmeck, S., et al.: MAPS: scaling privacy compliance analysis to a million apps. In: Proceedings on Privacy Enhancing Technologies, vol. 2019, p. 66 (2019)
27. Zimmeck, S., et al.: PrivacyFlash Pro: automating privacy policy generation for mobile apps. In: NDSS, vol. 2, p. 4 (2021)

Model-Based Meta-reinforcement Learning for Hyperparameter Optimization

Jeroen Albrechts, Hugo M. Martin, and Maryam Tavakol(✉)

Eindhoven University of Technology, Eindhoven, The Netherlands
m.tavakol@tue.nl

Abstract. Hyperparameter Optimization (HPO) plays a significant role in enhancing the performance of machine learning models. However, as the size and complexity of (deep) neural architectures continue to increase, conducting HPO has become very expensive in terms of time and computational resources. Existing methods that automate this process still demand numerous evaluations to find the optimal hyperparameter configurations. In this paper, we present a novel approach based on model-based reinforcement learning to effectively improve sample efficiency while minimizing resource consumption. We formulate the HPO task as a Markov decision process and develop a predictive dynamics model for efficient policy optimization. Additionally, we employ the `Deep Sets` framework to encode the state space, which is then leveraged in meta-learning for transfer of knowledge across multiple datasets, enabling the model to quickly adapt to new datasets. Empirical studies demonstrate that our approach outperforms alternative techniques on publicly available datasets in terms of sample efficiency and accuracy.

1 Introduction

The performance of machine learning models often relies heavily on the careful tuning of their hyperparameters. Techniques based on grid search and random search were widely used for finding the optimal hyperparameter configurations and proved valuable for many years [2]. However, as models have grown in complexity and size, the performance of these methods has significantly declined. Consequently, the focus of research has shifted toward developing automated approaches that find the best hyperparameter settings within a constrained time or resource budget, which is known as Hyperparameter Optimization (HPO).

Prominent techniques for addressing the HPO problem are mainly based on Bayesian optimization [2,23], multi-armed bandits [3,13,20], or a combination of both [8]. Although promising, these methods suffer from high computational costs and increased sample complexity as the number of hyperparameters grows. Reinforcement Learning (RL) offers an alternative due to its ability to optimize over longer horizons. Model-based RL methods, in particular, are highly sample-efficient, learning a model of the environment to improve performance with fewer

V. Julian et al. (Eds.): IDEAL 2024, LNCS 15346, pp. 27–39, 2025.
https://doi.org/10.1007/978-3-031-77731-8_3

real interactions [6]. Recent RL-based methods have improved sample efficiency for hyperparameter optimization [12,14–16,24], but these techniques either discretize the hyperparameter space, decreasing the pool of solutions, or model the problem in ways that limit the agent's ability to plan over a longer horizon.

In this paper, we propose a model-based RL approach for the HPO problem that enables knowledge transfer across datasets via meta-learning. Inspired by [12] and [15], we model the problem as a Markov decision process (MDP), where the search for the optimal set of hyperparameters unfolds as sequential decision-making. The dynamics model is trained to approximate the performance of hyperparameter configurations, and the state space is embedded using the `Deep Sets` [25] architecture. We further incorporate a novel clipping technique and adapt the policy to maximize the best-found reward rather than the cumulative reward, to align with the objective of the HPO problem. Moreover, leveraging meta-learning with a certain similarity measure to compare various datasets enables both the predictive model and the policy to train with initial knowledge, accelerating the learning process and increasing sample efficiency. Meta-learning has been successfully applied to RL scenarios [17] and has proven effective in increasing convergence speed in the HPO context [15,16,24].

Our main contributions are: (i) a novel model-based RL approach for HPO, (ii) a meta-learning scheme to transfer parameters across datasets, (iii) a new state representation using `Deep Sets`, (iv) a modified objective function, and (v) an evaluation of our method on six public datasets, demonstrating superior solution quality and sample efficiency compared to state-of-the-art techniques.

2 Related Work

Before automated hyperparameter selection, grid search and random search [2] were common. However, grid search becomes impractical for large and continuous spaces, while random search lacks a learning component, leading to poor performance in complex tasks. Subsequently, Bayesian optimization techniques emerged to address the HPO problem by constructing surrogate Bayesian models from validation results [2,19,23]. Among these, tree-structured Parzen estimator (TPE) [2] enhances alternative methods by utilizing a tree-based model instead of Gaussian-like surrogates, allowing to handle various variable types. Still, TPE struggles in large-scale search spaces.

On the other hand, methods utilizing multi-armed bandits have shown promising performance in practical scenarios [3]. Hyperband iteratively narrows down the search space using the successive halving algorithm [13]. While this approach requires less computational resources and maintains a flexible balance between exploration and exploitation, it samples configurations randomly, leading to reduced efficiency. HyperUCB [20] improves upon this by employing the upper confidence bound (UCB) algorithm [1] to leverage knowledge from previous evaluations. Alternatively, BOHB [8] combines the strengths of both Bayesian optimization and Hyperband, offering an efficient and scalable approach to optimal hyperparameter selection. More recently, FastHO [26] showed improvements on BOHB by using the progressive multi-fidelity technique.

RL-based approaches to HPO aim to reduce the dimensionality of the search space by turning the search problem into a sequential decision-making task, selecting hyperparameter values one at a time rather than all at once [14–16,22,24]. Wu et al. [22] adapt this approach to generate one hyperparameter configuration per iteration, with validation loss computed at the end of the iteration, either directly or via a surrogate model. Liu et al. [14] improve this method by embedding the model into a meta-learning framework, leading to enhanced performance across different datasets. Other techniques incorporate meta-learning into policy optimization, utilizing additional context inputs [15], dataset representations [24], or experience variables [16] to guide decision-making. However, these augmentations often focus only on specific evaluated configurations, overlooking potentially useful information from other tested settings during training. Alternatively, Hyp-RL [12] presents a deep Q-network with a recurrent neural architecture for policy optimization, exploiting meta-features for pre-training on multiple datasets. While promising, this approach is limited to discrete hyperparameter values and lacks permutation invariance. We address these issues by planning over a longer configuration horizon to achieve optimal performance.

3 Model-Based Meta-RL Framework

3.1 Preliminary

Let M denote a machine learning model with n hyperparameters to tune. The potential values for the i-th hyperparameter are determined within a domain Λ_i, leading to a hyperparameter space $\Lambda = \Lambda_1 \times \Lambda_2 \times \cdots \times \Lambda_n$. M_λ represents the model M with a selected hyperparameter configuration $\lambda \in \Lambda$. The hyperparameter space consists of various value types, such as real values or categorical values. Given the dataset $\mathcal{D}$ which is divided into the training set $\mathcal{D}_{train}$ and validation set $\mathcal{D}_{valid}$, the objective of the HPO problem is defined as λ^*, which can be expressed via $\lambda^* = \arg\min_{\lambda\in\Lambda} \mathcal{L}(M_\lambda|\mathcal{D}_{train}, \mathcal{D}_{valid})$, where $\mathcal{L}$ indicates the loss of model M on $\mathcal{D}_{valid}$ when trained using hyperparameters λ on $\mathcal{D}_{train}$.

3.2 MDP Formulation

To model HPO in an RL framework, we characterize the problem as a Markov decision process (MDP). An MDP provides a mathematical framework for modeling sequential decision-making and is defined by a tuple $(\mathcal{S}, \mathcal{A}, T, r, \gamma)$, where $\mathcal{S}$ and $\mathcal{A}$ represent the state and action spaces, respectively, $T(s'|s, a)$ denotes the transition distribution of next state s' under an action a performed in s, $r(s, a)$ is the associated reward, and $\gamma \in (0, 1)$ is the discount factor. The objective is to find an optimal policy $\pi^* : \mathcal{S} \to \mathcal{A}$ that maximizes the cumulative rewards: $\pi^* = \arg\max_\pi \mathbb{E}_\pi \left[\sum_{t=0}^{\infty} \gamma^t r(s_t, a_t)\right]$. We set the action space $\mathcal{A}$ to be the set of all hyperparameter configurations Λ. At each timestep t, an action a_t corresponds to selecting a hyperparameter setting from this space, $a_t = \lambda_t \in \Lambda$. Subsequently, the reward is derived from the validation loss of the model on $\mathcal{D}_{valid}$, given configuration λ_t, after training the model on $\mathcal{D}_{train}$. Hence, the reward only depends

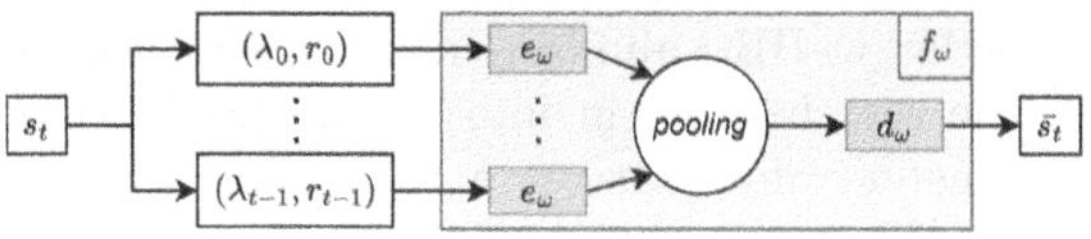

Fig. 1. State encoding via `Deep Sets`.

on the current action and is expressed as $r(\lambda_t) = 1 - \mathcal{L}(M_{\lambda_t} | \mathcal{D}_{train}, \mathcal{D}_{valid})$, which we also denote as r for simplicity.

Furthermore, we construct the state space to exploit as much information from past evaluations as possible. Prior RL-based techniques either form hyperparameter configurations incrementally or incorporate the recent or optimal configurations in defining the state [14–16,22,24], which might overlook the information gained from other model evaluations. Inspired by [12], we include the set of all previous tested hyperparameter configurations and their rewards in defining the state at time t, $s_t = \{(\lambda_1, r_1), (\lambda_2, r_2), \ldots, (\lambda_{t-1}, r_{t-1})\}$. However, instead of using an RNN architecture for state encoding, which is *not* permutation-invariant, we employ `Deep Sets` [25] to learn a representation $\bar{s}_t$ of s_t independent of previous actions' order.

3.3 State Encoding

The importance of effective state encoding in our approach is twofold. First, the state space consists of sets of varying sizes at each time which requires an efficient embedding into a fixed-size latent space to be incorporated in the underlying RL algorithm. Second, adopting a unified representation for states will facilitate the meta-learning process across diverse datasets. Therefore, we employ `Deep Sets` [25] to learn a robust state representation.

`Deep Sets` is a deep neural architecture designed to encode sets of objects into fixed-size vectors while guaranteeing permutation invariance. In this framework, each set element is mapped into a vector using a shared encoder. Then, the resulting vectors are merged through a "pooling" operator, such as *mean* or *sum*, combining all set items into a fixed-size vector. Finally, the output of the pooling operation feeds into a fully connected neural network, generating an output representation of constant length. We denote the overall process as f_ω. The encoding of a state s_t into $\bar{s}_t$ is depicted in Fig. 1.

We employ this framework to learn the state representation $\bar{s}_t = f_\omega(s_t)$ in our problem, which is carried out as a standalone optimization process in the initial step of meta-learning. We train f_ω using the learned latent vectors for a classification task, assigning each state to its corresponding dataset, and the network parameters ω are optimized with the cross-entropy loss.

3.4 Predictive Model

Model-based RL techniques train a dynamics model, allowing for the generation of artificial rollouts for policy optimization, aiming to enhance sample efficiency over model-free methods. In this section, we outline the design of our dynamics model, drawing inspiration from the MBPO algorithm [11].

The environment's dynamics consists of the transition function $T(s_{t+1}|s_t, a_t)$ and the reward r_t. T is deterministic in our scenario and can be derived from the actual action and reward, i.e., $s_{t+1} = \{s_t \cup (\lambda_t, r_t)\}$, thus estimating r_t alone is sufficient in order to determine the one-step dynamics. Following the approach in [11], we employ an ensemble of k models characterized by parameters $\{\psi_1, \psi_2, \ldots, \psi_k\}$, each initialized differently with the aim of reducing the variance of predictions. Every model in the ensemble is then trained via maximum likelihood, $\psi_i \leftarrow \arg\max_\psi \mathbb{E}_{\mathbb{B}_0}[\log p_{\psi_i}(r_t|\bar{s}_t, \lambda_t)]$, where $p_{\psi_i}(r_t|\bar{s}_t, \lambda_t) = \mathcal{N}(\mu_{\psi_i}(\bar{s}_t, \lambda_t), \Sigma_{\psi_i}(\bar{s}_t, \lambda_t))$, and $\bar{s}_t = f_\omega(s_t)$ indicates the encoded fixed-size representation of s_t, as mentioned in the previous section.

The predictive model is optimized using the replay buffer $\mathbb{B}_0$ collected during the meta-training procedure (see Sect. 3.7). However, this model may not generalize well to unseen datasets, thus we implement transfer learning by fine-tuning the last few layers of the pre-trained dynamics model on the new dataset. Hence, combining transfer learning with an ensemble model improves the predictive model's accuracy to learn a more precise reward from less data.

3.5 Policy Optimization

We employ Twin-Delayed Deep Deterministic policy gradient (TD3) [9] for policy optimization. TD3 is an actor-critic approach comprising a policy network π with parameters ϕ and two value networks Q, with parameters θ_1 and θ_2, collectively denoted as $\theta = \theta_1 \cup \theta_2$. The target networks parameters are represented by ϕ', θ'. TD3 is executed within the HPO procedure on a batch $\mathbb{B}_p$ of transitions sampled via the predictive model as well as Prioritized Experience Replay [18].

Furthermore, two modifications are applied to the target y of the Bellman error to adapt TD3 to our HPO setting: (i) replacing the sum of discounted rewards with the maximum reward and (ii) applying Q-value clipping. The first adjustment shifts the focus from maximizing the cumulative reward per episode to maximizing the highest obtained reward in an episode, as the objective of HPO is to find the best performing hyperparameter configuration. Hence, inspired by [10], we use a *max* operation instead of a *sum* when computing y. The second adaptation involves a novel Q-value clipping method developed for our scenario. The parameters of the critic networks are initialized with meta-parameters which may significantly overestimate Q-values. Higher Q-values reduce the effect of rewards when obtaining the target y, leading to low TD error δ and thus slow convergence. To address this, we cap the Q-values exceeding the maximum observed reward r^* recorded on the current dataset. These two adaptations lead to the modified target: $y \leftarrow \max\left[r, \min[r^*, \min_{j=1,2} \gamma Q_{\theta'_j}(\bar{s}, \lambda)]\right]$.

3.6 HPO Procedure

The HPO procedure outlined in Algorithm 1 illustrates the process of learning the optimal hyperparameter configuration λ^* for an unseen dataset $\mathcal{D}$ following

Algorithm 1: HPO procedure

input : Meta-parameters ϕ_0, θ_0, ψ_0, dataset $\mathcal{D}$ to tune, ω
1 $\phi, \phi', \theta, \theta', \psi \leftarrow \phi_0, \phi_0, \theta_0, \theta_0, \psi_0$
2 Reset λ^*, replay buffers $\mathbb{B}, \hat{\mathbb{B}}$, model errors $\mathcal{E}$, rewards $\mathcal{R}$
3 Initialize $\mathbb{B}$ with randomly sampled episodes using $\mathcal{D}$
4 **for** $e \leftarrow 1$ **to** E **do**
5   **if** *time to update the predictive model* **then**
6         $\psi_i \leftarrow \arg\max_\psi \mathbb{E}_{(\bar{s},\lambda,r)\sim\mathbb{B}}[\log p_{\psi_i}(r|\bar{s}, \lambda)] \quad \forall i = 1, \ldots, k$
7   Reset $\mathcal{E}$
8     $s_1 \leftarrow \{(\lambda_0, r_0)\}$ with λ_0 randomly sampled, and r_0 its reward
9     $\bar{s}_1 \leftarrow f_\omega(s_1)$
10   **for** $t \leftarrow 1$ **to** T **do**
11   Select action $\lambda_t \leftarrow clip(\pi_\phi(\bar{s}_t) + \epsilon, [\alpha_{low}, \alpha_{high}]), \quad \epsilon \sim \mathcal{N}(0, \sigma)$
12   Evaluate λ_t on $\mathcal{D}$ to obtain reward r_t
13   **if** $r_t > \max(\mathcal{R})$ **then** $\lambda^* \leftarrow \lambda_t$
14         $\bar{s}_{t+1} \leftarrow f_\omega(s_{t+1}) \quad$ where $s_{t+1} \leftarrow \{s_t \cup (\lambda_t, r_t)\}$
15   Append $(\bar{s}_t, \lambda_t, r_t, \bar{s}_{t+1})$ to $\mathbb{B}$ with max. priority
16   Append $|\hat{r}_t - r_t|$ to $\mathcal{E}$, with $\hat{r}_t \sim p_{\psi_i}(.|\bar{s}_t, \lambda_t), \quad \psi_i$ is randomly selected
17   Append r_t to $\mathcal{R}$
18   **if** mean$(\mathcal{E}) <$ std$(\mathcal{R})$ **then**
19   Generate a batch of data and add it to $\hat{\mathbb{B}}$
20   **if** *time to update the policy* **then**
21   Sample batch $\mathbb{B}_p \subset \mathbb{B}$ according to PER priorities
22   Sample batch $\hat{\mathbb{B}}_v \subset \hat{\mathbb{B}}$ s.t. $|\hat{\mathbb{B}}_v| = v|\mathbb{B}_p|$
23   Update ϕ, ϕ', θ, θ' with TD3 using $\mathbb{B}_p \cup \hat{\mathbb{B}}_v$
24 **return** λ^*

the meta-training phase (see Sect. 3.7). For this procedure, the parameters of the dynamics model, policy, and value networks are initialized with the meta-trained parameters, denoted by ϕ_0, θ_0 and ψ_0, respectively (line 1).

This algorithm maintains two replay buffers $\mathbb{B}$ and $\hat{\mathbb{B}}$. The former stores transitions from $\mathcal{D}$ and is initialized by randomly sampling hyperparameters and their validation loss on $\mathcal{D}$ (line 3), while the latter preserves transitions generated by the learned dynamics model branching from actual data stored in $\mathbb{B}$. We ensure that $\hat{\mathbb{B}}$ contains admissible transitions by generating data with the model only when the *mean of its absolute errors*, logged in $\mathcal{E}$, is lower than the *standard deviation of the rewards* derived from the dataset, logged in $\mathcal{R}$ (line 16–19). This approach allows for automatic adjustment to the uncertainty in reward predictions and has shown promising empirical results.

The algorithm iterates over E episodes of T steps. Within every episode, s_t consists of all evaluated configurations in the episode along with their corresponding rewards. s_t is then encoded into the representation $\bar{s}_t$ and added to the replay buffer (line 14–15). While the first state is randomly sampled (line 8–9), subsequent steps in the episode follow the policy π_ϕ, employing TD3 clipping

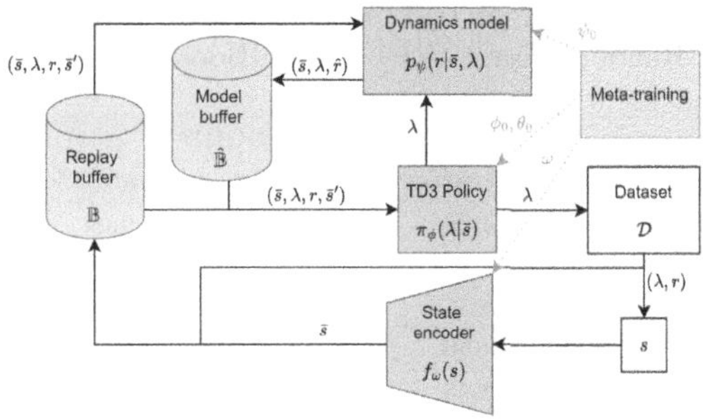

Fig. 2. The overall RL-based HPO pipeline.

technique (line 11–12). Additionally, model updates occur at specified intervals and when the length of the replay buffer $\mathbb{B}$ reaches a sufficient size. Consequently, all data in $\mathbb{B}$ is used to train the ensemble predictive model and $\mathcal{E}$ is cleared as the errors no longer accurately reflect the model's performance (line 5–7).

Moreover, periodically, we retrieve two mini-batches of data, one from the replay buffer $\mathbb{B}$ using PER priorities, and the other from the artificial data $\hat{\mathbb{B}}$, selected randomly but capped in size by a coefficient v. These batches are combined and leveraged to update the parameters of the actor and critic networks according to the modified TD3 update described in Sect. 3.5.

3.7 Meta Training

Meta-learning aims to acquire knowledge from prior tasks to facilitate learning of new tasks. In the context of HPO, each dataset is treated as a distinct task, and the meta-training process operates on a set of datasets, called meta-datasets, enabling accelerated learning when applying the procedure to new datasets.

In our setting, meta-training involves iteratively applying the HPO procedure described in Sect. 3.6 to each meta-dataset, without resetting the parameters of the networks between iterations. The initialization values for the actor (ϕ_0), critic (θ_0), and dynamics model (ψ_0) networks as well as the batch $\mathbb{B}_0$ are obtained from pre-trained models to aid learning in the subsequent iterations. The meta-training algorithm utilized in this phase closely resembles the HPO procedure in Algorithm 1, with three minor adjustments as outlined below.

First, we do not use a dynamics model, as time efficiency is not prioritized during the meta-training phase. Second, a meta-buffer $\mathbb{B}_0$ is maintained, storing all transitions $(\bar{s}, \lambda, r, \bar{s}')$ across all meta-datasets, which serves for training the predictive model (see Sect. 3.4). The third modification addresses exploding action values by introducing action normalization [21]. During meta-training, carrying over parameters that lead to extreme hyperparameter values to a new dataset may result in limited exploration capabilities, as exploration relies on Gaussian noise centered on the policy. While this issue is less likely during testing in the HPO procedure, where the policy is trained briefly on a single dataset, it often leads to divergence and impedes policy learning during meta-training.

Table 1. Performance of different HPO methods with **fixed sample size** on 6 test datasets using both Random Forest (RF) and XGBoost (XGB), averaged over 5 runs.

Dataset		MBMRL-HPO		BOHB		TPE	
		Accuracy	std (%)	Accuracy	std (%)	Accuracy	std (%)
Teaching Assistant	RF	**0.696**	0.49	0.631	9.04	0.613	4.89
	XGB	**0.671**	0.44	0.657	1.48	0.660	0.95
Maternal Health Risk	RF	**0.851**	0.07	0.826	4.16	0.830	1.83
	XGB	**0.859**	0.06	0.858	0.19	**0.859**	0.22
Cervical Cancer	RF	**0.873**	0.11	0.871	0.00	0.871	0.00
	XGB	**0.875**	0.14	0.873	0.09	0.873	0.22
Obesity	RF	**0.966**	0.10	0.960	1.21	0.960	0.37
	XGB	0.969	0.12	0.973	0.04	**0.975**	0.13
Musk	RF	0.893	0.59	**0.898**	1.33	0.892	0.64
	XGB	**0.926**	0.30	0.918	0.14	0.919	0.26
Android Malware	RF	0.991	0.08	**0.992**	0.14	0.992	0.12
	XGB	**0.993**	0.00	0.993	0.05	0.993	0.05

Given a hyperparameter configuration λ of dimension n, we compute the average absolute values of its components as $G = \frac{\|\lambda\|_1}{n}$. Accordingly, if $G > 1$, normalization is applied, resulting in a normalized action vector $\frac{\lambda}{G}$ for policy learning. Putting all components of our approach together, we illustrate the overall HPO pipeline, including the meta-training, in Fig. 2.

4 Empirical Study

4.1 Experimental Setups

We carry out experiments to evaluate the performance of our approach, employing two popular classification techniques, Random Forest (RF) [4] and Extreme Gradient Boosting (XGBoost) [5]. In these experiments, we aim to tune five hyperparameters for RF and ten for XGBoost. We further select sixteen publicly available datasets for classification tasks [7], allocating ten sets for meta-training and six sets for testing, with performance evaluation conducted via 10-fold cross-validation. Moreover, we compare our approach, denoted as **MBMRL-HPO**, against two top-performing state-of-the-art methods, **TPE** [2] and **BOHB** [8].

To measure the performance of HPO methods, we analyze both their computational efficiency and the quality of their optimal solution. The latter is determined by the accuracy of classification for a given hyperparameter configuration, while we impose a budget on the algorithm's computations in order to gauge the former. Initially, we set the budget based on the number of hyperparameter configurations to be evaluated, enabling us to assess sample efficiency in different methods (Sect. 4.2). In the remaining experiments, we enforce a time constraint, which is more in line with real-world situations (Sect. 4.3).

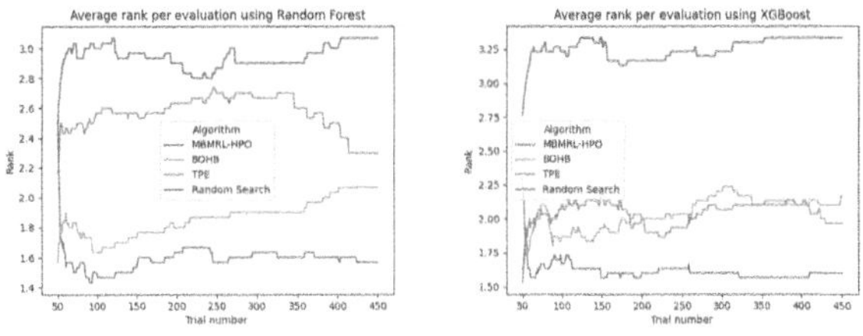

(a) Random Forest (b) XGBoost

Fig. 3. Average rank of HPO methods over all runs and datasets for 450 iterations of hyperparameter evaluation. Lower ranks indicate better performance.

We select the HPO procedure's hyperparameters once and keep them fixed throughout all experiments for better generalizability across datasets and classification algorithms. We also study the effect of the hyperparameters on the performances and conduct an ablation study showing that all components of our approach, particularly the meta-learning aspect, positively impact performance. These results are omitted from the paper due to space constraints.

4.2 Performance Evaluation with Fixed Sample Size

We first limit the budget to 450 as the number of hyperparameter configurations to assess. For MBMRL, we randomly select the initial 50 evaluations, used to populate the buffer $\mathbb{B}$. Table 1 shows the performance of the HPO algorithms on the six test datasets in terms of accuracy of classification techniques.

The results demonstrate that our approach consistently outperforms TPE and BOHB on most datasets for both classification algorithms. Specifically, it achieves superior performance in 9 out of the 12 experiments, and exhibits the lowest standard error in 8 experiments. These outcomes indicate that our algorithm not only requires fewer samples compared to its competitors but also offers greater reliability and consistency. Figure 3(a) and (b) further support this finding in terms of average rank of different algorithms. In this context, the rank of a method on a dataset is determined by its relative performance compared to other methods using the best hyperparameter configuration found on that dataset up to that point. Additionally, we include random search as a baseline in the plots, displaying notably inferior performance as expected.

4.3 Performance Evaluation Under Time Limit

In the second series of experiments, we allocate a fixed amount of time for methods to execute and deliver their best-found hyperparameter configuration. Accordingly, we conduct the experiments across three test datasets, "Teaching Assistant", "Musk", and "Android Malware", each evaluated under 5-, 15-, and 30-min time limits, resulting in a total of 9 setups per classification method. These datasets represent different scales of data size, with 150, 476, and 4464

Table 2. Performance of HPO methods **under time limit** (5, 15, and 30 min) on 3 test datasets, averaged over 3 runs. K indicates the number of evaluations.

Dataset–running time (min)		MBMRL-HPO			BOHB			TPE		
		Accuracy	std (%)	K	Accuracy	std(%)	K	Accuracy	std (%)	K
Teaching Assistant–5	RF	**0.693**	1.36	619	0.6	6.85	1219	0.671	1.24	1447
	XGB	0.671	0.46	437	0.676	2.46	1022	**0.689**	0.91	1262
Teaching Assistant–15	RF	**0.698**	1.19	1657	0.622	5.82	3724	0.691	0.45	3992
	XGB	0.682	0.46	1223	0.682	1.66	3216	**0.702**	0.45	3325
Teaching Assistant–30	RF	**0.702**	0.45	3049	0.644	3.41	7378	0.698	0.45	7106
	XGB	0.691	0.45	2266	0.682	1.66	6398	**0.709**	0.44	5829
Musk–5	RF	0.887	0.19	255	**0.893**	1.02	301	0.892	0.48	684
	XGB	**0.924**	0.11	285	0.917	0.1	255	0.918	0.11	360
Musk–15	RF	**0.899**	1.19	653	0.894	0.89	976	0.897	0.67	1883
	XGB	**0.927**	0.32	773	0.920	0.22	848	0.920	0.18	1048
Musk–30	RF	0.901	1.06	1384	0.894	0.89	1965	**0.904**	0.57	3715
	XGB	**0.929**	0.18	1446	0.922	0.39	1725	0.922	0.11	2067
Android Malware–5	RF	**0.991**	0.05	87	0.987	0.41	64	**0.991**	0.16	110
	XGB	**0.993**	0.05	51	0.992	0.04	52	0.991	0.05	57
Android Malware–15	RF	0.991	0.05	283	0.991	0.3	246	**0.992**	0.05	339
	XGB	**0.993**	0.03	160	0.992	0.02	131	0.992	0.02	179
Android Malware–30	RF	0.991	0.05	610	0.991	0.3	526	**0.993**	0.08	680
	XGB	**0.993**	0.03	317	**0.993**	0.02	285	**0.993**	0.01	359

datapoints respectively, leading to varying evaluation times. In each experiment, we record the highest performance achieved by each method at 5-s intervals, e.g., amounting to 360 measurements for a 30-min execution. Furthermore, we determine the total number of performed queries for evaluating the hyperparameter configurations in each experiment, represented by K.

Table 2 summarizes the performance of various HPO techniques based on the accuracy of classification models and the number of performed evaluations K. The results indicate that our approach achieves the highest performance in over half of the setups, while requiring fewer or a similar number of evaluations compared to BOHB and TPE. This improvement is evident in the two smaller datasets, "Teaching Assistant" and "Musk". Despite TPE and BOHB evaluating more configurations in the same timeframe, MBMRL-HPO performs better or on par in most cases with fewer evaluations, highlighting its sample efficiency. This outcome also extends to the larger "Android Malware" dataset, where evaluations per configuration take longer, resulting in reduced K. Nevertheless, the performance gap is less pronounced across all methods as classification results remain excellent for this dataset. The increased runtime from 5 to 30 min yields less improvement in MBMRL-HPO compared to BOHB or TPE in all experiments, emphasizing the advantage of our approach within time constraints.

Additionally, Fig. 4 displays the average ranks of three HPO algorithms during a 30-min runtime. For XGBoost (b), our proposed approach outperforms both other methods, particularly after 300 s, clearly separating from TPE. On

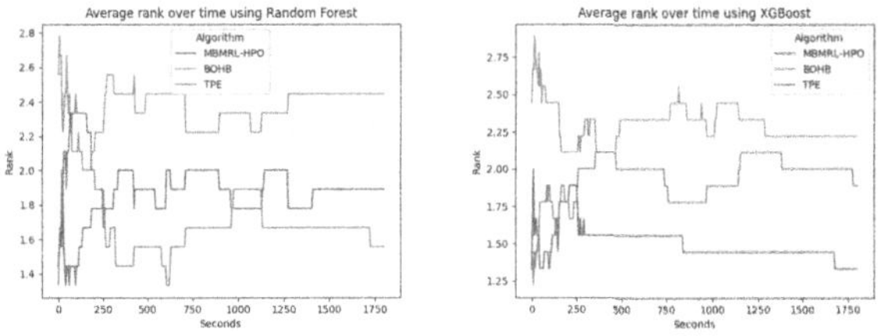

(a) Random Forest (b) XGBoost

Fig. 4. Average ranks of HPO methods over all runs and datasets with 30-min limit.

the other hand, MBMRL-HPO struggles to maintain its initial strong performance compared to TPE in Random Forest experiments (a). This is due to the fact that TPE can test more configurations in the same amount of time, especially with smaller datasets or faster underlying algorithms, which is the case for Random Forest. This pattern is also visible in Table 2, where generally all HPO methods are able to test more configurations for Random Forest than for XGBoost. Despite this, MBMRL-HPO shows superior performance in the early stages of the process, demonstrating the effectiveness of meta-learning within tight time constraints (less than 5 min). Besides, our approach excels in more complex tasks and larger datasets, such as tuning XGBoost on the large dataset "Android Malware" where MBMRL-HPO obtains the best result.

5 Conclusions

In this paper, we introduced an innovative model-based reinforcement learning approach to efficiently address the HPO problem. Our method employs meta-learning to accelerate the HPO process by transferring knowledge across various datasets. We leveraged `Deep Sets` to learn a unified representation of the state space, and integrated reward maximization and Q-value clipping to enhance the performance of TD3. By conducting a comparative evaluation with state-of-the-art HPO techniques, we demonstrated the superior performance of our approach in several classification tasks, attributed to its high sample efficiency.

References

1. Auer, P.: Using confidence bounds for exploitation-exploration trade-offs. J. Mach. Learn. Res. **3**, 397–422 (2002)
2. Bergstra, J., Bardenet, R., Bengio, Y., Kégl, B.: Algorithms for hyper-parameter optimization. In: Advances in Neural Information Processing Systems 24 (2011)
3. Bouneffouf, D., Rish, I., Aggarwal, C.: Survey on applications of multi-armed and contextual bandits. In: 2020 IEEE Congress on Evolutionary Computation (CEC), pp. 1–8. IEEE (2020)
4. Breiman, L.: Random forests. Mach. Learn. **45**(1), 5–32 (2001)

5. Chen, T., et al.: XGBoost: extreme gradient boosting. R package version 0.4-2 **1**(4), 1–4 (2015)
6. Chua, K., Calandra, R., McAllister, R., Levine, S.: Deep reinforcement learning in a handful of trials using probabilistic dynamics models. In: Advances in Neural Information Processing Systems 31 (2018)
7. Dua, D., Graff, C.: UCI machine learning repository (2017). http://archive.ics.uci.edu/ml
8. Falkner, S., Klein, A., Hutter, F.: BOHB: robust and efficient hyperparameter optimization at scale. In: International Conference on Machine Learning, pp. 1437–1446. PMLR (2018)
9. Fujimoto, S., Hoof, H., Meger, D.: Addressing function approximation error in actor-critic methods. In: International Conference on Machine Learning, pp. 1587–1596. PMLR (2018)
10. Gottipati, S.K., et al.: Maximum reward formulation in reinforcement learning. arXiv preprint arXiv:2010.03744 (2020)
11. Janner, M., Fu, J., Zhang, M., Levine, S.: When to trust your model: model-based policy optimization. In: Advances in Neural Information Processing Systems 32 (2019)
12. Jomaa, H.S., Grabocka, J., Schmidt-Thieme, L.: Hyp-RL: hyperparameter optimization by reinforcement learning. arXiv preprint arXiv:1906.11527 (2019)
13. Li, L., Jamieson, K., DeSalvo, G., Rostamizadeh, A., Talwalkar, A.: Hyperband: a novel bandit-based approach to hyperparameter optimization. J. Mach. Learn. Res. **18**(1), 6765–6816 (2017)
14. Liu, X., Wu, J., Chen, S.: Efficient hyperparameters optimization through model-based reinforcement learning and meta-learning, pp. 1036–1041 (2020)
15. Liu, X., Wu, J., Chen, S.: A context-based meta-reinforcement learning approach to efficient hyperparameter optimization. Neurocomputing **478**, 89–103 (2022)
16. Liu, X., Wu, J., Chen, S.: Efficient hyperparameters optimization through model-based reinforcement learning with experience exploiting and meta-learning. Soft. Comput. **27**(13), 8661–8678 (2023)
17. Nagabandi, A., et al.: Learning to adapt in dynamic, real-world environments through meta-reinforcement learning. arXiv preprint arXiv:1803.11347 (2018)
18. Schaul, T., Quan, J., Antonoglou, I., Silver, D.: Prioritized experience replay. arXiv preprint arXiv:1511.05952 (2015)
19. Snoek, J., Larochelle, H., Adams, R.P.: Practical Bayesian optimization of machine learning algorithms. In: Advances in Neural Information Processing Systems 25 (2012)
20. Tavakol, M., Mair, S., Morik, K.: HyperUCB: hyperparameter optimization using contextual bandits. In: Cellier, P., Driessens, K. (eds.) ECML PKDD 2019, Part I. CCIS, vol. 1167, pp. 44–50. Springer, Cham (2020). https://doi.org/10.1007/978-3-030-43823-4_4
21. Wang, C., Wu, Y., Vuong, Q., Ross, K.: Striving for simplicity and performance in off-policy DRL: output normalization and non-uniform sampling. In: International Conference on Machine Learning, pp. 10070–10080. PMLR (2020)
22. Wu, J., Chen, S., Liu, X.: Efficient hyperparameter optimization through model-based reinforcement learning. Neurocomputing **409**, 381–393 (2020)
23. Wu, J., Chen, X.Y., Zhang, H., Xiong, L.D., Lei, H., Deng, S.H.: Hyperparameter optimization for machine learning models based on Bayesian optimization. J. Electron. Sci. Technol. **17**(1), 26–40 (2019)

24. Wu, J., Liu, X., Chen, S.: Hyperparameter optimization through context-based meta-reinforcement learning with task-aware representation. Knowl.-Based Syst. **260**, 110160 (2023)
25. Zaheer, M., Kottur, S., Ravanbakhsh, S., Poczos, B., Salakhutdinov, R.R., Smola, A.J.: Deep sets. In: Advances in Neural Information Processing Systems 30 (2017)
26. Zhu, G., Zhu, R.: Accelerating hyperparameter optimization of deep neural network via progressive multi-fidelity evaluation. In: Lauw, H.W., Wong, R.C.-W., Ntoulas, A., Lim, E.-P., Ng, S.-K., Pan, S.J. (eds.) PAKDD 2020. LNCS (LNAI), vol. 12084, pp. 752–763. Springer, Cham (2020). https://doi.org/10.1007/978-3-030-47426-3_58

Towards Sustainable Precision: Machine Learning for Laser Micromachining Optimization

Luis Correas-Naranjo[1], Miguel Camacho-Sánchez[1], Laëtitia Launet[1], Milena Zuric[2], and Valery Naranjo[1](✉)

[1] Instituto Universitario de Investigación en Tecnología Centrada en el Ser Humano, HUMAN-tech, Universitat Politècnica de València, Valencia, Spain
vnaranjo@upv.es
[2] Fraunhofer Institute for Laser Technology ILT, Steinbachstraße 15, 52074 Aachen, Germany

Abstract. In the pursuit of sustainable manufacturing, ultra-short pulse laser micromachining stands out as a promising solution while also offering high-precision and qualitative laser processing. However, unlocking the full potential of ultra-short pulse lasers requires an optimized monitoring system capable of early detection of defective workpieces, regardless of the preprocessing technique employed. While advances in machine learning can help predict process quality features, the complexity of monitoring data necessitates reducing both model size and data dimensionality to enable real-time analysis. To address these challenges, this paper introduces a machine learning framework designed to enhance surface quality assessment across diverse preprocessing techniques. To facilitate real-time laser processing monitoring, our solution aims to optimize the computational requirements of the machine learning model. Experimental results show that the proposed model not only outperforms the generalizability achieved by previous works across diverse preprocessing techniques but also significantly reduces the computational requirements for training. Through these advancements, we aim to establish the baseline for a more sustainable manufacturing process.

Keywords: Laser Processing · Laser Structuring · Process Monitoring · Machine Learning · Sustainable Manufacturing

1 Introduction

In recent years, the evolution of laser machining techniques has transformed the manufacturing field. Among these innovative methods, ultra-short pulsed (USP) laser structuring has positioned itself as a competitive approach for achieving precise and high-quality laser micromachining across a wide range of materials.

As sustainability is becoming a priority in modern manufacturing practices, USP laser processing offers a more sustainable alternative to traditional manufacturing processes, with the absence of process or waste chemicals. Incorporating online monitoring into the laser processing can also help optimize the entire

V. Julian et al. (Eds.): IDEAL 2024, LNCS 15346, pp. 40–50, 2025.
https://doi.org/10.1007/978-3-031-77731-8_4

process, to drastically decrease wasted workpieces while increasing machine productivity.

Achieving an effective process monitoring of laser processing that minimizes workpiece waste requires a real-time system to enhance the quality assurance of micromachining. In recent years, the advent of machine learning (ML) methods has facilitated the application of innovative automatic algorithms to a wide variety of fields, including manufacturing [2,9,12]. Knaak et al. [9] introduced a random forest algorithm for assessing laser welding quality with data from two cameras, while Amini et al. [2] monitored selective laser melting on previously selected K-means clusters, and showed that more than 50 parameters impacted the printing quality. In a different approach, McDonnell et al. [12] used neural networks to optimize laser parameters before laser processing. Despite these advancements, monitoring USP laser processing remains a significant challenge due to the demanding requirements of high spatial resolutions and processing speeds [16], hence the need to optimize ML models to enable real-time implementation.

Simultaneously, current algorithms for USP laser process monitoring struggle to achieve promising results across inherently different preprocessing techniques such as milling, grinding, or polishing. In [10], Leyendecker et al. developed an ML solution to predict surface roughness, but results showed to be strongly dependent on workpiece properties and preprocessing techniques. This sensitivity to preprocessing techniques underscores the need for a versatile and effective system capable of meeting the diverse demands of micromachining processes.

In light of these challenges, this paper proposes an ML-based approach to address the challenges in USP laser micromachining, aiming to enhance results across multiple preprocessing techniques. Considering roughness as a key feature that reflects the overall process quality, the proposed approach focuses on predicting the final surface roughness after laser structuring, utilizing initial laser parameters and photodiode sensor data as inputs. Recognizing the high dimensionality of sensor data, this research also aims to strategically reduce input features to optimize monitoring ML models. This not only tackles the computational cost associated with training but also facilitates the application of lighter ML models, aiming for efficient real-time monitoring of the USP laser process.

2 Material

In this work, we leveraged a multi-sensor dataset for process monitoring provided by the Fraunhofer Institute for Laser Technology (ILT). As depicted in Fig. 1, the dataset comprises 60 experiments conducted on one of the lasers, each associated with a different workpiece and diverse preprocessing techniques, including milling, grinding, polishing, wire-EDM, and die-EDM. Within each experiment, 99 distinct sub-experiment samples were generated, each gathering information about laser parameters, initial and final roughness, along with sensor emission data for every layer. This sensor data was obtained from four photodiodes strategically mounted around the focusing lens, capturing infrared (IR), acoustic, laser reflection, and visible light (VIS) emissions. To generate the dataset, laser parameters were defined prior to processing for the samples of each

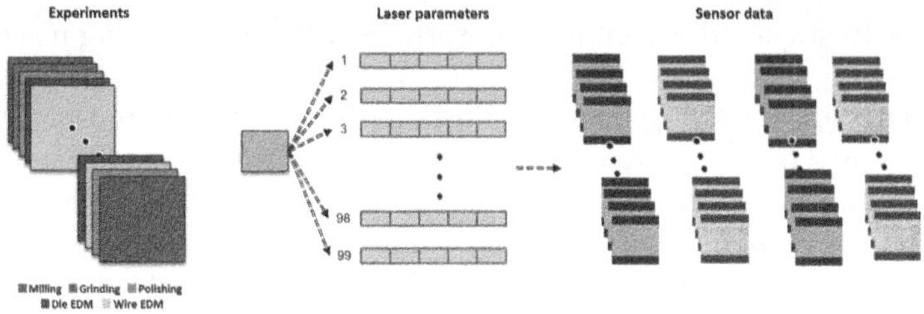

Fig. 1. Data structure for the different experiments leveraged in this project. (Left) Set of workpiece experiments with diverse preprocessing techniques; (Center) For each workpiece, 99 sub-experiments are carried out with varying laser parameters; (Right) Resulting sensor data for each available sub-experiment, consisting of 5684 layers in total per experiment.

experiment, thus resulting in a set of 99 parameters that were consistently used across all experiments. Specifically, these varying parameters correspond to the following: pulses per burst, pulse fluence (J/cm^2), laser power (W), and number of layers (laser scans over a particular area).

Building on this dataset, our approach aims to predict the final roughness of a sample regardless of the preprocessing applied (e.g., milling, grinding, etc.) before surface processing with the laser, while reducing input features for optimized monitoring.

3 Methods

The proposed pipeline is depicted in Fig. 2 and consists of four stages defined as follows: (A) Data preprocessing for both laser parameters and sensors signal; (B) Baseline model definition; (C) Feature importance selection; (D) Applicability with ML models.

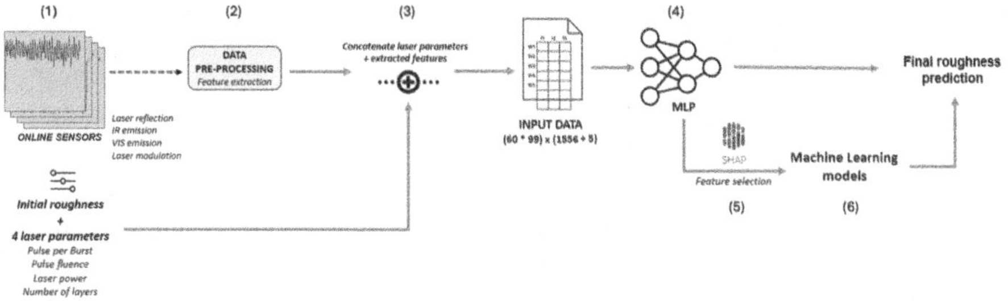

Fig. 2. Overall ML-based pipeline for optimized surface roughness prediction, (1) leverage data from the laser parameters and online sensors, (2) reduce sensor data input with feature extraction (see Fig. 3), (3) combine available input data to feed our models, (4) train a multi-layer perceptron (MLP) model for roughness prediction, (5) leverage the MLP weights to select most relevant features and, (6) by reducing the dimensionality of features, apply less computationally intensive ML models to predict roughness.

3.1 Data Preprocessing

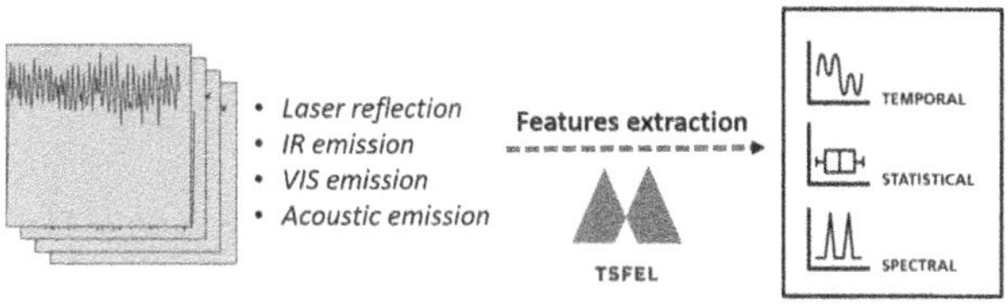

Fig. 3. Data preprocessing of the input sensor data that performs feature extraction using TSFEL, extracting characteristics from different domains.

To transform the information collected by the sensors into useful input data for the model, we used a Python package designed for feature extraction on time series data, the Time Series Feature Extraction Library (TSFEL) [3]. This enabled us to obtain characteristics from the statistical, temporal, and spectral domains of the available signals, as illustrated in Fig. 3.

As a result of this feature extraction, 389 features were extracted across the three domains for each sensor: 14 from the temporal domain, 40 from the statistical domain, and 335 from the spectral domain. To optimize real-time monitoring and reduce computational complexity, we specifically chose to extract characteristics solely from the last layer of the laser processing for each sample processed. This layer is expected to contain the most relevant information for predicting final surface roughness.

After extracting these features and assigning them with their corresponding laser parameters, we apply a min-max normalization to help the model better manage the values during training, thereby improving its generalization ability. In this way, each parameter and each feature is normalized based on its minimum and maximum values, confining the values between 0 and 1.

Finally, as a last data preprocessing step to prepare for model training, we apply a training-evaluation split to the dataset, where 80% is allocated for training and 20% for evaluation. This allows us to reliably evaluate the model with unseen data from the training process, determining if the model can truly generalize to new data. In addition, a cross-validation with $k = 4$ is applied on the training partition, as a resampling procedure to evaluate machine learning models on the limited data samples available.

3.2 Training Phase

Model Architecture. The proposed solution leverages the multi-layer perceptron (MLP) [13], a model based on neural networks capable of correctly predicting a key property in the subsequent study by establishing non-linear relationships between the different laser parameters and characteristics extracted from the sensor signals. Its training process consists of a forward propagation algorithm followed by a backpropagation process adjusting the weights of the connections to minimize the loss or error function.

Loss Function. We used the Mean Squared Error (MSE) to optimize models, a function that measures the error between each element in the input x and target y as follows:

$$\mathrm{MSE} = \frac{1}{N}\sum_{i=1}^{N}(y_i - \hat{y}_i)^2$$

where N is the total number of data points in the dataset, y_i is the ground truth, and $\hat{y}_i$ is the predicted value.

Hyperparameter Optimization. In the training phase of the model, we perform a thorough fine-tuning of model parameters to maximize its performance. Traditionally, this fine-tuning process relied on a trial-and-error approach, exploring values known to typically deliver satisfactory results, often employing a brute force grid search. However, in this study, we used the tree-structured Parzen estimator [15], a Bayesian optimization algorithm that leverages probabilistic modeling to intelligently guide the search for optimal hyper-parameter tuning. This algorithm is the foundation of the Optuna [1] library, an open-source hyperparameter optimization framework engineered to automate the hyperparameter search process. Employing this tool, we successfully identify the optimal values for the batch size, learning rate, L1 and L2 regularizations, and determine the most effective architecture for the MLP model. We also explore various optimizers and activation functions.

Feature Reduction. This project is situated within the context of an industrial production line, where efficiency and precision are of paramount importance. However, achieving real-time implementation requires a significant reduction in the computational costs of the monitoring system. To strike a balance between these two factors, we employ an explainability method to meticulously pinpoint the features that have the most significant impact on the model training process. This approach allows us to streamline the number of parameters involved, thereby reducing computational costs. For increased transparency and interpretability, we aim to achieve a reproducible solution for broader applications, specifically identifying influential features.

Therefore, to maximize the feature selection explainability, we leverage the SHapley Additive exPlanations (SHAP) [11] package, a robust Python library grounded in Game Theory. SHAP assigns a Shapley value to each input feature within the predictive context, quantifying the contribution of the respective feature towards a specific prediction decision. With this method, we aim to add valuable insights into the inner workings of the model.

4 Experiments

4.1 Experimental Settings

Training Experiments. To rigorously assess the effectiveness of the proposed approach, we conducted a series of experiments aimed at evaluating model performance under various conditions. The experiments were structured to provide comprehensive insights into the adaptability of the model across different laser processing techniques (e.g., milling, grinding, polishing), aiming to optimize its performance as a baseline for future real-time monitoring.

- **Baseline model with laser parameters** as input features. This experiment serves as a baseline to assess model performance without integrating sensor data, focusing on the influence of laser parameters on the predictive capabilities of the model.
- **Model with laser parameters and sensor data** as input, aiming to evaluate the performance improvement achieved by incorporating additional information from sensor measurements.
- **Dimensionality reduction model application.** To optimize the computational requirements, additional ML architectures are compared after reducing the number of input features with the SHAP library based on the best-performing model so far.

Training Settings. To facilitate a comprehensive comparison of model performance across preprocessing techniques, the training process was carried out in two distinct settings for each approach, individually and in combination. In the individual setting, a separate model was trained for each preprocessing method (i.e., milling, grinding, polishing, die-EDM and wire-EDM), with each preprocessing technique identified by its unique name. In the combined setting, samples from all preprocessing techniques were gathered to train a global model. As for the model architecture, it remained consistent for all training sessions (i.e., different settings) within each approach.

Evaluation Metrics. To objectively assess the effectiveness of the proposed models, we used the R-Squared (R2) and Root Mean Squared Error (RMSE) as reference metrics, respectively defined as follows:

$$R^2 = 1 - \frac{\sum (y_i - \hat{y}_i)^2}{\sum (y_i - \bar{y})^2}, \qquad RMSE = \sqrt{\frac{\sum_{i=1}^{N} \|y_i - \hat{y}_i\|^2}{N}}$$

Given y_i as the ground truth, $\hat{y}_i$ as the model predictions, and N as the total number of observations, the model fits better when R2 is closer to 1 and RMSE approaches 0.

4.2 Results

Baseline Model with Laser Parameters. When training the MLP model with the 4 laser parameters and the initial surface roughness, the optimal architecture determined using the Optuna library resulted in an MLP with a single hidden layer composed of three neurons. As presented in Table 1, the results for this experimental setup highlight the necessity for further refinement to ensure robust performance across diverse preprocessing scenarios. While the combined performance is satisfactory when training solely with laser parameters, reaching an R2 metric on the test set of 0.9710 and 0.9463 for milling (a) and combined (f) settings, respectively, its generalizability and adaptability to all preprocessing techniques are not consistent. This observation holds particular significance given that in a real-world scenario, data variability would be substantially higher.

Table 1. Results across all settings for the model trained with laser parameters only.

	Train R2	Test R2	Train RMSE	Test RMSE
Milling (a)	0.9726	0.9710	1.0786	1.0799
Grinding (b)	0.4026	0.2285	0.8585	0.8596
Polishing (c)	0.4730	0.3749	0.1627	0.1773
Die-EDM (d)	0.9231	0.9137	0.6145	0.6547
Wire-EDM (e)	0.5425	0.4771	0.3067	0.2859
Combined (f)	0.9514	0.9463	1.0742	1.0539

By incorporating the information extracted from sensor signals, we aim to improve the model's robustness and observe whether the results show improvement across various preprocessing techniques. In this new setup, the model's input now consists of 1561 input neurons, combining the initial 5 parameters with the 1556 features extracted from sensor data. This change alters the structure of the model compared to the previous configuration, thus resulting in an optimized model architecture of two hidden layers with 443 and 19 neurons, respectively, using the Optuna optimization package. Results obtained with this updated configuration are presented in Table 2, comparing results obtained in [10], highly sensitive to the preprocessing technique, with ours.

Table 2. Comparative table between the results obtained in [10] (above) and ours (below).

	Train R2	Test R2	Train RMSE	Test RMSE
[10] *approach on feature set G*				
Milling (a)	0.9881	0.9925	0.6748	0.5560
Grinding (b)	0.8757	0.6693	0.4144	0.6746
Polishing (c)	0.9683	0.8169	0.0377	0.1065
Die-EDM (d)	0.9928	0.9733	0.1666	0.3749
Wire-EDM (e)	0.7850	0.2339	0.1863	0.5133
Combined (f)	X	X	X	X
Our solution				
Milling (a)	0.9993	0.9812	0.1693	0.8682
Grinding (b)	0.9866	0.8227	0.1222	0.5180
Polishing (c)	0.9958	0.8931	0.0146	0.0733
Die-EDM (d)	0.9997	0.9649	0.0379	0.4177
Wire-EDM (e)	0.9996	0.8225	0.0094	0.1666
Combined (f)	0.9971	0.9811	0.2628	0.6264

In comparing our results with the model trained with laser parameters only (Table 1), incorporating sensor data demonstrates notable improvements in model generalization across all preprocessing techniques. Specifically, previously low results for preprocessing techniques such as grinding (b), polishing (c), and wire-EDM (e) now show a significant improvement, with the R2 metric for grinding (b), for example, increasing from 0.4026 to 0.9866 on the train set and 0.2285 to 0.8227 on the test set. These outcomes underscore the effectiveness of incorporating sensor data in enhancing model performance across diverse preprocessing scenarios. Figure 4 shows a scatter plot comparing the ground truth values to the predicted ones for this improved model, where each dot represents an individual data point's actual and predicted values. The alignment of dots along the diagonal indicates a close match between predicted and actual values, affirming the high performance of the prediction model.

In comparing our approach with the one presented in [10], ours shows significant improvement across all preprocessing techniques. Notably, while results for the grinding process (b) in [10] achieved a test R2 of 0.6693, our approach demonstrates a performance of 0.8227 in the same metric. These comparisons are broken down in Table 2.

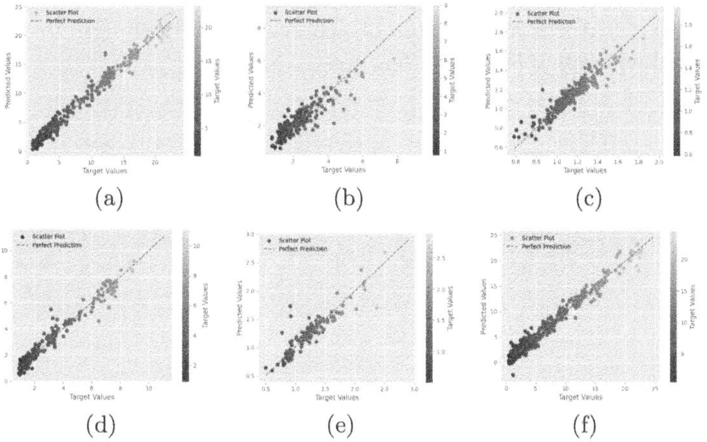

Fig. 4. Scatter plots corresponding obtained on the evaluation set, representing data point predictions with respect to ground truth.

Dimensionality Reduction Model Application. The top-performing baseline model was used as a foundation to identify most relevant features with the aim of optimizing the computational requirements of USP monitoring.

The Shapley values were extracted with SHAP from the models presented in Table 2, to identify which input neurons contributed the most information to the model. Through ablation experiments, we studied the optimal number of sensor features to find a tradeoff between feature reduction and performance. This analysis is depicted in Fig. 5-(b), comparing results with the original 1556 sensor features against significantly reduced sets, with 20 features identified as a minimum yielding good metrics. Further reduction in the number of features had a negative impact on model performance for most preprocessing techniques. The obtained Shapley results for these 20 most impacting features are exemplified in Fig. 5-(a), using the top-performing base model for the grinding preprocessing.

This substantial reduction in input feature dimensions paves the way for deploying lighter ML models, ultimately aiming to enable future optimized real-time monitoring. Leveraging the 20 selected sensor features, together with the 5 laser parameters and initial surface roughness, we conducted additional trainings using diverse ML models, encompassing extremely randomized trees (ET) [7], Random Forest (RF) [4], Decision Trees (DT) [14], CatBoost (CB) [6], LightGBM (LGBM) [8] and XGBoost (XGB) [5]. The ET reached the best results on all preprocessing techniques except for polishing, where CB slightly outperformed the ET. Results for the top-performing model, i.e., the ET, are detailed in Table 3, and demonstrate that lighter ML models trained with considerably reduced sensor features are capable of achieving results similar to those of an MLP trained with all 1556 sensor features.

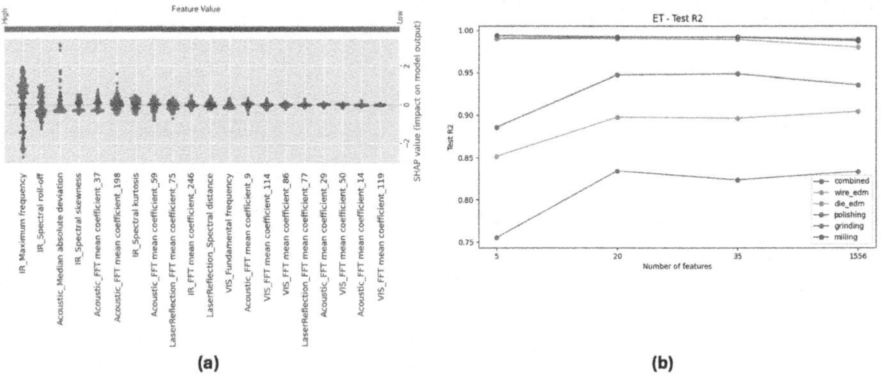

Fig. 5. (a): Example of the 20 selected Shapley feature values obtained for grinding preprocessing. (b): R2 test results on all preprocessing techniques, for different thresholds of selected features extracted with SHAP [11]. Points on the extreme right of the figure refer to reference results obtained without reducing features.

Table 3. Results obtained for each of the processes after selecting the most important features, with the top-performing model (ET).

	Train R2	Test R2	Train RMSE	Test RMSE
Milling (a)	1.0000	0.9956	0.0000	0.4170
Grinding (b)	1.0000	0.8186	0.0000	0.5237
Polishing (c)	0.9971	0.9473	0.0119	0.0515
Die-EDM (d)	1.0000	0.9924	0.0000	0.1936
Wire-EDM (e)	1.0000	0.9243	0.0000	0.1087
Combined (f)	1.0000	0.9934	0.0000	0.3696

5 Conclusion

In this paper, we proposed a machine learning approach for laser micromachining optimization that is effective across diverse preprocessing techniques. Our experiments demonstrated that sensors are a valuable source of information in predicting the final surface roughness. To address the impracticality of extracting a large number of features in a real-world setting, we conducted an explainability study to select the most influential features. By strategically reducing the dimensionality and computational cost of the input data, we were able to apply simpler machine learning models such as random forest without compromising performance across preprocessing techniques. Future work will involve testing this solution in our real-time system and aiming to achieve comparable predictions by focusing on earlier layers of the process.

Acknowledgments. Luis Correas, Miguel Camacho-Sánchez and Laëtitia Launet received funding from Horizon Europe, the European Union's Framework Programme for Research and Innovation, under Grant Agreement No. 101057457 (METAMORPHA).

References

1. Akiba, T., Sano, S., Yanase, T., Ohta, T., Koyama, M.: Optuna: a next-generation hyperparameter optimization framework. CoRR abs/1907.10902 (2019). http://arxiv.org/abs/1907.10902
2. Amini, M., Chang, S.: Process monitoring of 3D metal printing in industrial scale. In: International Manufacturing Science and Engineering Conference, vol. 51357, p. V001T01A035. American Society of Mechanical Engineers (2018)
3. Barandas, M., et al.: TSFEL: time series feature extraction library. SoftwareX **11**, 100456 (2020). https://doi.org/10.1016/j.softx.2020.100456
4. Breiman, L.: Random forests. Mach. Learn. **45**(1), 5–32 (2001). https://doi.org/10.1023/A:1010933404324
5. Chen, T., Guestrin, C.: XGBoost: a scalable tree boosting system. CoRR abs/1603.02754 (2016). http://arxiv.org/abs/1603.02754

6. Dorogush, A.V., Gulin, A., Vorobev, A., et al.: Fighting biases with dynamic boosting. CoRR abs/1706.09516 (2017). http://arxiv.org/abs/1706.09516
7. Geurts, P., Ernst, D., Wehenkel, L.: Extremely randomized trees. Mach. Learn. **63**(1), 3–42 (2006). https://doi.org/10.1007/s10994-006-6226-1
8. Ke, G., et al.: LightGBM: a highly efficient gradient boosting decision tree. In: Proceedings of the 31st International Conference on Neural Information Processing Systems, NIPS 2017, pp. 3149–3157. Curran Associates Inc., Red Hook (2017)
9. Knaak, C., Thombansen, U., Abels, P., Kröger, M.: Machine learning as a comparative tool to determine the relevance of signal features in laser welding. Procedia CIRP **74**, 623–627 (2018)
10. Leyendecker, L., Zuric, M., Nazar, M.A., Johannes, K., Schmitt, R.H.: Predictive quality modeling for ultra-short-pulse laser structuring utilizing machine learning. Procedia CIRP **117**, 275–280 (2023). https://doi.org/10.1016/j.procir.2023.03.047
11. Lundberg, S.M., Lee, S.I.: A unified approach to interpreting model predictions. In: Guyon, I., et al. (eds.) Advances in Neural Information Processing Systems, vol. 30. Curran Associates, Inc. (2017). https://proceedings.neurips.cc/paper_files/paper/2017/file/8a20a8621978632d76c43dfd28b67767-Paper.pdf
12. McDonnell, M.D., et al.: Machine learning for multi-dimensional optimisation and predictive visualisation of laser machining. J. Intell. Manuf. **32**, 1471–1483 (2021)
13. Popescu, M.C., Balas, V., Perescu-Popescu, L., Mastorakis, N.: Multilayer perceptron and neural networks. WSEAS Trans. Circ. Syst. **8**, 579–588 (2009)
14. Rokach, L., Maimon, O.: Decision trees. In: Maimon, O., Rokach, L. (eds.) Data Mining and Knowledge Discovery Handbook, vol. 6, pp. 165–192. Springer, Boston (2005). https://doi.org/10.1007/0-387-25465-X_9
15. Watanabe, S.: Tree-structured parzen estimator: understanding its algorithm components and their roles for better empirical performance. arXiv preprint arXiv:2304.11127 (2023)
16. Zuric, M., Nottrodt, O., Abels, P.: Multi-sensor system for real-time monitoring of laser micro-structuring. J. Laser Micro/Nanoeng. **14**(3), 245–254 (2019)

Association Rules Mining with Auto-encoders

Théophile Berteloot(✉), Richard Khoury, and Audrey Durand

Department of Computer Science and Software Engineering, Université Laval, Québec, Canada
theophile.berteloot.1@ulaval.ca,
{richard.khoury,audrey.durand}@ift.ulaval.ca

Abstract. Association rule mining is one of the most studied research fields of data mining, with applications ranging from grocery basket problems to explainable classification systems. Classical association rule mining algorithms have several limitations, especially with regards to their high execution times and number of rules produced. Over the past decade, neural network solutions have been used to solve various optimization problems, such as classification, regression or clustering. However there is still no efficient way to mine association rules using neural networks. In this paper, we present an auto-encoder solution to mine association rule called ARM-AE. We compare our algorithm to FP-Growth and NSGAII on three categorical datasets, and show that our algorithm discovers high support and confidence rule set and has a better execution time than classical methods while preserving the quality of the rule set produced.

Keywords: Association rules mining · Auto-encoder · Knowledge discovery

1 Introduction

Association rule mining (ARM) was first introduced by Agrawal [1] to solve the grocery basket problem, and since then it has found numerous applications in Knowledge Discovery in Database (KDD) problems ranging from financial analysis [19] to medical diagnostics [2]. An association rule (AR) is an implication of the form $A \Rightarrow C$, which can be read as "if antecedent A is true then consequent C must be true", where A and C are sets of different items (itemsets) in a database. An AR is defined by its antecedent, its consequent and two measures [11]. The first one is the support, which is the proportion of rows in the dataset where both the antecedent and the consequent appear. The second measure is the confidence, the conditional probability to observe the consequent given an observation of the antecedent.

The most widely-used mining strategies, such as the Apriori algorithm [1] and other exhaustive strategies [3,4,20], typically work by first mining frequent

V. Julian et al. (Eds.): IDEAL 2024, LNCS 15346, pp. 51–62, 2025.
https://doi.org/10.1007/978-3-031-77731-8_5

itemsets, then combining those itemsets to produce association rules. However, all these algorithms face the same problems: the number of rules they produce increases exponentially with the number of items in the database, and thus it becomes impossible for a human to sort through the rules returned to pick out the best ones [10]. Their execution time also become an issue with massive datasets [10]. Finally, these algorithms need support and confidence thresholds in order to efficiently search through the solution space, and those thresholds need to be carefully chosen: low values can lead to long execution times and an overabundance of rules, while high values cause the algorithm to miss interesting rules.

Researchers have tried to address these issues using top-k algorithms [9], metaheuristics [23] such as NSGAII [17], parallelism [12], and GPU [6]. Meanwhile, deep learning has been used in research from image analysis [16] to natural language processing [21] and have demonstrated a formidable capacity to find underlying relationships between input data [14]. Auto-encoders (AE) are a kind of neural network often used to find clusters [18] or to learn representations [27]. AE are trained in a supervised way using a dataset as an input and the same dataset as a target.

Thus, AE seems applicable to ARM, because while trying to rebuild the original dataset, the AE will learn links between columns in the dataset; we then can use those links to mine AR.

The contributions of this paper are therefore twofold:

- We propose a new approach to mine AR using AE, which we call ARM-AE.
- We compare our approach to FP-Growth, an exhaustive state-of-the-art algorithm, and NSGAII, a commonly used metaheuristic, based on execution time, number of rules produced, average support and confidence of the rules produced, and coverage (the percentage of FP-Growth rules that our approach finds). We use three categorical datasets taken from the popular UCI repository [7], ranging from a small dataset to a massive one. In all cases, ARM-AE is able to efficiently mine a limited number of high-quality rules, without the need to specify any mandatory thresholds.

The rest of this paper is organized as follows. In Sect. 2 we present the literature related to the usage of neural networks to solve ARM problems. In Sect. 3 we present our ARM-AE model. We conduct experiments with it, using FP-Growth and NSGAII as benchmarks, in Sect. 4, and discuss the results in Sect. 5. Finally, we provide a synthesis of our results and concluding remarks in Sect. 6.

2 Related Work

Several previous works have used deep learning within the context of ARM. For example, AR and a deep neural network have been used previously to predict car accidents based on tweets [26]. More specifically, AR were mined using Apriori on the tweets' tokens and used as input for a deep belief network to classify whether a tweet is about a car accident or not. AR and convolutional neural networks (CNNs) have also been used together for classifying reviews of three Spanish monuments and extracting the aspects that visitors didn't like to help improve the visitors' experience [24]. CNNs were first used to extract the aspects of monuments people felt strongly about. Then they were clustered into groups with the same meaning but represented by different words. Finally AR were mined to summarize sentiment paired with aspects. As a last example, Apriori has been used together with deep learning to predict which anti-cancer drugs should be given to an individual patient [25]. In this work, AR were mined from a pharmacogenomics dataset, and the extracted features were used to train a deep neural network.

While a lot of projects, such as the examples cited previously, combine both ARM and deep learning, few actually perform ARM using deep learning. One of them is [13], in which a neural network and a genetic algorithm were used to mine AR. The authors used a self-organizing map (SOM) to mine clustered pattern from the data, and then used a genetic algorithm to create the AR based on these patterns. In [15], a convolutional neural network was used to mine AR on medical datasets with complex attributes, then a probability estimation method validated the AR found.

None of the systems above use deep-learning to mine ARs directly, they either use deep learning to mine frequent itemsets or they use it alongside of ARM algorithms. Testing itemset combinations to find good ARs is a very time-consuming task. Our proposed ARM-AE approach mines ARs directly and thus doesn't spend time trying different combinations.

3 ARM-AE: Association Rule Mining with Auto-encoders

In this paper, we consider a dataset with binary attribute values. We define an item as a column with value 1 in the dataset, and an itemset as a group of items with no duplicates. In order to compare different rules we use the support and the confidence. The support is defined as the proportion of rows in the dataset D where both A and C appear: $Supp(A, C) = |A \cap C|/|D|$. The confidence is the conditional probability to observe the consequent given an observation of the antecedent: $Conf(A, C) = |A \cap C|/|A|$.

We propose to leverage the representation power of auto-encoders to directly extract ARs from the dataset. An AE consists of two parts, first the encoder which consists of several neuron layers of decreasing size, and second the decoder which consists of several layers of increasing size. The training goal of an AE is for the encoder to generate a compressed representation of the input, and for the decoder to generate a transformed version of the original input. Training for ARM-AE doesn't differ from a classical AE training. The input and the requested output of the AE are each rows of the dataset. The AE will thus learn to encode an input itemset in relation to similar itemsets, and the decoded output will be the similar itemsets, with items having a value between 0 and 1 representing how likely they are to belong to that itemset. Given that our goal is to learn a representation of the entire dataset, there is no need for evaluation or test datasets, and we train our model on the entire dataset. During training the AE will learn to recreate the input binary row, thus after training if an itemset is used as input, the AE will recreate it, but not perfectly. Items which appear frequently with the input itemset in the dataset will have higher output value than items which never appear with the input itemset in the dataset.

The core of our ARM-AE approach is presented in Algorithm 1. Our algorithm tries to mine a set of $N \times M$ rules, or N rules of maximum antecedent length M along with the most interesting subsets of the antecedent itemset, for each consequent. In ARM-AE, each item in the dataset becomes a consequent (line 5), though it would be trivial to change that line to focus the algorithm on a specific subset of items of interest. The consequent is composed of only one item, which is the most common use-case for ARM.

Algorithm 1. Association Rule Mining Auto-Encoder.

```
1: function ARM-AE(N, M, L)              ▷ N number of maximum-length rules per
   consequent, M maximum number of antecedents per rule, L similarity threshold
2:     Consequents ← items
3:     AE ←trained Auto-encoder
4:     Rules ← ∅
5:     for ∀Consequent ∈ Consequents do
6:         ConsequentRules ← ∅
7:         for I = 1 to N do
8:             Antecedents ← ∅
9:             for J = 1 to M do
10:                InputArray ← Antecedents ∪ Consequent
11:                AEOutput ← AE.forward(InputArray)
12:                AEOutput ← Sort(AEOutput)
13:                for Antecedent ∈ AEOutput do
14:                    if Antecedent ≠ Consequent and Antecedent ∉ Antecedents and
   ComputeSimilarity(ConsequentRules,Antecedents ∪ Antecedent) ≤ L then
15:                        Antecedents ← Antecedents ∪ Antecedent
16:                        ConsequentRules ← ConsequentRules ∪ Antecedents
17:                        Exit loop of line 13 (go to line 9)
18:                    end if
19:                end for
20:            end for
21:            Rules ← Rules ∪ ConsequentRules
22:        end for
23:    end for
24:    return Rules
25: end function
```

Initially, the consequent is the only input for the AE (lines 8–11). The AE gives as output a score between 0 and 1 for each item in the dataset; the higher the score, the higher the probability that this item appears often with the AE input. This output is sorted in decreasing order, and the item with the highest value that isn't the consequent, isn't already part of the antecedent, or would not make the antecedent more similar than a threshold L to a previously-discovered one as described in Sect. 3.2, is added to the antecedent of the rule and this new rule is added to the set of AR for that consequent (lines 12–16). The antecedent is appended to the consequent as input to the AE (line 10) and the loop repeats, adding one item at a time to the antecedent until it reaches the maximum length of M items (line 9). Then another rule is generated from scratch for that consequent until N rules are found (line 7) before moving on to the next consequent (line 5). The complete code of ARM-AE is available on GitHub.[1]

3.1 AE Training and Hyperparameters

As explained previously, there are three main hyperparameters to our algorithm, namely the number of maximum-length rules produced for each consequent N, the maximum number of items in the antecedent M, and the maximum similarity threshold L. There are also the hyperparameters for the AE itself, such as the number of layers and of neurons per layer, the number of training epochs, the learning rate, the activation function, the loss function, and the optimizer. These are typical for all AE architectures.

3.2 Similarity

An important step is computing the similarity between a new antecedent and those already found for a consequent, at line 14 of Algorithm 1. Without this step, starting from the same consequent, the AE would always discover the same antecedent. Imposing a maximum level of similarity between a new antecedent and those already discovered forces the AE to explore the solution space and to generate a variety of different AR for each consequent.

For our implementation of ARM-AE, the similarity metric we use is the maximum overlap between the items in the antecedent of the candidate rule and of existing rules for the same consequent, expressed as a percentage of the length of the existing rule. We only consider the similarity to rules of the same length as the candidate or longer, since otherwise an existing short and general rule would prevent the discovery of longer and more specific candidate rules. We can note however that this specific similarity function is not necessary for ARM-AE to function; another problem-specific similarity function could be used instead.

3.3 Illustrative Example

In order to better understand how mining Association Rules (AR) with an autoencoder works, we present an application of a simplified version of ARM-AE on a toy dataset. Our toy dataset is as follows $\begin{bmatrix} A\,0\,0\,0\,1\,1\,1\,1\,1\,1\,0\,0\,1 \\ B\,1\,1\,1\,0\,0\,0\,0\,0\,0\,1\,1\,0 \\ C\,1\,1\,1\,1\,1\,1\,1\,1\,1\,0\,0\,0 \\ D\,0\,0\,0\,0\,0\,0\,0\,0\,0\,1\,1\,1 \end{bmatrix}^T$.

[1] https://github.com/TheophileBERTELOOT/ARM-AE.

We can manually compute the itemsets with the highest support (how often each item appears with another item) for each item: Our itemsets are [(A, C), (B, C), (B, D),(A, D)]. All other combinations do not exist in our toy dataset.

To demonstrate how ARM-AE operates, we use a simple one-layer network, trained as an autoencoder. We used a learning rate of 10^{-2} and trained the model for 50 epochs. We can first analyze the results of our algorithm:

- Using A as input, here is the output of the network: $\begin{bmatrix} A & 0.96 \\ B & -0.30 \\ C & 0.94 \\ D & 0.78 \end{bmatrix}$ The highest score is obviously A, due to the autoencoder training process, but the second highest is for column C, and the third highest is for column D.
- Using B as input, the output is: $\begin{bmatrix} A & -0.41 \\ B & 0.98 \\ C & 0.95 \\ D & 0.78 \end{bmatrix}$ The highest is B as expected, but the second highest is C, and D is higher than A.
- Using C as input, the output is: $\begin{bmatrix} A & 0.79 \\ B & 0.76 \\ C & 0.94 \\ D & -0.11 \end{bmatrix}$ In order, the highest are C then A then B then D.
- Using D as input, the output is: $\begin{bmatrix} A & 0.77 \\ B & 0.75 \\ C & -0.67 \\ D & 0.98 \end{bmatrix}$ In order, the highest are D then A and B.

Our system correctly identified the most frequent itemsets [(A, C), (B, C), (B, D), (A, D)] and itemsets that do not exist have the lowest score.

To further showcase how our algorithm works, we analyze the evolution of the biases and weights of our neural network in Fig. 1. The weight wAB represents the weight of the first input of neuron A. The bias bA represents the bias of neuron A.

We can see that throughout the training, the weights representing the links between frequent items mainly increase or stay the same. On the other hand, the weights representing the links between items that do not appear together decrease or stay the same. For the biases, their evolution is here to modulate the evolution of the weights. Considering the evolution of these weights and biases, and that a neuron performs a weighted sum of its inputs and adds its bias, this explains how an auto-encoder can find association rules in a dataset.

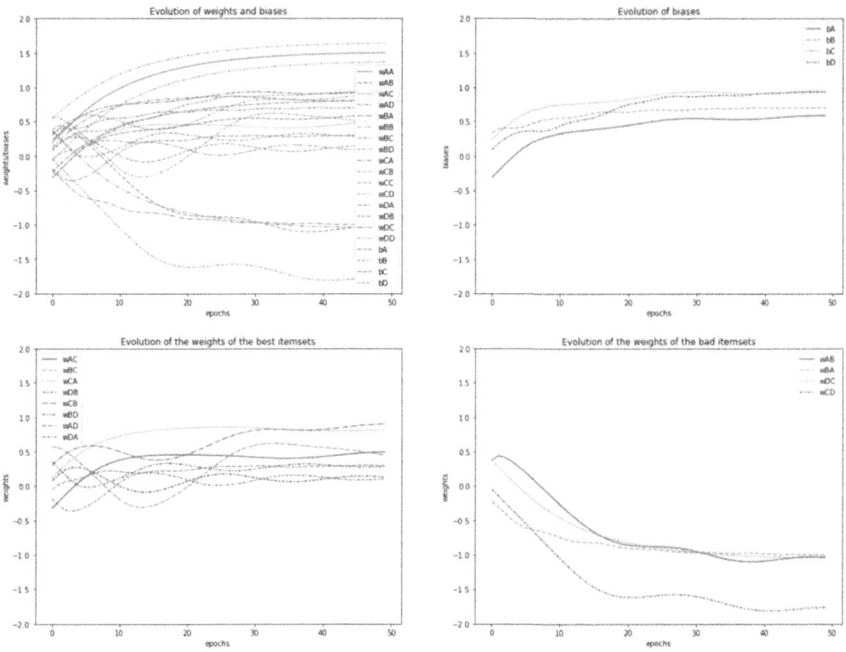

Fig. 1. Evolution of the weights and bias of our autoencoder trained on the toy dataset

4 Experiments

4.1 Datasets

We tested ARM-AE using three categorical datasets (Chess, Nursery, Plants) taken from the popular UCI repository [7]. Those datasets were selected because they have a wide range of number of rows (from 3.196 to 34,781) and items (from 32 to 75). Each dataset is preprocessed by replacing each categorical attribute with a one-hot set of binary attributes, with one column for each category value.

4.2 Hyperparameters

For the architecture of our AE, both the encoder and decoder are made up of three fully connected layers. The number of neurons for each layer is the number of different items in the dataset. This architecture has shown the best results in our experiments. The neurons use the tanh function as their activation function. The learning rate is 10^{-3}. The MSELoss is our loss function and we used the Adam optimizer.

We set the algorithm to look for rules with a maximum of two items in the antecedent ($M = 2$) and for two rules per consequent ($N = 2$). The similarity threshold $L = 0.5$, so there will be a maximum of one item overlap between the antecedents of the two pairs of rules. In order to set the number of training epochs, we use the training loss when it becomes lower than a threshold then we take that epoch as our final model. We tested several values for this threshold, and find that the best value is 0.1 which in practice means 1 to 5 epoch.

4.3 Benchmarks

We use FPGrowth [3] as one of our two benchmarks, as it is one of the fastest and most popular state-of-the-art exhaustive ARM algorithms. We used the implementation found in the Mlxtend library [22]. The support and confidence thresholds for each dataset are picked considering our time and computation ressources. In order to have a fair comparison with ARM-AE we selected a subset of rules among the thousands proposed by FPGrowth. For each item considered as consequent we selected the two rules with the highest support with one antecedent and the two rules with the highest support with two antecedent. If our method is perfect it should produce the same output as the subset selected from the output of FPGrowth. Our second benchmark is the NSGAII [5] algorithm, chosen because genetic algorithms and domination concepts are a popular trend in ARM [23] and NSGAII is a commonly-used baseline for meta-heuristic. We used the DEAP library [8] to implement our NSGAII. For our benchmark algorithm NSGAII, we use a population of 100 individuals, a crossover ratio of 0.9 and a mutation probability of 0.01. We set a maximum of 2 antecedents and 1 consequent in the rules. We keep the 300 best rules found across all generations for the plants and chess datasets and the 150 best for the nursery dataset, so results are comparable with ARM-AE and FP-Growth outputs. To set the number of generations we check if the average support and average confidence of the best rules found so far increase more than a threshold. If both decrease or increase less than the threshold then the NSGAII algorithm stops. We use a threshold of 0.01.

4.4 Metrics

We compute several metrics to compare ARM-AE and our benchmarks. The first one is the number of rules found by ARM-AE that have a support greater than 0. A rule that has a support greater than 0 is one that actually exists in the dataset. But since ARM-AE does not compute the support of the rules during its search, unlike FP-Growth and other exhaustive algorithms, it offers no guarantee in that respect, thus the importance of this metric. The second metric is the percentage of rules found by FP-Growth that are also found by our algorithm and NSGAII. Since FP-Growth performs an exhaustive search for all the best rules, this will indicate how close to complete the ARM-AE and NSGAII rule sets are. Recall however that we set ARM-AE's hyperparameters to generate four AR per consequent with only one or two items in the antecedent. Next, we compare the execution times for all three algorithms. For our algorithm we include the time spent to create the auto-encoder, the time to train it and the time spent to find the rules but not the time spent to compute the support and confidence. Finally we compare the average support and confidence of the rules produced by the three algorithms.

5 Results and Analysis

We trained and tested our AE and NSGAII independently 10 times with each dataset, in order to account for differences in the network's random initialization and the stochastic nature of meta-heuristics. We report the average results over all 10 runs. The FPGrowth algorithm, having no random initialization, was only run once.

The first metric to consider is the percentage of rules produced by ARM-AE that have a support greater than 0. Our experiments indicate that up to 4% of the rules produced actually don't exist in the dataset. Since ARM-AE is required to generate a number of rules for each consequent, some very-low-support or zero-support rules will be included when no high-support rules exist for a consequent. Without a computation of support, ARM-AE cannot detect or filter out these cases. This is a limitation of our approach. However, since such rules account for less than 4% of the extracted rule set, the amount of noise included is not a major issue.

The next metrics compare the AR set mined by ARM-AE, NSGAII and FP-Growth. First we consider the proportion of FP-Growth's rule set that is mined by ARM-AE and NSGAII. FP-Growth exhaustively mines all rules with one consequent and one or two antecedents that are above the set support and confidence threshold. For its part, ARM-AE discovers between 26% and 44% of that set, as is shown in Table 1. Regarding NSGAII, it mines up to 8% of the rules set mined by FP-Growth. That can be explained by the highly stochastic aspect of genetic algorithms and because NSGAII focuses on non-dominated rules where FP-Growth focuses on high support and high confidence rules. In non-dominated mining, if a rule has a very low support but a very high confidence and no other rules have a higher support with at least the same confidence, then it may not be kept. Table 1 also shows that the average support and confidence of the rules mined by our three algorithms. The results show that the rules found by ARM-AE have a support and confidence almost as high as those mined by FP-Growth, and in fact the confidence is higher in the case of the Plants dataset. That is despite ARM-AE including up to 4% of zero-support rules in its results, as mentioned previously. This shows that, although ARM-AE mines less than half the rules found by FP-Growth, it mines a high-quality subset of the rules that exist in the database. NSGAII mines a rule set with very high confidence and lower support than the two other algorithms. ARM-AE mines a rule set that is more similar to the one found by FP-Growth than NSGAII. We could expect ARM-AE rule set to have a very high support and confidence given that it mine less rules, but given how the algorithm is design, it's forced to create rules with non-frequent items, given that we ask it to find at least 4 rules for each items even the less frequent ones. Also to compute the average support and confidence display in Table 1 for FP-Growth we selected for each possible consequent the best two rules with 1 antecedent and the best two rules with 2 antecedents. The average support and confidence of FP-Growth is thus the maximum our algorithm can obtain. That explain why support and confidence of ARM-AE rule set and FPGrowth rule set are approximately the same.

Table 1. Coverage, support, confidence, execution time and number of rules per algorithm and dataset.

Dataset	ARM-AE					NSGAII					FP-Growth			
	coverage	support	confidence	Execution (s)	Rules	coverage	support	confidence	Execution (s)	Rules	support	confidence	Execution (s)	Rules
Chess	0.26	0.49	0.50	1.04	296	0.02	0.33	0.71	95.0	241	0.51	0.54	35.2	278,823
Nursery	0.38	0.10	0.33	0.5	124	0.08	0.06	0.38	99.0	150	0.12	0.37	1.4	18,140
Plants	0.44	0.08	0.75	1.22	280	0.02	0.05	0.78	979.0	261	0.09	0.70	114	301,918

Finally, we compare the performance of the three algorithms in terms of execution time and total number of rules generated, in Table 1. We consider first the execution time for the entire algorithm, including building and training the AE in the case of ARM-AE. As ARM-AE does not need to compute the support and confidence to produce rules, we don't include this time in the execution time of the algorithm. The most time-consuming step in every ARM algorithm is to compute the support and confidence of rules. FP-Growth uses a Fp-tree to speed up the mining of frequent itemsets, but then must compute the support and confidence of all the pairs of itemsets. NSGAII computes the support and confidence for a large part of the population at each generation explaining why it is slower than FP-Growth. By contrast, ARM-AE computes the support and confidence only for the subset of rules mined. For the three datasets, ARM-AE is much more efficient than FP-Growth and NSGAII in terms of execution time. In terms of number of rules, ARM-AE and NSGAII have the advantage of generating a small set of rules whose size is controlled by the algorithm's hyperparameters. By contrast, the exhaustive search of FP-Growth generates tens to hundreds of thousands of rules. As mentioned previously, this is one of the main weaknesses of algorithms such as this one [10]: it is impossible for a human to read and understand such a massive set of rules, and additional filtering or ranking algorithms must thus be applied in post-processing before the rules are suitable for human consumption. ARM-AE and NSGAII do not suffer from that issue.

6 Conclusion

In this paper, we introduced a new deep learning architecture to mine association rules, the Association Rules Mining Auto-Encoder(ARM-AE) algorithm. We compared this new algorithm to the popular state-of-the-art FP Growth algorithm and to the state-of-the-art genetic algorithm NSGAII on three categorical datasets. Our results show that ARM-AE produces a set of rules with an average support and confidence close to that of FP-Growth and with higher support than NSGAII, but with a much smaller execution time and by generating a much smaller number of rules than FP-Growth, which makes the results much more usable for a human being. Moreover, ARM-AE has hyperparameters to control the number of rules produced for each consequent, the maximum number of items in the antecedent of the rules, and the similarity between the rules, which again makes the algorithm much more user-friendly. The main drawbacks of ARM-AE stem from the fact it does not compute support and confidence for its rules, which means it cannot filter out zero-support rules and cannot give an importance ranking of the rules it mined, but these will be addressed in future works.

Acknowledgments. This research was made possible by the support of the INSPQ, as well as the financial support of the Canadian research funding agencies CIHR and NSERC.

Disclosure of Interests. The authors have no competing interests to declare that are relevant to the content of this article.

References

1. Agrawal, R., Srikant, R., et al.: Fast algorithms for mining association rules. In: Proceedings of the 20th International Conference on Very Large Data Bases, VLDB, vol. 1215, pp. 487–499. Citeseer (1994)
2. Alkeshuosh, A.H., Moghadam, M.Z., Al Mansoori, I., Abdar, M.: Using PSO algorithm for producing best rules in diagnosis of heart disease. In: 2017 International Conference on Computer and Applications (ICCA), pp. 306–311. IEEE (2017)
3. Borgelt, C.: Keeping things simple: finding frequent item sets by recursive elimination. In: Proceedings of the 1st International Workshop on Open Source Data Mining: Frequent Pattern Mining Implementations, pp. 66–70 (2005)
4. Christian, B.: An implementation of the FP-growth algorithm. In: Proceedings of the 1st International Workshop on Open Source Data Mining: Frequent Pattern Mining Implementations, pp. 1–5 (2005)
5. Deb, K., Pratap, A., Agarwal, S., Meyarivan, T.: A fast and elitist multiobjective genetic algorithm: NSGA-II. IEEE Trans. Evol. Comput. **6**(2), 182–197 (2002)
6. Djenouri, Y., Bendjoudi, A., Djenouri, D., Comuzzi, M.: GPU-based bio-inspired model for solving association rules mining problem. In: 2017 25th Euromicro International Conference on Parallel, Distributed and Network-Based Processing (PDP), pp. 262–269. IEEE (2017)
7. Dua, D., Graff, C.: UCI machine learning repository (2017). http://archive.ics.uci.edu/ml
8. Fortin, F.A., De Rainville, F.M., Gardner, M.A.G., Parizeau, M., Gagné, C.: DEAP: evolutionary algorithms made easy. J. Mach. Learn. Res. **13**(1), 2171–2175 (2012)
9. Fournier-Viger, P., Wu, C.-W., Tseng, V.S.: Mining top-K association rules. In: Kosseim, L., Inkpen, D. (eds.) AI 2012. LNCS (LNAI), vol. 7310, pp. 61–73. Springer, Heidelberg (2012). https://doi.org/10.1007/978-3-642-30353-1_6
10. Fund, I.: Comparing association rules and deep neural networks on medical data. Ph.D. thesis, University of Houston (2019)

11. Geng, L., Hamilton, H.J.: Choosing the right lens: finding what is interesting in data mining. In: Guillet, F.J., Hamilton, H.J. (eds.) Quality Measures in Data Mining. SCI, vol. 43, pp. 3–24. Springer, Cham (2007). https://doi.org/10.1007/978-3-540-44918-8_1
12. Heraguemi, K.E., Kamel, N., Drias, H.: Multi-swarm bat algorithm for association rule mining using multiple cooperative strategies. Appl. Intell. **45**(4), 1021–1033 (2016)
13. Kishor, P., Sammulal, P.: Association rule mining using an unsupervised neural network with an optimized genetic algorithm. In: Kumar, A., Mozar, S. (eds.) ICCCE 2018. LNEE, vol. 500, pp. 657–669. Springer, Singapore (2019). https://doi.org/10.1007/978-981-13-0212-1_67
14. LeCun, Y., Bengio, Y., Hinton, G.: Deep learning. Nature **521**(7553), 436–444 (2015)
15. Li, X., Li, D., Deng, Y., Xing, J.: Intelligent mining algorithm for complex medical data based on deep learning. J. Ambient. Intell. Humaniz. Comput. **12**(2), 1667–1678 (2021)
16. Lore, K.G., Akintayo, A., Sarkar, S.: LLNet: a deep autoencoder approach to natural low-light image enhancement. Pattern Recogn. **61**, 650–662 (2017)
17. Martín, D., Rosete, A., Alcalá-Fdez, J., Herrera, F.: QAR-CIP-NSGA-II: a new multi-objective evolutionary algorithm to mine quantitative association rules. Inf. Sci. **258**, 1–28 (2014)
18. Min, E., Guo, X., Liu, Q., Zhang, G., Cui, J., Long, J.: A survey of clustering with deep learning: from the perspective of network architecture. IEEE Access **6**, 39501–39514 (2018)
19. Nair, B.B., Mohandas, V., Nayanar, N., Teja, E., Vigneshwari, S., Teja, K.: A stock trading recommender system based on temporal association rule mining. SAGE Open **5**(2), 2158244015579941 (2015)
20. Pei, J., Han, J., Lu, H., Nishio, S., Tang, S., Yang, D.: H-Mine: fast and space-preserving frequent pattern mining in large databases. IIE Trans. **39**(6), 593–605 (2007)
21. Ramponi, A., Plank, B.: Neural unsupervised domain adaptation in NLP—a survey. arXiv preprint arXiv:2006.00632 (2020)
22. Raschka, S.: MLxtend: providing machine learning and data science utilities and extensions to Python's scientific computing stack. J. Open Source Softw. **3**(24), 638 (2018). https://doi.org/10.21105/joss.00638
23. Telikani, A., Gandomi, A.H., Shahbahrami, A.: A survey of evolutionary computation for association rule mining. Inf. Sci. **524**, 318–352 (2020)
24. Valdivia, A., et al.: What do people think about this monument? Understanding negative reviews via deep learning, clustering and descriptive rules. J. Ambient. Intell. Humaniz. Comput. **11**(1), 39–52 (2020)
25. Vougas, K., et al.: Deep learning and association rule mining for predicting drug response in cancer. A personalised medicine approach, p. 070490. BioRxiv (2017)
26. Zhang, Z., He, Q., Gao, J., Ni, M.: A deep learning approach for detecting traffic accidents from social media data. Transp. Res. Part C Emerg. Technol. **86**, 580–596 (2018)
27. Zhuang, F., Cheng, X., Luo, P., Pan, S.J., He, Q.: Supervised representation learning: transfer learning with deep autoencoders. In: Twenty-Fourth International Joint Conference on Artificial Intelligence (2015)

Using Contrastive Learning to Map Stylistic Similarities in Narrative Writers

María Valero-Redondo[1], Javier Huertas-Tato[2], Sergio D'Antonio Maceiras[2], Alejandro Martín[2](✉), and David Camacho[2]

[1] Universidad de Córdoba, Córdoba, Spain
182varem@uco.es
[2] Universidad Politécnica de Madrid, Madrid, Spain
{javier.huertas.tato,sergio.dantonio,alejandro.martin, david.camacho}@upm.es

Abstract. Exploring how literary authors influence each other and how intertextual connections can be found among them is a complex problem that is handled manually by experts. This research field, intertextuality, seeks to understand how writers relate to each other and how the works of one writer have reflected on the work of another. This analysis usually provides insight into the historical, cultural, biographical, and literary contexts that shape literary works. From a computational point of view, the approaches related to this problem have mainly dealt with problems of style classification, known as authorship attribution, but none have so far dealt with the problem of intertextuality. This paper proposes a novel approach based on contrastive learning to generate authorship embeddings that encapsulate the stylistic signatures of narrative writers and that can be projected into graphical representations to discern similarities between a dataset of literary writers. The embeddings generated with this approach aim to represent authorship styles are created and evaluated using a dataset of books from Project Gutemberg. During the evaluation, we perform a book and chunk-level evaluation, showing good performance in both cases. Finally, we present different graphical representations and provide a deep analysis of the relations that arise between the writers.

Keywords: Contrastive learning · Intertextual analysis · Language Models · Transformer

1 Introduction

The study of intertextuality has long been a topic of interest, providing information on historical, cultural, biographical and literary contexts that shape literary works. However, quantitatively capturing the essence of an author's style remains a challenging issue, particularly in the case of literature, where the nuances of language and narrative techniques come into play. This research proposes a contrastive learning-based approach to generate authorship embeddings that

V. Julian et al. (Eds.): IDEAL 2024, LNCS 15346, pp. 63–74, 2025.
https://doi.org/10.1007/978-3-031-77731-8_6

effectively encapsulate the stylistic signatures of writers. The goal is to build a meaningful representation in the form of graphs that can lead to a better understanding of the closeness between the authors and their narratives.

Understanding how different authors influence each other is essential for the analysis of the evolution of literature. Narrative is an ideal subject for such comparative analysis. The main challenge is to measure and compare the distinct stylistic features of each author quantitatively, a task that has usually relied on subjective and qualitative methods. Our goal in this research is to leverage contrastive learning to develop a model that not only reveals the unique stylistic elements of each author but also highlights their similarities and differences. Employing language models to create representations of authorship can enhance our understanding of the intertextual links within narrative, providing a useful instrument to help literary critics carry out these analyses.

The use of language models and contrastive learning allows to seek for a multi-dimensional representation of authors' style, including varied characteristics such as word choice, sentence structure, narrative techniques, or thematic elements. By training a model through contrastive learning to differentiate between the writing styles of various authors, it is possible to generate a vectorial representation to later build graphical visualizations where relations among authors can be studied.

Training and evaluation of the approach have been performed on a 15,000-book dataset from Project Gutemberg. We applied different thresholds to assess the effects of the number of authors and books per author considered. Besides, we also tested the performance when assessing individual chunks of the authors' work (limited by the input size of the architecture) and also entire books. For the architecture, we used an ELECTRA model in its small version, an efficient architecture that replaces the BERT MLM pre-training task for a replacement of tokens with plausible alternatives.

The paper is organized as follows. Section 2 describes the state-of-the-art regarding authorship embeddings and approaches in the literary area; Sect. 3 describes the methodology, including the contrastive learning approach followed. Section 4 describes the evaluation performed; Sect. 4 shows specific graphical representation and a deep evaluation of the relations found between writers and, finally, Sect. 5 shows the conclusions of the work.

2 Related Work

The concept of intertextuality was first introduced by Julia Kristeva in the 1960s [2], revolutionizing the way literary works are analyzed and interpreted. Intertextual analysis involves examining the relationships between texts and understanding how the themes, stylistic features, and narrative techniques of one writer's work influence those of another.

This approach has been instrumental in uncovering multifaceted dialogues between authors in different eras and literary movements. Historically, intertextual analysis in literary studies was predominantly qualitative, relying on the meticulous and subjective scrutiny of literary scholars. Pioneers in this field, such as Roland Barthes and Mikhail Bakhtin, emphasized the importance of viewing texts in a broader literary and cultural context, recognizing the interconnectedness of literary creation. Their work laid the foundation for understanding how authors engage in a continuous conversation with their peers and predecessors, consciously or subconsciously embedding references and motifs from other works.

With the advent of the digital humanities, computational methods started to be applied to intertextual analysis. Early computational approaches focused on specific aspects, such as thematic similarities, citation patterns, and stylistic mimicry. It was not until the development of Deep Learning architectures such as the Transformer [7,14] and language models that computational representations could be built to embed all these features.

This is, nonetheless, a field of research that is still in its infancy. Little research has yet studied the application of language models to assist in this task. From the NLP field, the closest research to this field is the study of authorship attribution. Following this research line, Ai et al. introduced Contra-X [1], a novel approach for authorship attribution. It involves fine-tuning pretrained language models with a contrastive objective to learn author-specific representations. Contrastive learning was also used in other research showing promising results [9,10].

Sundararajan et al. [12] explore the linguistic elements that define the author's style. This research analyzes the role of syntax and lexical choices (nouns, verbs, adjectives, and adverbs) in stylometry. The study demonstrates that syntax is important for cross-genre attribution, while additional lexical information benefits single-domain and cross-topic attributions. Furthermore, research indicates the strong influence of common and proper nouns on attribution, underlining the impact of topic interference.

As far as research in the field of Authorship Attribution is concerned, a large number of proposals come from competitions such as PAN. Examples include the work by Chen et al. [5], centered on the detection of changes in writing style for the identification of authors in multiauthor documents.

Siamese architectures have also proven to be powerful tools in these tasks. In PAN 2021, Tyo et al. [13] described an approach using a Siamese network with pre-trained BERT encoders.

Among all the approaches proposed so far, one of the most promising involves contrastive learning techniques. Guo et al. [8] proposed a novel contrastive learning method specifically designed for authorship verification. Unlike classical sample-based contrastive learning, this method efficiently uses sample pairs to compute the loss. The building of general representation through embeddings was also studied by Rivera et al. [11], particularly in the context of cross-domain transfer.

With the focus on intertextuality, very little research has yet addressed the combination of language models in this field. Burns et al. [4] proposed to approach intertextual profiling in Latin Literature using word embeddings extracted with word2vec. Although the approach shows the ability to extract

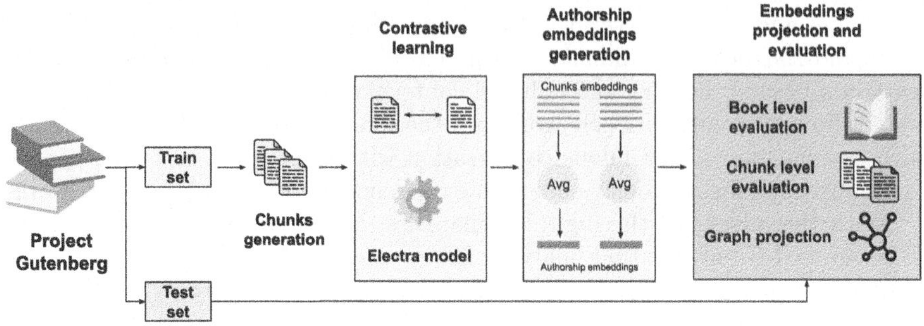

Fig. 1. Workflow of the methodology employed to build authorship embeddings for narrative writers.

style-determining features, our approach based on contrastive learning and a more advanced model, ELECTRA, produces embeddings of much greater stylistic richness. A specific use of a more advanced language model was proposed by Bamman and Burns in 2020, presenting Latin BERT [3], however this was a specific training of BERT without the focus on intertextuality.

3 Methodology

This section describes the methodology followed in this research to build meaningful representations of authorship for literary writers. Different subsections describe the data used, the training approach, how embeddings are built, and how meaningful graphical representations are finally created. Figure 1 summarizes the workflow followed in this research.

3.1 Dataset

Due to the focus of this research, we applied our analysis on the well-known *Project Gutenberg*. In particular, we used a 15,000 book dataset extracted from Kaggle[1], with books already labeled according to authors and titles. We removed books written by various authors because this is outside the scope of this research. After this, the dataset contained 13282 books from 4297 different writers. We then applied a different threshold to control the effects of the number of books considered for each author. Table 1 shows the different divisions created from this dataset, according to the minimum number of books per author to consider. This will allow to evaluate the performance of the language model according to the size of the space of authors, where a higher number implies the need for a finer analysis of the feature space to differentiate between a larger number of styles.

After preparing the different divisions of the dataset, every book was divided into chunks, according to the input size of the architecture. As we present in the

[1] https://www.kaggle.com/datasets/mateibejan/15000-gutenberg-books.

following subsection, the model used limits the input to 512 tokens, which are reduced to 510 after removing the additional start and end tokens.

Table 1. Threshold applied in the dataset according to the minimum number of books per author.

Min. Number books per author	No. Authors	Total no. books	No. text chunks
50	18	1,728	239,537
40	26	2,060	293,149
30	46	2,733	390,958
20	82	3,596	551,788
10	220	5,415	807,244

3.2 Contrastive Learning

In this research, we employ contrastive learning to develop authorship embeddings using the ELECTRA model. A contrastive learning-based approach is a powerful method to generate embeddings able to differentiate between similar and dissimilar data points, in this case, text segments from different authors. This approach is particularly effective for learning high-quality representations in a feature space where similar items are brought closer, and dissimilar items are pushed apart.

Regarding the specific model used, we used the ELECTRA model [6]. This architecture, proposed by Google and Standford researchers in 2020, proposed a different pre-training task than BERT. Instead of making use of replacement with the label `[MASK]`, they looked for a more efficient mechanism consisting of replacing tokens with possible alternatives. In this way, during training, a discriminative model was in charge of predicting whether a token had been replaced by a generated one.

This approach requires a modification of the loss function without adding or altering the architecture of the model. In this new contrastive learning function, for each text chunk, a 'positive' sample is selected (another text by the same author) and a 'negative' sample (a segment by a different author). The loss function then evaluates these samples in the embedding space created by the model. The loss function L is calculated according to:

$$L = \frac{1}{N} \sum_{i=1}^{N} [\max(0, d(e_i, e_{pos}) - d(e_i, e_{neg}) + \text{margin})] \tag{1}$$

where e_i is the embedding of the i-th text chunk, e_{pos} is the embedding of a positive sample, a different text from the author e_i, e_{neg} is the embedding of a negative sample, that is, from a different author, $d(x, y)$ is the distance between two embeddings, and *margin* is an auxiliary parameter to express a specific objective separation between the embeddings of different authors. We used the Euclidean distance and a Rectified Linar Unit (ReLU) ensures that the output is a positive value or zero.

During training, we used AdamW as optimizer with a $learningrate = 5e^{-5}$, $batchsize = 40$ and a total of 3 epochs. We also tested the dependence on the training data set. Since the goal of this research is to create writer representations efficiently and with the capacity to generalize, we aim to reduce the size of the training data. To do so, we first use a 30% split for training and a 70% split for testing. Subsequently, we reverse the figures and use 70% for training to be able to quantify the loss of information.

3.3 Authorship Embeddings

The ELECTRA model fine-tuned through contrastive learning provides authorship embeddings for text chunks of 510 tokens. The final authorship embeddings are generated by aggregating and averaging the embeddings of text segments associated with each author. Let E_a be the set of embeddings generated for the author a, where $E_a = \{e_{a,0}, e_{a,1}, ..., e_{a,n}\}$ where n is the number of chunks obtained for the different books of that author a. The mean embedding is then calculated as

$$\bar{\mathbf{e}}_a = \frac{1}{n_a} \sum_{i=1}^{n_a} \mathbf{e}_{a,i} \tag{2}$$

4 Evaluation

The aim of this section is to assess the ability of the proposed approach to generate meaningful authorship embeddings in the context of literary writers. For this purpose, we trained the model with 30% of the data and with 70% of the data in two different experiments. It is important to note that the final goal of the approach is to generate authorship embedding generalizable to a larger number of writers, even if not seen during training. However, before showing this in the next section, we have tested the performance of the proposal using two different approaches.

- **Book level evaluation:** During the test phase, an embedding is generated for each text segment, while an averaged authorship embedding is generated from all embeddings belonging to the same author. Then, the cosine similarity distance is calculated against the authorship embeddings generated during the training phase. If the closest vector represents the same author, then the test example is considered correctly classified. In general terms, a book-level evaluation allows to evaluate the capabilities of the authorship embeddings to place vectors from similar authors close in the space.

- **Chunk level evaluation:** This is fine-grained evaluation of the approach. It means evaluating for each 510 token segment in the test split if the original author can be correctly inferred. This is a highly complex task, since it consists of finding the author of small portions of text and where the major patterns that define the author's style are key.

Table 2. Results obtained using different thresholds in the number of authors considered and the size of the test set.

Test split	Min. number books per author	No. authors	Books level evaluation				Chunk level evaluation			
			Acc.	f1	Prec.	Recall	Acc.	f1	Prec.	Recall
30%	50	18	100,00%	100,00%	100,00%	100,00%	90,11%	81,22%	83,30%	81,23%
30%	40	26	100,00%	100,00%	100,00%	100,00%	87,11%	74,47%	75,56%	75,79%
30%	30	46	100,00%	97,10%	96,74%	97,83%	81,56%	65,89%	68,50%	65,36%
30%	20	82	97,56%	93,70%	93,09%	95,12%	73,77%	50,70%	54,34%	51,00%
30%	10	220	86,36%	83,18%	81,67%	86,36%	56,95%	26,51%	28,64%	28,50%
70%	50	18	100,00%	100,00%	100,00%	100,00%	85,80%	73,76%	75,52%	74,61%
70%	40	26	100,00%	100,00%	100,00%	100,00%	82,44%	69,80%	70,10%	71,07%
70%	30	46	97,83%	100,00%	100,00%	100,00%	73,20%	59,55%	61,80%	59,34%
70%	20	82	95,12%	96,75%	96,34%	97,56%	66,57%	45,22%	47,35%	46,16%
70%	10	220	92,27%	89,77%	88,56%	92,27%	48,05%	33,52%	35,59%	35,76%

Table 2 shows the results in terms of accuracy, f1, precision and recall when using 30% or 70% of the data for training and depending on the number of authors considered. As expected, the book-level evaluation produces better results than inferring the author from individual chunks.

The results with 30 test data are very high even with 82 authors, reaching 95% recall and more than 80% recall with 220 authors. This shows that when all segments of the same writer are combined, it is possible to infer the original writer of the text with high accuracy. When, on the other hand, it is evaluated at the level of individual segments, the results are lower, but show a very high performance nonetheless. With 26 authors, a recall of more than 80% is obtained, while with 82 authors the value drops to 51%. When the number of authors is increased to 220, there is a significant reduction in performance.

However, it is interesting to observe the values when only 30% of the data are used for training and the remaining 70% is evaluated. In this case, an interesting fact is that although the performance figures are reduced with a low number of authors, they increase when all the authors are considered, 220. This translates into possible overfitting when many authors and a large amount of training data are used.

The results show the effects on the number of authors and different books in the quality of the representation space. Although 18 authors and 1,728 different books allows to obtain very positive results, when the number of authors raises 220 with 5,415 books, the performance is significantly reduced.

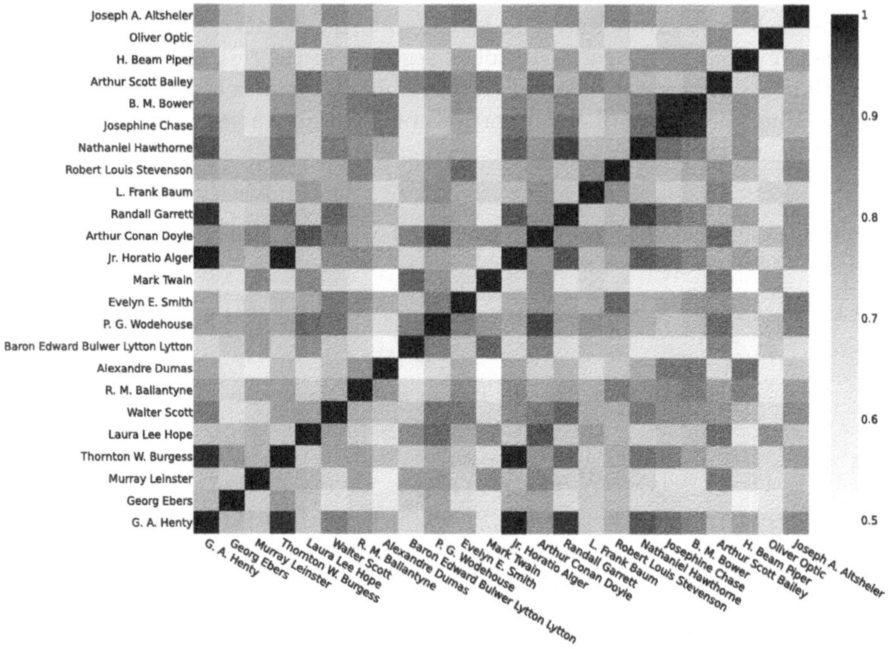

Fig. 2. Heatmap showing the degree of similarity based on the cosine distance for authors with more than 40 books in the dataset.

5 Authorship Embeddings Projection

Although the previous evaluation provided us with a way to quantify and measure the performance of the proposed approach through a classification task, the ultimate goal of this research is to explore the intertextual similarities that exist between writers and to take a step forward in the construction of new tools that will provide researchers with a powerful resource for detailed analysis of stylistic differences among authors.

For this purpose, we have created two different graphical representations. The first 2 consists of a heat map that shows the similarity of authorship between all authors with more than 40 books represented in the dataset. Many patterns of similarity can be seen. In general, the range of values, between 0.5 and 1.0, indicates that the construction of the embeddings has room for further separation. Nevertheless, the inherent similarity between the authors must also be taken into account, because of the literary nature of all of them, which also explains this closeness in the space of representation.

Figure 3 shows a 2-dimensional projection through PCA of the authorship embedding generated (in the test set) for each author. This representation, together with a graph with the closest similarity relations shown in Fig. 4, allows for a detailed analysis of the ability of the presented approach to find intertextual and stylistic differences and similarities between the writers.

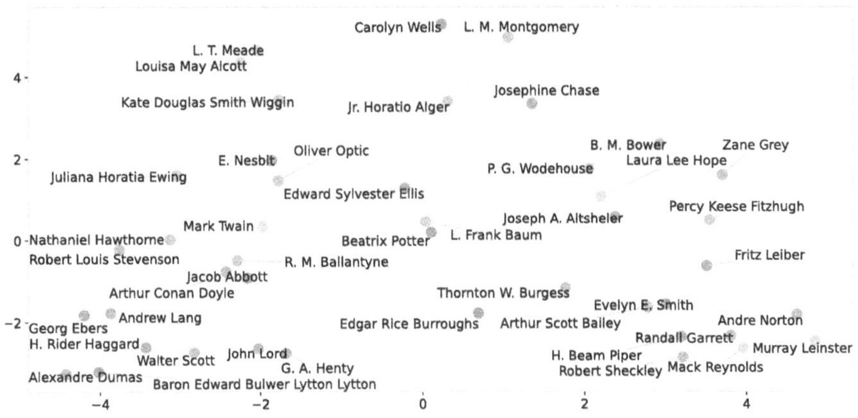

Fig. 3. Two-dimensional representation of the authorship embeddings through PCA.

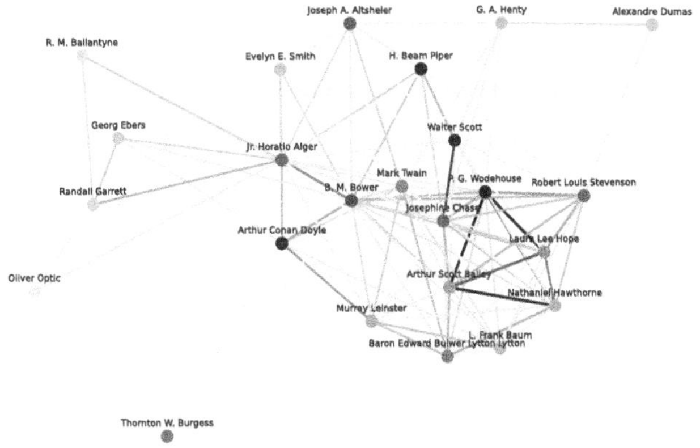

Fig. 4. Graph-based representation of the authorship similarity between writers. Only similarities higher than 0.8 were represented. The darkness of the edge indicates the degree of similarity.

We can address, for instance, the relation between B. M. Bower and Josephine Chase, placed close in both figures. Before delving into their intertextual similarities, it is essential to understand the context in which these authors wrote. Bertha Muzzy Bower, better known as B. M. Bower, was an American author who primarily wrote Western stories, flourishing in the early 20th century. Her works are characterized by their vivid depiction of ranch life, notably in the American West. Josephine Chase, on the other hand, is best known for her work under the pseudonym Carolyn Keene, under which she wrote the Nancy Drew series. Though writing in different genres, both authors contributed significantly to their respective literary fields, with Bower focusing on the Western genre and Chase on children's and young adult mystery novels. Both Bower and Chase employed narrative techniques that were ahead of their times, especially in the portrayal of strong female protagonists. Bower's works often featured women who were as capable and resilient as their male counterparts, which was a departure from the typical damsel-in-distress archetype prevalent in literature of that era. Similarly, Chase's Nancy Drew was a groundbreaking character, embodying intelligence, courage, and independence, qualities not commonly attributed to female characters in early 20th-century literature.

The theme of independence is a striking similarity between Bower and Chase. Bower's stories often revolved around the themes of self-reliance and the pioneering spirit, echoing the larger American narrative of westward expansion and exploration. Chase, through the character of Nancy Drew, explored the theme of independence in a different light, presenting a young woman who was intellectually independent and capable of solving mysteries on her own. The motif of the 'frontier' is prominent in Bower's work, symbolizing both a physical and metaphorical space where traditional norms could be challenged or redefined. In contrast, Chase's use of the 'hidden clue' motif served a similar purpose, allowing her young protagonist to step beyond societal expectations and engage in adventures and problem-solving.

In terms of tropes, Bower frequently used the 'lone hero' trope, albeit with a unique twist by occasionally making this hero a heroine. Chase, through the Nancy Drew series, used the 'amateur sleuth' trope to great effect, empowering her young readers, particularly girls, by portraying a strong, intelligent young woman solving complex mysteries.

While Bower and Chase wrote in different genres and for different audiences, their works exhibit remarkable similarities in their use of narrative techniques, themes, motifs, tropes, and symbolism. Both authors were pioneers in their respective fields, challenging societal norms and providing their readers with strong, independent female protagonists. Their contributions have had a lasting impact on American literature, and the parallels in their works offer a fascinating insight into the evolving nature of narrative storytelling in the early 20th century.

Another example is the relation between Arthur Conan Doyle and B. M. Bower, which, while hailing from distinct literary genres, detection fiction and westerns, respectively, exhibit a number of stylistic and narratological similarities that merit a detailed examination. Conan Doyle, renowned for his Sherlock Holmes series, and Bower, celebrated for her vivid portrayal of the American West, both demonstrate a mastery in crafting engaging narratives and complex characters, despite the superficial differences in their subject matter. Despite their distinct genres, Arthur Conan Doyle and B. M. Bower share several stylistic and narratological similarities. Their clear and descriptive narrative styles, the use of setting as a key element of their stories, and their methods of character development demonstrate a shared understanding of the importance of these elements in storytelling. The works of both authors are characterized by linear plot structures and a focus on justice and morality themes, reflecting their ability to engage readers not only with compelling narratives, but also with deeper thematic content. These similarities underscore the universality of certain narrative techniques and styles, transcending the boundaries of genre and era.

Another example is the stylistic and narratological similarities between Evelyn E. Smith and Randall Carrott. Both authors show a clear and engaging narrative style, with a focus on character development, thematic depth, and effective use of setting. Their ability to weave humor into serious themes, along with their use of dialogue and linear plot structures, highlights the universal aspects of storytelling that transcend genre boundaries. This comparative analysis not only sheds light on their individual narrative strengths, but also contributes to a broader understanding of mid-20th century literary trends.

6 Conclusion

In this paper, we have presented a first attempt to help in this task by using language models to construct vector representations of the styles of a set of writers. These embeddings are very useful for the creation of graphical representations such as those provided in this research, for example, a graph-based representation. The approach used, consisting of fine-tuning an ELECTRA model by means of contrastive learning, demonstrates excellent and promising capabilities to generate this space of representation, distancing authors with different styles, and bringing those with similar characteristics closer together. The evaluation, both at the level of predicting the authors from all their texts by averaging the individually generated embeddings, and at the level of text segments that maximize the input size of the ELECTRA model, shows that the vectors adequately construct a stylistic representation space.

Many possible avenues are ahead for the continuation of this work. As has been observed in the evaluation, when the number of writers increases, the space suffers a significant deterioration, and the possibility of predicting the actual writer of a text is reduced. This factor requires further research. On the one hand, to explain this effect and whether it is produced by particular sets of writers who show stylistic patterns that are more difficult to differentiate. On the other hand, it will be necessary to investigate other mechanisms for constructing this representation space, as well as more complex models with a greater number of parameters, although they will entail a greater learning process. Additional areas for future research could involve exploring the intertextual similarities between several works by the same author, reducing the number of works used for each writer, or seeking novel methods of training that target specific stylistic elements.

Acknowledgment. This work has been funded by the project PCI2022-134990-2 (MARTINI) of the CHISTERA IV Cofund 2021 program, funded by MCIN/AEI/10.13039/501100011033 and by the "European Union NextGenerationEU/PRTR"; by the Spanish Ministry of Science and Innovation under FightDIS (PID2020-117263GB-I00); by MCIN/AEI/10.13039/501100011033/ and European Union NextGenerationEU/PRTR for XAI-Disinfodemics (PLEC2021-007681) grant, by European Comission under IBERIFIER Plus - Iberian Digital Media Observatory

(DIGITAL-2023-DEPLOY-04-EDMO-HUBS 101158511); and by EMIF managed by the Calouste Gulbenkian Foundation, in the project MuseAI.

References

1. Ai, B., Wang, Y., Tan, Y., Tan, S.: Whodunit? Learning to contrast for authorship attribution. In: Proceedings of the 2nd Conference of the Asia-Pacific Chapter of the Association for Computational Linguistics and the 12th International Joint Conference on Natural Language Processing (Volume 1: Long Papers), pp. 1142–1157 (2022)
2. Allen, G.: Intertextuality. Routledge (2011)
3. Bamman, D., Burns, P.J.: Latin BERT: a contextual language model for classical philology. arXiv preprint arXiv:2009.10053 (2020)
4. Burns, P.J., Brofos, J.A., Li, K., Chaudhuri, P., Dexter, J.P.: Profiling of intertextuality in Latin literature using word embeddings. In: Proceedings of the 2021 Conference of the North American Chapter of the Association for Computational Linguistics: Human Language Technologies, pp. 4900–4907 (2021)
5. Chen, H., Han, Z., Li, Z., Han, Y.: A writing style embedding based on contrastive learning for multi-author writing style analysis. In: Working Notes of CLEF (2023)
6. Clark, K., Luong, T., Le, Q.V., Manning, C.: ELECTRA: pre-training text encoders as discriminators rather than generators. In: ICLR (2020). https://openreview.net/pdf?id=r1xMH1BtvB
7. Devlin, J., Chang, M.W., Lee, K., Toutanova, K.: BERT: pre-training of deep bidirectional transformers for language understanding. arXiv preprint arXiv:1810.04805 (2018)
8. Guo, M., Han, Z., Chen, H., Qi, H.: A contrastive learning of sample pairs for authorship verification. In: Working Notes of CLEF (2023)
9. Huertas-Tato, J., Martín, A., Camacho, D.: PART: pre-trained authorship representation transformer. In: Human-Centric Computing and Information Sciences, vol. 14, p. 39 (2024)
10. Huertas-Tato, J., Martín, A., Camacho, D.: Understanding writing style in social media with a supervised contrastively pre-trained transformer. Knowl.-Based Syst. **296**, 111867 (2024)
11. Rivera-Soto, R.A., et al.: Learning universal authorship representations. In: Proceedings of the 2021 Conference on Empirical Methods in Natural Language Processing, pp. 913–919 (2021)
12. Sundararajan, K., Woodard, D.: What represents "style" in authorship attribution? In: Proceedings of the 27th International Conference on Computational Linguistics, pp. 2814–2822 (2018)
13. Tyo, J., Dhingra, B., Lipton, Z.C.: Siamese BERT for authorship verification. In: CLEF (Working Notes), pp. 2169–2177 (2021)
14. Vaswani, A., et al.: Attention is all you need. In: Advances in Neural Information Processing Systems 30 (2017)

Automatic Classification of Signal and Noise in Functional Magnetic Resonance Imaging Scans Using Convolutional Neural Networks

Georgian Anghelescu[1(✉)], Camelia Chira[1], and Kristoffer N. T. Månsson[2,3]

[1] Faculty of Mathematics and Computer Science, Babeş–Bolyai University, Cluj-Napoca, Romania
{georgian.anghelescu,camelia.chira}@ubbcluj.ro

[2] Department of Clinical Psychology and Psychotherapy, Babeș-Bolyai University, Cluj-Napoca, Romania

[3] Centre for Psychiatry Research, Department of Clinical Neuroscience, Karolinska Institutet, Stockholm, Sweden
kristoffer.mansson@ki.se

Abstract. The integration of Artificial Intelligence (AI), particularly deep learning models like VGG16 and ResNet50, in the analysis of functional magnetic resonance imaging (fMRI) data has significantly advanced our understanding of brain functionality and the diagnosis of neurological disorders. This paper explores the application of Convolutional Neural Networks (CNNs) to enhance the accuracy and efficiency of fMRI data analysis, addressing challenges such as high dimensionality, noise, and the need for complex data preprocessing. Our study evaluates the performance of VGG16 and ResNet50 models and one 3D CNN on 2D and 3D fMRI datasets, highlighting the limitations of VGG16 in handling 2D data and demonstrating the superior performance of ResNet50 on balanced and unbalanced datasets. Additionally, we investigate the impact of using 3D data from the Human Connectome Project (HCP), achieving up to 98% accuracy on the validation set. The results indicate that CNNs can effectively replace traditional Independent Component Analysis (ICA) methods by leveraging their capability for automatic feature extraction and end-to-end learning.

Keywords: fMRI · Artificial Intelligence · Deep Learning · CNN · VGG16 · ResNet50

1 Introduction

fMRI is a non-invasive method crucial for understanding brain function. It tracks blood flow changes, providing insights into neural dynamics through Blood Oxygen Level Dependent (BOLD) imaging. The complexity and volume of fMRI data

V. Julian et al. (Eds.): IDEAL 2024, LNCS 15346, pp. 75–84, 2025.
https://doi.org/10.1007/978-3-031-77731-8_7

pose significant challenges, which AI, especially deep learning, can effectively address [1].

AI refers to the capability of machines to mimic human intelligence processes, including learning, reasoning, and self-correction. AI techniques, such as machine learning and deep learning, allow computers to analyze vast amounts of data, recognize patterns, and make informed decisions with minimal human intervention. These capabilities are particularly beneficial in handling complex and high-dimensional datasets like fMRI, where AI can enhance data analysis accuracy and efficiency by automating feature extraction and classification tasks [2].

The integration of AI in fMRI analysis has shown great promise in diagnosing neurological diseases and advancing our understanding of brain functionality. For instance, studies have demonstrated that deep learning methods can significantly improve the accuracy and efficiency of fMRI data interpretation, aiding in the diagnosis and treatment of disorders like Alzheimer's and depression. By leveraging AI, we can uncover patterns in brain activity that might elude human analysts, thus enhancing diagnostic capabilities and therapeutic strategies [3,4].

One major challenge when dealing with fMRI data is distinguishing between "signal" (true neural activity) and "noise" (artifacts and other non-neural signals). Thorough cleaning of the data to separate signal from noise is essential for accurate analysis [5].

Manually denoising fMRI data using ICA involves a multi-step process where the data is first decomposed into independent components using ICA algorithms, such as those provided by MELODIC (Multivariate Exploratory Linear Optimized Decomposition into Independent Components) [6]. Each component is then visually inspected by an expert who labels them as either "signal" (representing true neural activity) or "noise" (artifacts such as motion, physiological noise, or scanner-related issues). This classification is based on the spatial maps and time courses of the components, where noise components often exhibit characteristic patterns such as edges of the brain, ventricles, or high-frequency fluctuations. Once labeled, the noise components are removed or regressed out from the original fMRI data, leaving a cleaned dataset that better represents the underlying neural activity. This manual approach, while accurate, is time-consuming and subjective, requiring considerable expertise and effort for large datasets [5].

MELODIC is a tool from the FSL (FMRIB Software Library) used to automatically decompose fMRI data into spatially independent components. This process helps in identifying and separating signal from noise images, thus ensuring high data quality and providing valuable insights for training AI models. While MELODIC offers robust decomposition, it requires manual intervention to classify components, which can be time-consuming and subjective [6].

To automate the decision-making process, FIX (FMRIB's ICA-based Xnoiseifier) was developed. FIX uses a Random Forest classifier to automatically classify ICA components generated by MELODIC into signal and noise. This automation significantly reduces the time and effort required for data cleaning, making the process more efficient and objective. However, FIX is dependent on MELODIC to decompose the data into spatially independent components and the performance of classification is highly dependent on the quality of the training data. FIX achieved on conventional resting-state fMRI (rfMRI) 95% accuracy [5,7]. On new studies, without proper training, FIX does not perform very well because it relies on the quality and relevance of its training data. This limitation necessitated the exploration of more adaptive classifiers, leading to the development of CNNs classifiers.

CNNs offer a promising alternative by potentially enhancing the classification of fMRI data. CNNs can automatically learn complex spatial hierarchies of features, reducing the need for extensive manual preprocessing and feature engineering [8]. This capability makes CNNs particularly suited for handling high-dimensional fMRI data, enabling more efficient and accurate classification of signal and noise. By integrating feature extraction and classification into a single model, CNNs offer a more streamlined and potentially more accurate approach to ICA component classification, addressing the shortcomings of traditional machine learning methods used by FIX.

VGG16 and ResNet50 are well-established classifiers in the literature, commonly used for various image classification tasks. VGG16 is a deep network with 16 layers that has shown excellent performance in image classification tasks [9]. ResNet50, on the other hand, is a 50-layer network that introduces residual learning to address the vanishing gradient problem, allowing for the training of much deeper networks [10]. We will apply these classifiers to fMRI images to evaluate their performance and effectiveness in this domain, exploring their potential to enhance fMRI data analysis.

This paper aims to evaluate the effectiveness of AI models in classifying fMRI data by employing specific methodologies, particularly CNNs like VGG16, ResNet50, and an additional model specialized for 3D image classification. We analyze their performance on both 2D and 3D datasets, highlighting the challenges posed by high dimensionality and noise in fMRI data. Our study emphasizes the importance of precise hyperparameter tuning and the integration of 3D imaging data to accurately capture the spatial complexities of neural activity. By providing a thorough analysis of the current capabilities of AI in this field, the paper also outlines future research directions.

The rest of the paper is structured as follows: Sect. 2 presents the Methodology and Proposed Approach, detailing the data preprocessing steps in Subsect. 2.1, the development and training of CNN models in Subsect. 2.2, and the hyperparameter tuning process in Subsect. 2.3. Section 3 discusses the Experiments and Results, including the model performance for the 2D dataset in Subsect. 3.1, the model performance for the 3D dataset in Subsect. 3.2, and a combined discussion and exploration of ethical challenges in Subsect. 3.3. Finally, Sect. 4 provides the Conclusions and Future Work, summarizing the findings and outlining potential directions for further research.

2 Methodology and Proposed Approach

We utilized a meticulously curated dataset derived from fMRI scans of multiple volunteers. This dataset was enhanced through extensive scanning and a detailed manual labeling process using ICA, ensuring high-quality data for subsequent analysis. The dataset consisted of 2273 images, with 542 images classified as noise and 1731 as signal. The noise images represent artifacts and irrelevant data, while the signal images represent meaningful brain activity.

Additionally, we integrated data from the HCP that was used in FIX training, which includes 22,877 independent components formatted as 3D images with dimensions of (91, 109, 91), to enhance the comprehensiveness of our dataset. Among these images, 4,622 represent signal components, while the remaining are classified as noise [5,7].

2.1 Data Preprocessing

Our approach included normalization to ensure the reliability and consistency of the fMRI data by bringing all data to a comparable scale.

Additionally, for both datasets, we balanced them by removing some of the noise images to ensure equal quantities of components in each class. To make the training process more time-efficient, we consolidated the data into single .npy files. This approach reduces storage and memory usage and increases the training rate.

We experimented with multiple shapes for the input data as follows:

- **Simple 2D Data:** This is the data which went through ICA and was used in both its original unbalanced form and after balancing.
- **2D Data with Additional Channels:** The same 2D data was augmented with two additional channels, one representing the corresponding power spectrum and the other the time series. These datasets were also evaluated in both balanced and unbalanced forms.
- **3D HCP Data Treated as 2D:** The 3D data from the HCP was reduced in dimensions to fit the input requirements of VGG16 and ResNet50 models. These datasets were tested in both balanced and unbalanced forms.
- **3D HCP Data:** The original 3D HCP data was used in its balanced form.

These preprocessing steps are essential to prepare the fMRI data for effective training of deep learning models.

2.2 Development and Training of CNN Models

We developed AI models using CNNs, specifically VGG16 and ResNet50. VGG16 is a deep network with 16 layers that has shown excellent performance in image classification tasks [9]. ResNet50, on the other hand, is a 50-layer network that introduces residual learning to address the vanishing gradient problem, allowing for the training of much deeper networks [10].

Given the 3D nature of fMRI data, it is crucial to capture spatial dependencies across all three dimensions. Therefore, we designed a 3D CNN model structure, as outlined in Table 1, is designed to capture spatial dependencies across all three dimensions of fMRI data. It utilizes a series of convolutional and

Table 1. 3D CNN Model Structure

Layer	Type	Output Shape	Param #
conv3d	Conv3D	(None, 89, 107, 89, 32)	896
max_pooling3d	MaxPooling3D	(None, 44, 53, 44, 32)	0
dropout	Dropout	(None, 44, 53, 44, 32)	0
conv3d_1	Conv3D	(None, 42, 51, 42, 64)	55360
max_pooling3d_1	MaxPooling3D	(None, 21, 25, 21, 64)	0
dropout_1	Dropout	(None, 21, 25, 21, 64)	0
conv3d_2	Conv3D	(None, 19, 23, 19, 128)	221312
max_pooling3d_2	MaxPooling3D	(None, 9, 11, 9, 128)	0
dropout_2	Dropout	(None, 9, 11, 9, 128)	0
flatten	Flatten	(None, 114048)	0
dense	Dense	(None, 512)	58393088
dropout_3	Dropout	(None, 512)	0
dense_1	Dense	(None, 1)	513

pooling layers, followed by dropout layers to prevent overfitting, and concludes with fully connected layers for classification. This model architecture includes a total of 58,671,169 trainable parameters, highlighting its complexity and capacity to learn detailed features from the volumetric data.

The necessity of creating a model specifically for 3D data lies in its ability to capture more significant spatial information compared to 2D data, which flattens the volumetric data and loses critical spatial dependencies.

The training process for these models was conducted on high-performance computing systems equipped with GPUs to handle the computational demands. The models were trained on the preprocessed fMRI datasets using 80% split for training and 20% for validation to minimize overfitting and enhance generalizability.

2.3 Hyperparameter Tuning

Hyperparameter tuning is a critical step in the development of deep learning models. It involves adjusting parameters such as learning rate, batch size, and the number of epochs to optimize the model's performance. We used grid search and random search techniques to systematically explore the hyperparameter space and identify the optimal settings [11]. This process is computationally intensive but essential for achieving the best possible model performance.

3 Experiments and Results

We conducted experiments using two different approaches: training the well-known CNN models VGG16 and ResNet50 on the 2D dataset mentioned in the previous chapter, and training a custom-made 3D model on the HCP 3D dataset. These experiments aimed to evaluate the performance of these models on different data representations and to determine the advantages and limitations of each approach in classifying fMRI data.

3.1 Model Performance for the 2D Dataset

The performance of the VGG16 and ResNet50 models was evaluated on the preprocessed fMRI dataset. The results indicated that VGG16 was not complex enough to capture significant patterns from the 2D data. Despite the addition of channels for power spectrum and time series, the VGG16 model failed to produce meaningful results, highlighting its inadequacy for this type of data.

In contrast, ResNet50 demonstrated different behavior under various conditions. We experimented with different batch sizes, learning rates, and both balanced and unbalanced data. The results showed that ResNet50 reacted significantly to these changes:

- **Batch Sizes.** Varying the batch size impacted the training stability and convergence speed of ResNet50. Smaller batch sizes provided more stable updates but required more epochs to converge, while larger batch sizes converged faster but with less stability.
- **Learning Rates.** Different learning rates affected the training dynamics of ResNet50. A higher learning rate accelerated the training process but risked overshooting the optimal parameters, whereas a lower learning rate provided more precise updates but slowed down the training.
- **Balanced vs. Unbalanced Data.** ResNet50 performed better with balanced data compared to unbalanced data. Balancing the dataset ensured that the model did not become biased towards the majority class, leading to more generalizable performance.

However, the ResNet50 model only showed promising results in terms of training accuracy, not validation accuracy (Fig. 1 and 2). This discrepancy suggests potential overfitting, indicating that the model performed well on the training set but failed to generalize to the validation set. The potential reasons for this include:

- **Insufficient Training Data.** The dataset used for training might not have been large enough to provide the model with enough variety to generalize well.
- **2D Data Limitations.** Even with the additional channels for power spectrum and time series, treating the fMRI data as 2D might not capture the necessary spatial dependencies. The model might benefit more from directly using 3D data to fully leverage the volumetric information.

3.2 Model Performance for the 3D Dataset

When training on the 3D data from the HCP, the results were quite promising, with accuracy reaching up to 98% on the validation set. Initially, we conducted

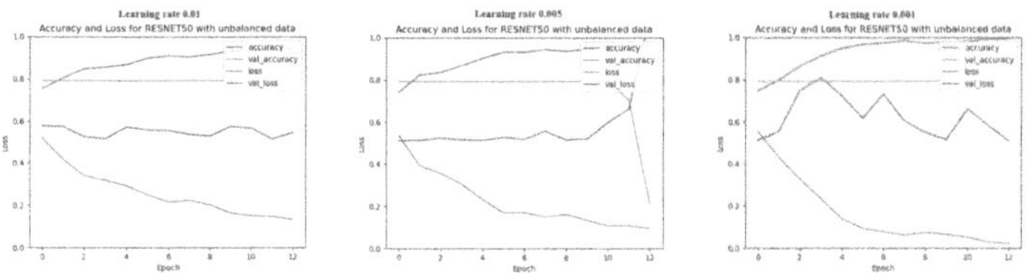

Fig. 1. ResNet50 Training with a Batch Size of 32 on an Unbalanced 2D Dataset

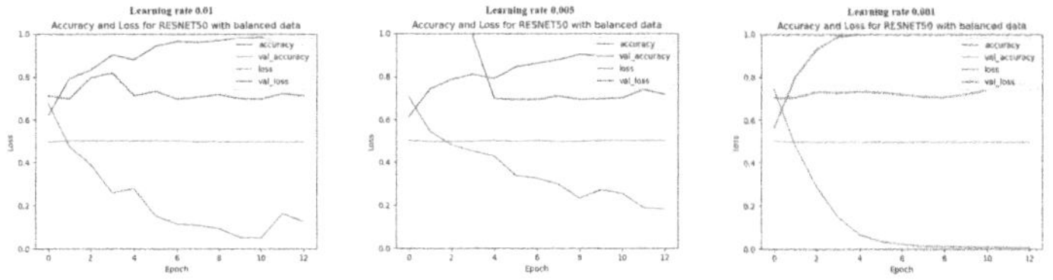

Fig. 2. ResNet50 Training with a Batch Size of 64 on a Balanced 2D Dataset

training for 10 epochs due to the time-consuming nature of processing the large dataset (Fig. 3). Observing potential for improvement, we extended the training by another 10 epochs with a mechanism to save the best model mid-training (Fig. 4).

Comparing this approach with the FIX melodic approach, several key differences and advantages of the CNN method emerged:

Feature Learning

- **Automatic Feature Extraction.** CNNs are excellent at learning spatial hierarchies of features automatically, which means they might discover more complex patterns in the data compared to manually engineered features used in the Random Forest approach in FIX [5,7].
- **Adaptability.** CNNs can potentially adapt to different types of features (spatial and temporal) from the fMRI data without requiring extensive manual feature engineering.

End-to-End Learning

- **Simplified Pipeline.** Using a CNN can simplify the processing pipeline since feature extraction and classification are integrated into a single model. This can reduce the need for separate steps of feature engineering and classification.

FIX first extracts various features from the ICA components generated by MELODIC. These features characterize both the spatial and temporal properties of each component. Examples of features include spatial properties (e.g., cluster size, edge fraction), temporal properties (e.g., power spectrum, time series autocorrelation), and various statistical measures [5,7].

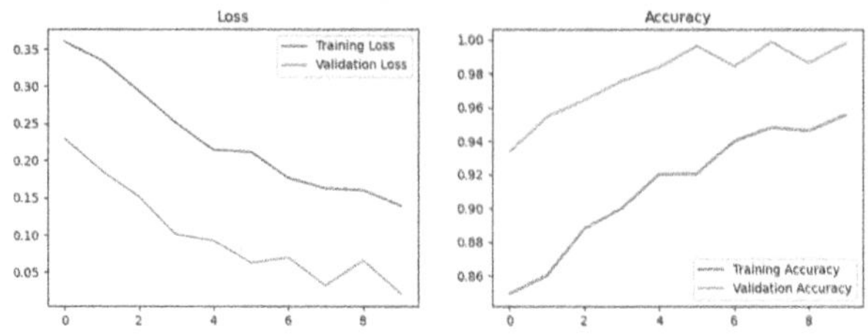

Fig. 3. Training and Validation Accuracy over first 10 Epochs of 3D CNN model training with HCP data

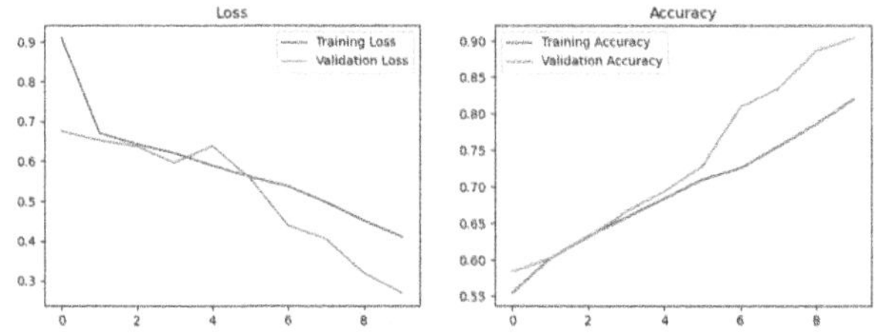

Fig. 4. Training and Validation Accuracy over last 10 Epochs of 3D CNN model training with HCP data

Our 3D CNN approach, on the other hand, directly processes the volumetric data, capturing complex spatial dependencies that are critical for accurate classification. The CNN model demonstrated superior performance in identifying relevant signal components compared to the FIX melodic method.

Overall, these results demonstrate that using 3D data significantly improves the model's performance and generalizability.

3.3 Discussion and Ethical Challenges

The results indicate that AI models can enhance the analysis of fMRI data, providing more precise and reliable interpretations of brain activity patterns. The use of advanced AI techniques such as transfer learning and integration of 3D imaging data can further improve the accuracy and efficiency of these models.

However, the integration of AI in fMRI data analysis also raises several ethical challenges. These include:

- **Data Privacy**: The use of sensitive medical data necessitates strict adherence to privacy laws and ethical guidelines to protect patient information.
- **Bias and Fairness**: AI models may inadvertently incorporate biases present in the training data, potentially leading to unfair or inaccurate conclusions. Ensuring fairness and transparency in AI decision-making is crucial.
- **Interpretability**: Deep learning models are often seen as "black boxes" due to their complex architectures. This lack of interpretability can be a barrier to their acceptance in clinical settings, where understanding the decision-making process is essential.

Addressing these ethical challenges requires a multidisciplinary approach, involving collaboration between AI researchers, medical professionals, ethicists, and legal experts. This collaborative effort is crucial to ensure that AI-driven fMRI data analysis is conducted responsibly and effectively, maximizing benefits while minimizing potential risks.

4 Conclusions and Fututre Work

The integration of AI, specifically CNNs, in the classification of fMRI data, has demonstrated significant potential in enhancing accuracy and efficiency. This paper explores the application of deep learning models, VGG16 and ResNet50, as well a custom 3D CNN on both 2D and 3D fMRI datasets.

Our experiments revealed that while VGG16 struggled with the complexity of 2D fMRI data, ResNet50 showed more promise under various conditions, including different batch sizes, learning rates, and balanced versus unbalanced datasets. However, ResNet50's performance was limited to training accuracy, with overfitting issues affecting its generalization to validation data. This highlighted the importance of dataset size and quality, as well as the limitations of treating fMRI data as 2D even with additional channels for power spectrum and time series.

The use of 3D data from the HCP yielded impressive results, with validation accuracy reaching up to 98%. This underscores the benefits of utilizing 3D data to fully capture spatial dependencies in fMRI images. The 3D CNN model's ability to automatically extract features and integrate feature extraction and classification into a single process demonstrated superior performance compared to traditional ICA methods, such as those employed in FIX melodic.

Comparative analysis with FIX melodic showed that while FIX effectively denoises fMRI data through manual feature engineering and hierarchical classifiers, CNNs offer a more adaptive and automated approach. CNNs excel in learning complex spatial hierarchies and adapting to various feature types without extensive manual preprocessing.

The findings indicate that AI models significantly outperform traditional methods in both depth and automation of feature extraction. Future work will focus on the development of predictive models for therapy response and the creation of a structural anomaly classifier. The predictive model aims to tailor therapeutic approaches based on initial fMRI scans and other relevant data, enhancing treatment efficiency and outcomes. The structural anomaly classifier will identify significant neurological abnormalities, ensuring the homogeneity of study participants and improving the reliability of neuroscientific studies. This research showcases AI's potential in revolutionizing fMRI data analysis and sets the groundwork for future advancements in the field.

To summarize, this research highlights the potential of CNNs to revolutionize fMRI data analysis, providing a robust alternative to traditional methods. Future work will focus on further optimizing CNN architectures, expanding dataset sizes, and exploring advanced data augmentation techniques to improve model robustness and generalizability. Continued exploration of 3D data representations and hybrid models could lead to even more significant advancements in the field.

References

1. Suhaimi, N.F.M., Htike, Z., Rashid, N.A.: Studies on classification of fMRI data using deep learning approach. ARPN J. Eng. Appl. Sci. **10**, 9748–9752 (2015)
2. Goodfellow, I., Bengio, Y., Courville, A.: Deep Learning. MIT Press, Cambridge (2016)
3. Yin, Z., Zhao, M., He, X., Zhang, Y.: Deep learning based framework for the detection of fMRI brain abnormalities. Neurocomputing **388**, 273–281 (2020)
4. Warren, J.D., Fletcher, P.C., Downey, L.E.: Functional MRI: current applications and future directions. Br. J. Radiol. **95**(1136), 20211017 (2022)
5. Griffanti, L., et al.: ICA-based artefact removal and accelerated fMRI acquisition for improved resting state network imaging. Neuroimage **95**, 232–247 (2014)
6. Beckmann, C.F., Smith, S.M.: Probabilistic independent component analysis for functional magnetic resonance imaging. IEEE Trans. Med. Imaging **23**(2), 137–152 (2004)
7. Salimi-Khorshidi, G., Douaud, G., Beckmann, C.F., Glasser, M.F., Griffanti, L., Smith, S.M.: Automatic denoising of functional MRI data: combining independent component analysis and hierarchical fusion of classifiers. Neuroimage **90**, 449–468 (2014)
8. Convolutional neural networks: an overview and application in radiology. Insights into Imaging. https://insightsimaging.springeropen.com/articles/10.1186/s13244-018-0673-2
9. Simonyan, K., Zisserman, A.: Very deep convolutional networks for large-scale image recognition. arXiv preprint arXiv:1409.1556 (2014)
10. He, K., Zhang, X., Ren, S., Sun, J.: Deep residual learning for image recognition. In: Proceedings of the IEEE Conference on Computer Vision and Pattern Recognition (CVPR), pp. 770–778 (2016)
11. Zhang, Z., et al.: A confusion matrix-based metric for evaluating classification models in machine learning. arXiv preprint arXiv:2007.08139 (2020)

How Resilient are Language Models to Text Perturbations?

Daniel Romero-Alvarado(✉), José Hernández-Orallo, and Fernando Martínez-Plumed

Universitat Politècnica de València, Valencia, Spain
dromalv@vrain.upv.es, jorallo@upv.es, fmartinez@dsic.upv.es

Abstract. Large language models typically rely on highly curated datasets that lack common irregularities such as typos and contractions, resulting in a mismatch between their training environments and real-world applications. This study evaluates the resilience of four prominent models in five different NLP tasks when confronted with perturbed inputs. We investigate three categories of perturbations: character-level, word-level and miscellaneous perturbations. By comparing performance on original and altered datasets, our results reveal a significant sensitivity to input perturbations across all models, with varying degrees of vulnerability depending on both the specific task and the type of perturbation. In particular, the XLNet model consistently shows superior robustness, while tasks involving grammatical coherence are most adversely affected.

Keywords: Language Models · Text Perturbations · Robustness

1 Introduction

Pre-trained Large Language Models (LLMs) such as BERT [9] and GPT [2], trained on large amounts of raw data, excel at various NLP tasks by extracting semantic and syntactic information. They have significantly advanced areas such as language translation, sentiment analysis and question answering, proving essential for real-world applications [2]. However, they struggle with the variability of real-world data, often relying on refined datasets that do not reflect the messiness of real-world use. For example, input during inference may contain more typos and spelling errors than during training, leading to performance degradation. These disruptions, which range from small character changes to more nuanced word substitutions, can particularly affect tasks requiring detailed word recognition or broader comprehension. Therefore, LLMs need to be able to manage and accurately interpret these disruptions across tasks (Fig. 1).

In this study, we investigate the robustness of LLMs by determining their performance stability when exposed to shifts in data distribution, observing how they handle perturbed inputs compared to normal inputs. Specifically, our goal is to identify the models that are most and least resilient to input perturbations, and to understand which NLP tasks are most susceptible to these changes.

V. Julian et al. (Eds.): IDEAL 2024, LNCS 15346, pp. 85–96, 2025.
https://doi.org/10.1007/978-3-031-77731-8_8

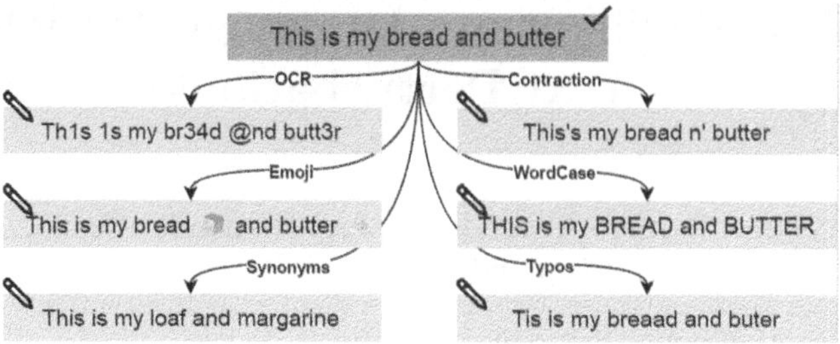

Fig. 1. Typos and perturbations are rare in controlled settings but common in real-world scenarios, causing major shifts in data distributions.

This analysis will guide the selection of optimal models for specific applications. We will also investigate the impact of different types of perturbations (such as character-level, word-level and syntactic changes) on model predictions. This will help refine our approaches to model evaluation and development. We will use different open-source LLMs and datasets, and employ over a dozen perturbation techniques such as insertion, deletion and synonym substitution to mimic realistic shifts in data distribution and to evaluate the adaptability of LLMs under different conditions.

This study offers three main contributions: (1) a detailed analysis of how four open-source, pre-trained BERT-based architectures handle different types and intensities of data perturbations across various NLP tasks; (2) a novel evaluation approach using Cohen's kappa that goes beyond typical character and word perturbations to include punctuation changes and semantically driven alterations, highlighting LLM vulnerabilities; and (3) comprehensive results revealing discrepancies in resilience to different perturbations, and identifying NLP tasks particularly vulnerable to input perturbations.

2 Related Work

Transformers, the most widely used NLP models [2], have significantly influenced the field. These models are pre-trained to generate token sequences and can be fine-tuned for specific tasks, allowing adaptability across various domains. Despite their capabilities, transformers can underperform on noisy, real-world data, often failing where humans would succeed. Studies reveal a notable performance gap between ideal benchmarks and real-world applications, highlighting challenges in handling minor errors such as typos [15].

Understanding model resilience to perturbations is essential for enhancing reliability. Moradi and Samwald [12] observed that LLM performance dropped by 8–18% with character- and word-level perturbations, varying by model and task complexity. Belinkov and Bisk [1] noted significant degradation in advanced neural machine translation systems under minor synthetic or natural noise. Jia and Liang [11] demonstrated that adversarial examples could compromise reading comprehension models. Ebrahimi et al. [10] revealed that character-level adversarial attacks could mislead text classification models and suggested training with such perturbed examples to boost robustness. Zhou et al. [28] enhanced lexical robustness through back-translation, generating paraphrases and increasing exposure to diverse linguistic phenomena.

Despite progress in assessing model robustness to text perturbations, significant limitations persist. Existing research mainly targets character- and word-level perturbations, overlooking other common real-world changes such as punctuation alterations and context-specific variations, such as social media tags. Additionally, the scalability and effectiveness of these models are hampered by using a fixed number of perturbations per sample, disregarding text length variation. Moreover, some studies limit the number of perturbed examples to mitigate meaning alteration caused by word changes, unintentionally favouring character-level robustness, which usually induces less semantic shift.

Recent works (e.g., [18]) have also evaluated the robustness of LLMs by comparing the dense representations of original and corrupted instances, to determine whether LLMs are able to *understand* the inputs rather than just *parsing* them.

To our knowledge, this paper is the first to offer a thorough, systematic evaluation of LLM robustness across various NLP tasks, tested against an extensive range of input perturbations going beyond traditional character and synonym-based changes to include punctuation variations, contractions, and other semantic modifications common in everyday language.

3 Methodology

NLP Tasks. In order to comprehensively test the applicability of our results to different NLP domains, we selected five supervised classification tasks using datasets from `HuggingFace`, all of which involved English-language texts. The datasets are summarised in Table 1 and detailed (including examples) in Appendix B in [16]. The training set size of each dataset is set to a maximum of 15,000 entries (for efficiency reasons), with unchanged validation and test sets (from the literature), except for HSOL, where the original dataset was randomly split into train-validation-test sets due to the absence of the latter two datasets.

Table 1. Summary of the five datasets used in our analysis.

Dataset	NLP Task	Size			#classes
		Train	Validation	Test	
CoLA [23]	Grammatical Coherence	8550	1040	1060	2
HSOL [7]	Hate Speech and Offensive Language	13878	5948	4957	3
MNLI [24]	Natural Language Inference	15000	9820	9800	3
SST2 [19]	Sentiment Analysis	15000	872	1820	2
PAWS [27]	Semantic Similarity	15000	8000	8000	2

Table 2. Summary of Perturbation Types, descriptions, and examples

Type	Perturbation	Description	Example (orig → perturbed)
Characterlevel	Keyboard	Substitutes characters with adjacent keys on a QWERTY keyboard	`hand` → `hanf`
	OCR	Mimics Optical Character Recognition errors by substituting similar-looking characters	`hello` → `hellu`
	Spelling Error	Introduces common spelling mistakes	`achieve` → `acheive`
	Typos	Implements random character insertions, deletions, or swaps	`random` → `andom`
Word-level	SwapSynWord Embedding	Replaces words with synonyms based on GloVE word embeddings	`It is nothing but caricature` → `It is nothing but an imitation`
	SwapSynWord Net	Substitutes words with synonyms from the WordNet database	`I'd like some pie, please` → `I'd like some cake, please`
Misc.	Contraction	Converts phrases to contracted forms	`They are coming, run!` → `They're coming, run!`
	InsertAdv	Adds adverbs before verbs	`He jumped the fence` → `He quickly jumped the fence`
	Prejudice	Substitutes geolocations to introduce bias	`I'm moving to Canada next month` → `I'm moving to Morocco next month`
	Punctuation	Alters punctuation marks	`The film was quite good` → `The film was quite good!`
	SwapNum	Changes numerical values	`3 people attended the meeting` → `15 people attended the metting`
	VerbTense	Changes the tense of verbs	`He started slow but ended up winning` → `He starts slow but ends up winning`
	Twitter	Adds mentions to usernames or short links	`God premise, bad ending` → `Good premise, bad ending @user789`
	WordCase	Converts text to upper case	`I'm not angry, not at all` → `I'M NOT ANGRY, NOT AT ALL`

Text Perturbations. We implemented three primary categories of perturbations to assess the resilience of our models (see Table 2 for a summary):

- **Character-level**: Modifying individual characters within words to simulate common typing errors and recognition issues. We controlled the *character perturbation percentage* (CP%), measuring the proportion of altered characters (e.g., a sentence with 20 characters and a CP% of 10% has around 2 characters affected). Perturbations were applied at intensities of 1%, 5%, and 10%, simulating real situations [20].
- **Word-level**: Replacing entire words to test the model's ability to handle context and meaning changes. We controlled the *word perturbation percentage* (WP%), so a sentence with 10 words and a WP% of 10% has 1 word altered. With sentences from 5 to 20 words, our experiments used WP% values of 10%, 20%, and 30%.
- **Miscellaneous** perturbations: Introducing changes in punctuation, contractions, and semantic elements to test the model's understanding and adaptation to non-standard language usage and structural variations.

When certain transformations, such as *SwapNum*, do not apply to all sentences—due to the lack of numerical digits—each perturbed sample is checked for differences from the original. After processing the dataset, the percentage of successfully perturbed samples is calculated. This metric assesses the results' robustness; for instance, if only 0.1% of the dataset is perturbed, drawing significant conclusions becomes challenging.

Even robust models can be vulnerable to specific perturbations. For example, under the *SwapNum* transformation, changing numerical details in a sentence (e.g., "this film scores 10 out of 10" to "this film scores 0 out of 10") affects sentiment analysis while preserving grammatical integrity. To implement these perturbations, we used the Python library `textflint` [22], which offers a wide range of general and task-specific text transformations.

Models. We explore a variety of BERT-inspired LLMs, each with different architectures or pre-training methods designed to overcome specific shortcomings of the original `BERT` model. Our study includes models implemented using resources from the `HuggingFace` transform library [25] and `PyTorch` [14]. The models evaluated are `DistilBERT` [17], `ELECTRA` [3], `XLNet` [26], and `Funnel Transformer` [6]. Each model was extended with a final classification layer to handle single label classification tasks, with further technical details provided in Appendix A in [16].

4 Experimental Setting

Model Finetuning. To optimise the performance of the models on specific datasets, we performed extensive training and fine-tuning on all models with different hyperparameters. We adjusted the initial learning rate (0.00001, 0.0005 and 0.0001) and the training duration (5, 7 and 10 epochs). Given the constraints of time and computational resources, we limited our hyperparameter optimisation to these two variables. Each training session was monitored, with accuracy and F1 score (which, unlike accuracy, remains reliable in the presence of imbalanced datasets) being recorded for both training and validation datasets. The best performing model configurations were identified using a weighted score that prioritised the validation score: $\mathrm{F1} = 0.2 \cdot \mathrm{F1}_{train} + 0.8 \cdot \mathrm{F1}_{validation}$ to account for potential overfitting and ensure generalisation. We used this custom scoring method to address potential biases in the model scores. Such biases can occur, for example, when the F1 score of the validation dataset appears disproportionately high due to randomness, especially with fewer training epochs. Further details regarding hyperparameter configuration in Appendix C in [16]. Detail regarding hardware, reproducibility and reporting results are in Appendix D in [16].

Evaluation. In order to identify discrepancies in model predictions, the test portion of each dataset was subjected to a series of perturbations as outlined previously, resulting in a collection of perturbed datasets that can now be accessed on HuggingFace[1].

---

[1] https://huggingface.co/DaniFrame.

Once the best combination of hyperparameters was found, each original (unaltered) test set was then fed into the model, returning label predictions. As many of the datasets used in this study are also used in popular benchmarks such as the GLUE benchmark [21], original test labels were often unavailable, and thus evaluation metrics such as accuracy or F1 score could not be computed directly. Instead, we measured performance using Cohen's kappa coefficient κ [4], which quantifies the degree of agreement between the predicted labels of the original and perturbed datasets, adjusting for agreement that occurs by chance (since we have different number of labels in some datasets).

When aggregating results across datasets and perturbations (Sect. 5.1), the proportion of perturbed examples (p) in each perturbed dataset was taken into account to calculate the *robustness score* defined as $\kappa \cdot p$, rather than relying solely on Cohen's kappa coefficient. This was done to prevent models from receiving inflated scores when the dataset had minimal perturbations.

5 Results

Next, each NLP task will be detailed, evaluating model scores and harmful perturbations. An aggregated analysis will follow.

Grammatical Coherence. The results for the CoLA dataset are shown in Table 3 (left). Cohen's kappa values range from 0.10 to 0.50, indicating low to moderate agreement above chance for character-level perturbations. The `ELECTRA` model consistently outperforms others, especially for keyboard and typos perturbations, remaining robust at 10% character modification. For word-level perturbations in the same table, `DistilBERT` is more robust, particularly as perturbations increase. However, high kappa values in the *Swap Synonym Word Embedding* perturbation are biased due to the small proportion of perturbed examples, misrepresenting the model's true performance. For other perturbations, `DistilBERT` remains robust. Notably, for the *WordCase* perturbation, models such as `ELECTRA` and `Funnel Transformer`, which do not distinguish case during tokenisation, show greater robustness compared to case-sensitive models such as `DistilBERT` and `XLNet`. The *InsertAdv* and *VerbTense* perturbations significantly degrade all models' performance, while the effects of *Punctuation* and *Twitter* are milder.

Hate Speech and Offensive Language. Table 3 (right) presents Cohen's kappa values for the HSOL task, ranging from 0.3 to nearly 0.9. This indicates better model robustness on HSOL compared to the CoLA task, due to lower linguistic knowledge requirements. `XLNet` shows slightly better robustness to most perturbations. Word-level and character-level perturbations follow a similar pattern, with all models faring better than in the CoLA task. `Funnel Transformer` outperforms other models in this setup. Other perturbations do not significantly impact model predictions, with some kappa values reaching up to 0.96.

Table 3. Cohen's kappa scores for each model and perturbation type for CoLA (left) and HSOL (right). The best score is highlighted. The last column shows the % of examples with at least one change in the text.

Grammatical Coherence (CoLA)						
Character level						
Perturbation	**CP%**	**DistilBERT**	**ELECTRA**	**Funnel Transf.**	**XLNet**	**% samples**
Keyboard	1	0.41	**0.47**	0.37	0.43	100.0
	5	0.28	**0.30**	0.20	0.23	100.0
	10	0.08	**0.12**	0.05	0.06	100.0
OCR	1	0.39	**0.44**	0.33	0.41	99.8
	5	0.21	**0.36**	0.25	0.29	99.8
	10	0.13	**0.17**	0.06	0.08	99.8
SpellingError	1	0.38	0.38	0.37	**0.39**	99.2
	5	**0.28**	0.26	0.18	0.27	99.2
	10	0.10	**0.10**	0.07	0.09	99.2
Typos	1	0.44	**0.48**	0.41	0.42	95.6
	5	0.32	**0.36**	0.24	0.32	97.6
	10	0.10	**0.16**	0.07	0.10	99.2
Word level						
Swap Synonym Word Embedding	10	0.89	**0.92**	0.87	0.87	39.7
	20	0.88	0.88	0.86	**0.89**	39.7
	30	0.88	0.89	0.88	**0.89**	39.7
Swap Synonym WordNet	10	**0.46**	0.32	0.32	0.37	99.8
	20	**0.39**	0.27	0.26	0.31	99.8
	30	**0.27**	0.16	0.16	0.15	99.8
Miscellaneous						
Contraction		**0.99**	0.98	0.96	0.98	16.7
InsertAdv		**0.36**	0.24	0.22	0.30	99.4
Prejudice		**1.00**	**1.00**	**1.00**	0.99	0.5
Punctuation		0.84	**0.85**	0.76	0.57	100.0
SwapNum		**1.00**	**1.00**	**1.00**	1.00	1.2
VerbTense		**0.38**	0.23	0.31	0.35	88.1
Twitter		**0.76**	0.74	0.64	0.76	100.0
WordCase		0.00	**1.00**	**1.00**	0.18	100.0

Hate Speech and Offensive Language (HSOL)						
Character level						
Perturbation	**CP%**	**DistilBERT**	**ELECTRA**	**Funnel Transf.**	**XLNet**	**% samples**
Keyboard	1	0.81	0.82	**0.84**	0.83	100.0
	5	0.62	0.62	0.65	**0.66**	100.0
	10	0.35	0.33	0.35	**0.40**	100.0
OCR	1	0.83	0.81	0.82	**0.86**	100.0
	5	0.66	0.59	0.60	**0.72**	100.0
	10	0.41	0.31	0.31	**0.50**	100.0
SpellingError	1	0.79	0.80	**0.82**	0.81	98.5
	5	0.59	0.61	**0.64**	**0.64**	98.6
	10	0.38	0.41	0.45	**0.46**	98.6
Typos	1	0.83	0.81	0.83	**0.86**	97.0
	5	0.60	0.60	0.61	**0.64**	99.4
	10	0.36	0.29	0.34	**0.39**	99.9
Word level						
Swap Synonym Word Embedding	10	0.97	**0.97**	**0.98**	**0.97**	31.1
	20	0.97	0.97	**0.98**	0.98	31.1
	30	0.98	0.97	**0.98**	0.97	31.1
Swap Synonym WordNet	10	0.69	0.75	**0.78**	0.71	99.1
	20	0.55	0.62	**0.67**	0.58	99.1
	30	0.42	0.51	**0.56**	0.46	99.1
Miscellaneous						
Contraction		0.99	0.99	**1.00**	1.00	26.5
InsertAdv		0.94	0.92	**0.96**	0.95	90.6
Prejudice		1.00	1.00	**1.00**	1.00	0.6
Punctuation		0.94	0.95	**0.96**	0.95	100.0
SwapNum		1.00	1.00	**1.00**	1.00	11.8
VerbTense		0.98	0.97	**0.99**	0.97	73.4
Twitter		0.93	0.93	**0.97**	0.95	100.0
WordCase		0.00	**1.00**	**1.00**	0.22	100.0

Natural Language Inference. The performance of different models under various perturbations for the MNLI task is in Table 4 (left). For character-level perturbations, kappa values range from 0.3 to 0.75, with `XLNet` being the most resilient, particularly at a 1% perturbation rate. `DistilBERT` and `ELECTRA` experience larger drops compared to Funnel Transformer. For word-level perturbations, kappa scores range from 0.6 to 0.75. `DistilBERT` is the most sensitive, whereas `ELECTRA` matches `XLNet` and surpasses Funnel Transformer. Other perturbation types show similar trends. `XLNet` consistently performs as well as or better than `Funnel Transformer`, except for *Twitter*-specific perturbations, where `XLNet` excels. For *WordCase* perturbations, kappa scores are higher than in Grammatical Coherence and Hate Speech tasks.

Sentiment Analysis. According to Table 4 (right), SST2 mirrors trends seen in HSOL task. No single model consistently outperformed others across different perturbations, though each, except `DistilBERT`, excelled in certain scenarios. `DistilBERT` remained stable at lower perturbations but struggled with higher ones. In contrast, `XLNet` excelled when 10% of characters were altered. For word-level perturbations, `Funnel Transformer` showed greater robustness than `XLNet`, regardless of perturbation size. The *Swap Synonym with Word Embedding* perturbation had minimal impact on all models, with a kappa coefficient of 1, indicating perfect agreement due to few perturbed samples. Miscellaneous perturbations found `ELECTRA` generally gave the best results, with `Funnel Transformer` close behind. `DistilBERT` was notably more vulnerable, although, except for the *WordCase* perturbation, all models produced consistent predictions.

Table 4. Cohen's kappa scores for each model and perturbation type for MNLI (left) and the SST2 (right). The best score is highlighted.

Natural Language Inference (MNLI)						
Character level						
Perturbation	**CP%**	**DistilBERT**	**ELECTRA**	**Funnel Transf.**	**XLNet**	**% samples**
Keyboard	1	0.63	0.74	0.73	**0.74**	99.9
	5	0.45	0.53	0.54	**0.57**	99.9
	10	0.18	0.22	0.29	**0.33**	99.9
OCR	1	0.63	0.73	0.72	**0.73**	99.8
	5	0.44	0.52	0.53	**0.55**	99.8
	10	0.21	0.24	0.32	**0.34**	99.8
SpellingError	1	0.68	**0.77**	0.75	0.74	98.7
	5	0.49	0.59	**0.60**	0.60	98.8
	10	0.28	0.36	**0.40**	0.39	98.9
Typos	1	0.65	**0.75**	0.72	0.73	93.3
	5	0.46	0.55	0.54	**0.56**	98.2
	10	0.22	0.27	0.33	**0.35**	99.6
Word level						
Swap Synonym Word Embedding	10	**0.91**	0.87	0.86	0.87	19.3
	20	**0.91**	0.87	0.85	0.87	19.3
	30	**0.91**	0.87	0.85	0.87	19.3
Swap Synonym WordNet	10	0.70	**0.76**	0.74	0.75	99.3
	20	0.62	**0.70**	0.69	0.69	99.3
	30	0.54	**0.63**	0.61	0.63	99.3
Miscellaneous						
Contraction	-	0.95	**0.98**	0.97	0.98	11.0
InsertAdv	-	0.63	**0.64**	0.62	0.63	94.7
Prejudice	-	**0.99**	0.99	0.99	0.99	0.5
Punctuation	-	0.83	**0.92**	0.90	0.90	100.0
SwapNum	-	**0.97**	0.97	0.97	0.97	3.8
VerbTense	-	0.90	**0.94**	0.92	0.92	68.7
Twitter	-	0.56	**0.84**	0.73	0.81	100.0
WordCase	-	0.35	**1.00**	**1.00**	0.49	99.9

Sentiment Analysis (SST2)						
Character level						
Perturbation	**CP%**	**DistilBERT**	**ELECTRA**	**Funnel Transf.**	**XLNet**	**% samples**
Keyboard	1	0.88	0.91	**0.93**	0.91	100.0
	5	0.69	0.75	**0.77**	**0.77**	100.0
	10	0.28	0.43	0.37	**0.47**	100.0
OCR	1	0.86	**0.93**	0.93	0.92	100.0
	5	0.67	**0.77**	0.75	0.74	100.0
	10	0.28	0.45	0.34	**0.47**	100.0
SpellingError	1	0.91	0.94	**0.94**	0.92	99.9
	5	0.70	0.80	**0.81**	0.78	99.9
	10	0.52	0.62	0.63	**0.64**	99.9
Typos	1	0.90	0.92	**0.93**	0.89	99.9
	5	0.69	**0.76**	0.75	**0.76**	100.0
	10	0.32	0.45	0.32	**0.46**	99.9
Word level						
Swap Synonym Word Embedding	10	1.00	1.00	**1.00**	1.00	7.4
	20	1.00	0.99	**1.00**	1.00	7.4
	30	1.00	1.00	**1.00**	1.00	7.4
Swap Synonym WordNet	10	0.88	0.92	**0.93**	0.91	99.4
	20	0.81	0.88	**0.89**	0.86	99.4
	30	0.73	0.82	**0.83**	0.79	99.4
Miscellaneous						
Contraction		0.98	**0.99**	0.99	0.98	27.0
InsertAdv		0.88	**0.94**	0.93	0.89	90.7
Prejudice		**1.00**	**1.00**	**1.00**	**1.00**	0.0
Punctuation		0.94	0.97	**0.97**	0.95	100.0
SwapNum		**1.00**	1.00	**1.00**	**1.00**	4.1
VerbTense		0.94	**0.98**	0.96	0.95	71.3
Twitter		0.93	**0.96**	0.96	0.94	100.0
WordCase		0.00	**1.00**	**1.00**	0.35	100.0

Semantic Similarity. Table 5 shows Cohen's kappa scores for the PAWS task. The `XLNet` model consistently outperforms others, especially as perturbation increases from 1% to 5% and 5% to 10%. `XLNet` maintains superior performance in most conditions, clearly surpassing `Funnel Transformer`, and significantly outperforming `ELECTRA` and `DistilBERT` with minimal perturbations. For word-level perturbations, *Swap Synonym Word Embedding* has a notable influence due to higher sample proportions. `XLNet` again excels, showing minimal performance degradation regardless of perturbed words percentage. For other perturbations such as *VerbTense*, *Twitter* and *Punctuation*, `ELECTRA` and `Funnel Transformer` show better kappa values. Overall, the Semantic Similarity pattern differs from other NLP tasks, with models such as `DistilBERT` suffering significantly under *WordCase* perturbation.

5.1 Aggregated Analysis

To evaluate model effectiveness across scenarios, we analyse aggregated results by models, NLP tasks, and perturbations (see Fig. 2). The scores account for the proportion of perturbed samples as explained in Sect. 4, normalising to prevent inflation from datasets with fewer perturbations, such as *Contraction* or *Prejudice*. The *WordCase* perturbation is excluded from overall calculations due to its unique tokenisation effects on `DistilBERT` and `XLNet`.

`XLNet` proved most robust with an average kappa score of 0.487, with `ELECTRA` and `FunnelTransformer` close behind with kappa scores of 0.472 and 0.464 each.

Table 5. Cohen's kappa scores for each model and perturbation type for the PAWS. The best score is highlighted.

Semantic Similarity (PAWS)						
Character level						
Perturbation	CP%	DistilBERT	ELECTRA	Funnel Transf.	XLNet	% samples
Keyboard	1	0.66	0.70	0.75	**0.85**	100.0
	5	0.13	0.08	0.22	**0.37**	100.0
	10	0.04	0.03	0.10	**0.17**	100.0
OCR	1	0.69	0.65	0.77	**0.84**	100.0
	5	0.21	0.08	0.28	**0.38**	100.0
	10	0.14	0.09	0.30	**0.31**	100.0
SpellingError	1	0.73	0.77	0.82	**0.87**	100.0
	5	0.30	0.30	0.43	**0.55**	100.0
	10	0.24	0.29	0.42	**0.51**	100.0
Typos	1	0.67	0.70	0.79	**0.85**	100.0
	5	0.16	0.12	0.26	**0.44**	100.0
	10	0.07	0.06	0.20	**0.32**	100.0
Word level						
Swap Synonym Word Embedding	10	0.46	0.37	0.44	**0.50**	72.1
	20	0.46	0.37	0.44	**0.50**	72.1
	30	0.46	0.38	0.45	**0.50**	72.1
Swap Synonym WordNet	10	0.76	0.84	0.86	**0.88**	100.0
	20	0.62	0.72	0.76	**0.80**	100.0
	30	0.59	0.71	0.74	**0.78**	100.0
Miscellaneous						
Contraction		1.00	**1.00**	1.00	5.01	27.0
InsertAdv		0.49	0.64	**0.77**	99.75	90.7
Prejudice		0.96	0.98	**0.98**	2.86	0.0
Punctuation		**0.96**	0.96	0.95	100.00	100.0
SwapNum		0.75	0.98	**0.98**	40.15	4.1
VerbTense		**0.98**	0.97	0.96	76.41	71.3
Twitter		0.87	**0.95**	0.94	100.00	100.0
WordCase		**1.00**	**1.00**	0.64	100.00	100.0

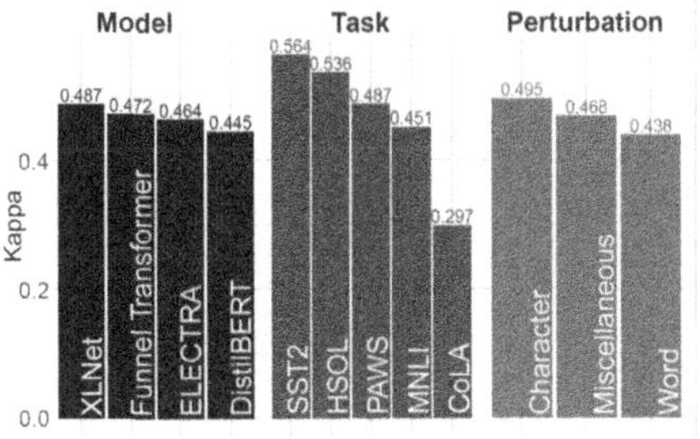

Fig. 2. Average kappa score per model, NLP task and perturbation.

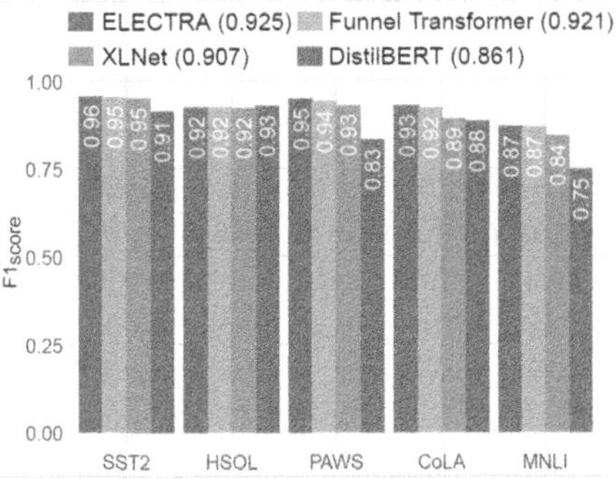

Fig. 3. $F1$ scores for the different (fine-tuned) models across different tasks (no perturbation). Model's average performance in the legend.

Our findings challenge the assumption that higher initial model accuracy directly correlates with greater robustness across tasks [8]. Despite strong initial scores from models such as `ELECTRA` and `FunnelTransformer` (see Fig. 3), these did not consistently translate to robustness as measured by kappa scores, indicating a need for deeper investigation into robustness factors beyond initial performance.

In NLP tasks, the Corpus of Linguistic Coherence (CoLA) proved the most difficult, with a score of 0.297. No model achieved 0.50 in this dataset, particularly with character-level perturbations, indicating potential over-fitting to specific grammatical structures. This results in significant performance drops when typographical errors occur. Conversely, tasks such as Sentiment Analysis (SST2) and Hate Speech Identification (HSOL) were more robust, scoring 0.536 and 0.564 respectively. Tasks requiring logical reasoning, such as Semantic Similarity and Natural Language Inference, scored around 0.48.

Initially, word-level perturbations were considered the most damaging, but results showed that character-level perturbations had more disruptive effects. Further analysis suggested discrepancies due to less common perturbations, such

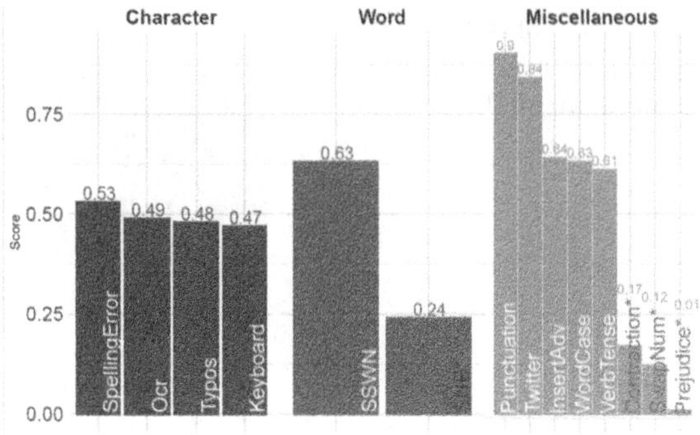

Fig. 4. Average kappa score for character-level, word-level and miscellaneous perturbations. Values in grey* indicate that the mean proportion of perturbed samples in the dataset is less than 50%.

as *Contraction*, *Prejudice*, *SwapNum* for miscellaneous perturbations, and *SwapSynonymWordEmbeddings* for word-level perturbations, which scored lower. These outliers reduced the average kappa values for their categories.

Excluding these outliers, character-level perturbations proved the most challenging (average kappa = 0.495), while word-level and miscellaneous types demonstrated higher robustness with values of 0.632 and 0.723, respectively. Common spelling mistakes caused kappa scores of 0.4 to 0.5, even with 10% character misspelling. Miscellaneous perturbations had varied impacts (Fig. 4): adding punctuation, links, and usernames resulted in high model scores (0.8–0.9), whereas inserting adverbs before verbs and capitalising letters made models less robust.

Overall, `XLNet` and `ELECTRA` are robust to severe perturbations, excelling in grammatical coherence and semantic similarity tasks. However, `DistilBERT` is more sensitive to higher levels of perturbation.

6 Conclusions

Our evaluation of LLM robustness under various textual perturbations provides valuable insights into their performance in realistic conditions. We found a significant performance decrease with increased perturbation intensity, especially for character-level perturbations, which were more challenging than word-level and miscellaneous ones.

`XLNet` stood out as the most robust model across most tasks and perturbation types, making it a strong candidate for applications needing high resilience to input noise. However, its higher computational cost may limit its use in certain scenarios. Models such as `ELECTRA` and `Funnel Transformer` also handled input perturbations well with lower computational demands, showing their value in resource-limited environments.

Grammatical coherence tasks highlighted all models' sensitivity to linguistic inaccuracies, while, tasks such as sentiment analysis and hate speech detection showed better robustness, indicating these models' potential in subjectively interpreting text.

Future research should explore nuanced perturbation types, such as semantic drifts and culturally specific text modifications [13], and extend robustness evaluations to multilingual and low-resource languages [5].

Acknowledgements. We acknowledge support from: CIPROM/2022/6 (FASSLOW); IDIFDER/2021/ 05 (CLUSTERIA) funded by Generalitat Valenciana; the EC H2020-EU grant agreement No. 952215 (TAILOR); US DARPA HR00112120007 (RECoG-AI); FISCALTICS (I+D+i PID2022-140110OA-I00), granted by MICIU/ AEI/10.13039/501100011033 and by ERDF, EU; and Spanish grant PID2021-122830OB-C42 (SFERA) funded by MCIN/AEI/10.13039/501100011033 and "ERDF A way of making Europe".

References

1. Belinkov, Y., Bisk, Y.: Synthetic and natural noise both break neural machine translation. In: International Conference on Learning Representations (2018)
2. Brown, T., et al.: Language models are few-shot learners. In: NeurIPS 33, pp. 1877–1901 (2020)
3. Clark, K., Luong, M.T., Le, Q.V., Manning, C.D.: ELECTRA: pre-training text encoders as discriminators rather than generators. In: International Conference on Learning Representations (2020)
4. Cohen, J.: A coefficient of agreement for nominal scales. Educ. Psychol. Measur. **20**(1), 37–46 (1960)
5. Conneau, A., et al.: Unsupervised cross-lingual representation learning at scale. arXiv:1911.02116 (2019)
6. Dai, Z., Lai, G., Yang, Y., Le, Q.V.: Funnel-transformer: filtering out sequential redundancy for efficient language processing. In: NeurIPS (2020)
7. Davidson, T., Warmsley, D., Macy, M., Weber, I.: Automated hate speech detection and the problem of offensive language. In: Eleventh International AAAI Conference on Web and Social Media (2017)
8. Demontis, A., et al.: Why do adversarial attacks transfer? Explaining transferability of evasion and poisoning attacks. In: 28th USENIX Security Symposium (USENIX Security 2019), pp. 321–338 (2019)
9. Devlin, J., Chang, M.W., Lee, K., Toutanova, K.: BERT: pre-training of deep bidirectional transformers for language understanding. arXiv:1810.04805 (2018)
10. Ebrahimi, J., Rao, A., Lowd, D., Dou, D.: HotFlip: white-box adversarial examples for text classification. In: Proceedings of the 56th Annual Meeting of the Association for Computational Linguistics (Volume 1: Long Papers), pp. 31–36 (2018)
11. Jia, R., Liang, P.: Adversarial examples for evaluating reading comprehension systems. In: Conference on Empirical Methods in NLP, pp. 2021–2031 (2017)
12. Moradi, M., Samwald, M.: Evaluating the robustness of neural language models to input perturbations. arXiv:2108.12237 (2021)
13. Morris, J.X., Lifland, E., Yoo, J., Grigsby, J., Jin, D., Qi, Y.: TextAttack: a framework for adversarial attacks, data augmentation, and adversarial training in NLP. In: Proceedings of the 2020 Conference on Empirical Methods in NLP: System Demonstrations, pp. 119–126 (2020)

14. Paszke, A., et al.: Automatic differentiation in PyTorch. In: NIPS-W (2017)
15. Ribeiro, M.T., Wu, T., Guestrin, C., Singh, S.: Beyond accuracy: behavioral testing of NLP models with checklist. arXiv:2005.04118 (2020)
16. Romero-Alvarado, D., Hernández-Orallo, J., Martínez-Plumed, F.: Supplementary material: how resilient are language models to text perturbations? (2024). https://github.com/Daniframe/TFG-GCD.git
17. Sanh, V., Debut, L., Chaumond, J., Wolf, T.: DistilBERT, a distilled version of BERT: smaller, faster, cheaper and lighter. In: NeurIPS EMC2 Workshop (2019)
18. Singh, A., Singh, N., Vatsal, S.: Robustness of LLMs to perturbations in text (2024). https://arxiv.org/abs/2407.08989
19. Socher, R., et al.: Recursive deep models for semantic compositionality over a sentiment treebank. In: Proceedings of the 2013 Conference on Empirical Methods in NLP, pp. 1631–1642 (2013)
20. Volk, M., Furrer, L., Sennrich, R.: Strategies for reducing and correcting OCR errors. In: Sporleder, C., van den Bosch, A., Zervanou, K. (eds.) Language Technology for Cultural Heritage. TANLP, pp. 3–22. Springer, Heidelberg (2011). https://doi.org/10.1007/978-3-642-20227-8_1
21. Wang, A., Singh, A., Michael, J., Hill, F., Levy, O., Bowman, S.R.: GLUE: a multi-task benchmark and analysis platform for natural language understanding. arXiv:1804.07461 (2018)
22. Wang, X., et al.: TextFlint: unified multilingual robustness evaluation toolkit for NLP. In: 59th ACL: System Demonstrations, pp. 347–355 (2021)
23. Warstadt, A., Singh, A., Bowman, S.R.: Neural network acceptability judgments. Trans. Assoc. Comput. Linguist. **6**, 625–641 (2018)
24. Williams, A., Nangia, N., Bowman, S.R.: A broad-coverage challenge corpus for sentence understanding through inference. In: Proceedings of the 2018 Conference of the North American Chapter of the Association for Computational Linguistics: Human Language Technologies, Volume 1 (Long Papers), pp. 1112–1122 (2018)
25. Wolf, T., et al.: Transformers: state-of-the-art NLP (2020). https://huggingface.co/transformers
26. Yang, Z., Dai, Z., Yang, Y., Carbonell, J., Salakhutdinov, R.R., Le, Q.V.: XLNet: generalized autoregressive pretraining for language understanding. In: NeurIPS 32 (2019)
27. Zhang, Y., Baldridge, J., He, L.: PAWS: paraphrase adversaries from word scrambling. In: Proceedings of the 2019 Conference of the North American Chapter of the Association for Computational Linguistics: Human Language Technologies, vol. 1, pp. 1298–1308 (2019)
28. Zhou, X., Zhang, M., Huang, H., Han, J.: Improving lexical robustness in text classification. In: Proceedings of the Annual Conference of the North American Chapter of the Association for Computational Linguistics (2019)

Emotional Sequential Influence Modeling on False Information

Debashis Naskar(✉), Subhashis Das, and Sara Rodríguez González

BISITE, Department of Computer Science and Automation, USAL, Salamanca, Spain
{debashis,subhashis,sara}@usal.es

Abstract. The extensive dissemination of false information in social networks affects netizen's social lives, morals, and behaviours. When a neighbour expresses strong emotions (e.g., fear, anger, excitement) based on a false statement, these emotions can be transmitted to others, especially through interactions on social media. Therefore, exploring the mechanism that explains how an individual's emotions change under the influence of a neighbour's false statement is a practically important task. In this work, we systematically examining the public's personal, interpersonal, and historical emotional influence based on social context, content, and emotional-based features. The contribution of this paper is to build an emotionally infused model called the Emotional-based User Sequential Influence Model+(E-USIM+) to understand users' temporal emotional propagation patterns and predict future emotions against false information.

Keywords: Emotion Detection · False Information · BERT · OSNs

1 Introduction

The existence of false information in Online Social Networks (OSNs) can trigger strong emotions and arouse netizens' feelings of fear and sadness. Emotional contagion paired with false information intentionally misguides the public by disseminating social media postings. Emotions are indeed regarded as a dominant driver of human behaviour, and understanding the role of emotions in spreading false information is crucial for developing effective strategies to combat their negative impact on social media [4]. False information infused with strong emotional elements such as fear, outrage, or excitement are more likely to be shared and spread rapidly across social networks. In this regard, Zaeem et al. [9] observed the significant association between negative sentiment with fake news and positive sentiment with true news. Emotions have a contagious quality and leading individuals to share content that resonates with their emotional state. Researchers [7] believe that emotions embedded in online rumours are essential determinants of the spreading diffusion dynamics in social media. Recently, a study [2] explored how emotionally framed content triggers emotional response posts by social media users and how corresponding users' response posts affect their sharing behaviour on social media.

V. Julian et al. (Eds.): IDEAL 2024, LNCS 15346, pp. 97–102, 2025.
https://doi.org/10.1007/978-3-031-77731-8_9

In addition to understanding emotional relationships in disseminating false information, capturing internal and external influence plays an important role in social networks. The work in reference [10] proposed influence graph in order to significantly differentiate between intentional and unintentional fake news spreaders. In social networks, existing works have been implemented differently in capturing emotional relationships as well as understanding different influential features on social conversation. For example, significant work [1] has been conducted by capturing personal and social identities within social discourse to better understand opinion behaviours. However, to the best of our knowledge, systematically understanding personal and social influential patterns with dynamic emotion detection over rumour/non-rumour has not yet been explored. On the other hand, it is also essential to capture the user's historical influential pattern in order to understand the dynamic nature of the conversation.

To exhibit this work, we propose an emotionally infused model called *Emotional-based User Sequential Influence Model+*(E-USIM+). The purpose of this work is to identify users' future emotions on false information by systematically examining different influential features along with emotional ones and their temporal propagation patterns. By taking advantage of our proposed model, the user's updated emotion can be predicted by taking into account personal, interpersonal and historical influential patterns on social conversation. The effectiveness of the proposed approach is demonstrated by analysing different topics on publicly available Twitter (currently known as X) datasets.

2 Methodology

The ultimate goal of the proposed framework is to detect public emotions in rumour/non-rumour conversations by adopting the BERT and GRU methods. Figure 1 illustrates the overall architecture of the proposed model, which consists of four components, namely *feature-based temporal social network*, *emotion-based user influence network*, *semantic feature extraction* and *emotion classification on false information*. All components are connected to three layers, i.e., the input layer, the fusion layer and the output layer. First, the input layer obtains the user's personal interests, initial emotional propagation structure and set of tweets' words along with structural information. Second, in the fusion layer, the GRU and BERT models are used to obtain the feature vector of the user's temporal internal emotional influential information and semantic feature information represented as $C_{u,i}(t)$ and $F_u(i)$. These two feature vectors are fused to obtain a final feature vector set $H_{u,i}(t) = (C_{u,i}(t) \oplus F_u(i))$, where $\oplus$ indicates the vector concatenating operation. Finally, the output layer is passed through the *Softmax* function to predict the emotion classification as an output class label on false information.

2.1 Feature-Based Temporal Social Network

To effectively study emotional influence and behavioural impact on social media, we have formulated *feature-based temporal social network* based on temporal

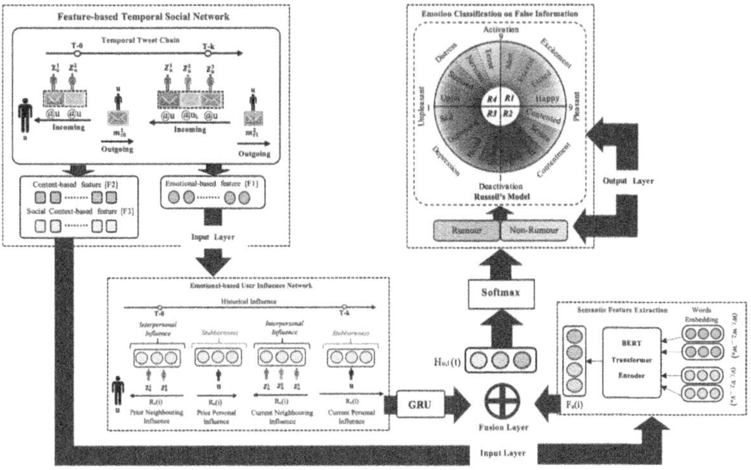

Fig. 1. The Overall Architecture of Our Proposed Model

conversation by identifying three distinct features, i.e., *emotional-based features [F1]*, *content-based features [F2]* and *social context-based features [F3]*.

Given a direct graph $G = (V, E)$, where each vertex $u \in V$ represents a user and each edge $(u, v) \in E$ represents a user u following to another user v. The each user $u \in V$, his/her posting sequence contains $PS_u = {<}M_u(i), S_u(i), t_u(i){>}$ where a tweet message $M_u(i)$ with emotion state $S_u(i)$ at time $t_u(i)$ posted by u. Taken into account G, we considered a collection of all external set for each user $u \in V$, which denoted as a $Z_u = \{v|(u, v) \in E\}$. The total number of users is N and size of the neighbor set Z_u is $n(u)$. For every participating u's data, we have estimated the *incoming* tweets posted by v that the user u has received between the time of two consecutive *outgoing* tweets is referred to as t_0 and t_1. Additionally, we also include the incoming tweets posted by v before posting u's first tweet at time t_0. Given a user u, as we construct a typical temporal tweet chain or tuple chain of u's posted a message at each timestamp $t_u(i)$, where $\downarrow$ denotes (`incoming`) tweets and $\uparrow$ denotes the (`outgoing`) tweets.

$$u = <.., \downarrow M^C_{t_{0-1}}>, \uparrow M^C_{t_0}, <\downarrow M^C_{t_{0+1}}, ..>, \uparrow M^C_{t_1}, <\downarrow M^C_{t_{1+1}}, ...>, \uparrow M^C_{t_2}, ...$$

where the user u posts his first tweet at t_0 denoted by the tuple $\uparrow M^C_{t_{0-i}}, i = 1, 2, ..$ and $C \in \{tweet, reply, retweet\}$. Similarly, incoming tweets between the Z_u's tweets are denoted by the tuple $<\downarrow M^C_{t_{k+0}}, \downarrow M^C_{t_{k+1}}, ..>$.

2.2 Emotional-Based User Influence Network

The *emotional-based feature [F1]* is determined using Russell's circumflex model of affect [8]. We also have adopted well-grounded emotional lexicons EmoLex [5] to increase the coverage of the emotional words in texts. Each tweet message

$M_u(i)$ contains the emotional state R_i^C and the core affect map identified sixteen states ($R_u(i) \in \{1, 2, 3, ..., 16\}$). To reduce the number of emotional states, we consider four quadrants defined in Russell's model and named them as *excitement* ($R1$), *contentment* ($R2$), *depression* ($R3$), and *distress* ($R4$). All the incoming set of emotion tweets from neighbor are denoted by the tuple $< \downarrow R^C_{t_{k+0}}, \downarrow R^C_{t_{k+1}}, ..>$ and fall between the user's personal emotion tweets $\uparrow R^C_{t_k}$ and $\uparrow R^C_{t_{k+1}}$.

$$u = <.., \downarrow R^C_{i,t_{0-1}}>, \uparrow R^C_{i,t_0}, < \downarrow R^C_{i,t_{0+1}}, ..> \uparrow R^C_{i,t_1}, < \downarrow R^C_{i,t_{1+1}}, ..>, \uparrow R^C_{i,t_2}, ..$$

At each time stamp, four crucial emotion influence factors are consider for shaping one's future dynamics emotion, i.e., prior neighboring $PN_{u,i}(t)$, prior personal $PP_{u,i}(t)$, current neighboring $CN_{u,i}(t)$ and historical emotion influence $HI_{u,i}(t)$. Therefore, future emotion state $S_u(i)$ is classified at the next timestamp t_u based on personal emotion sequence $\uparrow PR_{u,i}(t) = \langle \uparrow R^C_{i,t_0} \ldots, \uparrow R^C_{i,t_k} \rangle$ sent by the user u at t and all the neighboring emotion of the messages $\downarrow NR_{u,i}(t) = \langle \downarrow R^C_{i,t_{0-1}} \ldots, \downarrow R^C_{i,t_k} \rangle$ received from $t-1$ to t. To proposed the sequential influence model, we already taken place personal and neighboring emotion influence, i.e., derived by $X_{u,i}(t) \in \langle PN_{u,i}(t), PP_{u,i}(t), CN_{u,i}(t) \rangle$. To update u's next emotion at $t_u(i+1)$, the historical emotion influence $HI_{u,i}(t-1)$ could be replaced by the temporal internal emotion state $C_{u,i}(t-1)$. In our sequential influence model, we acknowledge the GRU to prove more appropriate and compact for dropping irrelevant information and updating the relevant information with affordable computation cost. This unit is formally expressed as follows:

$$\lceil r_{u,i}(t) = \sigma\left(W_r\left[C_{u,i}(t-1), X_{u,i}(t)\right] + b_r\right) \tag{1}$$

$$\lceil g_{u,i}(t) = \sigma\left(W_g\left[C_{u,i}(t-1), X_{u,i}(t)\right] + b_g\right) \tag{2}$$

$$\widetilde{C}_{u,i}(t) = \tanh\left(W_c\left[\lceil r_{u,i}(t) \cdot C_{u,i}(t-1), X_{u,i}(t)\right] + b_c\right) \tag{3}$$

$$C_{u,i}(t) = \lceil g_{u,i}(t) \cdot \widetilde{C}_{u,i}(t) + (1 - \lceil g_{u,i}(t)) \cdot C_{u,i}(t-1) \tag{4}$$

where, $W_r, W_g, W_c \in R^{c \times d_w}$, and $b_r, b_g, b_c \in R^c$ and d_w is the dimension of emotion state region.

2.3 Semantic Feature Extraction

We have employed a BERT in our proposed framework to extract semantic features information from *content-based feature [F2]* and *social context-based features [F3]* of the user interactions. The multi-head self-attention mechanism of the BERT model significantly resolves the issue of ignoring meaningful context and extracted feature vectors which accurately represent the literal meaning of the text.

The tweet content within the length n is represented as a set of words $Q=\{w_1, w_2, ..., w_n\}$ and the corresponding social structural feature vector set $K= \{v_1, v_2, ..., v_n\}$ is obtained through the BERT pre-processing model. The initialized text word vectors and structural information are submitted to the

Transformer, where feature vector $(Q, K) \in T$ learned by the multi-headed self-attention layer to fed the feed-forward neural network, and the learned tweet semantic representation vector F_u is obtained as follows:

$$F_u(i) = max(0, TW_a + b_a) \tag{5}$$

where W_a and b_a are the corresponding weight matrix and bias terms of the feed-forward neural network.

2.4 Emotion Classification on False Information

For the purpose of significantly detecting u's future emotion state $S_u(i)$ on rumour/non-rumour, we formulate the problem as a predicting probability distribution by given final fusion information $H_{u,i}(t)$ and the output layer for the E-USIM+ is a *softmax* function to output the probabilities of four distinct emotional region over false information.

$$P(S_u(i)|H_{u,i}(t)) = Softmax(VH_{u,i}(t) + b) \tag{6}$$

where $V \in R^{c \times d_w}$, and $b \in R^c$. R^c is the number of emotional regions for all u.

Table 1. Performance comparison between the E-USIM+ with other Models.

Method	Rumour					Non-Rumour				
	Acc	F-R1	F-R2	F-R3	F-R4	Acc	F-R1	F-R2	F-R3	F-R4
DeGroot	0.5198	0.4239	0.4811	0.5010	0.5144	0.5802	0.6319	0.5708	0.5213	0.4817
Voter	0.5445	0.4286	0.4927	0.5102	0.5024	0.5934	0.6236	0.6199	0.5390	0.4944
E-USIM	0.6234	0.4982	0.5393	0.6198	0.6305	0.6688	0.7027	0.6910	0.6210	0.5403
E-USIM+	**0.6728**	**0.5699**	**0.5893**	**0.6316**	**0.7012**	**0.7187**	**0.7393**	**0.7245**	**0.6612**	**0.5947**

3 Experimental Evaluation

We have used a publicly available dataset [3] with 111 events with tweet IDs and user information (60 rumours and 51 non-rumours) to particularly focus on predicting users' future emotions on false information by analyzing user influential features. To understand the dynamic nature of the users, we filtered out from the raw datasets the users who sent 0 or only one message and tweets that did not show any emotion. The final figure after filtering comprises 14,752 users (4876 rumours and 9876 non-rumours) over a total of 1,92,359 tweets posted. For validation of our model, we compare it with other baseline methods, **Degroot** and **Voter** model, along with our simplified version of the **E-USIM** model [6].

Given the sequence of u's emotion, we split the data into 90% of training and the remaining 10% as test dataset according to the temporal sequence. Accuracy (Acc) measures the percentage of correctly predicted emotions among all testing instances. We also compare models in terms of F-measure on *R1-Excitement*,

R2-Contentment, *R3-Depression*, and *R4-Distress* separately. Table 1 shows that E-USIM+ almost outperform in all evaluation metrics as compared to other baseline models for both rumour and non-rumour conversations. The result demonstrates that integrating historical information into an influential model is better for predicting someone's emotions than other influential models.

4 Conclusion

This paper proposes an emotionally infused model, i.e., E-USIM+, to detect emotion against false information by systematically examining different influential patterns and semantic feature information. Our proposed model has the ability to learn emotional influence and also detect future emotions over false information. The experiments demonstrated that the proposed model outperforms other compared influence models when updating users' future emotions.

Acknowledgements. This project has received funding from the EU's Horizon 2020 research and innovation programme under the MSCA grant agreement No. 101034371.

References

1. Chen, C., Wang, Z., Li, W.: Modeling opinion influence with user dual identity. In: Proceedings of the 2017 ACM on Conference on Information and Knowledge Management, Singapore, pp. 2019–2022 (2017)
2. Hosseini, A.S., Staab, S.: Emotional framing in the spreading of false and true claims. In: Proceedings of the 15th ACM Web Science Conference 2023, Austin, TX, USA, pp. 96–106 (2023)
3. Kwon, S., Cha, M., Jung, K.: Rumor detection over varying time windows. PLoS ONE **12**(1), e0168344 (2017)
4. Liu, Z., Zhang, T., Yang, K., Thompson, P., Yu, Z., Ananiadou, S.: Emotion detection for misinformation: a review. Inf. Fusion **107**, 102300 (2024)
5. Mohammad, S., Turney, P.: Emotions evoked by common words and phrases: using mechanical turk to create an emotion lexicon. In: Proceedings of the NAACL HLT 2010 Workshop on Computational Approaches to Analysis and Generation of Emotion in Text, pp. 26–34. ACL, Los Angeles (2010)
6. Naskar, D., Onaindia, E., Rebollo, M., Singh, S.R.: Predicting emotion dynamics sequence on Twitter via deep learning approach. In: Proceedings of the 18th International Conference on Advances in Mobile Computing & Multimedia, pp. 20–24 (2020)
7. Pröllochs, N., Bär, D., Feuerriegel, S.: Emotions in online rumor diffusion. EPJ Data Sci. **10**(1), 51 (2021)
8. Russell, J.A.: A circumplex model of affect. J. Pers. Soc. Psychol. **39**(6), 1161 (1980)
9. Zaeem, R.N., Li, C., Barber, K.S.: On sentiment of online fake news. In: 2020 IEEE/ACM International Conference on Advances in Social Networks Analysis and Mining (ASONAM), pp. 760–767 (2020)
10. Zhou, X., Shu, K., Phoha, V.V., Liu, H., Zafarani, R.: "This is fake! Shared it by mistake": assessing the intent of fake news spreaders. In: Proceedings of the ACM Web Conference 2022, pp. 3685–3694 (2022)

CSSDH: An Ontology for Social Determinants of Health to Operational Continuity of Care Data Interoperability

Subhashis Das(✉), Debashis Naskar, and Sara Rodríguez González

BISITE Research Group, Department of Computer Sciences and Automation, University of Salamanca, 37007 Salamanca, Spain
{subhashis,debashis,sara}@usal.es

Abstract. The rise of digital platforms has led to an increasing reliance on technology-driven, home-based healthcare solutions, enabling individuals to monitor their health and share information with healthcare professionals as needed. However, creating an efficient care plan management system requires more than just analyzing hospital summaries and Electronic Health Records (EHRs). Factors such as individual user needs and social determinants of health, including living conditions and the flow of healthcare information between different settings, must also be considered. Challenges in this complex healthcare network involve schema diversity (in EHRs, personal health records, etc.) and terminology diversity (e.g., ICD, SNOMED-CT) across ancillary healthcare operations. Establishing interoperability among various systems and applications is crucial, with the European Interoperability Framework (EIF) emphasizing the need for patient-centric access and control of healthcare data. In this paper, we propose an integrated ontological model, the *Common Semantic Data Model for Social Determinants of Health (CSSDH)*, by combining ISO/DIS 13940:2024 ContSys with WHO Social Determinants of Health. CSSDH aims to achieve interoperability within the Continuity of Care Network.

Keywords: Continuity of care · EHR · OWL · Social Determinants of Health

1 Introduction

The rise of digital platforms and services is helping to bridge the gap between organizational requirements and citizens' needs, particularly in health and social care. This trend underscores the value of remote monitoring and home-based solutions, driving a growing demand for digitized services. However, relying solely on Electronic Health Records (EHRs) is insufficient; understanding user needs and the social context of care is essential. Key considerations include purchasing agility, socioeconomic factors, digital literacy, and consent for accessing healthcare information. Securely accessing information across settings requires a robust

V. Julian et al. (Eds.): IDEAL 2024, LNCS 15346, pp. 103–109, 2025.
https://doi.org/10.1007/978-3-031-77731-8_10

healthcare management architecture, including access controls and data modeling.

According to the World Health Organization (WHO, 2024), poorly designed and ineffective EHRs often create a significant gap between what policymakers, health professionals, and researchers know and what they need to know to improve population health. Data from various sources, such as primary care, acute care, home care, national registries, and vital statistics systems, can be easily modified or expanded to meet the informational needs of rehabilitation.

The 2024 report, *Operational Framework for Monitoring Social Determinants of Health Equity (SDHE)*, published by the WHO [12], highlights that inequities in COVID-19 exposure, illness, and death emphasize the need to transform public health data and information systems. These systems should be equity-oriented to effectively monitor SDHE [12].

Significantly, Social Determinants of Health (SDH) are estimated to contribute to 30% to 55% of health outcomes. Exploring the relationship between SDH and outcomes is becoming increasingly important in healthcare as we strive to enhance continuous patient care. And if there is no provision to capture this information in the existing EHR system, we will not be able to monitor the patient's overall health. The recent initiative by the US Department of Health also gave emphasis to Capturing and recording SDH data under the Healthy People 2030 initiative.

In this paper, we proposed an Integrated Ontology-based Information Model called the Common Semantic Data Model for Social Determinants of Health (CSSDH) by combining ISO 13940: System of concepts to support continuity of care (ContSys) and WHO SDH. This model enables continuity of care data interoperability. The paper is structured as follows: related work is presented in Sect. 2, the overall methodology is described in Sect. 3, and results and evaluation, along with a discussion on future work, are covered in Sect. 4.

2 Related Work and Shortcoming

In the current literature, there are few initiatives aimed at formalizing a model that enables the semantic interoperability of continuity of care and incorporates social care aspects into the EHR model. For instance, the Health at Home [13] project provided a conceptually integrated framework to include social care within the patient journey. Cantor and Thorpe [1] highlighted key challenges and the need to integrate SDH data into EHR systems from a population health perspective. From an ontology engineering perspective, only two ontologies are available on BioPortal: the Social Determinants of Health Ontology (SOHO) [10] and the Social Prescribing Ontology [9], both directly related to SDH. However, they are primarily lists of terms in OWL format and do not qualify as formal ontologies that can be used as an Information Model for data integration purposes [5]. On the other hand, the SDoHo ontology [2] is overloaded with classes not exclusive to SDH, such as demographic information, age, etc. A few local initiatives aim to build universal screening for SDH alongside EHRs [11], but no final information model or publicly accessible tool exists. To our knowledge, there are no integrated formal ontological models that incorporate SDH into healthcare information models. In our previously published work [3,4], we introduced a formal ontological model based on the ContSys standard, aligned with the top-level ontology DOLCE [6]. Figure 1 shows a partial view of the ContSys ontology class visualization using the WebProtégé tool. The upper part of the figure, in green, includes DOLCE classes such as *mentalObject*, *stative*, and *event*, while the other classes are domain-specific and derived from ISO/DIS 13940:2024 ContSys. In this work, we extended our model to include concepts related to SDH.

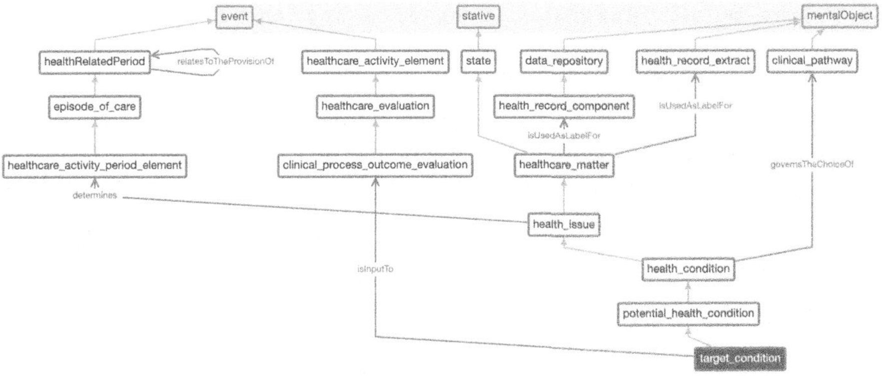

Fig. 1. target condition → health Condition alignment with DOLCE Top-level

3 Methodology

In this paper, we describe the development of a formal ontological schema for social factors influencing patients' continuous clinical outcomes, called the Common Semantic Data Model for Social Determinants of Health (CSSDH). Below, we describe the method used to develop CSSDH.

3.1 Modelling Concepts, Object Properties and Data Properties

In this work, we used four main resources for concept collection: the International Classification of Diseases (ICD)-11 Chapter 24, the SOHO ontology [10], the Social Prescribing Ontology [9], and the SDH value set from the Gravity project. We have reused most of the terminology from the SOHO ontology and organized it based on the OntoClean methodology [8]. For example, "Lay Off From Job" within the *Economic Stability* category is modeled as a data property associated with the class *subject of care* to collect relevant information. We have extended our model with additional properties, such as "Medical Services

Not Available at Home", "Problems Associated with Exposure to Radiation" and "Problems Associated with Exposure to Tobacco Smoke", borrowed from ICD-11 to make data collection more comprehensive and granular.

3.2 Implementation

CSSDM ontology model created using Protégé, an open-source ontology editing tool developed by Stanford Center for Biomedical Informatics Research. The main objective of the CSSDM model is to capture patients' SDH within the EHR system while keeping patient privacy. CSSDM should be capable of answering Competency Questions (CQ) [7] like

```
CQ1: Give me all patients lives in a low-income area.
CQ2: Provide a list of all patients who are laid off
from their job and whose houses are also crowded.
```

CSSDM ontology has 171 classes including subject of care (i.e. patient), care professional, observation, hospital appointment, and referral. 141 relations (i.e. object properties) We have added a total of 210 data properties among them 171 are related to SDH. The rationale behind collecting SDH information only as true or false (i.e.boolean datatype) as this type of data is sensitive personal information with significant privacy and security considerations [12]. Figure 2 depicted the *subject of care* class and property restrictions on the class hierarchy panel in Protégé and on the left panel it detailed SDH-related properties (i.e. social and community factors, Neighbourhood and built-in environment) associated with the subject of care.

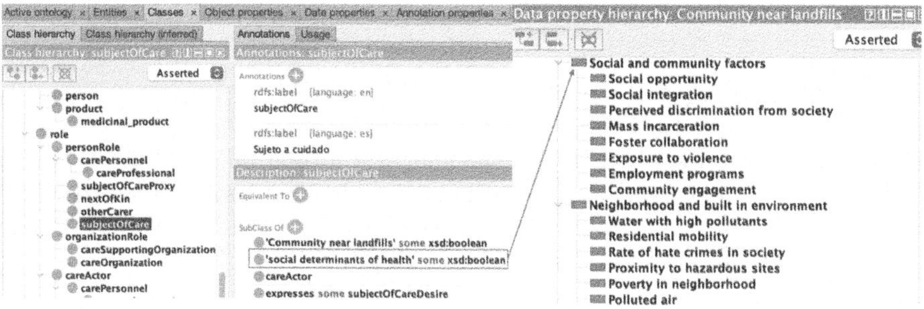

Fig. 2. Subject of Care SDH

4 Results and Discussion

Evaluation: Consistency of the CSSDM ontology is checked by the HermiT reasoner, a Protégé plugin. It identifies subsumption relations between classes. On the other hand Description Logic (DL) based query revealed that model is logically sound in retrieving information. For example, if policymakers would like to know *all patients who are laid off from their jobs as well as their home is crowded* (CQ2), CSSDM model can retrieve that Fig. 3 depicted such an answer.

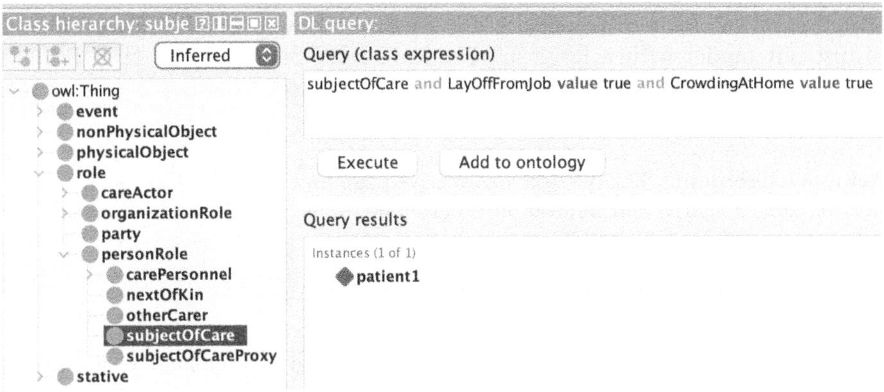

Fig. 3. Description Logic (DL) query execution with HermiT reasoner

We also have checked CSSDM ontology using oops ! ontology pitfall scanner and have obtained no pitfall badge.

We evaluated our ontology-driven CSSDM by modeling the Competency Queries elicited in Sect. 3.2 as `SPARQL` queries and assessing the returned answers. To address *CQ1*, we associated the data properties *LayOffFromJob* and *CrowdingAtHome* with the class *subjectOfCare* and set their data types as `Boolean`. A snapshot of the `SPARQL` query structure is provided.

```
PREFIX rdf: <http://www.w3.org/1999/02/22-rdf-syntax-ns#>
PREFIX owl: <http://www.w3.org/2002/07/owl#>
PREFIX rdfs: <http://www.w3.org/2000/01/rdf-schema#>
PREFIX xsd: <http://www.w3.org/2001/XMLSchema#>
PREFIX  coc: <http://purl.org/net/for-coc#>
PREFIX fhir: <http://hl7.org/fhir/>
SELECT ?subjectOfcare
?forename ?Lay_off_from_job ?CrowdingAtHome
WHERE { ?subjectOfcare fhir:Patient.forename ?forename.
    ?subjectOfcare   coc:Lay_off_from_job ?Lay_off_from_job.

OPTIONAL {?subjectOfcare
coc:Crowding_at_home ?CrowdingAtHome}
}
```

CSSDM aims to provide a formal OWL representation of the continuity of care domain, addressing identified gaps and overcoming the limitations of loosely available SDH control vocabularies. We have extensively reused existing ontologies and standards vocabularies to facilitate patient data capture in the context of SDH. CSSDM is designed to protect patient privacy while still providing healthcare professionals with the necessary information about social factors affecting the subject of care.

We have created classes and properties that offer flexibility for data capture and integration, as well as supporting semantic reasoning. Our future plan is to test our model with a large dataset from NHS Scotland and HSE Ireland, MIMIC-III free clinical Dataset, and a regional dataset from the Recoletas Red Hospital using our existing collaboration.

Acknowledgement. This project has received funding from the European Union's Horizon 2020 research and innovation programme under the Marie Skłodowska-Curie grant agreement No. 101034371.

References

1. Cantor, M.N., Thorpe, L.: Integrating data on social determinants of health into electronic health records. Health Aff. **37**(4), 585–590 (2018)
2. Dang, Y., et al.: Systematic design and data-driven evaluation of social determinants of health ontology (SDoHO). J. Am. Med. Inform. Assoc. **30**(9), 1465–1473 (2023)
3. Das, S., Hussey, P.: ContSOnto: a formal ontology for continuity of care. In: pHealth 2021, pp. 82–87. IOS Press (2021)
4. Das, S., Hussey, P.: HL7-FHIR-based ContSys formal ontology for enabling continuity of care data interoperability. J. Personalized Med. **13**(7), 1024 (2023)
5. Dhayne, H., Haque, R., Kilany, R., Taher, Y.: In search of big medical data integration solutions-a comprehensive survey. IEEE Access **7**, 91265–91290 (2019)
6. Gangemi, A., Guarino, N., Masolo, C., Oltramari, A., Schneider, L.: Sweetening ontologies with DOLCE. In: Gómez-Pérez, A., Benjamins, V.R. (eds.) EKAW 2002. LNCS (LNAI), vol. 2473, pp. 166–181. Springer, Heidelberg (2002). https://doi.org/10.1007/3-540-45810-7_18
7. Grüninger, M., Fox, M.S.: The role of competency questions in enterprise engineering. In: Rolstadås, A. (ed.) Benchmarking—Theory and Practice. IAICT, pp. 22–31. Springer, Boston (1995). https://doi.org/10.1007/978-0-387-34847-6_3
8. Guarino, N., Welty, C.A.: An overview of OntoClean. In: Staab, S., Studer, R. (eds.) Handbook on Ontologies. IHIS, pp. 201–220. Springer, Heidelberg (2009). https://doi.org/10.1007/978-3-540-92673-3_9
9. Jani, A., et al.: Using an ontology to facilitate more accurate coding of social prescriptions addressing social determinants of health: feasibility study. J. Med. Internet Res. **22**(12), e23721 (2020)
10. Kollapally, N.M., Chen, Y., Xu, J., Geller, J.: An ontology for the social determinants of health domain. In: 2022 IEEE International Conference on Bioinformatics and Biomedicine (BIBM), pp. 2403–2410. IEEE (2022)

11. Meyer, D., Lerner, E., Phillips, A., Zumwalt, K.: Universal screening of social determinants of health at a large US academic medical center, 2018. Am. J. Public Health **110**(S2), S219–S221 (2020)
12. World Health Organization: Operational framework for monitoring social determinants of health equity. World Health Organization (2024)
13. Pecoraro, F., Luzi, D., Pourabbas, E., Ricci, F.L.: A system of concepts to support the integration of health and social care and assistive domotics services: the health@ home project. Inform. Health Soc. Care **46**(3), 333–344 (2021)

Padel Two-Dimensional Tracking Extraction from Monocular Video Recordings

Álvaro Novillo(✉), Víctor Aceña, Carmen Lancho, Marina Cuesta, and Isaac Martín De Diego

Rey Juan Carlos University, Data Science Laboratory,
C/ Tulipán, s/n, 28933 Móstoles, Spain
{alvaro.novillo,victor.acena,carmen.lancho,
marina.cuesta,isaac.martin}@urjc.es
http://www.datasciencelab.es

Abstract. This study introduces a novel framework for the automatic two-dimensional tracking of padel games using monocular recordings. By integrating advanced Computer Vision and Deep Learning techniques, our algorithm detects and tracks players, the court, and the ball. Through homography, we accurately project detected player positions onto a two-dimensional court, enabling comprehensive tracking throughout the game. We tested the proposed algorithm using amateur video recordings of padel games found in literature. This approach remains user-friendly, cost-effective, and adaptable to various camera angles and lighting conditions. This makes it accessible to both amateur and professional players and coaches, providing a valuable tool for performance analysis. Additionally, the proposed framework holds potential for adaptation to other sports with minimal modifications, further broadening its applicability.

Keywords: Computer Vision · Padel · Deep Learning · Court Detection · Homography · Applied Intelligence · Tracking Data

1 Introduction

The unique combination of speed, agility, and strategy in padel has attracted many players, gaining immense popularity worldwide. This surge in popularity has also motivated an increase in scientific research on this dynamic racquet sport, encompassing studies that quantify player movements and shots [12] to those examining performance, bio-mechanical, physical, and physiological factors [3,10]. The integration of *Computer Vision (CV)* and *Artificial Intelligence (AI)* into such studies has revolutionized the analysis and coaching of padel games [6,13,14]. This technology allows for precise tracking of the movements and techniques of athletes, enabling an in-depth analysis of their performance.

V. Julian et al. (Eds.): IDEAL 2024, LNCS 15346, pp. 110–120, 2025.
https://doi.org/10.1007/978-3-031-77731-8_11

The application of *CV* and *AI* has been particularly beneficial in professional sports, where high-end cameras record high-resolution and high-frame-rate videos. These videos are then combined with image processing for referee assistance or data collection, leading to the development of sophisticated sport video tracking systems like Hawk-Eye [2], widely used in popular sports such as tennis, baseball, and basketball. Similarly, in padel, commercial *AI* solutions that use *CV* and machine learning to track the ball and provide real-time analytics are being developed. These systems, such as AIBall [1] and PadelCourt [5], offer a consumer-friendly approach to these complex technologies.

However, such professional *CV* tracking systems require significant resources, special multi-camera setups, and are not affordable for individuals or amateurs. The potential to develop a comparable system based on everyday technology would be immensely beneficial for the average user, democratizing access to advanced performance analysis tools.

To adress that need, the present study introduces an innovative framework for automatic player and ball detection using simple monocular recordings of padel games. Our cutting-edge algorithm seamlessly integrates several state-of-the-art *CV* techniques to accurately identify and track key elements in the video, including players, the court, and the ball. From this data, we can extrapolate player positions onto a two-dimensional court, enabling precise and automatic two-dimensional tracking of padel games.

Our approach stands out for its user-friendly design, cost-effectiveness, and adaptability to different camera angles and lighting conditions. This makes it an invaluable tool not only for professional players and coaches seeking detailed performance insights but also for amateur enthusiasts looking to enhance their game with advanced analytics. By making high-level tracking technology accessible to a wider audience, we aim to revolutionize the way padel is played and analyzed at all levels.

The paper is organized as follows. Section 2 discusses the basic principles of the proposed framework. In Sect. 3, we provide a detailed account of the experiments and results for real padel game videos. Section 4 concludes the work and suggests further improvements.

2 Methodology

Our framework is based on a three-module process. First, the court is detected, defining the Region *Of Interest (ROI)* to be studied. Then, *CV* techniques are applied to detect players, and finally, *Deep Learning (DL)* is used to detect the ball. Figure 1 contains a summarized flowchart diagram of the algorithm. The aforementioned modules are represented by triangles, while user inputs are represented by ovals placed above the pipeline. The subsequent stages of the workflow are represented by boxes. The output of the algorithm is represented as an oval below the chart.

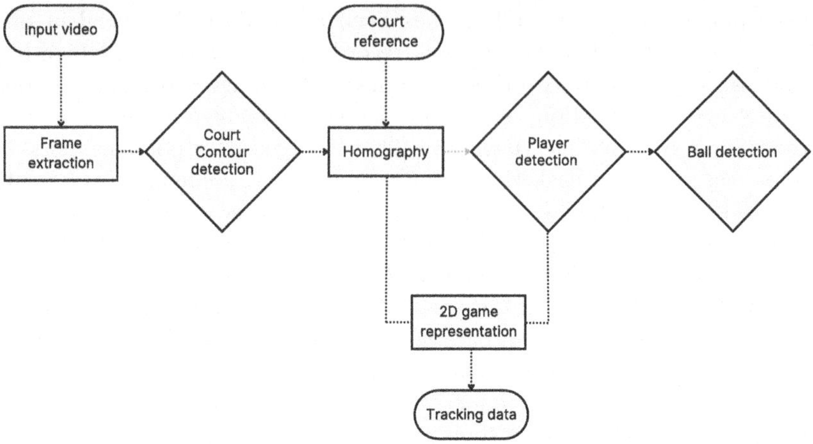

Fig. 1. Flowchart of the proposed framework. The primary modules are represented by triangles, while user inputs are represented by ovals placed above the pipeline. The following stages of the workflow are depicted as boxes, while the algorithm's output is illustrated as an oval beneath the chart.

2.1 Court Detection Module

The basic concept of the proposed court detection method is simple. By using elementary image preprocessing techniques in conjunction with contour detection [15], the goal is to automatically delineate the court within the image. Once the court contour is detected, its four corners are extracted, and linear interpolation is performed between them. Then, the intersections between these lines are estimated as the court boundaries. Using the estimated points, *four-point homography estimation* [9] is applied to a reference court. This involves mapping the court in the image to a reference image of the court provided by the user[1]. After this process, the *ROI* of the video is defined, where both the players and the ball are detected, and a homography matrix is obtained that allows mapping the detected position of the players to a two-dimensional representation of the game. The algorithm of this module contains the following steps:

Mask Generation. The image is first segmented using K-means clustering with $K = 2$, as shown in Fig. 2a. $K = 2$ seems to be the most convincing choice, since the entire court is likely to be detected as one of the clusters. It is then assumed that the court is placed in the most central regions of the image, since the camera placement in professional padel videos follows a de facto standard, consisting of a high angle shot at approximately 7.6 m above the court floor. Under this assumption, the clusters within a 5×5 square of pixels centered in the middle of the frame are examined. The frequency of each cluster in such region is then measured. After that, the most frequent cluster in this area is the one assumed

[1] The reference image should be a grayscale template of the padel court.

to be the court. Using the selected cluster, the image is then transformed into a two-dimensional array, where pixels belonging to the identified cluster are assigned a value of 1, and the remaining pixels are assigned a value of 0. For further clarification, see Fig. 2b. This process generates a binary mask that is then used to detect the location of the court.

Refinement of the Mask. To improve the performance of our court detection mask, a series of morphological transformations are applied, including frame averaging and blurring. These transformations are applied to a batch of five randomly sampled frames, from which binary masks are extracted. The resulting arrays are then averaged to produce an average mask, see Fig. 2c.

Contour Detection. Once a refined binary mask has been generated, contours within this mask are identified using the approach proposed in [15]. The contour detection algorithm identifies the boundaries of the highlighted regions (regions equal to 1 in the binary mask). Among the detected contours, it is assumed that the largest contour corresponds to the court. This assumption is based on the fact that the court is the most significant and prominent feature in the image. See Fig. 2d.

Extracting Extreme Points. From the identified court contour, the algorithm extracts the extreme points: the topmost, bottommost, leftmost, and rightmost points. These points serve as estimates for the corners of the court. To improve the estimation of the court borders, linear interpolation is performed using the four extracted corners from the contour, and then the court borders are determined as the intersection of these lines. To ensure the points form a coherent quadrilateral representing the court, the intersection points are ordered. This ordering step ensures that the points are in the correct sequence: top-left, top-right, bottom-right, and bottom-left. Figure 2e.

Homography Transformation. Using the extracted reference boundaries of the court, we can compute *four-point homography matrix* to map an input court reference image to the detected court region. In projective geometry, a $2D$ point $x \in \mathbb{R}^2$ is represented in homogeneous coordinates as $\boldsymbol{X} \in \mathbb{P}^2$, where $\mathbb{P}^2$ denotes the 2 dimensional projective space. The homogeneous coordinate vector $\boldsymbol{X}$ is a 3-dimensional vector $\boldsymbol{X} = [x, y, 1]^T$. Thus, a two-dimensional four-point homography is a transformation between two planes in the projective space $\mathbb{P}^2$. This transformation can be represented by a 3×3 matrix H. The relationship between a point $\boldsymbol{X}$ in the original plane and its image $\mathbf{X}'$ in the transformed plane is given by $\boldsymbol{X}' = H\boldsymbol{X}$. Here, $\boldsymbol{X} \in \mathbb{P}^2$ and $\boldsymbol{X}' \in \mathbb{P}^2$ are homogeneous coordinate vectors. In matrix form, this can be written as: $\begin{bmatrix} x' \\ y' \\ w' \end{bmatrix} = \begin{bmatrix} h_{11} & h_{12} & h_{13} \\ h_{21} & h_{22} & h_{23} \\ h_{31} & h_{32} & h_{33} \end{bmatrix} \begin{bmatrix} x \\ y \\ 1 \end{bmatrix}$ where $[x, y, 1]^T$ is the homogeneous coordinate vector of a point in the original

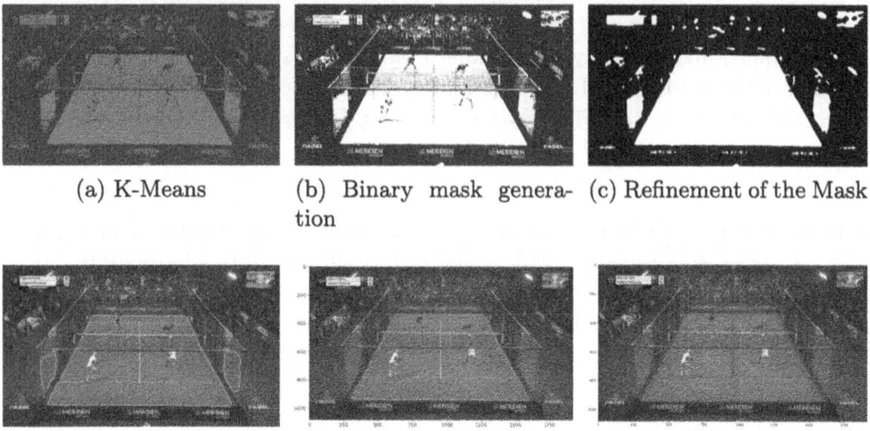

(a) K-Means (b) Binary mask generation (c) Refinement of the Mask

(d) Contour Detection (e) Extracting Extreme Points (f) Homography Transformation

Fig. 2. Court detection process on a professional game recording extracted from *Youtube*. After loading the image, K-means clustering with $K = 2$ is applied (2a). Then, in a 5×5 square of pixels centered in the middle of the image, the most frequent cluster is inspected, and a binary mask is created where all pixels belonging to this cluster are set to 1 and the rest to 0 (2b). This is followed by blurring and frame averaging of the mask (2c). Contour detection is then applied to this mask to extract all contours in the image (2d). Assuming that the court contour is the one that has the maximum area, the yard contours are detected. The detected boundary is outlined in green. In red, the court contour is smoothed by fitting straight lines through the upper-left, upper-right, lower-right, and lower-left corners of the detected contour (2e). Finally, the court corners are estimated by the intersections of these lines, and *4-point homography* is applied to map a reference court onto the image (2f).

plane, and $[x', y', w']^T$ is the homogeneous coordinate vector of the corresponding point in the transformed plane. H is the *homography matrix*, a 3×3 matrix but with 8 DoF (degrees of freedom) as it is estimated up to a scale. In our case, the computed homography matrix maps the detected court area in the image to a reference image of the court. Figure 2f represents the *four-point homography* for the professional game taken as a reference. Using this method, the court is successfully detected, along with its lines and different regions.

2.2 Player Detection Module

To efficiently detect and track the players within the field, we use YOLOv8's object detection model [8]. YOLOv8 was chosen for its high accuracy, real-time performance, and tracking capabilities. To ensure that only players inside the court are considered, the player detections are filtered by verifying that the center of each bounding box is inside the *ROI*.

Then, in order to obtain two-dimensional tracking data of the aforementioned players, we employ the homography matrix computed in the court detection module. This matrix allows the transformation of the coordinates of the detected players from the image perspective to the top-down, two-dimensional view of the court. For each player detection, since there is no information on the distance and height at which the recording camera is placed, the position of the player is estimated as the center of the bottom of the detection bounding box, since it can be assumed to be coplanar with the court. The homography matrix is then applied to such point, yielding the transformed coordinates on the two-dimensional real court.

2.3 Ball Detection Module

Finally, to detect the ball, TrackNet [4], a *DL* network for tracking high-speed and small objects in sports applications, is implemented. TrackNet is particularly effective for ball detection in racquet sports due to its ability to handle fast-moving objects and provide accurate position tracking. The TrackNet system consists of a *Convolutional Neural Network (CNN)* and a *Deconvolutional Neural Network (DeconvNet)* [11]. It generates a heat map showing the position of the object by processing successive frames.

While the ball can be detected in the video, estimating its position in a two-dimensional real court from the extracted positions in the video is not straightforward. Since there is no prior information about the camera specifications, it is not possible to perform deformation corrections. In addition, the exact placement of the camera, both in distance and height with respect to the court, is unknown, making it impossible to compute the actual trajectory of the ball in real-world coordinates. Monocular video recordings do not provide any information about the depth of the scene. Consequently, the aforementioned methodology allows the direct transformation of only the ball bounces into a two-dimensional representation of the game, since they can be assumed to be coplanar points in the plane. Future work will involve the simulation of the ball trajectory in the two-dimensional representation based on the obtained tracking and the detected bounces.

3 Experiment and Results

To evaluate the accuracy of the framework, we have implemented it on the public amateur padel dataset proposed in [7]. The source code is available at https://github.com/URJCDSLab/DS_Padel.git for reproducibility. The dataset consists of videos taken from different camera angles using different devices. The camera angles range from close to the de facto standard in professional video streaming to court-level views of the games. For this study, we will apply the proposed algorithm using three different perspectives, thus covering all professional and other state-of-the-art camera angles: *main camera view*, *second camera view*, and a *court-level shot.*

For the *main camera view*, a Panasonic AG-UX180 4K professional camcorder was utilized. The camera was positioned at approximately 6 m above the court plane and 10 m from the back wall, thus obtaining a camera angle that is close to the de facto standard in professional video streaming. This angle minimizes the impact of the occlusion of the mesh panel and structural elements.

The *second camera view* is positioned at a slightly higher position. A GoPro Hero9 Black was utilized on a tripod positioned above a ceiling panel structure with a stepped ceiling. This video shows a lamp hanging from the ceiling which occludes a part of the top-left court.

For the *the court-level recordings*, a tripod-mounted smartphone camera was employed at a height of approximately 1.45 m and a distance of 3.88 m from the court.

In Fig. 3, the efficacy of the algorithmic court detection module is evaluated in each of the employed perspectives. It can be observed that our algorithm performs adequately in both the *main* and *secondary* camera views, yet fails to accurately capture the court position at the court level. After applying the homography, there are slight deviations from the real position of the court lines, introducing uncertainty to the later extracted two-dimensional tracking.

Since the court placement has been successfully estimated in both the *main* and *secondary* views, we extract both the player and the ball positions from the videos, to then use the homography matrix previously computed to generate two-dimensional tracking of both perspectives.

Figure 4 contains both the players and ball detection in both views: *main* view, and *secondary* view. In both cases, the same frame is being displayed. It can be seen that all objects are properly tracked, and there are no significant differences between the two views. In both cases, two-dimensional tracking of the game can be successfully extracted using the previously computed homography matrix. Although the results are comparable, discrepancies between the two perspectives are evident. In the *main* view, the homography is able to accurately capture the vertical line, while having some problems with the first horizontal line. Meanwhile, in the *secondary* view, the main vertical line is not correctly detected, with some slight deviations.

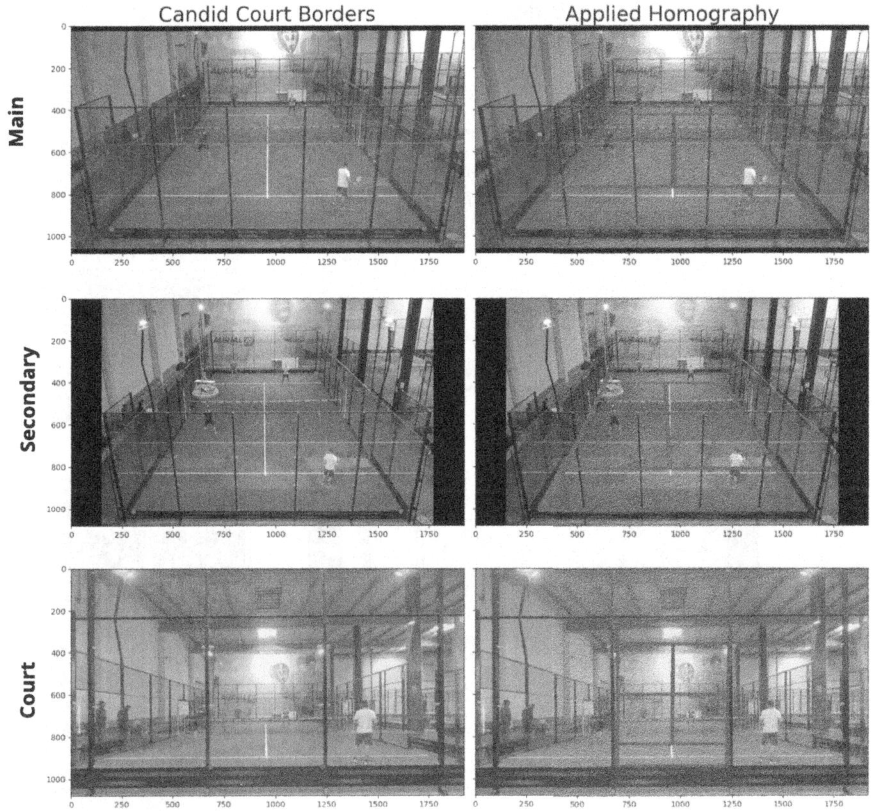

Fig. 3. Court detections evaluated in each of the employed perspectives. The algorithm performs adequately in both the *main* and *secondary* camera views, close to the de facto standard from professional video recordings. However, the proposed method is unable to capture the court position at the court level.

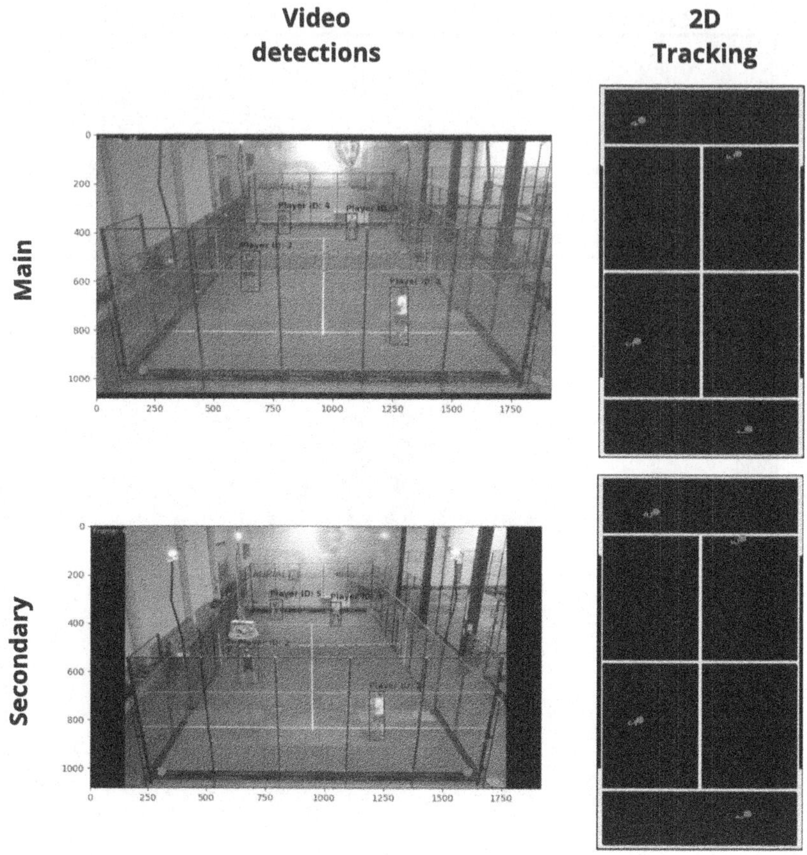

Fig. 4. Players and ball detection in both *main* view and *secondary* view. In both cases, the same frame is being displayed. The extracted two-dimensional tracking from the game of such frame is represented in the right column.

4 Conclusions

The proposed framework effectively detects and tracks both players and the ball in the video footage. Testing the algorithm on the existing data in the literature revealed that it performs optimally when the camera is positioned at standard angles commonly used in professional video streaming. However, the algorithm struggles with court-level views due to the limited field of view, which hinders accurate court mapping.

Despite this limitation, our framework offers a low-cost solution that benefits both amateurs and professionals. Its adaptable design holds potential for application to other sports with minimal modifications, such as tennis, squash, and badminton, where tracking player movements and ball trajectories is equally crucial for performance analysis.

Future work will focus on several key areas to enhance the system. First, creating an annotated dataset specifically for court identification would improve the training of the court detection system. Additionally, detecting more key points of the court before applying homography will allow for accurate court-level detections, even when the court is not fully visible.

Second, developing an algorithm to detect ball bounce positions and simulating the ball's trajectory based on real-world coordinates will enable us to reconstruct the ball's trajectory in two dimensions, which is crucial for a thorough tactical analysis. In addition to tracking player positions, incorporating information about the types of shots made by players will add significant detail, allowing for a comprehensive technical and tactical game analysis.

By implementing these enhancements, the system will provide greater accuracy and utility in sports analysis, contributing significantly to tactical evaluation and player development. Furthermore, this would enable the creation of an automatically produced datasets of padel matches, including annotations for courts, players, and the ball, complementing the work in the state of the art.

Acknowledgements. This research has been supported by grants from Rey Juan Carlos University (Ref: C1PREDOC2020) and the Spanish Ministry of Science and Innovation, under the Knowledge Generation Projects program: XMIDAS (Ref: PID2021-122640OB-100).

References

1. Aiball: Aiball - it's not all about technique (2024). https://www.aiball.io. Accessed 20 June 2024
2. Bal, B., Dureja, G.: Hawk eye: a logical innovative technology use in sports for effective decision making. Sport Sci. Rev. **XXI** (2012). https://doi.org/10.2478/v10237-012-0006-6
3. Garcia-Gimenez, A., Pradas de la Fuente, F., Castellar Otin, C., Carrasco Paez, L.: Performance outcome measures in Padel: a scoping review. Int. J. Environ. Res. Public Health **19**(7), 4395 (2022)
4. Huang, Y.C., Liao, I.N., Chen, C.H., İk, T.U., Peng, W.C.: Tracknet: a deep learning network for tracking high-speed and tiny objects in sports applications (2019)
5. Inc., I.: Padelcourt - artificial intelligence for padel (2023). https://padelcourt.app. Accessed 20 June 2024
6. Javadiha, M., Lacasa, E., Ric, A., Andújar, C., Susin, A.: Estimating player positions from padel high-angle videos: accuracy comparison of recent computer vision methods. Sensors **21**(10, article 3368), 1–17 (2021). https://doi.org/10.3390/s21103368, http://hdl.handle.net/2117/345963
7. Javadiha, M.: Padelvic: multicamera videos and motion capture data of an amateur padel match (2024). https://doi.org/10.17398/2952-2218.2.89, https://github.com/UPC-ViRVIG/PadelVic
8. Jocher, G., Chaurasia, A., Qiu, J.: Ultralytics YOLO, January 2023. https://github.com/ultralytics/ultralytics
9. Luo, Y., et al.: A review of homography estimation: advances and challenges. Electronics **12**, 4977 (2023). https://doi.org/10.3390/electronics12244977

10. Martín-Miguel, I., Escudero-Tena, A., Muñoz, D., Sánchez-Alcaraz, B.J.: Performance analysis in padel: a systematic review. J. Hum. Kinet. **89**, 213–230 (2023). https://doi.org/10.5114/jhk/168640
11. Noh, H., Hong, S., Han, B.: Learning deconvolution network for semantic segmentation. In: 2015 IEEE International Conference on Computer Vision (ICCV), pp. 1520–1528 (2015). https://doi.org/10.1109/ICCV.2015.178
12. Priego Quesada, J.I., Olaso Melis, J., Llana Belloch, S., Pérez Soriano, P., González García, J.C., Sanchis Almenara, M.: Padel: a quantitative study of the shots and movements in the high-performance. J. Hum. Sport Exerc. JHSE **8**(4), 925–931 (2013). https://doi.org/10.4100/JHSE.2013.84.04
13. Santiago, C.B., Sousa, A., Estriga, M.L., Reis, L.P., Lames, M.: Survey on team tracking techniques applied to sports. In: 2010 International Conference on Autonomous and Intelligent Systems, AIS 2010, pp. 1–6. IEEE (2010)
14. Shih, H.C.: A survey of content-aware video analysis for sports. IEEE Trans. Circuits Syst. Video Technol. **28**(5), 1212–1231 (2017)
15. Suzuki, S., be, K.: Topological structural analysis of digitized binary images by border following. Comput. Vis. Graph. Image Process. **30**(1), 32–46 (1985). https://doi.org/10.1016/0734-189X(85)90016-7, https://www.sciencedirect.com/science/article/pii/0734189X85900167

Drowsiness Detection Using Vital Sign Sensors and Deep Learning on Smartwatches

Vitor Augusto da Rosa Pereira[1(✉)], Rafael Alceste Berri[2], and Fernando Santos Osório[1]

[1] University of São Paulo, São Paulo, São Carlos 13566-590, Brazil
{vitor.arp,fosorio}@usp.br
[2] Federal University of Rio Grande, Rio Grande, Rio Grande do Sul 96203-900, Brazil
rafaelberri@furg.br

Abstract. Drowsy drivers contribute to high rates of road accidents, leading to numerous fatalities and injuries. The dangers of drowsiness extend beyond the roads, affecting workplaces that require continuous attention and focus. This work aims to develop a non-intrusive, real-time system capable of assessing the user's drowsiness level by processing physiological signals collected by smartwatch sensors. The study utilized photoplethysmography (PPG) and heart rate data from a smartwatch, labeled with an adapted Karolinska Sleepiness Scale (KSS). Deep Learning techniques were used to fine-tune hyperparameters and train a 1-dimensional Convolutional Neural Network to reach better prediction performance on drowsiness data. Besides, a 5-fold cross-validation technique was used to evaluate performance and create an ensemble, enhancing prediction robustness through election method. The results show significant prediction accuracy using the ensemble on test data (91.41%). The final pre-trained model was integrated into the smartwatch, creating a real-time drowsiness detection system. This system alerts the user through sound and vibration feedback, preventing them from falling asleep and reducing the risk of accidents.

Keywords: Drowsiness · Smartwatch · Photoplethysmography (PPG) · Artificial neural networks · Deep learning · Ensemble model

1 Introduction

Drowsiness is one of the main causes of fatalities on highways. In Brazil, about 20% of traffic accidents are related to drowsiness, with lack of rest compromising driving at the same level as alcohol ingestion [2]. Yet in Brazil, between 2014 and 2019, around twenty thousand accidents where the driver slept at the wheel were registered, causing 1,788 fatalities and 19,450 injured people [1]. The dangers of drowsiness extend beyond roads, impacting workplaces as well. A study [4] found that the 1-hour daylight saving time shift in the U.S. reduced sleep by 40 min on average, leading to a 5.7% increase in workplace injuries in the mining sector.

V. Julian et al. (Eds.): IDEAL 2024, LNCS 15346, pp. 121–132, 2025.
https://doi.org/10.1007/978-3-031-77731-8_12

Advancements in health technology have enabled the monitoring and storage of health data on small devices, leading to the development of wearables like smartwatches. These devices track physiological signals such as heart rate, blood pressure, oxygen saturation, and sleep quality using sensors like PPG, ECG, and body composition sensors. The market for wearables has grown, particularly in the biomedical sector [6,13].

Diverse methods for drowsiness detection have been developed, with many previous works utilizing physiological-based measurements, machine learning techniques, and video and subjective reports as ground truth for this purpose. Lee et al. [12] used ECG and PPG sensors from wearable devices to detect driver drowsiness. Data was labeled using videos from the drivers, pre-processed and used to generate recurrence plots. Features like determinism, divergence, and entropy were extracted to train KNN, SVM, and random forest models. Additionally, recurrence plots were directly used in a CNN, which achieved the highest accuracy of 70%. Kundinger et al. [10,11] used heart rate data (HR) from wearables, PPG and ECG to detect driver drowsiness. Data were labeled using KSS scale self-ratings and videos from driving simulation. Statistical features like max, min, mean, and std were used to train KNN, SVM, Random Forest, and MLP models. KNN achieved 98% accuracy for heart rate and 92% for PPG, while Random Tree and Random Forest reached 97% accuracy with ECG data. Rundo et al. [15,16] used a PPG sensor embedded in a simulator driving wheel to detect driver drowsiness. Data were labeled as drowsy or alert using EEG sensors. These labeled data were then input into neural networks, with MLP achieving 95% accuracy and 1D CNN reaching 98.5% accuracy. Zhang et al. [17] propose a drowsiness detection method based on remote PPG measurements captured by a camera distant from the user. The study compares this method with HR estimation using wearable ECG. Both remote PPG and wearable HR data are input into a 1D CNN model, achieving an accuracy of 0.70 AUC for remote PPG data and 0.71 AUC for wearable HR data.

Physiological-based methods have led to positive results. Even so, none of the previous research offers a solution for real-time drowsiness detection. Some approaches rely on intrusive devices, such as medical-grade ECG monitors or equipment integrated directly into vehicles, like PPG sensors embedded in the steering wheel, restricting their use to this specific application. Considering this, this work aims to develop a non-intrusive real-time drowsiness detection system based on the processing of physiological signals obtained from sensor readings of a smartwatch. To achieve this, a deep learning model capable of detecting drowsiness from smartwatch data has been developed. It is also essential to seek and implement algorithms that allow the predictive model to operate directly on the smartwatch, allowing real-time functionality.

This paper is organized as follows. Section 2 details the methodology to develop the drowsiness detection system. Section 3 presents and discusses the obtained results. Finally, Sect. 4 summarizes the conclusions and future works from this research.

2 System Proposal

The proposal of this paper is a non-intrusive real-time system designed for detecting drowsiness using a smartwatch. Figure 1 shows the overall system diagram, which underlies into four stages: Data Acquisition, Model Generation, Prediction, and Alert System.

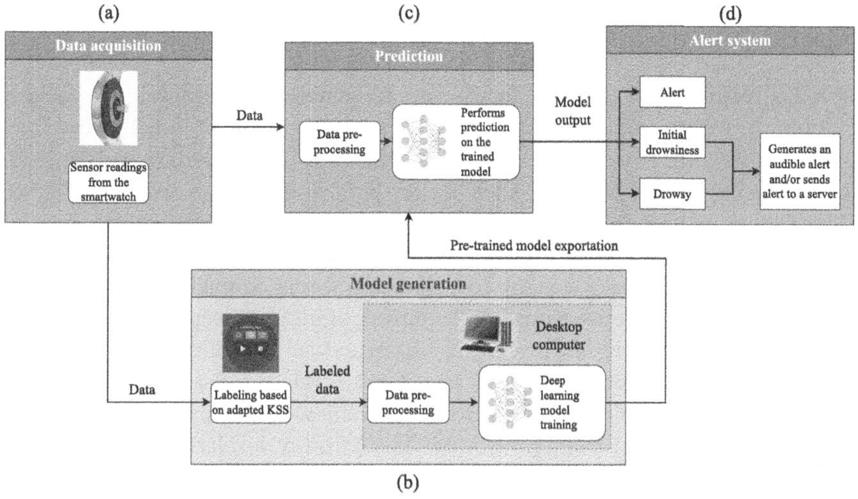

Fig. 1. Diagram of the proposed system.

Samsung Galaxy Watch4 smartwatch was utilized, which stands out for having sensors that are relevant for the development of work, alongside the Android Wear OS operating system, which simplifies app development. This smartwatch offers three variations of the PPG (Photoplethysmography) sensor signal [14], an optical sensor that measures vital signs at the user's wrist. These variations are green, red, and infrared, each generating a different raw analog-to-digital value, with data acquisition rates of 25 Hz for the green PPG and 100 Hz for the red and infrared PPGs. Additionally, the watch includes a heart rate sensor that records data at 1 Hz in beats per minute (BPM).

2.1 Data Acquisition

The Data Acquisition module shown in Fig. 1-(a) consists of reading the data provided by the smartwatch sensors. The sensor reading is conducted continuously for both the heart rate sensor and the three variations of the PPG. This module provides the collected data for the Model Generation (b) and for the Prediction (c) module.

2.2 Deep Learning Model Generation

In the Model Generation module (Fig. 1-(b)), data from the sensors are collected and labeled by the watch user. Subsequently, they are exported to a computer, pre-processed and used to train the deep learning model. This specific module is highlighted in orange because it is used only once, exclusively to generate the predictive model. After this step, the drowsiness detection system relies only on the modules colored in cyan. The following subsections will detail each step of the model generation process.

Data Labeling. The samples for the model generation module were collected on different days and times, including nighttime periods to induce drowsiness, and during the day, covering both alert and drowsy states. All data were collected with the individual at rest.

This work is based on an adaptation of the subjective sleepiness scale KSS [18], aiming to simplify the data labeling process by reducing the number of classes for drowsiness assessment to just three levels equivalent to the original scale: 3 - Alert; 6 - Some signs of sleepiness; and 8 - Sleepy, but some effort to keep awake. The original levels were assigned to the following labels: Alert, Initial Drowsiness, and Drowsy. At the Alert level, the user is fully attentive with no signs of drowsiness. The Initial Drowsiness level applies when the user starts feeling tired and slightly drowsy. The Drowsy level is marked when the user feels very fatigued and close to falling asleep. Drowsiness level labeling must be in sync with the user's perception of their current state, so that upon noticing an increase or decrease, a reassignment to the appropriate level should be done. Consequently, each sensor reading is linked to a label corresponding to the level of drowsiness.

Data Pre-processing. After gathering and labeling the data, it is then transferred from the smartwatch to a computer, where a data processing step is performed (Fig. 1-(b)). Invalid readings, such as zero, null, and negative values, are removed. Besides that, it was necessary to perform a step to remove outliers. In this work, thresholds of -3 and 3 in the statistical measure z-score [3] were used for this removal. Data normalization is also required, scaling the data from 0 to 1. Finally, data was grouped into windows of adjacent readings and associated with a single class. The sliding window approach was adopted for both inputs and outputs (labels). To determine the output of each window, the Mode function was applied to the vector containing the labels.

Deep Learning Model Training. The deep learning model adopted was a one-dimensional convolutional neural network. Its architecture is shown in Fig. 2.

Grid search [5] was used to optimize the hyperparameters of the neural network. In Fig. 2, the parameters marked with a question mark (?) were determined by applying this technique. The parameters are: First convolutional layer filters

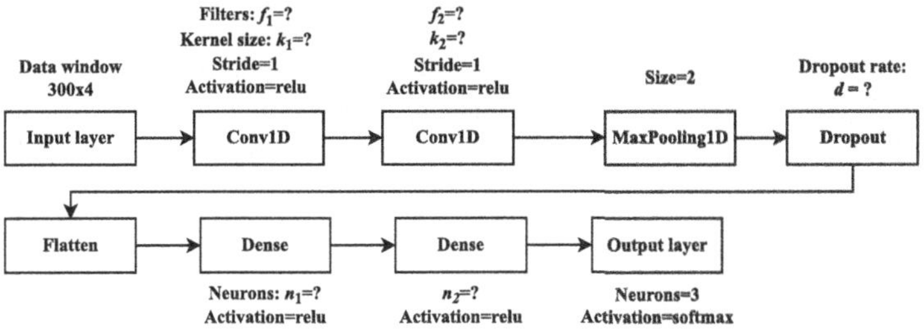

Fig. 2. Proposed 1-D CNN architecture.

(f_1); Second convolutional layer filters (f_2); Size of the kernel of the first convolutional layer (k_1); Size of the kernel of the second convolutional layer (k_2); Number of neurons in the first dense layer (n_1); Number of neurons in the second dense layer (n_2); dropout rate (d); and batch size (b). To select an appropriate number of epochs for training and avoid overfitting [8], the performance of Accuracy and Loss metrics is monitored through graphs, both in the training and validation sets, throughout the training epochs. For the final model training, k-fold cross validation technique was employed [9]. An ensemble [7] was formed from the created models through a voting process, where the final prediction is determined based on the majority of the obtained predictions.

2.3 Prediction and Alert System

The Prediction module (Fig. 1-(c)) uses the models generated in the previous module to make inferences from the smartwatch's sensor readings. Then, the pre-trained neural networks are exported in TensorFlow Lite format and loaded onto the smartwatch. Once the models are implemented on the smartwatch, the sensor data is used for prediction. However, upon receiving the data from the Data Acquisition module, pre-processing is necessary, following the same process used in the training stage (see Sect. 2.2 - data pre-processing). Finally, the data is used in the Ensemble, which through a voting process, infers the final output from the five models. In tie situations, the first vote identified by the algorithm is considered.

The resulting output is then utilized in the Alert System (Fig. 1-(d)), which is also integrated into the smartwatch and assumes the responsibility of verifying if the model output indicates any level of drowsiness. If the model classifies the data as "Initial Drowsiness" or "Drowsy", a notification is sent to inform the user that he is in a drowsy state.

3 Results and Discussion

3.1 Dataset

A total of 19 data collection sessions were conducted at different times and on different days. The data was collected at a sampling rate of 20 Hz, with an average duration of 4 h per session. Subsequently, these sessions were concatenated into a single dataset[1] and pre-processed, resulting in 4,890,260 sensor readings. As shown in Table 1, each reading includes a heart rate value along with values for each PPG variation. Additionally, each reading has a corresponding label (column *drowsiness*).

Table 1. Sample of acquired and labeled data.

Index	heartRate	ppgGreen	ppgRed	ppgIR	drowsiness
0	54.0	1584091.0	5970731.0	6388383.0	0.0
1	54.0	1584091.0	5971202.0	6392174.0	0.0
2	54.0	1581111.0	5971295.0	6391469.0	0.0
3	54.0	1579343.0	5972599.0	6396137.0	0.0
4	54.0	1579321.0	5971906.0	6392898.0	0.0
...	...	...	...	...	...
524998	85.0	1962633.0	4921448.0	5109467.0	1.0

The data were then normalized and structured into data windows, resulting in a total of 16,300 input windows. Each one has a format of 300 rows (data points) by 4 columns (features) and represents 15 s of sensor reading time, as the data is collected at a frequency of 20 Hz. The choice of a 300-data-point window size was based on empirical reasoning. Additionally, 16,300 outputs were obtained, with one output corresponding to each data window. Among the 16,300 samples, 7,019 belong to the Alert class, 4,557 to the Initial Drowsiness class, and 4,724 to the Drowsy class. Figure 3 graphically presents examples of data windows from the 3 classes. These samples have their normalized value on the y-axis, window indices on the x-axis, and the plotted curves represent the features.

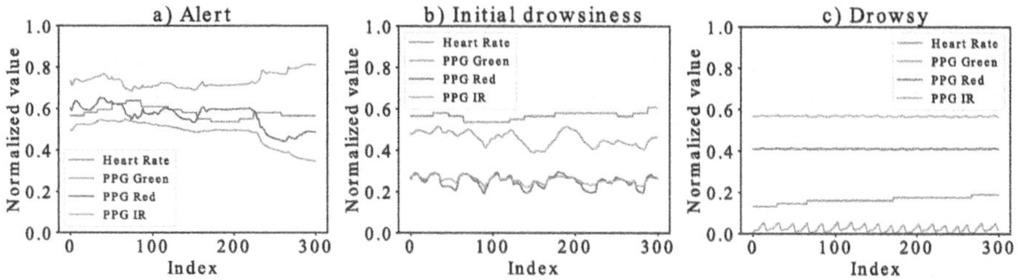

Fig. 3. Data windows for the Alert (a), Initial drowsiness (b) and Drowsy (c) classes.

An analysis of the pre-processed data grouped by class was conducted. As shown in Table 2, it is observed that as the drowsiness level increases, the average heart rate (HR) decreases. The same pattern is observed in the standard

[1] The pre-processed dataset, without the normalization and window structuring steps, is available at https://www.kaggle.com/datasets/vitoraugustx/drowsiness-dataset.

deviation of HR, which decreases as drowsiness levels increase. This indicates that heart rate variability is lower at higher levels of drowsiness.

Both *ppgRed* and *ppgIR* columns show a similar pattern to HR, with their averages decreasing as drowsiness increases. However, no specific trend was observed regarding the standard deviation of the PPG columns.

Table 2. Mean and standard deviation of the columns for samples of each class.

a) Class Alert		
Column	Mean	Std
heartRate	86.12	11.02
ppgGreen	1998717.39	397460.64
ppgRed	5799788.22	358897.25
ppgIR	5886285.55	296717.13

b) Class Initial drowsiness		
Column	Mean	Std
heartRate	79.82	8.78
ppgGreen	2160249.95	507568.43
ppgRed	5618291.80	412224.83
ppgIR	5775197.09	455042.23

c) Class Drowsy		
Column	Mean	Std
heartRate	64.68	6.79
ppgGreen	2101197.91	417292.01
ppgRed	5436226.71	303707.27
ppgIR	5448053.13	439913.47

3.2 Neural Network Hyperparameter Optimization

The variables applied in the grid search algorithm (see Sect. 2.2), and their possible hyperparameter combinations, were: f_1: [32, 64, 128]; f_2: [32, 64, 128]; k_1: [3, 5, 9]; k_2: [3, 5, 9]; n_1: [32, 64, 128]; n_2: [16, 32, 64]; d: [0.25, 0.4, 0.5]; b: [32, 64, 128]. The training was limited to 10 epochs to reduce training time and was conducted using the complete dataset with the 2-fold cross-validation technique to evaluate performance. At the end of the grid search, the best configuration for the convolutional neural network was selected, achieving a validation accuracy of 86.17%. The network architecture can be seen in Fig. 2. The best hyperparameters were: f_1: 32; f_2: 128; k_1: 3; k_2: 3; n_1: 128; n_2: 64; d: 0.25; and b: 128.

3.3 Choosing the Ideal Number of Epochs for Training

Figure 4 shows the Loss and Accuracy plots of the model trained for 400 epochs using the best parameters obtained (see Sect. 3.2). For this, the complete dataset was utilized, divided into a ratio of 80% for training and 20% for validation. In the left plot, it can be observed that the validation loss decreases until a certain number of epochs and then starts to increase, while the training loss continues to decrease, indicating the point of overfitting. In the right plot, it is noted that the accuracy curve on the validation set stabilizes and no longer increases, while the training accuracy continues to increase and tends to 100%, also indicating overfitting. It is concluded that the stopping point for the final model training is reached at 50 epochs, as observed in the highlighted regions in red in Fig. 4, where the lowest validation loss and the stabilization point of validation accuracy are found.

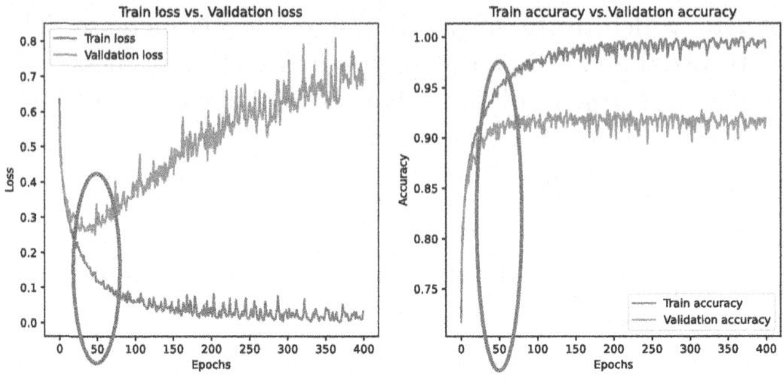

Fig. 4. Left: region of lowest validation loss; Right: region of validation accuracy stabilization.

3.4 Cross-Validation

A 5-fold cross-validation approach was adopted, using 80% of the dataset for cross-validation and the remaining 20% for testing. The cross-validation set included 13,040 samples, divided into 5,634 for the Alert class, 3,657 for Initial Drowsiness, and 3,749 for Drowsy. The test set comprised 3,260 samples, distributed as 1,385 for Alert, 900 for Initial Drowsiness, and 975 for Drowsy. The folds were divided so that the labels were balanced between them, with an average of 1,228.8, 731.4, and 749.8 samples per fold for Alert, Initial Drowsiness, and Drowsy classes, respectively.

The architecture used in each of the five models is defined in Sect. 2.2, using the best hyperparameters obtained from the experiments in Sect. 3.2. Training was conducted for 50 epochs, as defined in the experiments of Sect. 3.3. To evaluate the training of the 5 models, predictions were made on the validation sets and the test set data. The average accuracy of the 5 models on the training folds was 95.22%, with a standard deviation of 0.37%. The average accuracy on the validation folds was 89.58%, with a standard deviation of 1%. In the test set, the accuracy remained at 89.73% with a standard deviation of 0.49%. These results are shown in Fig. 5-(a). The accuracy of each model, along with the mean and standard deviation, are presented in Table 3. Both the validation and test sets showed that the models had similar accuracy levels, resulting in a low standard deviation.

Table 3. Models accuracy table.

Dataset	Models					Mean	Std
	1	2	3	4	5		
Train	95.19	95.91	94.78	95.07	95.16	95.22	0.37
Validation	89.46	90.30	90.99	89.61	91.22	90.31	0.71
Test	90.77	91.01	90.83	90.31	90.00	90.58	0.37

With the ensemble creation, accuracy of the combined models was evaluated on the test data. The experiments resulted in an accuracy of 91.41%. As observed in Fig. 5-(b), there was a slight improvement (¡1%) in generalization when using the ensemble technique, as its accuracy was better than the average accuracy of the 5 models on the test data (90.58%).

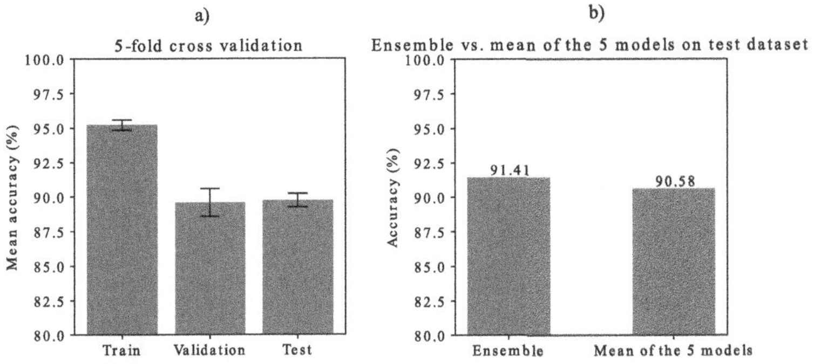

Fig. 5. Average accuracy and standard deviation of the 5 models in cross-validation (training and validation) and in the test set (a). Prediction accuracy of the ensemble vs. mean of the 5 models on the test data (b).

Figure 6 displays the confusion matrix of the ensemble on test data. Both in the individual models and in the ensemble, the lowest accuracy rate was observed in the Initial Drowsiness class. A significant rate of false positives and false negatives was observed between the Alert and Initial Drowsiness classes, indicating that the data classified as Alert or Initial Drowsiness share similar characteristics and have patterns that are not easily distinguishable by the neural network. These results may also reflect the difficulty users face in collecting and labeling data, specifically in discerning the physiological difference between being alert and being initially drowsy.

3.5 Drowsiness Levels Real-Time Prediction

After the final training of the deep learning models on a computer (see Fig. 1), the models and their weights were exported. These were then integrated into the smartwatch, enabling real-time inferences.

Figure 7 presents a schematic illustrating the operation of the developed drowsiness detection system. A drowsy user wearing the smartwatch has their vital signs collected and processed by the artificial neural network, which returns an output of the "Drowsy" class, for example. Once drowsiness is detected, the smartwatch sends a notification with sound, vibration, and a visual alert.

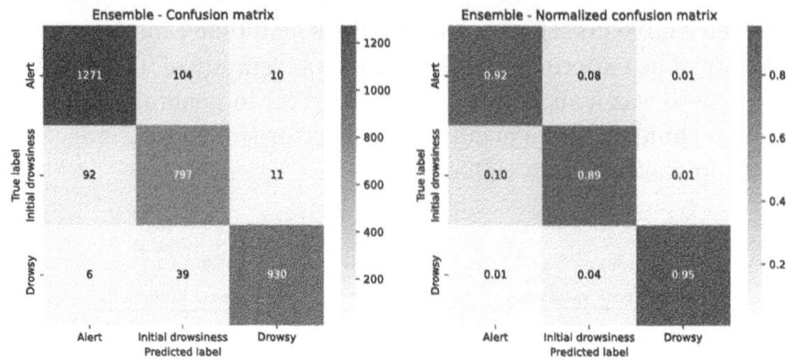

Fig. 6. Left: confusion matrix; Right: normalized confusion matrix. Both representing the ensemble on the test data.

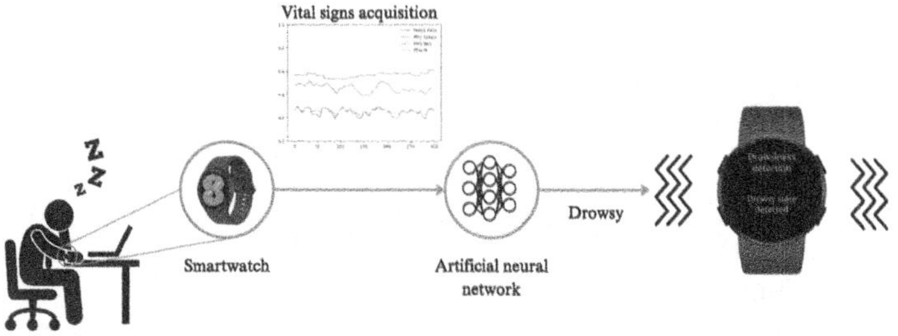

Fig. 7. Diagram illustrating the operation of the drowsiness detection system.

4 Conclusions and Future Work

In this study, a drowsiness detection system was developed using vital sign sensors in the Samsung Galaxy Watch4 smartwatch. Heart rate and photoplethysmography data were used, labeled based on the adapted KSS, processed, and utilized to train a one-dimensional Convolutional Neural Network, which underwent an optimization phase for the selection of the best hyperparameters.

The 5-fold cross-validation technique was applied to evaluate the system's performance and create an ensemble model. The results indicate notable prediction accuracy values in the test data using the ensemble (91.41%). The average accuracy of the five models resulted in 90.58% in the test data and 90.31% in the validation folds, indicating an improvement in prediction with the ensemble. The final model's performance was consistent across all classes, with Alert and Drowsy achieving the highest accuracies (92% and 95%, respectively), while Initial Drowsiness presented the lowest accuracy (89%).

The real-time drowsiness detection system was implemented on the watch, using data collection and processing algorithms. It monitors vital signs, detects drowsiness, and alerts the user with sound and vibration, reducing accident risk. Unlike traditional Advanced Driver-Assistance Systems (ADAS), the proposed approach offers a significant advantage as a wearable device, allowing continuous use with the person rather than being restricted to vehicles. Compared to existing literature, the solution offers advantages like real-time functionality and statistical robustness, achieved through cross-validation, ensemble methods and Deep Learning techniques.

In future work, it is considered to improve data collection and labeling. One option is to use only two classes: "not drowsy" and "drowsy." Experiments could label data using all KSS scale levels, grouping them into classes [10]. Another approach is to adopt non-subjective labeling methods, such as video-based labeling (e.g., eye closure, yawning) [11,12], or EEG sensors to monitor brain activity [16], aiming to eliminate the dependence on the subjective state of drowsiness experienced by the user. Future work will also evaluate the model's performance using intra-personal (same person) vs. inter-personal (different people) data, considering physiological variations like cardiac activity and drowsiness symptoms. Customizing the model through an initial adjustment phase and personalized training could better adapt it to each individual's characteristics. Finally it is intended to study how input window size impacts inference time and model accuracy, aiming to minimize window size without compromising accuracy, thereby enabling faster inference and better real-time performance.

Acknowledgments. We thank PPG-CCMC (PPG-CCMC is the acronym for *Programa de Pós-Graduação em Ciências de Computação e Matemática Computacional* in Portuguese.), University of São Paulo and Federal University of Rio Grande, which made this work possible. This study was financed in part by the *Coordenação de Aperfeiçoamento de Pessoal de Nível Superior - Brasil* (CAPES) - Finance Code 001 and iTEC/FURG.

References

1. ABRAMET: Problemas na saúde do motorista deixaram mais de 200 mil feridos e 12,4 mil mortos, entre 2014 e 2019, no brasil (Outubro 2019). https://abramet.com.br/noticias/saude-no-transito-abramet/
2. ABRAMET: Sono e direção não combinam. dormir ao volante está entre as principais causas de mortes no trânsito. Jornal Medicina de Tráfego **2**, 1–8 (2019)
3. Aggarwal, V., Gupta, V., Singh, P., Sharma, K., Sharma, N.: Detection of spatial outlier by using improved z-score test. In: 2019 3rd International Conference on Trends in Electronics and Informatics (ICOEI), pp. 788–790. IEEE (2019). https://doi.org/10.1109/ICOEI.2019.8862582
4. Barnes, C.M., Wagner, D.T.: Changing to daylight saving time cuts into sleep and increases workplace injuries. J. Appl. Psychol. **94**(5), 1305–1317 (2009). https://doi.org/10.1037/a0015320
5. Bergstra, J., Bengio, Y.: Random search for hyper-parameter optimization. J. Mach. Learn. Res. **13**(null), 281–305 (2012). https://www.jmlr.org/papers/volume13/bergstra12a/bergstra12a.pdf
6. Chandel, R.S., Sharma, S., Kaur, S., Singh, S., Kumar, R.: Smart watches: a review of evolution in bio-medical sector. Mater. Today Proc. **50**, 1053–1066 (2022). https://doi.org/10.1016/j.matpr.2021.07.460, 2nd International Conference on Functional Material, Manufacturing and Performances (ICFMMP-2021)

7. Dietterich, T.G.: Ensemble methods in machine learning. In: Multiple Classifier Systems, pp. 1–15. Springer Berlin Heidelberg, Berlin, Heidelberg (2000). https://doi.org/10.1007/3-540-45014-9_1
8. Hawkins, D.M.: The problem of overfitting. J. Chem. Inf. Comput. Sci. **44**(1), 1–12 (2004). https://doi.org/10.1021/ci0342472
9. Kohavi, R.: A study of cross-validation and bootstrap for accuracy estimation and model selection. In: Proceedings of the 14th International Joint Conference on Artificial Intelligence - Volume 2, pp. 1137–1143. IJCAI'95, Morgan Kaufmann Publishers Inc., San Francisco, CA, USA (1995). https://dl.acm.org/doi/10.5555/1643031.1643047
10. Kundinger, T., Riener, A.: The potential of wrist-worn wearables for driver drowsiness detection: a feasibility analysis. In: Proceedings of the 28th ACM Conference on User Modeling, Adaptation and Personalization, pp. 117–125. UMAP '20, Association for Computing Machinery, New York, NY, USA (2020). https://doi.org/10.1145/3340631.3394852
11. Kundinger, T., Sofra, N., Riener, A.: Assessment of the potential of wrist-worn wearable sensors for driver drowsiness detection. Sensors **20**(4) (2020). https://doi.org/10.3390/s20041029
12. Lee, H., Lee, J., Shin, M.: Using wearable ECG/PPG sensors for driver drowsiness detection based on distinguishable pattern of recurrence plots. Electronics **8**(2) (2019). https://doi.org/10.3390/electronics8020192
13. Nasir, S., Yurder, Y.: Consumers' and physicians' perceptions about high tech wearable health products. Procedia Soc. Behav. Sci. **195**, 1261–1267 (2015). https://doi.org/10.1016/j.sbspro.2015.06.279, world Conference on Technology, Innovation and Entrepreneurship
14. Reisner, A., Shaltis, P.A., McCombie, D., Asada, H.H., Warner, D.S., Warner, M.A.: Utility of the photoplethysmogram in circulatory monitoring. Anesthesiology **108**(5), 950–958 (2008). https://doi.org/10.1097/ALN.0b013e31816c89e1
15. Rundo, F., Spampinato, C., Battiato, S., Trenta, F., Conoci, S.: Advanced 1d temporal deep dilated convolutional embedded perceptual system for fast car-driver drowsiness monitoring. In: 2020 AEIT International Conference of Electrical and Electronic Technologies for Automotive (AEIT AUTOMOTIVE), pp. 1–6 (2020). https://doi.org/10.23919/AEITAUTOMOTIVE50086.2020.9307400
16. Rundo, F., Spampinato, C., Conoci, S.: Ad-hoc shallow neural network to learn hyper filtered photoplethysmographic (ppg) signal for efficient car-driver drowsiness monitoring. Electronics **8**(8) (2019). https://doi.org/10.3390/electronics8080890
17. Zhang, Y., Tsujikawa, M., Onishi, Y.: Sleep/wake classification via remote PPG signals. In: 2019 41st Annual International Conference of the IEEE Engineering in Medicine and Biology Society (EMBC), pp. 3226–3230 (2019). https://doi.org/10.1109/EMBC.2019.8857097
18. Åkerstedt, T., Gillberg, M.: Subjective and objective sleepiness in the active individual. Int. J. Neurosci. **52**(1–2), 29–37 (1990). https://doi.org/10.3109/00207459008994241, pMID: 2265922

Benchmarking Out of the Box Open-Source LLMs for Malware Detection Based on API Calls Sequences

Ciprian-Alin Simion[1,2], Gheorghe Balan[1,2(✉)], and Dragoş Teodor Gavriluţ[1,2]

[1] Faculty of Computer Science, "Al.I. Cuza" University of Iasi, Iasi, Romania
gbalan@bitdefender.com
[2] Bitdefender Laboratory, Iasi, Romania

Abstract. The rise of generative AI has created several scenarios where older technologies (or human intervention) can be replaced by an agent that relies on LLMs. In this paper, we evaluate if an LLM is suited for malware detection and in what scenarios. For this task, we compare 4 open source models [LLama2-13B, Mistral, Mixtral and Mixtral-FP16] using a set of 20000 malware and 20000 benign files for which we provide behavioral information as a list of API calls sequences. The goal is to identify scenarios where these types of models can be successfully used for malware detection.

1 Introduction

LLMs[1] and generative AI are being utilized across multiple industries, including cyber-security. And while the immediate usage (in terms of information visibility, better support, recommendation or assistance for Security Operation Centers) is clear, it is still debatable if such models can be successfully used to replace the current malware detection technologies that every security product has.

From this point of view, there are several aspects that need to be evaluated: **reliability** (a detection test executed in identical conditions should yield the same results), **accuracy** and **false positive rate**. These parameters are required in particular for scenarios where detection is performed locally (on the client endpoint) and real-time protection (blocking access to a resource while it is being scanned) is needed. Some scenarios (such as real-time protection) will also require a reduced **inference time**. As such, we will evaluate this metric as well to see how well a model can be used in these cases.

This paper focuses on evaluating if LLM models (as they are today), can be used to replace malware detection and for what scenarios they are best suited for.

This study looks into the out-of-box capabilities of various open source models (without any fine tuning) and attempts to evaluate their ability to identify malicious content. It is also important to highlight that LLMs (while generic)

[1] Large Language Models.

V. Julian et al. (Eds.): IDEAL 2024, LNCS 15346, pp. 133–142, 2025.
https://doi.org/10.1007/978-3-031-77731-8_13

can not be used for all problems and in some scenarios a non-generative AI method combined with feature engineering might be more practical.

In terms of malware detection, most security products use some sort of behavior analysis (either for malicious pattern identification) or for feature engineering. This concept relies on sandboxes, emulators or various types of API hooking mechanisms to extract API calls and/or system events and infer a behavior that will further be used to detect a threat. While in some cases, various behavioral patterns (such as encrypting multiple files) are good enough to provide a reliable detection, in others, a more complex form of correlation and understanding is required. This technique is generally sound as a malware has to do something malicious (or to exhibit some malicious behavior) in order to achieve its purpose. With this is mind, we are wondering if having access to the raw data (in our case the API calls sequence) provides sufficient information for an LLM to infer the maliciousness of a file.

The rest of the paper is organized as follows: Sect. 2 presents similar research around this topic, Sect. 3 describes several scenarios where the use of LLMs will be evaluated, Sect. 4 shows how we organized our evaluation experiment, Sect. 5 details our results and finally, Sect. 6 draws some conclusions based on our experiments.

2 Related Work

As shown in [1], LLMs are used for both defensive and offensive functions. In this paper, the authors have analyzed over 30 papers for defensive functions and over 30 for offensive practices. The ones focused on offensive applications, target actions like Reconnaissance, Initial Access, Execution, Defense Evasion or Credential Access [2]. On the other hand, the defensive applications include Identification, Protection, Detection, Response and Recovery.

One very interesting research focused on detecting Android APK[2] files and proposed a malware detecting framework named MalBERT [3]. The dataset used for their experiment was comprised of 10K malign files and 12K benign files. One task was to classify the files into malware or clean and the second task they pursued was to classify each file into groups like spyware, dropper, clicker, etc. based on their behaviour. The authors compared their proposed framework with other models like LSTM[3], XLNet [4], RoBERTa [5] and DistilBERT [6]. After running all tests, the best performing model on both tasks was BERT [7], with 97% accuracy on the binary classification problem and 91% on the multi classification task.

Another interesting paper is [8] where J. Chen et al. proposed a framework (SLAM - Sliding Local Attention Mechanism) that used Windows API[4] sequences as input. Besides the raw API sequences, they also embedded the category of each API in a 2 dimensional input vector. The dataset they used

[2] Android Package Kit.
[3] Long Short Term Memory.
[4] Application Programming Interface.

was comprised of 110K benign samples and 27K malign samples. In SLAM, the processing of samples goes as follows: splitting the input vector, initialising the local attention window from the split, training the CNN[5], aggregating the results and lastly, applying softmax on the aggregated results. When compared with Random Forest, Attention CNN LTSM (ACLM) [9] and two-stream CNN Attention Model (TCAM) [4] on a 10-fold cross validation, the authors showed that the SLAM framework performed best with 97% average accuracy, followed by TCAM with 92%.

In this research paper [10], the authors propose a teacher-student system to obtain smaller models that still perform at the same level as their bigger counterpart. The problem they are tackling is that of web content filtering. Their research shows how one could distill the knowledge of a model with 770 million parameters into one 175 times smaller and still maintain the original accuracy. The models compared were BERT [7], eXpose [11] and GPT-3 Babbage.

With multiple LLMs arising that focus on generating code, some focus on analysing code as well. This paper [12], proposes a new framework that uses GPT-2, BERT and LSTM to detect vulnerabilities in C, C++ and Java code. As in the previous paper mentioned, this framework also employs the use of Knowledge Distillation in a teacher-student setup to keep their models small. The results of their framework were compared to VulDeBERT [13] and LSTM. The best results were obtained when using ChatGPT-2 on the SARD [14] dataset: 93.59%.

Rando et al. propose a framework called PassGPT in [15]. Their goal in this research paper is to show how PassGPT outperforms current password generation methods based on Generative Adversarial Networks (GANs). Besides this, they also introduce a new concept named *guided password generation* which implies generating passwords w.r.t. arbitrary constraints. For their experiment, they used the datasets of leaked passwords from LinkedIn and RockYou which are the same used in their previous research (PassGAN [16]). Their initial proposed model, PassGPT, is based on the GPT-2 architecture. Secondly, PassVQT, adds the vector-quantization of the latent space on top of this architecture. After fine tuning these models on the aforementioned datasets, they evaluated the models against PassGAN and PassGAN+. Their results show that PassGPT(41.86%) surpasses both GAN based methods (15.09%, 23.33%) when tasked to generate password from the RockYou test set with 10^9 guesses, while the PassVQT (21.46%) only outperformed the vanilla PassGAN.

3 Problem

In this section we design an experiment that tries to evaluate how well suited a LLM model is for malware detection, based on several parameters:

[5] Convolutional Neural Network.

- **Detection Rate**: (most of the current security solution have more then 98% detection rate[6]. Thus, it would be expected for such a model to have similar (if not better) results.
- **False Positive Rate**: having too many false positives can lead to either an alert fatigue scenarios (if we are dealing with solutions that rely on security analytics such as EDR[7] or XDR[8]) or to a severe usage impediments (in case of protection scenarios where access to a malicious file is blocked)
- **Inference time**: this is relevant in particular for scenarios where the scanning is performed locally (especially if associated with a protection technology - where access to a file will be blocked until that file is scanned). If the time needed to scan a file is too long, the entire OS will be severely slow down to a point where it will freeze.

It should also be mentioned that LLMs are in particular good with texts (or anything that can be translated into a human readable form). As such, the first experiments tried were based on script malware [17] rather than on binary files. However, for this experiment we will try to address the detection capability of an LLM based on API calls (that reflect the behavior and can be expressed in a more readable language) extracted from binary files.

Since one of our scenarios relies on running the models locally, we decided to use only LLMs that could be run locally (and NOT a service like ChatGPT that implies a cloud connectivity and potential privacy issues). It is also important to notice that there are certain scenarios where cloud connectivity will not be possible (e.g., when protecting private data - such as classified documents, medical databases, etc.).

3.1 Emulators

A static emulator is used to obtain the list of API Calls. This is done by simulating the functionality of an Operating System but without affecting the real host OS. Hence, an emulator usually implements different OS API Calls and a Virtual File System in order to keep track of file changes performed by a sample. The strength of the emulator system comes from its low resource usage compared to a virtualization system. Moreover, the changes performed in the emulated Virtual File System (VFS) can be easily analyzed or hooked in order to trick the sample under scrutiny to perform all its execution paths (e.g.: being in an emulator, we can simulate different errors, such as file creation error in order to analyze what happens when a specific file is not present on the operating system - this might be useful in find a kill switch[9]). This can be implemented by using different configurations. In a default configuration, an implemented API just returns a dummy response and its call is saved in the file state.

[6] https://www.av-comparatives.org/latest-tests/.
[7] Endpoint Detection and Response.
[8] Extended detection and response.
[9] https://www.theguardian.com/technology/2017/may/13/accidental-hero-finds-kill-switch-to-stop-spread-of-ransomware-cyber-attack.

Even if it has the previously mentioned advantages, an emulator has certain limitations:

- each emulator should implement a maximum number of emulated instructions; this is needed for both real-time scanning (where the analysis time is limited because the sample execution is halted) and emulation efficiency; however, some optimisations must be implemented, as an attacker might use large loops with different API Calls in order reach the maximum number of emulated instructions.
- there are specific APIs or instructions which can not be implemented (due to their architecture dependency and complexity); hence, using such an instruction, the process is stopped; of course those APIs/instructions can be skipped but the obtained results may be inconsistent.
- maybe the most difficult step, is to implement the usage of third-party libraries; to do so, those libraries should exists a priori in VFS; if such a library is not found, the emulation is stopped.

Getting past those limitations by implementing different optimization techniques is a measure of emulation quality that differentiate emulators from one another [18,19]. For this paper, we have used an custom designed emulator provided by a security company that implements checks for several anti-emulation techniques used by malware.

4 Experiment

4.1 Databases

Our database consists of 20000 malicious files and 20000 benign ones collected in the last 6 months. These samples were emulated and the API call sequence for each one of them was collected. The resulted sequences (after additional preprocessing) were further submitted to different LLM models.

The total number of uniquely identified APIs (from both class samples) is 4231. When analysing the samples, we found that the maximum length of the extracted API sequences is 6999 (as restricted by emulator implementation). The median sequence length for benign class is 156, while for malicious files the emulator managed to extract a median of 507.

After processing a sample, the emulator would yield a sequence of IDs corresponding to a list of API calls. Here is an example of such a sequence for a sample with hash *917dcae22d56d22cf418aa3b5fba2ff4*: *1225, 1313, 1329, 1182, 1329, 1182, 1020, 1182, 1023, 1329, 1182, 1329, ...*.

Because the LLMs are known to work better with natural language, or something closer to that, we translated the sequence of IDs above into one of API names using a mapping table similar to this (Table 1):

Table 1. Index to Function Name mapping

Index	Library	Function name
1020	kernel32	FlsAlloc
1023	kernel32	FlsSetValue
1182	kernel32	GetProcAddress
1225	kernel32	GetSystemTimeAsFileTime
1313	kernel32	IsProcessorFeaturePresent
1329	kernel32	LoadLibraryExW

4.2 Models

As stated before, the models used to conduct our experiment are: LLama-13b [20], Mistral [21], Mixtral and Mixtral-FP16 [22]. Each of these models were hosted in a cloud environment that exposed an API. The models were pulled from the HuggingFace[10] repository and are pre trained for general purposes and not fine tuned for the task of malware detection.

4.3 Setup

For this experiment we started by crafting the prompts for the models. Each prompt was built out of three parts: prefix, api_seq and suffix. After trying out some variants where we asked it to respond with percentages, categories or qualifiers, we observed best results by asking it to yield a digit on an 1 to 10 scale. Even so, we still had to rerun prompts for some samples where the model could not respond (time out, error) or was delusional (it's response was unusable). When it came to the api_seq, we stripped each API call from the sequence of its prefixing library name and only left the function name. For example, if we had a sequence like the one presented in Subsect. 4.1, after removing the library names, we would end up with a sequence like this:

GetSystemTimeAsFileTime, IsProcessorFeaturePresent, LoadLibraryExW, GetProcAddress, LoadLibraryExW, GetProcAddress, FlsAlloc, GetProcAddress, FlsSetValue, LoadLibraryExW, GetProcAddress, LoadLibraryExW, GetProcAddress, FlsAlloc, GetProcAddress, FlsGetValue, GetProcAddress, FlsSetValue, ...

We then regrouped them in a sequence separated by commas. One common limitation when using LLM is the length of the prompt (the model's capacity). Because we only used one shot prompts (without using the context from a previous prompt) and the maximum length of a prompt was 30 (prefix), 70000 (api_seq) and 300 (suffix), we did not reach the prompt maximum limit of any of the models.

For each call to the models, we gather the textual response and the duration of the call. After we obtained an answer for each sample, we parsed the results

[10] https://huggingface.co/.

using regular expressions in two steps. The first step implied trying to match one of the following expressions to the text response: "(\d{1}) out of 10", "it a (\d{1})", "it an (\d{1})", "as (\d{1})", "as a (\d{1})", "as an (\d{1})", "at a (\d{1})", "at an (\d{1})", "with a (\d{1})", "with an (\d{1})", "choose (\d{1})", "Answer (\d{1})", "associated with malware ((\d{1}))". If none of these matched the response, we tried matching more generic regular expressions with a broader coverage like "(\d{1})", "([1|2|3|4|5|6|7|8|9|10]{1})". If we still did not get a match, we considered it an error and skipped the sample from the results.

For each model, we calculated average response time, accuracy, detection rate and false positive rate.

5 Results

After all processing we have obtained usable results for each model as follows:

- llama2-13b - 19958 malign, 19996 benign;
- mistral - 17380 malign, 17362 benign;
- mixtral - 19281 malign, 19303 benign
- mixtral-fp16 - 8258 malign, 13935 benign

In order to compute detection rate and false positive rate, we needed a threshold value that would separate the two classes (benign, malign) on the 1 to 10 scale. We chose values 5,6,7,8,9 to act as class borders, considering everything above the value or equal to it, as being malign and everything below the value as being benign.

Table 2 depicts the accuracy of the models while Table 3 shows detection rate and false positive rate for each model across all threshold values. All hashes and results can be found on a public repository[11].

Table 2. LLM Accuracy computed for different threshold values (T)

T	LLama2-13B	Mistral	Mixtral	Mixtral-fp16	Average
5	49%	47%	50%	48%	50%
6	49%	47%	51%	49%	50%
7	50%	48%	51%	52%	53%
8	50%	50%	58%	66%	56%
9	50%	51%	67%	72%	50%

Both Tables 2 and 3 also show the average column. This was computed by adding the digits obtained from the models' responses and dividing them by their number. It is important to highlight that in some cases not all models yielded

[11] https://github.com/alinsimion/llm_api_seq_det.

Table 3. LLM Detection Rate and False Positive Rate computed for different threshold values (T)

	LLama2-13B		Mistral		Mixtral		Mixtral-fp16		Average	
T	Det	FPR	Det	FPR	Det	FPR	Det	FPR	Det	FPR
5	99%	99%	82%	88%	97%	96%	97%	94%	97%	98%
6	99%	99%	73%	79%	96%	94%	96%	90%	92%	90%
7	49%	48%	59%	63%	95%	92%	94%	84%	65%	64%
8	0%	0%	33%	33%	84%	67%	82%	46%	17%	0%
9	0%	0%	18%	15%	55%	21%	55%	11%	0%	0%

a usable result, resulting in only one, two or three results. This being said, the results obtained for the average column may not reflect the exact average of the other 4 columns.

As for response times, the total amounts of time needed for inference for each model, as well as the average amount for a single response for each model, are presented in Table 4.

Table 4. LLM one Request Average duration and Total Duration for each Model

	Malicious		Benign	
Model	Req Avg.	Total	Req Avg.	Total
Llama2-13b	14 s	81 h 04 m	13 s	70 h 46 m
Mistral	9 s	49 h 54 m	12 s	68 h 25 m
Mixtral	11 s	39 h 12 m	10 s	37 h 45 m
Mixtral-FP16	33 s	54 h 26 m	35 s	63 h 39 m

6 Conclusions

The results show that the tested models are not well suited for malware detection (based on the sequences of API calls). This is likely the result of the contents of the training set (used to train the four LLMs that we have evaluated). Current open source malware databases contain signatures and various type of hashes so it is unlikely that a model will have lots of data to infer from (in our case with sequences of API calls).

Another conclusion is that these models are not suited for real time detection as they require several seconds to provide an answer, while current security products do so in milliseconds. We should also point out that a security product architecture might rely on a cloud service and a filter that extracts metadata locally and sends them to the cloud service only in some scenarios (e.g. an unknown process is being executed). This technique, combined with the fact the a cloud service might rely on dedicated hardware could be practical enough to use LLMs for some scenarios. There is however a matter of cost and bandwidth that still needs to be taken under consideration.

Furthermore, we can observe that the accuracy is close to 50% in many cases and this makes the model unreliable (even as a second opinion).

Additional tests are still needed that involve fine-tuning a model (for this specific task) and also a comparison with regular (non-generative) machine learning models that are now used by many security products. It's also important to highlight that a fine-tuning task will be time demanding as it will first require the extraction of all API calls from multiple malware and benign files.

References

1. Motlagh, F.N., Hajizadeh, M., Majd, M., Najafi, P., Cheng, F., Meinel, C.: Large language models in cybersecurity: state-of-the-art, 2024
2. Mitre corporation 2024. mitre attack. https://attack.mitre.org/matrices/enterprise/. Accessed 15 Apr 2024
3. Rahali, A., Akhloufi, M.A.: Malbert: using transformers for cybersecurity and malicious software detection, 2021
4. Yang, Z.: XLNet: generalized autoregressive pretraining for language understanding, 2020. all
5. Liu, Y., et al.: A robustly optimized bert pretraining approach, Roberta (2019)
6. Sanh, V., Debut, L., Chaumond, J., Wolf, T.: Distilbert, a distilled version of bert: smaller, faster, cheaper and lighter, 2020
7. Devlin, J., Chang, M.-W., Lee, K., Toutanova, K.: Bert: pre-training of deep bidirectional transformers for language understanding, 2019
8. Chen, J., et al.: Slam: a malware detection method based on sliding local attention mechanism. Secur. Commun. Netw. **1–11**, 2020 (2020)
9. Shiqi, L.: Android malware analysis and detection based on attention-CNN-LSTM. J. Comput. 31–43 (2019)
10. Voros, T., Bergeron, S.P., Berlin, K.: Web content filtering through knowledge distillation of large language models, 2023
11. Saxe, J., Berlin, K.: Expose: a character-level convolutional neural network with embeddings for detecting malicious URLs, file paths and registry keys, 2017
12. Omar, M.: Detecting software vulnerabilities using language models, 2023
13. Kim, S., Choi, J., Ahmed, M.E., Nepal, S., Kim, H.: Vuldebert: a vulnerability detection system using bert. In: 2022 IEEE International Symposium on Software Reliability Engineering Workshops (ISSREW), pp. 69–74 (2022)
14. Zhou, X., Verma, R.M.: Vulnerability detection via multimodal learning: datasets and analysis. In: Proceedings of the 2022 ACM on Asia Conference on Computer and Communications Security, ASIA CCS '22, pp. 1225–1227, New York, NY, USA, 2022. Association for Computing Machinery (2022)
15. Rando, J., Perez-Cruz, F., Hitaj, B.: PassGPT: Password modeling and (guided) generation with large language models. In: Tsudik, G., Conti, M., Liang, K., Smaragdakis, G. (eds.) Computer Security – ESORICS 2023. ESORICS 2023. LNCS, vol. 14347, pp. 164–183. Springer, Cham (2024). https://doi.org/10.1007/978-3-031-51482-1_9

16. Hitaj, B., Gasti, P., Ateniese, G., Perez-Cruz, F.: PassGAN: a deep learning approach for password guessing. In: Deng, R.H., Gauthier-Umaña, V., Ochoa, M., Yung, M. (eds.) ACNS 2019. LNCS, vol. 11464, pp. 217–237. Springer, Cham (2019). https://doi.org/10.1007/978-3-030-21568-2_11
17. Introducing virustotal code insight: empowering threat analysis with generative AI
18. Choi, S., Chang, T., Yoon, S.W., Park, Y.: Hybrid emulation for bypassing anti-reversing techniques and analyzing malware. J. Supercomput. **77**, 471–497 (2020)
19. Martignoni, L., Paleari, R., Roglia, G.F., Bruschi, D.: Testing CPU emulators. In: International Symposium on Software Testing and Analysis (2009)
20. Rozière, B., et al.: Open foundation models for code, Code llama (2024)
21. Jiang, A.Q., et al.: Mistral 7b, 2023
22. Jiang, A.Q., et al.: Mixtral of experts, 2024

Multimodal Visio-Lingual Content Analysis to Detect Fake Content on Reddit

Adrián Girón(✉), Javier Huertas-Tato, and David Camacho

Universidad Politécnica de Madrid, Madrid, Spain
adrian.giron@upm.es

Abstract. The spread of misinformation across online social networks poses a significant threat to political stability, societal harmony, and economic integrity. This paper tackles the challenge of analyzing online posts, which often consist of diverse heterogeneous modalities like text or image. We introduce a multimodal architecture designed to accurately identify manipulated content. Our approach leverages a large-scale, multimodal dataset encompassing images, captions, comments, and metadata for each post. A novel encoding strategy is employed to capture both the semantic content and hierarchical structure of the comments. Based on CLIP, an early fusion technique is applied to process and merge the hidden representations. This method effectively combines multimodal and unimodal data according to their respective information channels. The performance is evaluated against the *Fakeddit* dataset, achieving a competent accuracy in binary classification tasks (0.9506 acc_{test}) with significant hardware limitations, and outperforming the SotA in more complex categorization tasks, with 0.9509 acc_{test} and 0.9371 acc_{test} in 3-way and 6-way label classification, respectively.

Keywords: Multimodal · Misinformation · *Fakeddit*

1 Introduction

Online Social Networks have become an information hub for modern society. The ration of information has increased with the number of users, becoming highly complex to distinguish legitimate content from manipulations. Misinformation is a complex problem with catastrophic consequences (i) in the political landscape, seen during the *Brexit* in the United Kingdom, where misleading data about economic consequences or immigration were claimed (ii) in the public safety, the *Capitol Hill attack* in USA, where Donald Trump made unfounded claims of voter fraud (iii) and as a sanitary risk, with all the disseminated misinformation during the COVID-19 pandemic.

Behind this content, there is a considerable social engineering effort, becoming undetectable for a large percentage of the population. They contain certain

V. Julian et al. (Eds.): IDEAL 2024, LNCS 15346, pp. 143–154, 2025.
https://doi.org/10.1007/978-3-031-77731-8_14

elements that enable high virality [18], using the user itself for the dissemination and composing a network of hoaxes too complex to counteract manually.

These hoaxes can be seen in any format. From misleading headlines to images that have been manipulated or even generated by AI. In this context, it is AI itself that can help us detect this content automatically, being able to recognize intricate patterns in the complex composition of hoaxes. In addition, multimodal systems are able to simultaneously pay attention to several information channels, detecting the combined semantics of modalities (Fig. 1).

Fig. 1. Examples of Reddit posts. Left picture is not manipulated, the right picture has been manipulated. They were extracted from *Fakeddit* [14].

We benchmark our work with *Fakeddit*, which contains multimodal posts from *Reddit*. These posts contain an image, its caption, the tree of comments and some metadata. We aim to demonstrate that a multimodal model is a better option for hoax detection than an unimodal perspective. Through an early fusion approach we merge different information channels when determining the veracity of a post. The architecture is based on CLIP [16] as an encoder for both images and its captions. We encode, in a novel way, the tree of comments as text using semantic particles that symbolize interactions between users and feed them to DeBERTa [6].

2 Related Work

2.1 Multimodal Fusion

A frequent approach for modality combination is early fusion. This method involves extracting separate representation for each modality to combine them to obtain the multimodal features. Other approaches in the case of image-text classification employ a visual encoder such a *CNN*/*ViT* for the visual content and text transformer for the textual information. In this context, Wang et al. [20] propose an early fusion approach using RoBERTa [12] to encode the concatenation of a news headline and its body and a ViT [2] to extract the features of the accompanying image. Then, each of the representations and its concatenation are classified adding the losses of each head.

These techniques have a structural problem. The representations of each modality are contained in a different dimensional space, which means that merging operations other than concatenation (sum, average, dot-product) are meaningless. We propose to use CLIP [16] as a multimodal encoder since it does ensure this property by approximating by contrastive learning images and captions in the same semantic space.

2.2 Encoding of Graph-Structured Semantic Data

Typically, comments in social networks follow a nested response tree format. Graph-structured data whose nodes are unstructured information like text, usually require the use of graph neural networks who follow a recursive neighborhood aggregation scheme. These models usually require large computational resources, for this reason, especially in a demanding multimodal context, strategies are needed to simplify the training process. Xie et al. [21] create a *Stance Reasoning Network*. They use *BERT* for all post-response pairs and apply self-attention between them like if they were a fully-connected graph, however, this approach scales quadratically with the number of comments per post. Kirchknopf et al.

We encode the interactions directly into a string in a *T5* manner [17] preserving both the semantics and the structure. Then, we take advantage of the great versatility of a SotA model like DeBERTa [6] to obtain the embeddings. Thus, this approach is much less time and resource consuming, obtaining, however, very competitive results.

2.3 Fake News Detection

There exist multiple approaches in terms of the technique or architecture employed or the type and number of modalities taken into account. Li et al. [11] propose using an ensemble of SotA text-transformers. Yuan et al. [22] address multimodal content moderation using *AM3*, a visio-lingual transformer. Peiguang et al. [10] present *EMEAF*, a multimodal architecture that aligns image-detected elements with their caption keywords.

In [14], Nakamura et al. introduce the multimodal dataset of *Fakeddit* which includes the image, caption and metadata of *Reddit* posts, furthermore, they propose using a *BERT + ResNet* early fusion architecture. Kirchknopf et al. [9] use ResNet101-v2 [5] and Inception-v3 [19], projecting the modalities into a small dimension to merge them with metadata. In [15], Stefanos-Iordanis et al. address image and text misalignment, proposing the use of *CLIP*. In [4], Hamed et al. obviate all non-textual content of the posts and hydrate *Fakeddit* with emotion analysis.

3 Methodology

Verifying posts on social networks is a difficult task mostly due to their format. Data is highly heterogeneous and is distributed along several information

Table 1. Modalities comparation to different SotA papers who approach multimodality. We test to incorpore metadata but it leads to worse results. We think this is due to the dimensional difference between modalities.

Approach	Image	Text	Comments	Metadata	Others
Li et al. [11]	-	x	-	-	-
Wang et al. [20]	x	x	-	-	-
Yuan et al. [22]	x	x	-	-	-
Peiguang et al. [10]	x	x	-	-	-
Nakamura et al. [14]	x	x	-	-	-
Kumar et al. [8]	x	x	-	-	-
Stefanos-Iordanis et al. [15]	x	x	-	-	-
Nadeem et al. [13]	x	x	-	x	-
Giachanou et al. [3]	x	x	-	x	-
Xie et al. [21]	x	x	x	-	-
Hamed et al. [4]	x	-	-	-	Emotions and sentiments
Kirchknopf et al. [9]	x	x	x	x	-
Ours	x	x	x	x*	-

channels, that is, the problem becomes multimodal. However, the multimodality hypothesis is put forward which states that even though the modalities do not share the same channel, they do share the intent. In other words, if one is able to combine and take advantage of all the points of view, more information can be extracted than by taking them into account with a unimodal approach (Table 1).

For each post, three different modalities are taken into account, these being the image of the post, its text (caption and comments) and some metadata (up- and downvotes, number of comments and score). To process and merge the different modalities, an early fusion based architecture will be used, that is, each modality will be processed by an encoder and then the representations will be combined.

Commonly, the general idea when combining image and text is usually the same, using a text transformer for the text and some SotA computer vision model for the image. Subsequently, merge the representations and/or project them in order to classify them. This approach is standard procedure as seen in related works. However, we seek to go a step further and use CLIP [16] as an encoder so that the representations obtained from both modalities are not only of quality but are intrinsically related. CLIP has the same architecture as it is composed of two completely independent encoders, the difference being that they have been trained by contrastive learning to project images and text in the same dimensional space. This model obtains better results than some many-shot SotA models even when performing classification at zero-shot (encoding the labels as text). Taking advantage of this property, we expect to obtain a better multimodal representation of the image and its caption than in the traditional way.

CLIP is used to encode image and caption. Specifically, the *clip-vit-large-patch14* version is used, which differs from other versions using a Vision Transformer (ViT) [2] with 24 attention layers, depth of 1024 dimensions and 16 attention heads, since in Radford et al. [16] is seen to have the best performance. In the case of caption, a text transformer of 12 attention layers, hidden dimension of 768 dimensions and 12 attention heads is used. We encode captions and comments separately for two reasons (i) the caption shares a semantic relationship with the image and comments do not have to (ii) the maximum sequence size of the CLIP text encoder is limited and would truncate most of the comments.

In order to obtain a quality representation of the comments, DeBERTa [6] is used with very good results in the state of the art thanks to the use of disentangled attention. In multiple NLP tasks it manages to outperform well-stablished models such as BERT [1] or RoBERTa [12], which is why it is chosen. Finally, in the case of metadata, since there is no specific encoder, a multilayer perceptron will suffice. It should be noted that both CLIP and DeBERTa are based on their pre-trained versions.

The architecture proposed to address multimodal fusion is composed of different branches, each of them in charge of combining a specific channel. There are 4 branches (i) image processing (ii) metadata processing (iii) textual information processing (see the joint meaning of caption and comments) (iv) multimodal branch that combines all channels. In this way, both unimodal and multimodal contributions are taken into account.

These branches can be divided into 3 clearly differentiated sections as can be seen in the Fig. 2. The first of the sections has been named encoding module and is responsible for processing each of the modalities with the relevant encoder and then (if it is not already) projecting through a feed-forward layer to a 512-dimensional vector. Then there is the projection module in charge of merging the different modalities according to their information channel and projecting these representations to a smaller dimension, in this case 128. Finally, a classification head will concatenate all the representations into a single 512-dimensional vector containing both the multimodal and unimodal load, projecting it to a two-neuron linear output layer that will produce the binary logits for the classification problem.

It should be noted that the activation function chosen was the GELU. In addition, regarding regularization, it has been empirically seen that a dropout of 0.1 after the encoders and layernorm in the layers of the projection module helped the convergence of the model.

Apart from the previously explained architecture, some with less modalities or even unimodal ones have been designed in the same way (branch-based). These can be understood as the one shown in Fig. 2 eliminating the modalities and branches that are not involved. These are all unimodal (images, captions, comments and metadata), as well as combinations such as images and captions; captions and comments (text-based modalities) and images, captions and comments (bypassing metadata).

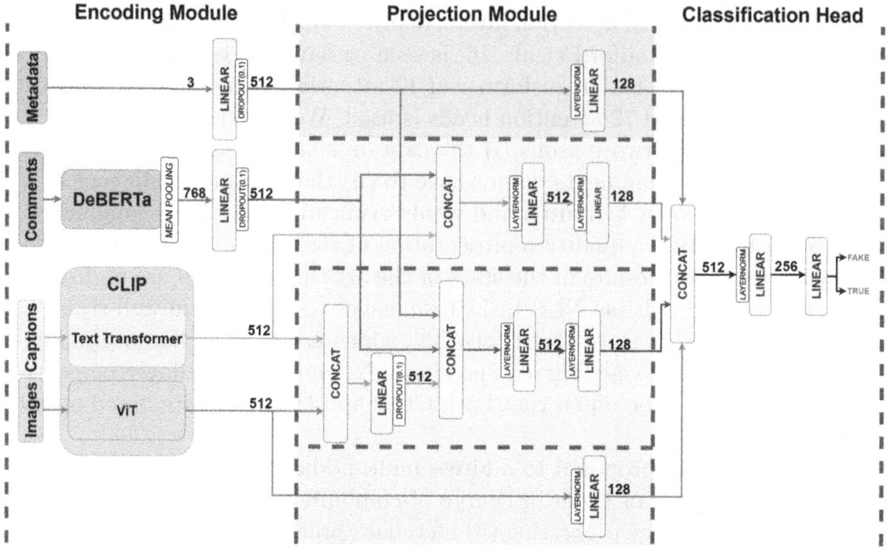

Fig. 2. Multimodal architecture designed to process and combine all modalities. Vertically, the network can be separated into three modules: (i) encoding module, (ii) projection module and (iii) classification head. In addition, horizontally the projection module is separated according to the information channel through which the modality is transmitted, from top to bottom (i) metadata, (ii) text, (iii) multimodal, (iv) image.

4 Experimental Setup

This section provides an overview of the dataset used to train and test the proposed architecture, the hardware available and the evaluation metrics we used to assess the performance.

4.1 Dataset

We used *Fakeddit* [14], a dataset with 700,000 multimodal posts from *Reddit* ranging from memes to news; containing links to images, captions and metadata. We use the *PushShift API* to hydrate the dataset with the comment tree associated with each post.

When downloading the images, we noticed that there were corrupt images. We encountered *"image is not available"* style images and placeholder images of the download domain. We fine-tuned an *Efficient-NetB0* architecture to classify these corrupted images from non-corrupted ones and, by means of its inference, the dataset is automatically filtered. Finally, the images are resized to 224×224 pixels, subsequently normalizing their pixel values to between [0, 1].

The comments, as in other social networks, have a graph format, specifically a tree format. In order not to increase the complexity of the model, it is proposed

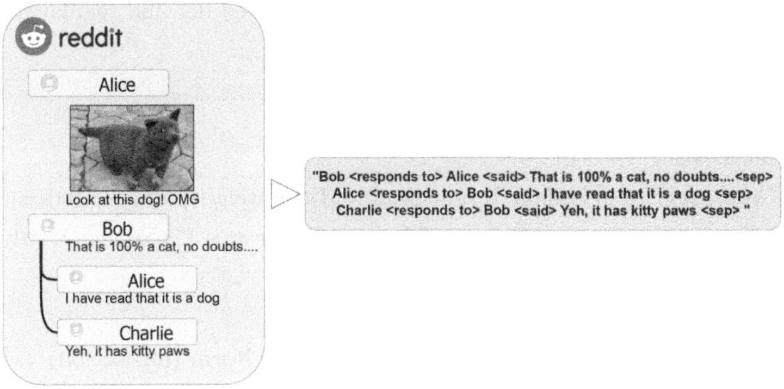

Fig. 3. The tree of comments is embedded directly in a text writing the interactions as semantic particles.

to embed the structure directly in text and interpret it as such through direct connections. Each interaction would be given by three elements (i) sender, (ii) comment and (iii) receiver as it can be shown in Fig. 3.

It should be noted that in cases where there are no comments this field is filled in as *<no comments>*. Finally, all interactions are concatenated into a single text. In this way not only the meaning is maintained, but also the structure.

The metadata that *Fakeddit* provides are the score (*likes*), upvote ratio (*likes/dislikes*) and number of comments. In the metadata we locate three fundamental problems. First, a large presence of nulls resulting from the direct integration of the dataset *The PSBattles* (Heller et al., 2018) [7] that account for about one-fifth of the data; second, a large power-law distribution in number of comments and score. Finally, we found a potential bias as this dataset has been labeled in a straightforward manner based on the *subreddit* of provenance.

The score nulls were filled with 0 s and the logarithm in base 10 was applied to smooth the power law distribution. Finally, to avoid the bias, we first grouped the posts by *subreddit* and normalize each metadata between [0, 1] with the statistics of each group.

4.2 Training and Evaluation Techniques

In most experiments, the encoder weights were frozen to reduce computational cost. However, we occasionally unfroze the last transformer blocks to assess performance impact. A learning rate of 1e-5 was used for fine-tuning the encoders and 1e-3 for other layers. We used *AdamW* as the optimizer and *CrossEntropy* as the loss function. The *ReduceLROnPlateau* scheduler, which reduced the learning rate by 0.1 after 10 epochs without improvement, was employed. The batch size was 16, the largest multiple of 2 that fitted on the GPU. Training used *Cross Validation* for 30 epochs, and performance was evaluated using accuracy and F1-score (Table 2).

Table 2. Training hyperparameters, optimizer, scheduler and regularization.

Hyper-parameter	Value
Learning rate	1e−3 (1e−5 for finetuning)
Batch size	16
Optimizer	AdamW (amsgrad = True, weight_decay = 0.01)
Scheduler	ReduceLROnPlateau (factor = 0.1, patience = 10)
Loss function	Cross Entropy
Training Epochs	30
Regularization	Dropout (encoding) & LayerNorm (projection)

4.3 Equipment and Timing

The network training is done with a 24 GB VRAM NVIDIA GeForce RTX 3090 GPU. Experiments range from a few hours in the unimodal case to a day or more in the multimodal cases.[1]

5 Experiments and Results

Table 3 shows the different experiments as a function of the encoders used, the modalities involved and the last unfrozen transformer blocks of the encoders. Due to the available experimentation capacity, these experiments have been performed with a small sample of the dataset, namely 15,000 training and 5,000 validation posts, the metrics being the best validation metrics in each experiment.

Table 3. Our results for both, unimodal and multimodal experiments, encoders and unfrozen last transformer blocks.

Encoder(s)	Unfrozen	Images	Captions	Comments	Metadata	Accuracy	F1-score
CLIP, DeBERTa	2	x	x	x	-	**0.9310**	**0.9100**
CLIP, DeBERTa	1	x	x	x	-	0.9293	0.9097
CLIP, DeBERTa	0	x	x	x	-	0.9268	0.9053
CLIP	4	x	x	-	-	0.9031	0.8742
CLIP	2	x	x	-	-	0.8949	0.8622
CLIP	0	x	x	-	-	0.8808	0.8487
CLIP, DeBERTa	0	-	x	x	-	0.8957	0.8667
CLIP, DeBERTa-V3	0	-	x	x	-	0.8868	0.8552
CLIP, LongFormer	0	-	x	x	-	0.8899	0.8598
CLIP, DeBERTa	2	x	x	x	x	0.8585	0.8506
CLIP	0	-	x	-	-	0.8451	0.7998
DeBERTa	0	-	-	x	-	0.8248	0.7594
CLIP	0	x	-	-	-	0.8184	0.7727
MLP	-	-	-	-	x	0.6934	0.6917

[1] The code to replicate the experiments is available at https://github.com/adgiz05/multimodal-disinformation-detection.

From analyzing the unimodal experiments, it can be concluded that text-based modalities are better at distinguishing fakes from non-fakes. Even with just the caption, performance is more than acceptable. Comparing multimodal combinations, caption and comments (both with 0 layers unfrozen) perform better than image and caption. This suggests that comments provide more information than images when combined with captions. However, combining all three modalities significantly boosts model performance, indicating that even though images perform the worst unimodally, their contribution in a multimodal context is valuable.

Introducing metadata complicates model convergence significantly. In unimodal experiments, metadata performance was far inferior to other modalities. The initial hypothesis was that spurious artifacts existed between metadata and other modalities, but this was dismissed after testing all combinations with similar results. The alternative hypothesis suggests a complexity disparity: metadata, a small numerical vector of raw data, contrasts with the high-dimensional abstract representations of other modalities.

The column "*unfrozen*" indicates the number of transformer blocks unfrozen from each encoder. Hardware limitations prevented unfreezing more attention blocks. As more blocks are unfrozen, results improve, which makes sense because we are limiting the general knowledge that the encoders have thanks to the received pre-training. With higher computational capacity and complete fine-tuning of the encoders, performance could significantly improve.

Finally, it should be noted that as more modalities are incorporated, the gap between accuracy and F1-score decreases. In other words, multimodality not only enhances accuracy but also contributes to more robust models. Each modality offers a unique perspective, which, when combined, leads to more confident predictions.

Table 4. Training with 10 times more data and results with 2, 3 and 6-way labels.

Encoders	Labels	Val Acc	Val F1	Test Acc	Test F1
CLIP, DeBERTa	2	0.9517	0.9375	0.9506	0.9369
CLIP, DeBERTa	3	0.9510	0.9514	0.9509	0.9548
CLIP, DeBERTa	6	0.9373	0.9274	0.9371	0.9282

In order to faithfully compare our results with the state of the art, the best of the models (images, captions and comments with 2 unfrozen transformer blocks) is trained with about a quarter of the training dataset (150,000 posts) and evaluated against the full validation and test sets. It should be noted that, so far, all experiments have been for binary classification, however, *Fakeddit* is multi-label and allows for 3 and 6 classes to be classified in addition. Although this is not the main purpose of this paper, we investigate how the proposed architecture works in these cases. The results are shown in the Table 4.

Table 5. SotA comparison of our method. Hamed et al. [4] does not appear because they compute AUC and F1-score instead of accuracy.

Architecture	Acc. 2-way	Acc. 3-way	Acc. 6-way	Train sample size	Test sample size
Our architecture	0.9506	**0.9509**	**0.9371**	150,000	full set (~60,000)
Armin et al.	**0.9550**	-	-	full set (~560,000)	full set (~60,000)
Xie et al.	0.9663*	-	-	<13,400	<1,460
Nakamura et al.	0.8909	0.8890	0.8588	full set (~560,000)	full set (~60,000)

There are several papers with similar results. Hamed et al. [4] achieve better results in f1-score, however, the training of their model is done with a small sample of the data so it makes a great premise, but it is not comparable with the SotA. With Xie et al. [21] something similar happens, they take a sample of the dataset by choosing certain subreddits. It is worth remembering that this attribute proposes a large bias as it directly attacks the label of the posts. Finally, Armin et al. [9] achieve a result of about 0.5% more than ours in accuracy for the binary case (taking these into account the metadata). It is important to note that we reach a 99.53% of their performance in the binary case just using a quarter of the available train set and a much less time and resource consuming model (Table 5).

Analyzing only the text-based results (captions and comments), we can see how our model (see Fig. 3) outperforms by up to 4% the results of Armin et al. (0.8590) even without having fine-tuned the encoders for that particular experiment. These results are not comparable since different samples are used, however, the ablations obtained with the entire dataset hints that the difference may be even larger. This implies that the power of DeBERTa coupled with our coding strategy for the comments is capable of extracting extremely high quality representations.

6 Conclusion

In this paper a multimodal branch-based early fusion architecture was proposed. By taking into account different modalities that pursue the same objective, it is possible to predict misinformation not only more effectively, but also more robustly. Furthermore, it is found that the number of modalities need not be directly proportional to the model performance, as there may be some (in our case metadata) that hinder the fusion with the rest and, therefore, the convergence of the model. The ability of CLIP to represent in the same dimensional space images and texts proves to obtain a high quality representation of images and their captions in the field of misinformation, obtaining SotA results despite technical limitations.

Despite not being able to improve SotA in the binary task, it should be considered as an achievement considering the following limitations (i) the model is trained with only a quarter of the training data, (ii) the encoders are not fully fine-tuned, (iii) a very limited batch size, (iv) the use of one less modality. On the other hand, in 3-way and 6-way label tasks, our architecture achieves the state of the art versus Nakamura et al. [14].

In the future, it is expected to perform a greater fine-tuning of the encoders or increase the batch size utilizing better equipment. In addition, we plan to incorporate more modalities and solve the spurious artifacts issue with metadata. Also, test graph-based approaches to encode the comments. Finally, extrapolate the architecture to other media or multimodal misinformation datasets.

Acknowledgments. This work has been funded by Grant PLEC2021-007681 (XAI-DisInfodemics) and PID2020-117263GB-100 (FightDIS) funded by MCIN/AEI/10.13039/501100011033 and by "ERDF A way of making Europe", by the "European Union" or by the "European Union NextGenerationEU/PRTR", by grant PCI2022-134990-2 (MARTINI) of the CHISTERA IV Cofund 2021 program, funded by MCIN/AEI/10.13039/501100011033 and by the "European Union NextGenerationEU/PRTR", by Calouste Gulbenkian Foundation, under the project MuseAI - Detecting and matching suspicious claims with AI.

References

1. Devlin, J., Chang, M.W., Lee, K., Toutanova, K.: BERT: pre-training of deep bidirectional transformers for language understanding, p. 13 (2018). http://arxiv.org/abs/1810.04805
2. Dosovitskiy, A., et al.: An image is worth 16 × 16 words: transformers for image recognition at scale. In: International Conference on Learning Representations (2021). https://openreview.net/forum?id=YicbFdNTTy
3. Giachanou, A., Zhang, G., Rosso, P.: Multimodal fake news detection with textual, visual and semantic information. In: Sojka, P., Kopeček, I., Pala, K., Horák, A. (eds.) TSD 2020. LNCS (LNAI), vol. 12284, pp. 30–38. Springer, Cham (2020). https://doi.org/10.1007/978-3-030-58323-1_3
4. Hamed, S.K., Ab Aziz, M.J., Yaakub, M.R.: Fake news detection model on social media by leveraging sentiment analysis of news content and emotion analysis of users' comments. Sensors (Basel, Switzerland) **23**(4), 1748 (2023). https://doi.org/10.3390/s23041748
5. He, K., Zhang, X., Ren, S., Sun, J.: Deep residual learning for image recognition. In: 2016 IEEE Conference on Computer Vision and Pattern Recognition (CVPR), pp. 770–778 (2016). https://doi.org/10.1109/CVPR.2016.90
6. He, P., Liu, X., Gao, J., Chen, W.: Deberta: decoding-enhanced BERT with disentangled attention. CoRR abs/2006.03654 (2020). http://dblp.uni-trier.de/db/journals/corr/corr2006.html#abs-2006-03654
7. Heller, S., Rossetto, L., Schuldt, H.: The PS-battles dataset - an image collection for image manipulation detection. arXiv abs/1804.04866 (2018). https://api.semanticscholar.org/CorpusID:4877931
8. Kaliyar, R.K., Kumar, P., Kumar, M., Narkhede, M., Namboodiri, S., Mishra, S.: Deepnet: an efficient neural network for fake news detection using news-user engagements. In: 2020 5th International Conference on Computing, Communication and Security (ICCCS), pp. 1–6 (2020). https://doi.org/10.1109/ICCCS49678.2020.9277353

9. Kirchknopf, A., Slijepčević, D., Zeppelzauer, M.: Multimodal detection of information disorder from social media. In: 2021 International Conference on Content-Based Multimedia Indexing (CBMI), pp. 1–4 (2021). https://doi.org/10.1109/CBMI50038.2021.9461898
10. Li, P., Sun, X., Yu, H., Tian, Y., Yao, F., Xu, G.: Entity-oriented multi-modal alignment and fusion network for fake news detection. IEEE Trans. Multimed. **24**, 3455–3468 (2022). https://doi.org/10.1109/TMM.2021.3098988
11. Li, X., Xia, Yu., Long, X., Li, Z., Li, S.: Exploring text-transformers in AAAI 2021 shared task: COVID-19 fake news detection in English. In: Chakraborty, T., Shu, K., Bernard, H.R., Liu, H., Akhtar, M.S. (eds.) CONSTRAINT 2021. CCIS, vol. 1402, pp. 106–115. Springer, Cham (2021). https://doi.org/10.1007/978-3-030-73696-5_11
12. Liu, Y., et al.: Roberta: a robustly optimized BERT pretraining approach (2019). http://arxiv.org/abs/1907.11692
13. Nadeem, M.I., et al.: EFND: a semantic, visual, and socially augmented deep framework for extreme fake news detection. Sustainability **15**(1), 133 (2022). https://doi.org/10.3390/su15010133
14. Nakamura, K., Levy, S., Wang, W.Y.: Fakeddit: a new multimodal benchmark dataset for fine-grained fake news detection. In: Calzolari, N., et al. (eds.) Proceedings of the Twelfth Language Resources and Evaluation Conference, pp. 6149–6157. European Language Resources Association, Marseille, France (2020). https://aclanthology.org/2020.lrec-1.755
15. Papadopoulos, S., Koutlis, C., Papadopoulos, S., Petrantonakis, P.C.: Figments and misalignments: a framework for fine-grained crossmodal misinformation detection. CoRR abs/2304.14133 (2023). https://doi.org/10.48550/ARXIV.2304.14133
16. Radford, A., et al.: Learning transferable visual models from natural language supervision. In: International Conference on Machine Learning (2021). https://api.semanticscholar.org/CorpusID:231591445
17. Raffel, C., et al.: Exploring the limits of transfer learning with a unified text-to-text transformer. J. Mach. Learn. Res. **21**(140), 1–67 (2020). http://jmlr.org/papers/v21/20-074.html
18. Solovev, K., Pröllochs, N.: Moral emotions shape the virality of COVID-19 misinformation on social media. arXiv e-prints arXiv:2202.03590 (2022)
19. Szegedy, C., Vanhoucke, V., Ioffe, S., Shlens, J., Wojna, Z.: Rethinking the inception architecture for computer vision (2015)
20. Wang, Z., Shan, X., Zhang, X., Yang, J.: N24News: a new dataset for multimodal news classification. In: Proceedings of the Thirteenth Language Resources and Evaluation Conference, pp. 6768–6775. European Language Resources Association, Marseille, France (2022). https://aclanthology.org/2022.lrec-1.729
21. Xie, J., Liu, S., Liu, R., Zhang, Y., Zhu, Y.: SERN: stance extraction and reasoning network for fake news detection. In: 2021 IEEE International Conference on Acoustics, Speech and Signal Processing (ICASSP), ICASSP 2021, pp. 2520–2524 (2021). https://doi.org/10.1109/ICASSP39728.2021.9414787
22. Yuan, J., Yu, Y., Mittal, G., Hall, M., Sajeev, S., Chen, M.: Rethinking multimodal content moderation from an asymmetric angle with mixed-modality (2023)

MetaLIRS: Meta-learning for Imputation and Regression Selection

Işıl Baysal Erez(✉), Jan Flokstra, Mannes Poel, and Maurice van Keulen

EEMCS, University of Twente, 7500 AE Enschede, The Netherlands
i.baysalerez@utwente.nl

Abstract. Missing data is a prevalent problem in data science for many fields such as natural, social, and health sciences. Since most regression methods can not handle missing data directly, imputation methods are used in data pre-processing. Finding the best imputation method is non-trivial, however. Moreover, our results show that an independent choice for a best imputation method does not always result in the best predictive performance in the end; the combination matters. Furthermore, search-based approaches for finding a best-fitting imputer/regressor-pair can be computationally intensive. In this paper, we propose the MetaLIRS (**Meta** **L**earning **I**mputation and **R**egression **S**election) framework for developing resource-friendly ML-based recommendation models for method selection. With MetaLIRS, we constructed a proof-of-concept recommendation model based on 12 meta-features that achieves an accuracy of 63% for selecting the best-fitting imputer/regressor-pair. A data scientist can use this model for a quick resource-friendly recommendation on which imputation and regression method to use for their particular data set and task without the need for an expensive grid search among methods.

Keywords: Missing data · Imputation · Regression

1 Introduction

In pattern recognition problems, regression models play an important role in data analysis. These models are used in various real-world applications such as stock market predictions [16], house prices, retinol levels [34], etc. Regression models, however, are highly sensitive to data quality problems such as missing. Many models cannot be directly applied to data with missing values [4,41]. Moreover, missing data can negatively affect the predictive performance of regression models.

Early proposals for ad-hoc solutions to handle missing data for regression are listwise [33] and pairwise deletion of data samples [37]. These techniques, however, cause information loss and bias. A better method is imputation which replaces missing values with suitable estimated values [37]. There are various imputation method ranging from statistical [21,38] to machine learning [35,36]

V. Julian et al. (Eds.): IDEAL 2024, LNCS 15346, pp. 155–166, 2025.
https://doi.org/10.1007/978-3-031-77731-8_15

and deep learning [26,43] based. Filling in missing values separately and subsequently applying regression is known in literature as "impute-followed-by-regression" [20]. Another strategy is selecting regression models that can handle missing data directly, such as some ensemble methods based on Gradient Boosting Decision Tree (GBDT) of which Extreme Gradient Boosting (XGBoost) [8] and Light Gradient-Boosting Machine (LightGBM) [17] are common examples. Due to the "built-in" structures [9] of these models, they can be used directly for regression without the need for imputation.

A complication is that there is neither a single best imputation method nor a single best regression model for all tasks [40]. Typically, data analysts use a technique called "grid search" to try and evaluate all potentially appropriate imputation and ML methods. Furthermore, they often independently select a best-fitting imputation and ML method while certain combinations of imputer and regressor work better than others [30]. To ease the grid search-based selection of models, AutoML tools such as TPOT (Tree-based Pipeline Optimization Tool) [28] and AutoWeka [19] have been developed to automate and optimize the search for a best-fitting machine imputation method, learning method, and accompanying configurations [13]. However, AutoML's range of considered data imputation methods is limited [44], and its search-based approach is rather compute-intensive.

Meta-learning-based approaches have a high potential to increase the performance of AutoML systems [31]. Meta-learning (MtL) methodologies aim to understand the learning process and re-use knowledge from previous experiences. They can be used for algorithm selection and yield substantial advantages [2,5,14]. We propose the *MetaLIRS (Meta Learning Imputation and Regression Selection) framework* and use it to construct a recommender for selecting the best (imputer, regressor)-pair. The predictive model of the recommender is based on meta-features of the given data set and regression task, which requires low computation resources avoiding the need for a grid search. This work developed the recommendation model and extends our preliminary work [3].

We strive for a resource-friendly recommender that is applicable to a wide range of real-world scenarios. As opposed to similar approaches like [22,44], we consider a diverse set of imputation methods including more advanced imputation methods such as Imputation by Chained Equations (ICE), MissForest, Multi Layer Perceptron (MLP), and Generative Adversarial Imputation Nets (GAIN). Moreover, we support low, medium, and high missingness scenarios.

The main contributions of this paper are:

- MetaLIRS, a framework for developing resource-friendly ML-based recommendation models for selecting a best-fitting combination of imputation and regression method based on meta-features of a given data set and regression task. It is extensible with more meta-features, imputation and regression methods, data sets, etc.
- A proof-of-concept Random Forest-based recommendation model constructed with MetaLIRS based on extensive experimental training data from a wide

range of meta-features, data sets, imputation and regression methods, and missingness scenarios.
- Insight into the performance of imputation methods in low, medium, and high missingness scenarios, how they affect regression performance, and which data characteristics (meta-features) are most important in imputation method selection.

2 Related Work

Meta-learning approaches use machine learning algorithms to build relationships between the performance of algorithms and the characteristics of the data [5]. That allows them to use the knowledge gathered from these relationships to recommend the best algorithms for new datasets [5]. Moreover, many meta-learning approaches are utilized separately for a singular-task recommendation, such as regression or imputation [2,23].

In computer science, while meta-learning approaches for algorithm recommendation have been studied early on [32], few studies focus on algorithm selection for pre- and post-imputation in the presence of missing data. Regarding imputer recommendation, Zou et al. [45], an early study, evaluated eight statistical imputation methods (Deletion, Median/Mode Imputation (MMI), Group/Class MMI, EM Imputation, etc.) and one machine learning-based method (k-NN) for handling missing data. They assessed the performance of these methods using the C4.5 and ELEM2 classification algorithms. Zagatti et al. [44] developed a meta-learning system for recommendation of different data pre-processing methods such as data normalization, standardization, imputation, and data integration. Liu and Hauswirth [22] presented A Provenance Meta Learning Framework to select the imputation and classification methods. Their framework also includes basic mean/median imputation, they also add different k-NN variants imputers based on machine learning, and three classification methods (C4.5, KNN, and RIPPER). The framework generated If-Then rules based on the Rule Induction Engine model.

For the regressor's recommendation, Lorena et al. [23] adapted data complexity meta-features such as feature correlation measures, linearity measures, smoothness measures, and geometry, topology, and density measures for regression tasks. They used the SVR (Support Vector Regression) model as the meta-learner to recommend 14 different regression models. Aguiar et al. [2] extracted meta-features similar to previous work and extracted more features based on general and statistical information to find the most suitable solution for the multi-target regression models.

In addition to the studies above, a recently published paper [30] on the research indicates that since 2005, meta-learning approaches used for data pre-processing problems have included imputation methods based on statistical techniques and basic machine learning algorithms. However, given the increasing diversity of data, other machine learning and deep learning-based imputation techniques may also be necessary. Therefore, we conducted a broader analysis of imputation methods. Furthermore, as noted in the literature review study [30], a notable focus is selecting classification algorithms. Our study focuses on regression algorithms of varying complexities to identify and recommend the best imputation+regression model pair.

3 The MetaLIRS Framework

We present MetaLIRS (Meta Learning Imputation and Regression Selection) framework, and the aim is to find the best imputer+regressor pair with the model we built through this framework. We explain the MetaLIRS framework by describing its phases as depicted in Fig. 1.

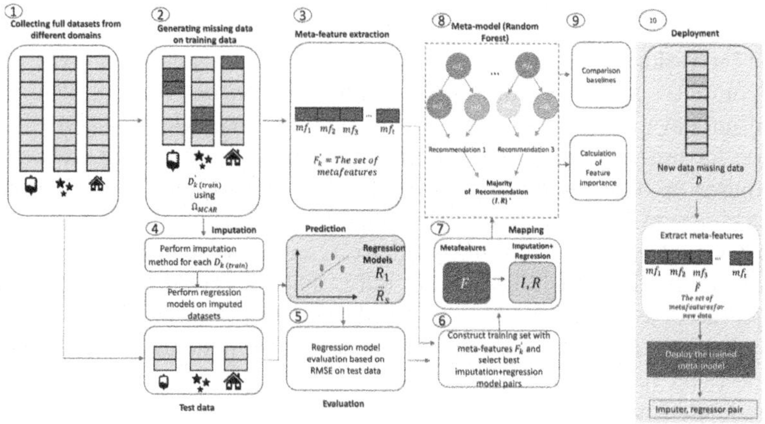

Fig. 1. An overview MetaLIRS (Meta Learning Imputation and Regression Selection) framework.

The MetaLIRS framework has two main phases: Learning and recommendation.

- **Learning phase:** Before setting the meta-database for the meta-model complete datasets (from steps 1 to 2 in Fig. 1) were selected from different domains such as health, business, and other fields. Missing datasets were generated from these datasets according to MCAR data mechanisms with different missing ratios: 1%, 5%, 10%, 30%, 50%, 70% on training data. Then (step 3 in Fig. 1), 12 meta-features (F'_k) were calculated using general features (number of features, ratio of dimensionality, missing level), statistical features (standard deviation, variance, correlation, covariance, kurtosis, skewness, 25th percentile, 75th percentile) from each missing data (for more details, see [1]). Six different imputation methods were employed to predict missing values and complete the datasets. These methods include statistical approaches such as

Mean imputation and ICE, machine learning methods like k-NN and MissForest, and deep learning methods such as MLP and GAIN.
Five regressors (Support Vector Regression (SVR), MLP, Adaptive Boosting (AdaBoost), XGBoost, and LightGBM were trained on completed datasets by filled several imputers, LightGBM and XGBoost were also trained directly with missing data. The best-performing imputation method and regression model pairs were determined according to the test dataset's root mean squared error (RMSE) values (from steps 4 to 5 in Fig. 1). Then, the mapping (from steps 6 to 7 in Fig. 1) involved constructing a meta-database using meta-features and the best imputer-regressor pair. The outcomes of these steps are integrated into a single meta-dataset and meta-model built in accordance with the above requirements. We select the Random Forest meta-model to recommend the best imputer+regressor pair. Random Forest is an ensemble technique composed of a series of decision trees, each of which is built using a random subset of features. It is preferred for meta-learning approaches because it exhibits low bias and variance [6], and the most popular and widely used, more robust [12], and has high predictive accuracy [2,30,44] among meta-learning models. The training stage of the Random Forest meta-model showed number 8 in Fig. 1. Comparison baselines and calculation feature importance values (step 9 in Fig. 1) performed. As baselines, we compared our meta-model to search-based AutoML tools such as TPOT [28] and Auto-sklearn [13], and we also considered the imputer+regressor pair that obtained the best overall performance. We calculated the importance of impurity-based features to assess the impact of meta-features.
- **The recommendation (deployment) phase:** The deployment stage (number 9 in Fig. 1) involves the recommendation process. For a new dataset, the set of meta-features are computed and fed to the Random Forest meta-model to recommend the best imputer+regressor pair.

4 Experimental Set-Up

4.1 Data Preparation Step for Meta-database

Datasets: We use 16 datasets from healthcare, biology, business, geography domains, and others, from OpenML [39] and UCI repositories [18]. More details about used datasets are given in Appendix Table 1 [1]. We applied scaling to the features by performing standardization before training/test splitting to prevent data leakage.

Missing Data Mechanism: MCAR (missing completely at random) data mechanism, where the probability of missing data is equal for all features, was simulated with six different missing ratios of 1%, 5%, 10%, 30%, 50%, 70%. To emulate the MCAR data mechanism, we sample from a Bernoulli distribution ($\mathcal{B}$), following similar settings as employed in [25].

Imputation Methods: We used various imputation methods (Mean [21], k-NN [36], ICE [37,38], MissForest [35], MLP [15], GAIN [43] and employed the default settings included in the scikit-learn package [29] for the implementations, except for GAIN imputation (We utilized the original source code provided by the authors [43]).

Regression Models: We selected five regression methods: MLP [15], AdaBoost [11], and SVM [10] algorithms, and used the scikit-learn package [29] to set up these regression models. Official repositories were used for XGBoost [8] and LightGBM (Light Gradient-Boosting Machine) [17]. We divided the dataset into a 75% training subset and a 25% testing subset. We used imputation methods after train and test splitting to prevent data leakage. We just emulated missing data on the training set. All experiments were performed independently 3 times using different train/test split settings. We demonstrated the hyperparameter settings used for the imputers and regressors in Appendix Table 4 [1] and Table 5 [1], respectively.

Evaluation Metrics: To assess the effectiveness of imputers and regression models, we used the RMSE metric.

4.2 Setting of Recommendation Stage

We used the Random Forest classification model [6] as a meta-model to recommend the best imputer+regressor pairs.

We adopted the Leave-One-Out Cross-Validation (LOOCV) strategy to evaluate the model's overall performance. LOOCV uses each instance sequentially as the test set, and the remaining observations are used as the training dataset. Thus, the model's performance is evaluated on each dataset, allowing a better estimate of the overall performance. The Random Forest model was trained using 30 different random initializations of the meta-model. We used the default values for hyperparameters of the classification model defined in the scikit-learn [29] library and used default hyperparameters. The extracted meta-features are listed in Appendix Table 6 [1]. All meta-features are numerical. For the meta-database label, we used the Synthetic Minority Oversampling Technique (SMOTE) to balance the minority class in the imbalanced dataset (only on train set) [7].

5 Results

We chose six widely used imputers and seven regression (XGBoost and LightGBM are used two times with imputation and without) models to validate the recommended meta-model. Three approaches were compared with our proposed RF meta-model: (1) majority class for regressor, (2) TPOT framework, and (3) Autosklearn framework (in Table 3). We also compared the performance of the meta-model classification with random selection (Table 5).

The number of times each algorithm was in each rank position is indicated in Table 1. For imputation results, MissForest showed 35 times the best results. MLP achieved the highest rank 25 times, making it the second-best performer. For the regression part, using the LightGBM without imputer ranked first in

40 out of 96 cases. In Table 2, we showed the number of times the imputer and regression models best rank based on RMSE values of regressor models for the MCAR data mechanism. Using the LightGBM regression model without imputer showed the best performance.

Table 1. Number of times each algorithm assumed each rank position for imputer and regression models. The lowest rank is better.

Rank	Imputation						Regression						
	Mean	ICE	k-NN	MissForest	MLP	GAIN	SVR	MLP	AdaB.	XGB.	LightG.	XGB (No imp.)	LightG.(No imp.)
1	12	20	4	35	25	-	6	9	-	10	17	14	40
2	10	21	17	18	29	1	2	1	1	14	44	14	20
3	9	22	20	14	25	6	22	21	18	4	9	14	8
4	21	24	28	10	9	17	23	9	19	12	8	22	6
5	21	8	24	10	3	30	8	10	10	32	12	9	12
6	36	1	3	9	5	42	23	4	4	18	2	13	10
7							12	42	42	6	2	10	-

The meta-model we recommended has consistently demonstrated very competitive performance across various datasets and levels of missing data in Table 3. It demonstrated better performance for the plasma dataset with a low level of missing data compared to majority voting and the TPOT framework. However, our meta-model showed the best results for high levels of missing data, except for the TPOT framework. Similarly, this meta-model showed results similar to those of majority voting and TPOT for the concrete dataset with low missing data levels. On the other hand, for high missing data levels, it performed similarly to TPOT, while Autosklearn achieved the best results. In Table 4, we compared training time for all models (imputer+regressor) for a 1% missing level. While the mean imputer showed the fastest performance, the MissForest imputer showed the longest computational time for the wine quality dataset. For the training time of regressors, the MLP was significantly longer compared to other regression models. The best-fitting imputer+regressor pair is Missforest with LightGBM in most cases. While TPOT had a more reasonable training time, Auto-sklearn took approximately 10 min to train and find the optimal model, despite being trained with data completed using a single imputation method for one missing level. A meta feature-based recommendation model can come to this recommendation directly, while a grid search-based method would have to evaluate much more costly imputation methods such as MissForest and regression methods such as MLP.

The results in Table 5 indicate that our recommendation model performs noticeably better than the random selection baseline technique. Our recommendation model can effectively recommend the best imputer+regressor pair with an accuracy of 0.63. If we had chosen randomly, we would have achieved an average accuracy of 0.20. The advantage of the recommendation model over the random model based on the trial and error approach is clear. Moreover, we analyze the feature importance of classification with the trained RF meta-model. Figure 2

Table 2. The number of times regression models and imputation methods achieve the best rank is based on the regression model's performance for all datasets. Experiments were performed on the 16 datasets under the MCAR data mechanism. Experiments run three different seeds.

Imputation Methods	Regression models				
	SVM	XGBoost	LightGBM	AdaBoost	MLP
No Imputation	-	-	24	-	-
	-	5	-	-	-
KNN	1	4	2	3	1
Mean	3	1	4	-	-
ICE	1	1	7	1	-
MissForest	1	3	10	5	-
MLP	1	3	7	1	-
GAIN	-	4	3	-	-

Table 3. Comparison of Methods on Different Datasets' RMSE values. Experiments run one seed. Lower RMSE is better.

Datasets	Meta-Model	Majority	TPOT	Autosklearn
Plasma [27] (%1)	232.94	236.97	276.30	221.25
Plasma (%60)	244.89	249.87	224.70	264.45
Concrete [42] (%1)	4.3	4.29	3.96	4.4
Concrete (%60)	9.8	8.87	9.81	7.9

Table 4. Imputation and regression models training time values (in seconds) for the wine-quality-red dataset for one seed and a 1% missing level. For the TPOT and Auto-sklearn tools, the median imputation training time and the regressors' training time were included.

Imputation	Time (s)	Regression	Time (s)	TPOT (s)	Autosklearn (s)
Mean	0.002	MLP	**38.75**	32.12	**605.96**
k-NN	0.019	SVM	0.69		
ICE	0.39	AdaBoost	1.42		
MissForest	**801.91**	XGBoost	0.01		
MLP	19.99	LightGBM	0.23		
GAIN	18.25	XGBoost (No imputation)	0.01		
-	-	LightGBM (No imputation)	0.04		

Table 5. The results of evaluation metrics for the classification model with MCAR data mechanism. For all metrics, higher is better. Results are presented with mean and standard deviation over 30 different random initializations of Random forest.

Mechanism	Accuracy	Precision	Recall	F1-score
MCAR	$0.63_{\pm 0.01}$	$0.64_{\pm 0.01}$	$0.63_{\pm 0.01}$	$0.63_{\pm 0.01}$

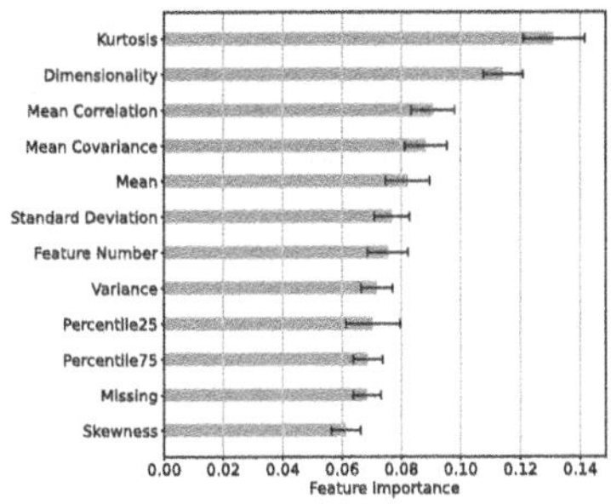

Fig. 2. Feature importance to induce a meta-model recommending a best imputer+regressor pair.

shows that average kurtosis and dimensionality are the most important features of the Random Forest (RF). It was also observed that, among these 12 features, the RF is least sensitive to missing values and average skewness.

6 Discussion and Future Work

We created a MetaLIRS framework to select the best imputer and regressor pair and developed a Random Forest meta-model within this framework. We collect complete datasets and generate missing data using various missing levels (1%, 5%, 10%, 30%, 50%, and 70%) in a meta-database to train the recommendation model. Future work involves expanding the database with more datasets, including larger ones.

Finding the optimum regression model is difficult due to the challenge of consistently guaranteeing the best imputation procedure across all levels of missing data. Thus, the RMSE values were used to label the meta-models. Investigating various ranking techniques could lead to a comprehensive meta-database in future studies.

We selected simple and statistical meta-features based on their easy computability and frequent usage in previous works [24]. Future studies may benefit from incorporating other measures, such as data-complexity, which are used in the literature [23].

We proposed a classifier based on the recommender model to select the best regressor+imputer pairs. The recommendation model is a classifier with imputer/regressor-pairs as classes. The training data obtained through extensive experiments is rather skewed as LightGBM turned out to be best in many cases. Therefore, we combined minority classes into one leaving five classes (for more details, see [1]).

During the comparison analysis between the meta-model and other models, we observed that the TPOT and Autosklearn frameworks sometimes had difficulty handling missing datasets due to their basic imputation setting (Median imputation).

Finally, our training set has been obtained from experimenting with out-of-the-box imputation and regression methods. Including hyperparameter optimization is an important avenue for future research.

7 Conclusion

In this paper, we proposed MetaLIRS, a framework for developing resource-friendly ML-based recommendation models for selecting a best-fitting combination of imputation and regression method based on meta-features of a given data set and regression task. With MetaLIRS, we constructed a recommendation model based on 12 meta-features that achieves an accuracy of 63% for selecting the best-fitting imputer/regressor-pair among 6 common and more advanced imputation methods, 5 regression methods, determined over a wide range of scenarios represented by 16 data sets, 6 MCAR missingness levels (1%, 5%, 10%, 30%, 50%, 70%). A data scientist can use this model for a quick resource-friendly recommendation on which imputation and regression method to use for their particular data set and regression task without the need of an expensive grid search among methods. In our experimentation, we observed that the best imputation methods for low-missingness scenarios turn out to be the worst for high-missingness scenarios and vice versa. Nevertheless, when selection is combined with regression method selection, LightGBM came out best in combination with only certain imputers or no imputation at all (LightGBM supports training with missing data). Most important meta-features turned out to be average kurtosis, then dimensionality. Finally, it was shown that it is important to view the selection of imputation and regression methods as a combined problem: independently selecting an imputation method first did not yield the best imputer/regressor pair in most scenarios, depending on the level of missingness.

Acknowledgments. This research has been partly funded by the Turkey's Ministry of National Education study-abroad program.

References

1. Appendix. https://shorturl.at/pwsNE
2. Aguiar, G.J., Santana, E.J., de Carvalho, A.C., Junior, S.B.: Using meta-learning for multi-target regression. Inf. Sci. **584**, 665–684 (2022)
3. Baysal Erez, I., Flokstra, J., Poel, M., van Keulen M.: Artificial intelligence in medicine. In: 22nd International Conference, AIME 2024, Salt Lake City, UT, USA (2024). https://aime24.aimedicine.info/papers-and-posters/, Poster Section
4. Bertsimas, D., Pawlowski, C., Zhuo, Y.D.: From predictive methods to missing data imputation: an optimization approach. J. Mach. Learn. Res. **18**(196), 1–39 (2018)
5. Brazdil, P., van Rijn, J.N., Soares, C., Vanschoren, J.: Metalearning: Applications to Automated Machine Learning and Data Mining. Springer, Cham (2022)
6. Breiman, L.: Random forests. Mach. Learn. **45**, 5–32 (2001)

7. Chawla, N.V., Bowyer, K.W., Hall, L.O., Kegelmeyer, W.P.: SMOTE: synthetic minority over-sampling technique. J. Artif. Intell. Res. **16**, 321–357 (2002)
8. Chen, T., Guestrin, C.: XGBoost: a scalable tree boosting system. In: Proceedings of the 22nd ACM SIGKDD International Conference on Knowledge Discovery and Data Mining, pp. 785–794 (2016)
9. Correia, A., Peharz, R., de Campos, C.P.: Joints in random forests. In: Advances in Neural Information Processing Systems, vol. 33, pp. 11404–11415 (2020)
10. Cortes, C., Vapnik, V.: Support-vector networks. Mach. Learn. **20**, 273–297 (1995)
11. Drucker, H.: Improving regressors using boosting techniques. In: ICML, vol. 97, p. e115. Citeseer (1997)
12. Feurer, M., Springenberg, J.T., Hutter, F.: Using meta-learning to initialize Bayesian optimization of hyperparameters. In: Proceedings of the 2014 International Conference on Meta-Learning and Algorithm Selection, vol. 1201, pp. 3–10. Citeseer (2014)
13. Feurer, M., Klein, A., Eggensperger, K., Springenberg, J., Blum, M., Hutter, F.: Efficient and robust automated machine learning. Adv. Neural Inf. Proc. Syst. **28** (2015)
14. Garouani, M., Ahmad, A., Bouneffa, M., Hamlich, M., Bourguin, G., Lewandowski, A.: Using meta-learning for automated algorithms selection and configuration: an experimental framework for industrial big data. J. Big Data **9**(1), 57 (2022)
15. Hinton, G.E.: Connectionist learning procedures. In: Machine Learning, pp. 555–610. Elsevier (1990)
16. Illa, P.K., Parvathala, B., Sharma, A.K.: Stock price prediction methodology using random forest algorithm and support vector machine. Mater. Today: Proc. **56**, 1776–1782 (2022)
17. Ke, G., et al.: LightGBM: a highly efficient gradient boosting decision tree. In: Advances in Neural Information Processing Systems, vol. 30 (2017)
18. Kelly, M., Longjohn, R., Nottingham, K.: The UCI machine learning repository. https://archive.ics.uci.edu. Accessed 16 Jan 2024
19. Kotthoff, L., Thornton, C., Hoos, H.H., Hutter, F., Leyton-Brown, K.: Auto-WEKA 2.0: automatic model selection and hyperparameter optimization in WEKA. J. Mach. Learn. Res. **18**(25), 1–5 (2017)
20. Le Morvan, M., Josse, J., Scornet, E., Varoquaux, G.: What's a good imputation to predict with missing values? In: Advances in Neural Information Processing Systems, vol. 34, pp. 11530–11540 (2021)
21. Little, R.J., Rubin, D.B.: Statistical Analysis with Missing Data. Wiley, Hoboken (1987)
22. Liu, Q., Hauswirth, M.: A provenance meta learning framework for missing data handling methods selection. In: 2020 11th IEEE Annual Ubiquitous Computing, Electronics & Mobile Communication Conference (UEMCON), pp. 0349–0358. IEEE (2020)
23. Lorena, A.C., Maciel, A.I., de Miranda, P.B., Costa, I.G., Prudêncio, R.B.: Data complexity meta-features for regression problems. Mach. Learn. **107**, 209–246 (2018)
24. Michie, D., Spiegelhalter, D.J., Taylor, C.C., Campbell, J.: Machine Learning, Neural and Statistical Classification. Ellis Horwood (1995)
25. Muzellec, B., Josse, J., Boyer, C., Cuturi, M.: Missing data imputation using optimal transport. In: International Conference on Machine Learning, pp. 7130–7140. PMLR (2020)
26. Nazabal, A., Olmos, P.M., Ghahramani, Z., Valera, I.: Handling incomplete heterogeneous data using VAEs. Pattern Recogn. **107**, 107501 (2020)

27. Nierenberg, D.W., et al.: Determinants of plasma levels of beta-carotene and retinol. Am. J. Epidemiol. **130**(3), 511–521 (1989)
28. Olson, R.S., Bartley, N., Urbanowicz, R.J., Moore, J.H.: Evaluation of a tree-based pipeline optimization tool for automating data science. In: Proceedings of the Genetic and Evolutionary Computation Conference, GECCO 2016, pp. 485–492. ACM, New York (2016). https://doi.org/10.1145/2908812.2908918
29. Pedregosa, F., et al.: Scikit-learn: machine learning in Python. J. Mach. Learn. Res. **12**, 2825–2830 (2011)
30. Pio, P.B., Rivolli, A., de Carvalho, A.C., Garcia, L.P.: A review on preprocessing algorithm selection with meta-learning. Knowl. Inf. Syst. **66**(1), 1–28 (2024)
31. Rakotoarison, H., Milijaona, L., Rasoanaivo, A., Sebag, M., Schoenauer, M.: Learning meta-features for AutoML. In: International Conference on Learning Representations, ICLR 2022 (spotlight) (2022)
32. Rice, J.R.: The algorithm selection problem. In: Advances in Computers, vol. 15, pp. 65–118. Elsevier (1976)
33. Schafer, J.L., Graham, J.W.: Missing data: our view of the state of the art. Psychol. Methods **7**(2), 147 (2002)
34. Spíndola Garcêz, L., et al.: Serum retinol levels in pregnant adolescents and their relationship with habitual food intake, infection and obstetric, nutritional and socioeconomic variables. Nutrients **8**(11), 669 (2016)
35. Stekhoven, D.J., Bühlmann, P.: Missforest–non-parametric missing value imputation for mixed-type data. Bioinformatics **28**(1), 112–118 (2012)
36. Troyanskaya, O., et al.: Missing value estimation methods for DNA microarrays. Bioinformatics **17**(6), 520–525 (2001)
37. Van Buuren, S.: Flexible Imputation of Missing Data. CRC Press (2018)
38. Van Buuren, S., Groothuis-Oudshoorn, K.: MICE: multivariate imputation by chained equations in R. J. Stat. Softw. **45**, 1–67 (2011)
39. Vanschoren, J., van Rijn, J.N., Bischl, B., Torgo, L.: OpenML: networked science in machine learning. SIGKDD Explor. **15**(2), 49–60 (2013). https://doi.org/10.1145/2641190.2641198
40. Wolpert, D.H., Macready, W.G.: No free lunch theorems for optimization. IEEE Trans. Evol. Comput. **1**(1), 67–82 (1997)
41. Woźnica, K., Biecek, P.: Does imputation matter? Benchmark for predictive models. arXiv preprint arXiv:2007.02837 (2020)
42. Yeh, I.C.: Modeling of strength of high-performance concrete using artificial neural networks. Cem. Concr. Res. **28**(12), 1797–1808 (1998)
43. Yoon, J., Jordon, J., Schaar, M.: Gain: missing data imputation using generative adversarial nets. In: International Conference on Machine Learning, pp. 5689–5698. PMLR (2018)
44. Zagatti, F.R., et al.: Metaprep: data preparation pipelines recommendation via meta-learning. In: 2021 20th IEEE International Conference on Machine Learning and Applications (ICMLA), pp. 1197–1202. IEEE (2021)
45. Zou, Y., An, A., Huang, X.: Evaluation and automatic selection of methods for handling missing data. In: 2005 IEEE International Conference on Granular Computing, vol. 2, pp. 728–733. IEEE (2005)

Pipeline for Semantic Segmentation of Large Railway Point Clouds

Hugo Gabrielidis[1,2(✉)], Filippo Gatti[1], and Stephane Vialle[2,3]

[1] CentraleSupélec, ENS Paris-Saclay, CNRS, LMPS - Laboratoire de Mécanique Paris-Saclay UMR 9026, Université Paris-Saclay, 91190 Gif sur Yvette, France
{hugo.gabrielidis,filippo.gatti}@centralesupelec.fr
[2] CNRS, LISN - Laboratoire interdisciplinaire des sciences du numérique, Université Paris-Saclay, 91405 Orsay, France
[3] CentraleSupélec, 91192 Gif sur Yvette, France
stephane.vialle@centralesupelec.fr

Abstract. This paper presents a novel deep-learning pipeline to segment large railway datasets with minimal manual annotation, notoriously time consuming. The pipeline adapts DINOv2 [11] for labeling point clouds, with tailored self-distillation pre-training and fine-tuning. The adopted transformer architecture successfully generalizes to multiple railway datasets, with a lightweight pipeline that outperforms manual labeling speed by a factor of 6, despite requiring a final segmentation check and correction. This groundbreaking achievement bridges the gap between the need for annotated point clouds in railway industry and the lack of publicly available annotated datasets.

Keywords: 3D Point Cloud · Deep Learning · Semantic Segmentation · Large Scale data

1 Introduction

Semantic Segmentation of Point Clouds in (PCSS) represents the task of assigning a semantic class to each point in the cloud, transforming it from a tensor of shape (N, c) the intensity to a tensor of shape $(N, 1)$ where the second dimension is a number associating the point to its class. PCSS is a critical operation in modern transportation infrastructure management [13], to enhance railway design and maintenance, including safety inspections and asset management, by reliably tracking the evolution of the infrastructure geometry. Efficient segmentation aids in detecting anomalies, planning repairs, and optimizing resource allocation. PCSS consists into annotating distinct segments in 3D unlabeled and noisy point clouds, identifying different railway structural and functional components, such as ground, track, vegetation, closure, building, catenary posts, catenary cables among others. In this sense, segmented point clouds are indispensable for railway design and maintenance, in order to track possible geometrical defects verify the viability of the track and ensure the safety of the train while traveling.

V. Julian et al. (Eds.): IDEAL 2024, LNCS 15346, pp. 167–179, 2025.
https://doi.org/10.1007/978-3-031-77731-8_16

Recent advancements in deep learning have shown great premises in improving image [9] and point clouds [10] segmentation performance. In the case of railway engineering, several studies showed good results in point cloud segmentation while using basic deep leaning architecture [12]. However, a single file corresponding to a single railway point cloud can contain millions of points, making it extremely time-consuming to manually label, approximately 1.5 h per file. Consequently, most studies focus on datasets with fewer than a hundred files. Additionally, these studies do not leverage any form of pre-training schemes. Therefore, the enormous amount of unlabeled point clouds available in railway engineering is yet to be exploited for segmentation purposes. Studies have shown that Transformers [15] are one of the best available choice to learn relevant features from unlabeled point clouds datasets [3,11].

The main objective of this paper is to present a fast automated segmentation pipeline tailored for railway infrastructure point clouds datasets. The main contributions of this work can be summarized as follows:

- The first end-to-end pipeline for railway point clouds segmentation is introduced, capable of labeling the feature sample using only a few samples.
- An adaptation of the Dinov2 [3] model is tailored to segment 3D point clouds data that achieves state-of-the-art performance on railway infrastructures datasets, proving that Transformers architecture with strong pre-training are rather efficient in the context of large point clouds semantic segmentation.

The remainder of the paper is organized as follows: Sect. 2 presents current related works. Section 3 outlines the pursued strategy to achieve state-of-the-art performance by targeting a consistent reduction of the labeling time. The different steps and issues we encountered are described and the pipeline is described in details. Section 4 details the conducted experiments. Finally, Sect. 5 summarizes our research and expends on future works.

2 Related Works

Recent works related to the methods and domain of our study can be divided into two main categories, namely: (1) Existing 3D LiDAR pre-training methods, (2) Railway LiDAR Point Cloud Semantic Segmentation, and (3) End-to-End pipeline for segmentation of unlabeled point clouds dataset

Pre-training: Given the abundance of unlabeled data available today, recent studies have focused on leveraging this data to pre-train large deep learning models, which can then be fine-tuned for various downstream tasks. For instance, in order to leverage the vast amount of text data, [5] proposed a Transformer model [15], called BERT, predicting a word randomly masked from a sentence during pre-training. In the image domain, DINO [3] demonstrates that Vision Transformer (ViTs), when pre-trained using self-distillation, reveals explicit information about an image's semantic segmentation, a feature that is less apparent while training ViTs in a supervised manner. Several studies on point clouds segmentation has already adapted akin training scheme, for their PointCAM model [14] the authors implemented self-distillation while using a network to generate the masks in an adversarial manner to generate more intricate masks. In [17] authors used a Transformer architecture for their self-distillation pre-training but added a decoder on top of their student network. However, to confirm their findings usual point cloud datasets such as ShapeNet [4] or ModelNet40 [16] are used. Those datasets contain point clouds with less than 50,000 points, which is way less than the millions of points usually presents in railways points clouds. Inspired by those works, in this paper, we aim to prove the efficiency of such self-pre-training method based on a transformer architecture in the context of semantic segmeNtation of large railway point clouds.

Railway LiDAR Point Cloud Semantic Segmentation: Despite the limited amount of deep learning approaches for Point Cloud Semantic Segmentation (PCSS) in the context of railway engineering, in recent years, a significant growth of general purpose approaches have been developed and successfully applied.

Several recent studies focus on the introduction of tailored datasets suited for deep learning on railway segmentation. The most stringent constraint is provided by the large spatial extent of point clouds referring to railway infrastructures. Typical railway point clouds often include more than 1 million points, making manual labeling of a large dataset of this type too time-consuming. According to the authors of the RailcloudHDF dataset [1], annotating a single point cloud takes around an hour and a half. Thereby the majority of railway dataset published nowadays doesn't include more than 100 point clouds [7]. Thus making it hard to train proper model that would be able to generalize well to other railway datasets. One interesting study focused on the creation of a Deep Learning architecture that would be fitted for the segmentation of Railways Point Clouds [8]. It introduces the cylindrical convolution as railway structure such as rail, catenary pole, catenary wire takes a cylindrical shape. This approach outperform classical models such as KPConv [2]. We aimed to demonstrate that, with appropriate pre-training, transformers can achieve state-of-the-art results in large-scale point cloud semantic segmentation, particularly in the context of railway segmentation. To the best of our knowledge Transformer architectures have never been applied in the context of semantic segmentation of railway data, but they will be at the heart of our approach and we will illustrate their ability to efficiently process large scale datasets.

End-to-End Pipeline for Segmentation of Unlabeled Point Clouds Dataset: Recent studies have shown significant progress in weakly supervised semantic segmentation of point clouds, reducing the need for extensive manual annotation. [18] achieved notable results with only 10% of points labeled. However, these kind studies used classical datasets and didn't address the challenge of unbalanced datasets common in railway point clouds. In this paper, we aim to create an End-to-End pipeline for railway point cloud semantic segmentation, with the goal of labeling an yet unannotated dataset while requiring the minimum amount of manual labeling.

3 Methodology

3.1 Pipeline Presentation

The objective of our research is to segment the current unlabeled The Société Nationale des Chemins de fer Français (SNCF) [French National Railway Company] dataset (see Sect. 3.2) and to develop a model that automates the process, thus greatly reducing the need for manual intervention in a more human-free pipeline. To this end, we have designed the pipeline illustrated in Fig. 1:

1. **Pre-training:** in order to leverage the vast amount of available yet unlabeled 3D point cloud, e.g., the LIDARHD dataset[1], we leveraged a transformer self-distillation pre-training scheme. Originally introduced in [3], this self-pre-training scheme encourages the transformer architecture to learn important features beforehand, leading to a faster and better convergence during the fine-tuning phase. This method solely uses unlabeled data, and has been proven capable of outperforming supervised training scheme allowing the transformer to expose relevant features in the attention heads to facilitate the downstream tasks [3]. Section 3.3 outlines details on the methods used for the pre-training.
2. **Fine-tuning:** We tried fine tuning the pre-trained model with several methods: unsupervised learning, weakly supervised learning and classical semantic segmentation fine tuning. As we first tried to create an end-to-end pipeline without the need of any labeling, we turned to the latest unsupervised GrowSP approach [19]. However, this method appears to not be well-suited in the case of extremely large point clouds as the metrics reached were judge insufficient. As unsupervised learning seems to not be adequate for our use-case, we decided to switch to classical fine tuning while still having as goal for our pipeline to need the minimal amount human intervention. To this end we make use of the SNCF labeled dataset (see Sect. 3.2).
3. **Classification on the SNCF dataset:** After the fine-tuning we apply our model to our target SNCF unlabeled dataset. We use the SNCF dataset as our primary labeling target. However, this pipeline can be generalized to other point clouds datasets across different domains.
4. **Correction:** After applying our model on the unlabeled SNCF dataset, we assessed its performance. One of the metrics we put an emphasis on was the time taken to correct the model error against the time taken to manually label a raw point. The correction process took us approximately 15 min per files (model segmentation and manual correction), while labeling a raw point cloud file usually takes more than one hour and half. Thus making the labeling

[1] Charles Gaydon, IGN. Myria3D: Deep Learning for the Semantic Segmentation of Aerial Lidar Point Clouds: https://github.com/IGNF/myria3d (2022).

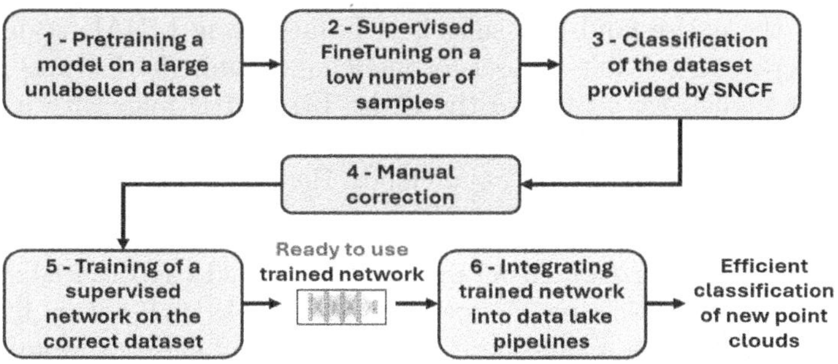

Fig. 1. Semantic Segmentation of Point Cloud Pipeline using few labeled samples

process 6 times faster. For instance, manual labeling of the SNCF unlabeled dataset (≈1350 point clouds) takes around 2025 h, against the 338 h deemed necessary to correct our pre-trained model. Manual annotation was conducted using the CloudCompare software (CloudCompare, 2022).

5. **Training a supervised network on the corrected dataset:** After labeling the whole dataset using our models to reduce time, we plan to re-train a final model on the whole labeled dataset (a combination of the previously unlabeled SNCF dataset, labeled at step 4, and the original SNCF labeled dataset, described in Sect. 3.2), which will become our final model. We will then compare the results of this final model with the one used in the first supervised training session.
6. **Integrating trained network into data lake pipelines:** The final model will be integrated into a data lake pipeline to automatically segment incoming point clouds from various SNCF projects. This process includes monitoring the model's performance to ensure its robustness against potential data drift as railway structure quickly evolves with times. The trained model on the complete labeled dataset is planned to be deployed on edge devices for real-time segmentation (leveraging knowledge distillation [6] and quantization).

We are presently focused on the correction phase (step 4) of the pipeline and are manually labeling our datasets by correcting the model errors. We plan to monitor the improvement in model performance as more labeled data become available.

3.2 Datasets

For our study, we considered several datasets for our pre-training and several test datasets to measure how efficient our pre-training approach is in the context of labeling large railways point cloud datasets. The datasets considered and the reasons for our choices are detailed below.

LIDARHD: A dataset created by the French National Institute for Geographic and Forestry Information (IGN). This institute produces and distributes 3D mapping of the entire land and subsoil of France using LIDAR scan, in `.las` format. To create a suitable dataset for modeling meaningful features using our Transformer model, we compared the entire LIDARHD map with a `.shp` file provided by SNCF to extract areas around railways. As a result, we collected a total of 10,000 unlabeled railway point clouds. Hereafter, we refer to these 10,000 files as the *LIDARHD dataset* (730 GB).

SNCF Unlabeled Dataset: In our project, we received a substantial dataset we needed to annotate, consisting of approximately 2000 LIDAR (`.las`) files. These files were collected from a rail-borne scan of a railway network around Nice in the south of France. The dataset, provided without any accompanying labels, included files of varying sizes. To ensure consistency with established conventions in other railway datasets [1,8], we opted to retain only point clouds containing over 100.000 points. Each file was manually inspected to determine whether it depicted a railway scene. Scenes from train stations and other irrelevant areas were excluded. As a result, we curated a collection of 1,365 files relevant to our task. We refer to those 1,365 files as the *unlabeled SNCF dataset* (135 GB).

SNCF Labeled Dataset: After unsuccessful attempts at obtaining significant results using unsupervised learning on the SNCF unlabeled dataset, we sought datasets containing railway point clouds created with similar LIDAR scanning methods. We requested the Saint-Etienne Railborne dataset from the Data3D team, associated with the author of [8], since those data were created using similar scans as the SNCF unlabeled dataset. This dataset contains approximately 40 labeled point clouds. We refer to this dataset as the *SNCF labeled dataset.*

Rail3D: As a new and easy to access dataset, the Rail3D [7] dataset is ideal to test the efficiency of our pre-training pipeline. This dataset concern three countries: Hungary, France, and Belgium, with a total length of almost 5.8 km and approximately 288 million points. We define the Rail3D dataset as the combination of the SNCF and Hungarian Mobile Laser Scanning (HMLS) datasets described in the Rail3D paper.

3.3 Self Supervised Pre-training and Fine Tuning

To exploit the large unlabeled datasets available as usual labeled railway datasets are either too small or unpublished because of privacy issue, we decided to use the 10,000 unlabeled point clouds extracted from the LIDARHD dataset. To this end, we train a transformer encoder using a self-distillation method. The transformer encoder, referred to as the *student*, is trained alongside a model of similar architecture called the *teacher* [11]. The teacher's weights are updated based on an exponential moving average (over the epochs) of the student's ones.

After implementing the pre-processing steps outlined in Sect. 4.1, the teacher network is provided with a sequence derived from the tokenization and embedding of the global crop. Concurrently, the student network receives a masked version of the encoding of the local crop. These inputs are then passed over to their respective networks, resulting in the generation of two latent representations

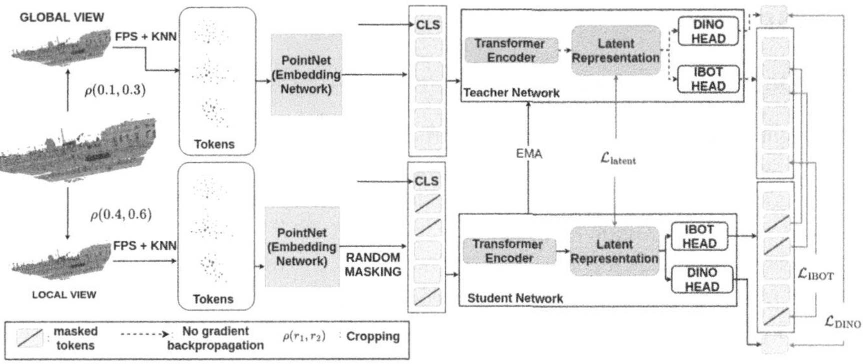

Fig. 2. Self supervised distillation pipeline applied to railway point cloud

of the point cloud. To compare the two latent representations, we incorporate three different losses inspired by [11], namely the L_{latent}, IBOT and DINO losses. L_{latent} is the mean square error directly computed on the latent representation of the two networks.

The IBOT loss, also known as the patch-level loss, compares features extracted by the student from the masked sequence, with the ones extracted by the teacher from the corresponding unmasked patch [20]. Both features are passed through a multi-layer perceptron (MLP) before computing their cross-entropy, witch read:

$$\mathcal{L}_{IBOT} = -\sum_{i=1}^{M} p_{ti} \log p_{si} \tag{1}$$

On the other hand, the DINO loss, also referred as to objective level loss, compares the two networks outputs class tokens (CLS), associated to their respective latent representations. The comparison is performed through a cross-entropy loss between the Multi Layer Perceptron (MLP) heads with a softmax activation applied to each of the two CLS, according to the following expression:

$$\mathcal{L}_{CLS} = -p_{t_{CLS}} \log p_{s_{CLS}} \tag{2}$$

The pre-training loss is a weighted sum of the three. The entire pre-training framework is schematized in Fig. 2. As railway point clouds are greatly imbalanced [1,8], a penalty is applied to the loss related to the least populated classes, namely $\rho = \sqrt[2.5]{\frac{N_{\max}}{N_c}}$, with Nc being the he percentage of points belonging to class $\mathcal{C}$ the number of points within the training sets. Each parameters is used as a coefficient for the class cross-entropy loss. Fine-tuning is performed using the trained student network, on top of which we add a segmentation head, i.e., a 1D convolutional network with ReLU activation and batch normalization layers, in order to transform the latent space encoding created by the student network.

4 Experiments

4.1 Data Preprocessing

Our point clouds processing was inspired from classical schemes adopted in the railwail industry see [1]. Each point cloud dataset was down-sampled to a 0.05 m spaced grid, addressing varying point density due to the scanned environment and the surveying wagon's speed.

The spatially varying intensity value $I(x, y, z)$ was normalized based on its highest possible value (65535), where coordinates were normalized on $[0, 1]^3$. As railways point clouds are deeply unbalanced [8], we applied data augmentation on files containing the least represented classes. For instance, point clouds in the SNCF labeled dataset were rotated, duplicated and white noise was added to the point clouds containing catenary pole, catenary wires and building classes. During the pre-training phase, two different views of augmented point clouds were instantiated, obtained by cropping the initial one. This cropping technique randomly chooses a center point within the cloud and retains its k nearest neighbors, with $k = \lfloor (1 - c)N \rfloor$ where N is the number of points in the point cloud and c the randomly chosen cropping ratio. This approach preserves local spatial relationships within the cropped region, offering different perspectives of the point cloud data at varying scales. The global view is obtained by cropping with a random ratio between 0.1 and 0.3, while the local view is obtained by cropping with a random ratio between 0.4 and 0.6. In order to adapt the input size to the transformer architecture, we utilized the furthest point sampling (FPS) algorithm to select 500 central points. Subsequently, we employed the K-nearest neighbors (KNN) algorithm to retrieve the 100 nearest neighbors for each central point, transforming the original input into a sequence of [500, 100, 4], as required by the transformer model. We also apply random masking to the student patches.

4.2 Experimental Setup: Time Requirements and Training Details

We deemed necessary to underline the computational cost of the present study, as it has rarely been discussed.

1. Pre-training: The majority of our efforts were focused on optimizing the pre-training schema, as this stage was pivotal for all subsequent studies. After implementing and evaluating various strategies, we ultimately selected the approach detailed in Sect. 3.3. The complete pre-training phase, conducted using Distributed Data Parallelism on three A100 GPUs of Paris-Saclay University super computing facility, required approximately 400 GPU hours, with the final training run lasting 24 h.
2. Fine-tuning: To prioritize the evaluation of our pre-training scheme's efficiency, we did not conduct hyper-parameter optimization for fine-tuning on the different test dataset. Both our model and the baseline were trained on three NVIDIA A100 GPUs for 24 h.

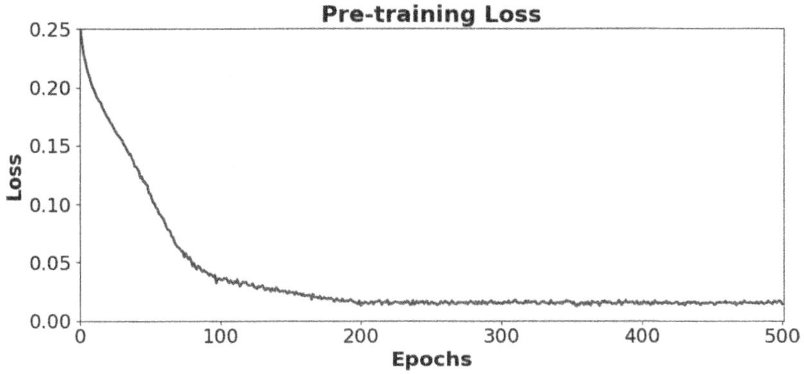

Fig. 3. Pre-training final loss curve

For each model tested, we employed a batch size of 8 and an Adam optimizer with an initial learning rate of $3e10^{-4}$, coupled with a cosine learning rate scheduler.

4.3 Results

Since the pre-training process performance cannot be directly be assessed by metrics, we adapted our pre-training hyperparameters each times the model performance were judged insufficient during the fine-tuning. In Fig. 3, we observe that, the testing loss curves achieves during the final pre-training of our model

In order to assess our model performance we took into consideration several metrics. The first relevant metrics to our task the hypothetical time taken to label the entire unlabeled SNCF dataset, comparing a manual labeling of the entire dataset against a labeling consisting of correcting the model error. The others more commonly used metrics in point clouds semantic segmentation are as follow:

- **Mean Overall Accuracy (mOA):** This metric takes into account the accuracy of all the classes and computes their mean.
- **Intersection over Union (IoU):** For each class we compute the overlap between the predicted segmentation and the ground truth segmentation.

$$\text{IoU} = \frac{\text{Nb. of points of a class correctly classified}}{\text{The total nb. of points that are classified as belonging to that class}}$$

- **Mean Intersection over Union (mIoU):** Mean of the IoU over the classes.

After our first fine-tuning we assessed the time taken to label 10 Point Clouds randomly taken from our SNCF unlabeled dataset of 1365 files. In order to manually annotate a railway point cloud we make use of the CloudCompare software (CloudCompare, 2022). It took us approximately 90 min to annotate a raw point cloud. Using our fine-tuned model first to annotate and then manually correct it's error drastically reduce this time to around 15 min per files, thus making the

(a) Segmentation truth

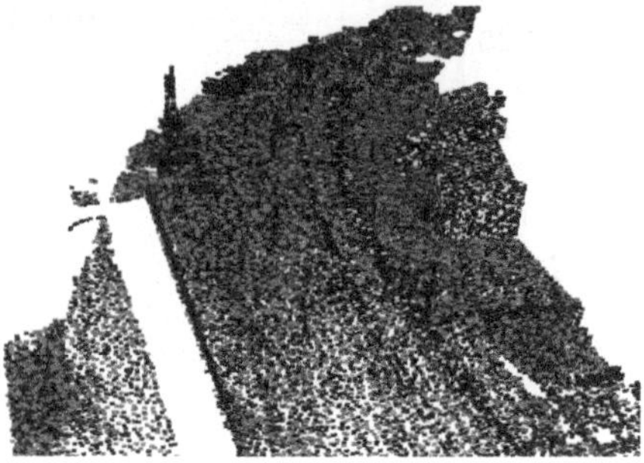

(b) GrowSP

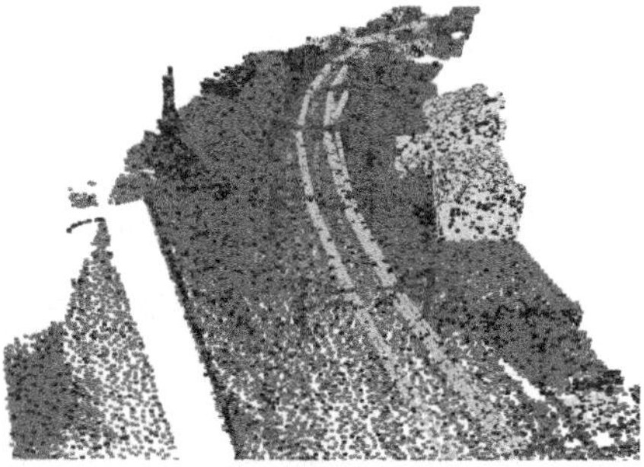

(c) PointNet++

(d) KPConv

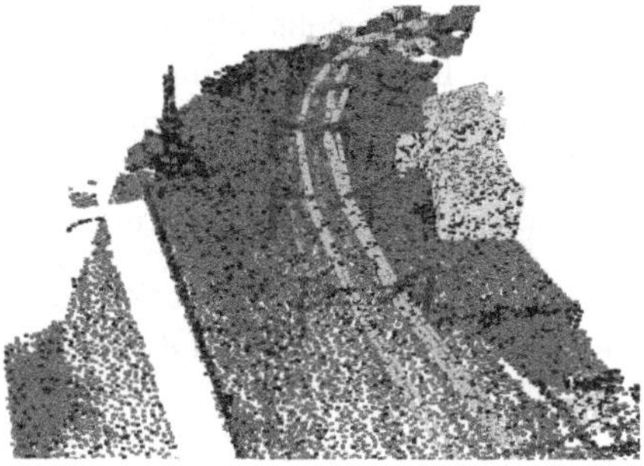

(e) Ours without pre-training

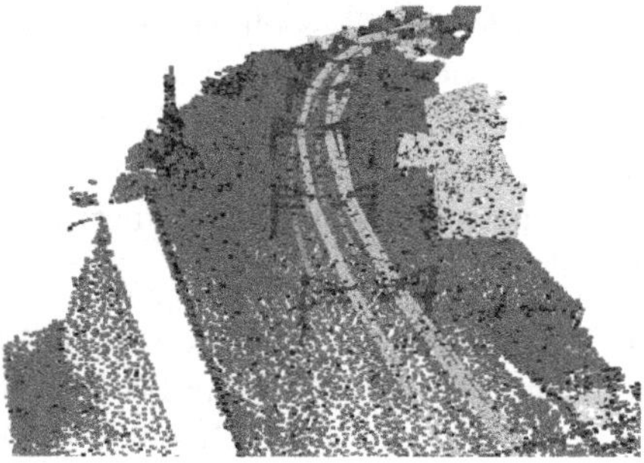

(f) Ours with pre-training

Fig. 4. Comparison of segmentation model on the SNCF labeled dataset, with: Ground in brown, Rail in bright green, Vegetation in green, Fence in red, Building in yellow, Catenary Pole in blue, Catenary Wire in cyan, Structures in grey, Environment in pink. (Color figure online)

process 6 times faster. In the context of an application to the unlabeled SNCF dataset, which contains 1365 files this would mean an improvement of $\approx$ 70 days of manual labeling.

Table 1. Comparison of performance between our approaches and other classical models on the SNCF labeled dataset, mOA (%) and mIoU (%)

Method	mOA	mIoU	Ground	Rail	Vegetation	Fence	Building	Cat. Pole	Cat. Wire	Struct.	Env.
GrowSP	30.3	23.4	48.3	35.6	43.4	5.1	21.3	14.1	16.2	22.9	4.0
PointNet++	87.4	71.5	89.3	77.6	84.2	35.2	70.4	82.9	92.4	85.1	26.1
KPConv	93.2	74.7	95.3	83.6	90.1	32.3	75.2	86.4	**96.2**	85.6	27.5
Ours (no pre-training)	87.2	71.1	90.7	75.3	87.5	27	75.8	86.1	87.3	84.7	25.3
Ours	**94.3**	**78.8**	**95.4**	**87.2**	**96.6**	**38.1**	**84.1**	**88.3**	93.2	**94.6**	**31.7**

Table 2. Comparaison of performance between our approaches and other classical model on the Rail3D dataset, mOA (%) and mIoU (%)

Method	mOA	mIoU	Ground	Vegetation	Rail	Poles	Wires	Signaling	Fences	Installation	Building
GrowSP	33.2	25.6	51.3	36.6	47.1	38.6	21.3	3.1	12.1	15.1	5.3
PointNet++	95.0	80.1	92.3	90.1	75.7	85.7	94.3	74.5	79.6	36.8	92.1
KPConv	96.2	85.3	91.2	92.6	75.9	**94.2**	**98.0**	76.5	**95.8**	47.6	95.8
Ours (no pre-training)	94.3	79.3	90.0	87.3	71.5	88.7	93.1	67.4	70.5	52.6	92.5
Ours	**98**	**89.6**	**93.7**	**98.0**	**88.4**	92.1	97.6	**87.4**	94.2	**56.4**	**98.7**

We then compared our pre-trained model (after step 1) with famous state of the art Deep Learning architecture such as PointNet++ and KPConv, we also documented the results we obtained while testing the unsupervised GrowSP [19] on those datasets. We compare those model on the following datasets: the labeled SNCF dataset and the Rail3D dataset. We split each dataset into a training and testing set with a 80%/20% ratio. The results are reported in Tables 1 and 2.

As seen in both results Tables 1 and 2 our approach outperforms classical point cloud segmentation network. In both cases the unsupervised GrowSP shows good results for an unsupervised approach but not satisfactory enough to allow faster labeling. The classic transformer architecture without pre-training already achieved correct results. We observe that using the self-distillation pre-training method leads a 7.7% and 10.3% improvement in mIoU metrics for the SNCF labeled dataset and the Rail3D dataset, proving the efficiency of this pre-training scheme.

In Fig. 4, the results of the different models on a point cloud from the SNCF labeled dataset are shown. The points marked in black indicate the model errors. Segmentation quality varies across methods. GrowSP produces noisy results with poor boundaries. PointNet++, the proposed method without pre-training and KPConv show improvement but still have inconsistencies. Our pre-trained method achieves the best results, closely matching the ground truth.

5 Conclusions

In this paper, we presented an end-to-end pipeline capable of segmenting large railway point cloud datasets using only a few labeled ones. Dinov2 architecture was adapted and pre-trained on LIDAR data. The pre-trained model was fine-tuned on multiple railway datasets, surpassing baseline performance in semantic segmentation tasks. The pre-training pipeline have proven itself well suited for semantic segmentation on large railway point cloud data, paving the way to labeling the entire SNCF dataset and embark the model onto edge devices.

Acknowledgment. We would like to extend our gratitude to the DATA3D team at SNCF for providing us with the SNCF-labeled dataset. Computations were performed on Ruche, Paris-Saclay University supercomputer facility.

References

1. Abid, M., Teixeira, M., Mahtani, A., Laurent, T.: RailCloud-HdF: a large-scale point cloud dataset for railway scene semantic segmentation. In: Proceedings of the 19th International Joint Conference on Computer Vision, Imaging and Computer Graphics Theory and Applications. SCITEPRESS - Science and Technology Publications, Rome, Italy (2024)
2. Boulch, A.: Convpoint: continuous convolutions for point cloud processing. Comput. Graph. **88** (2020)
3. Caron, M., et al.: Emerging properties in self-supervised vision transformers. In: 2021 IEEE/CVF International Conference on Computer Vision (ICCV) (2021)
4. Chang, A.X., et al.: Shapenet: an information-rich 3d model repository (2015)
5. Devlin, J., Chang, M.W., Lee, K., Toutanova, K.: BERT: pre-training of deep bidirectional transformers for language understanding. In: Proceedings of the 2019 Conference of the North American Chapter of the Association for Computational Linguistics: Human Language Technologies (Volume 1: Long and Short Papers) (2019)
6. Hinton, G.E., Vinyals, O., Dean, J.: Distilling the knowledge in a neural network. CoRR abs/1503.02531 (2015)
7. Kharroubi, A., Ballouch, Z., Hajji, R., Yarroudh, A., Billen, R.: Multi-context point cloud dataset and machine learning for railway semantic segmentation. Infrastructures **9**(4) (2024)
8. Manier, A., Moras, J., Michelin, J.C., Piet-Lahanier, H.: Railway LIDAR semantic segmentation with axially symmetrical convolutional learning. ISPRS Ann. Photogrammetry Remote Sens. Spat. Inf. Sci. **V-2-2022**
9. Minaee, S., Boykov, Y., Porikli, F., Plaza, A., Kehtarnavaz, N., Terzopoulos, D.: Image segmentation using deep learning: a survey. IEEE Trans. Pattern Anal. Mach. Intell. **44**(7) (2022)
10. Nguyen, A., Le, B.: 3D point cloud segmentation: a survey. In: 2013 6th IEEE Conference on Robotics, Automation and Mechatronics (RAM) (2013)
11. Oquab, M., et al.: DINOv2: learning robust visual features without supervision. Trans. Mach. Learn. Res. (2024)
12. Soilán, M., Nóvoa, A., Sánchez-Rodríguez, A., Riveiro, B., Arias, P.: Semantic segmentation of point clouds with pointnet and KPConv architectures applied to railway tunnels. ISPRS Ann. Photogrammetry Remote Sens. Spat. Inf. Sci. **V-2-2020**
13. Soilán, M., Sánchez-Rodríguez, A., del Río-Barral, P., Perez-Collazo, C., Arias, P., Riveiro, B.: Review of laser scanning technologies and their applications for road and railway infrastructure monitoring. Infrastructures **4**(4) (2019)
14. Szachniewicz, M., Kozłowski, W., Stypułkowski, M., Zięba, M.: Self-supervised adversarial masking for 3D point cloud representation learning (2023)

15. Vaswani, A., et al.: Attention is all you need. In: Advances in Neural Information Processing Systems, vol. 30. Curran Associates, Inc. (2017)
16. Wu, Z., et al.: 3D ShapeNets: A Deep Representation for Volumetric Shapes (2015). http://arxiv.org/abs/1406.5670
17. Zeid, K.A., Schult, J., Hermans, A., Leibe, B.: Point2vec for self-supervised representation learning on point clouds. In: Pattern Recognition (2024)
18. Zhang, Y., Li, Z., Xie, Y., Qu, Y., Li, C., Mei, T.: Weakly supervised semantic segmentation for large-scale point cloud (2022). https://arxiv.org/abs/2212.04744
19. Zhang, Z., Yang, B., Wang, B., Li, B.: Growsp: unsupervised semantic segmentation of 3D point clouds. In: Proceedings of the IEEE/CVF Conference on Computer Vision and Pattern Recognition (CVPR) (2023)
20. Zhou, J., et al.: iBOT: Image BERT Pre-training with Online Tokenizer (2022)

Preliminary Investigation on Machine Learning and Deep Learning Models for Change of Direction Classification in Running

Pranay Jaiswal[1], Abhishek Kaushik[1](✉), Fiona Lawless[1], Tiago Malaquias[2], and Fergal McCaffery[1]

[1] Dundalk Institute of Technology, Dundalk, Ireland
{pranay.jaiswal,abhishek.kaushik,fiona.lawless,fergal.mccaffery}@dkit.ie
[2] STATSports, Newry, Northern Ireland
t.malaquias@statsports.com

Abstract. The ability to detect, define, and classify Change of Direction (COD) movements during running plays a crucial role in sports science, as it has been widely used to assess athlete performance. Automating the process of COD classification during live games or training can provide real-time feedback. In this study, we evaluated Machine Learning (ML) and Deep Learning (DL) models for the classification of COD using accelerometers and gyroscope sensor data, and speed data were calculated from the Global Positioning System (GPS) sensor data. We hypothesized that DL algorithms classify COD better than ML classification algorithms. Comparative analysis showed that the best-performing DL and ML models showed similar behavior. Similarly, the statistical analysis observed no significant difference. This emphasized the importance of accurate model selection.

Keywords: Change of Direction · Machine Learning · Deep Learning · Sports · Running · Classification

1 Introduction

The evolution of wearable technology has significantly impacted athlete activity tracking and analysis, allowing multiple benefits to athletes and coaches. With the integration of advanced sensors and smart devices, monitoring athletes' activity has become more efficient, accurate, and discreet [33]. Wearable trackers, such as wristbands or specialized vests, provide real-time data from the activity using accelerometers, gyroscopes, magnetometers, and Global Positioning System (GPS) sensors. These sensors provide a comprehensive dataset that provides valuable insights into various activity metrics during practice sessions or live games, including steps taken, distance covered, calories burned, and movement intensity, enabling detailed analysis of athletes' game-play [10,11].

V. Julian et al. (Eds.): IDEAL 2024, LNCS 15346, pp. 180–191, 2025.
https://doi.org/10.1007/978-3-031-77731-8_17

Change of Direction (COD) movements are common across sports like football, basketball, rugby, hockey, and tennis. Because of their dynamic character, they pose a higher risk of injury [6,16,24]. The COD movement involves deceleration, change in momentum, and acceleration in the new direction [6]. On average, a player performs more than 700 COD movements during a competitive football game [5], making COD a crucial determining factor for an athlete's performance. This quantification of the COD movement heavily relied on manual annotation, which involves constant monitoring and labeling of the data, making this task costly, labor-intensive, and time-consuming [5,9,23,29].

Various Machine Learning (ML) and Deep Learning (DL) classification algorithms have been studied for Human Activity Recognition (HAR), suggesting their use could benefit COD classification [4,26,32]. However, there is still a significant research gap in COD classification; not many studies have been done for COD classification using ML and DL models. To the best of the authors' knowledge, an automated machine-learning approach that uses a GPS-derived COD angle as a feature for COD movement classification has been presented by Reilly et al. [28]. Although many studies have examined COD movements, they have not yet attempted to classify them [3].

1.1 Motivation

Optimizing COD performance enhances individual player skills and impacts team strategy and competitive outcomes in the dynamic world of sports [8,34]. Its classification while training or live games using wearable sensors is necessary to reduce the cost, time, and labor dependency. Automated approaches using COD detection algorithms with ML and DL classification models may offer solutions to mitigate this issue and provide real-time feedback to athletes and coaches.

This study aims to assess the accuracy of various ML and DL models in classifying COD movements using a novel approach for detecting COD movement while running on a new COD dataset. DL models are best suited for sequential data and perform better than ML models [18]. Our research hypothesis for this study is that DL algorithms outperform ML classification algorithms.

The Research Questions (RQ) to explore the hypothesis are given below:

1. How do ML and DL models perform on the COD dataset based on various evaluation metrics?
2. How do the best-performing ML and best-performing DL models compare in terms of accuracy when applied to COD?

The rest of the paper is organized as follows: Sect. 2 defines the methodology. Section 3 discusses the results obtained from our methodology. Section 4 discusses our research hypothesis, and Sect. 5 provides concluding remarks and future work.

2 Materials and Methods

The COD classification structured pipeline consists of five fundamental steps: data collection, COD window detection, data preprocessing, feature extraction,

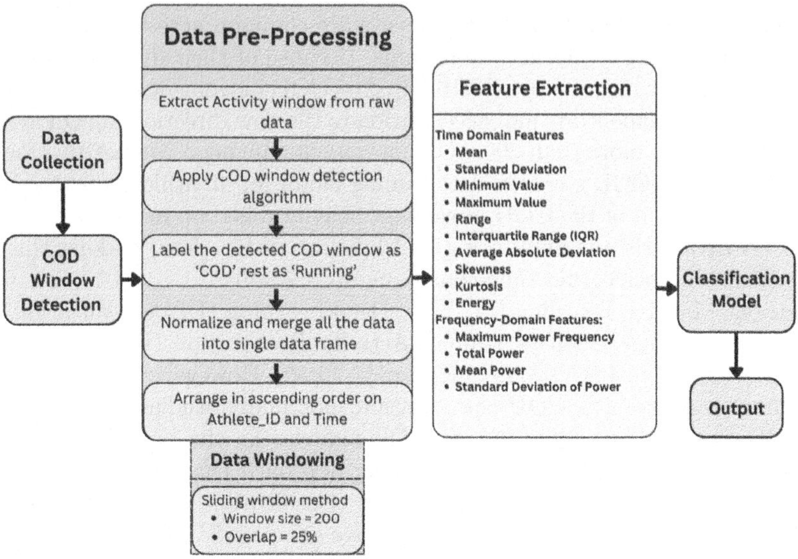

Fig. 1. Structured pipeline for COD classification.

and classification. As shown in Fig. 1, the following section provides a detailed description of each step.

2.1 Data Collection

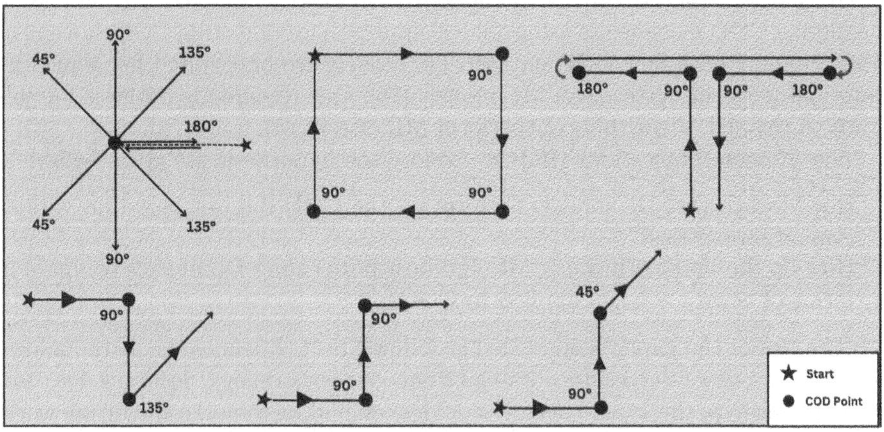

Fig. 2. Six Different COD drills used during data collection.

Data is collected in an open football field, and STATSports manages the data collection process. After collection, they provide the raw dataset for analysis. Seven athletes participated in the data collection process. Participants have to run six predetermined COD drills as shown in Fig. 2, in total consisting of a different number of repetitions of 4 different angular changes (three at 45°, eleven at 90°, three at 135°, and three at 180° COD events for each athlete throughout this process) wearing a commercially available tracking device[1]. A drone is positioned directly above the designated point of the COD area to capture all trials. This allows comparisons between different measurement devices and ensures the running paths are followed accurately. The tracking device data are extracted using STATSports software in a raw text file containing different sensor data. This study uses 'Speed' data from a GPS sensor sampled at 10 Hz, a triaxial 'Accelerometer,' and 'Gyroscope' data sampled at 100 Hz for analysis. As the athletes wear the tracking device during this data collection, each activity drill's start and end times are extracted using video analysis.

2.2 COD Window Detection

This COD window detection method uses the basic principles of deceleration and acceleration during COD movement [15]. First, the yaw angle, which measures the rotation angle around a vertical axis, is calculated using gyroscope data. A Gaussian filter is applied to both the yaw angle and the speed data to clean and smooth the data [17]. Significant changes in the filtered yaw angle that exceed 10° within a 1-s interval are identified, along with drops in speed below 10 m/s during the same period, indicating slowing down. Then, the speed increasing (acceleration) or decreasing (deceleration) is detected by analyzing changes in the first derivative of the filtered speed.

A short time window is created around the maximum changes in the yaw angle to look for patterns, such as a speed drop just before the maximum yaw angle change and specific acceleration and deceleration patterns before and after this change. The threshold for the time window is the entry speed greater than 3 m/s, and the duration of the window is less than 2 s. Finally, the angle of movement is calculated based on the largest change in the yaw angle within this window [17].

2.3 Data Preprocessing

Different data frames are initially extracted using the drill start and end time. Then, the COD detection algorithm is applied to identify the COD windows for each data frame, and then each data point is labeled as 'COD' activity for the detected window. In contrast, the rest of the data is labeled as a 'Running' activity for each participant's data. All the data is first normalized[2] and then merged into a single data frame where the 'Running' activity has 20358 and the

[1] (Apex Units, STATSport, Newry, Co. Down, Northern Ireland).

[2] (Min-Max Scaler function is used).

'COD' activity has 17066 rows of data, respectively. Then, arrange the data in ascending order based on the 'Athlete_ID' and 'Timestamp' columns [31].

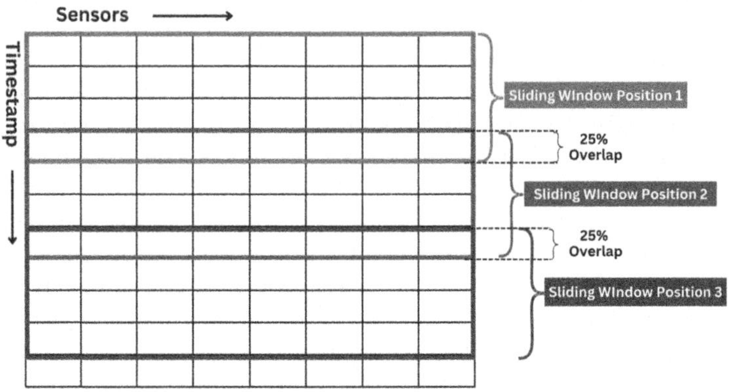

Fig. 3. Sliding window of size 200 data-points and 25% overlap.

Data Windowing. Data windowing is a fundamental preprocessing technique for effectively utilizing time-series data in ML and DL models. It involves segmenting the dataset into windows or segments to capture temporal patterns accurately. Based on observing the detected COD window, the average length is 140–170 data points. So, to cover one whole COD window in a single segment, we implement a sliding window method with a window size of 200 data points (equal to 2 s, as IMUs are sampled at 100 Hz) and a 25% overlap. This predetermined step size iterated over the dataset, as shown in Fig. 3. The accelerometer and gyroscope readings across multiple axes and speed measurements are retrieved and grouped into segments within each iterative cycle. The activity label for each segment is determined based on the 'mode' of activity observed within the window. The count of the events before and after windowing is shown in Figs. 4, 5 respectively, where initially, the 'Running' activity has 20358, and the 'COD' activity has 17066 data points, and after applying the window method 421 segments of 'Running' activity and 324 segments of 'COD' activity were formed [25].

2.4 Feature Extraction

To extract features from segmented data, we convert raw sensor readings into a set of descriptive features that represent the essential characteristics of the signal [14]. Time-domain features show patterns in the distribution and variability of sensor readings over time. Frequency-domain features detail the signal's frequency components using methods such as the Fourier or wavelet transform [2]. These characteristics enable precise representation of the underlying data for

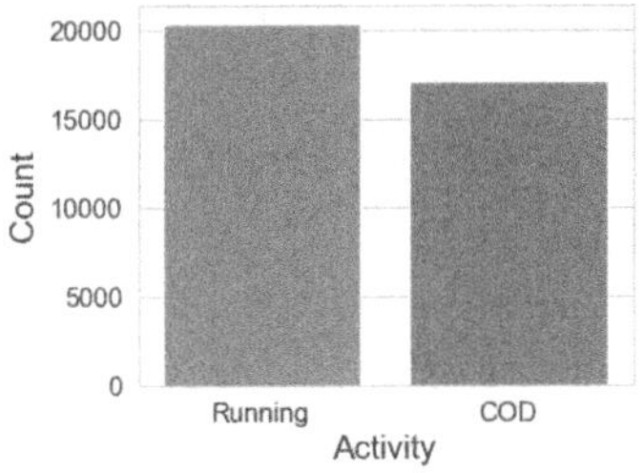

Fig. 4. Data points per label before windowing.

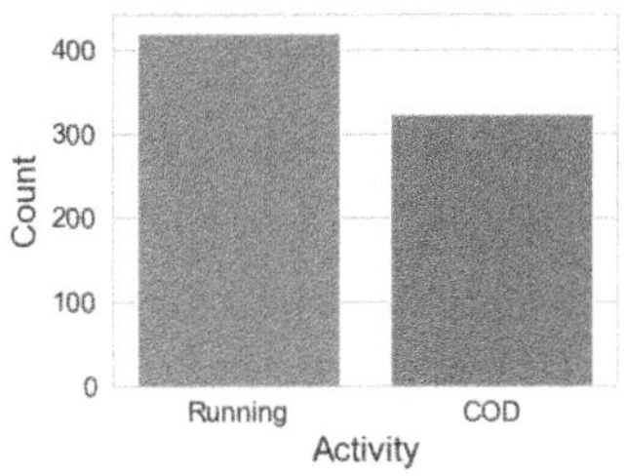

Fig. 5. Segments per label after windowing

analysis and modeling tasks by capturing crucial aspects of the signal's morphology, dynamics, and frequency content. Ten time-domain and four frequency-domain features were selected based on the literature [20]. These features were extracted from the segmented window of the triaxial accelerometer, gyroscope, and speed data, so 98 (7*(10+4)) time-domain and frequency-domain features for every window were used as the input for ML models.

DL models like Recurrent Neural Networks (RNNs) and Long Short-Term Memory (LSTM) can automatically learn time-domain and frequency-domain features from raw data. For instance, RNNs can directly extract features and capture temporal dependencies from sequential data in both the frequency and time domains [13]. This eliminates manual feature engineering and allows DL models to learn data representations best suited to the given task. We use segmented windows of raw sensor values from the triaxial accelerometer, gyroscope, and speed data as input for DL models.

2.5 Classification Models

After investigating the literature on HAR, we have selected 4 ML and 2 DL models for our analysis. These models were used due to their superior performance on online datasets, such as UCI-HAR, WISDOM, and PAMAP2. Hyperparameters for each model as shown in Table 1 are also employed based on the literature [12,19,21,28,30,35].

For the architecture of LSTM shown in Fig. 6 and RNN shown in Fig. 7 models, we use two hidden layers with 128, 64, and 128, 32 neurons, respectively, followed by dropout layers to mitigate overfitting by randomly deactivating some input units. Then, a dense, fully connected layer is used to analyze the features extracted by the hidden layers, followed by another dense layer, the final layer, to make predictions.

10-fold cross-validation is used to evaluate the classification models to ensure fair understanding and generalizability [22]. We measure the performance of each model by evaluating key classification metrics such as accuracy, precision, recall, and F1-score.

Table 1. Hyperparameters used for different ML and DL models

Model	Hyperparameters
Random Forest (RF)	criterion = 'entropy' n_estimators=100
Support Vector Machine (SVM)	kernal = 'linear'
Logistic Regression (LR)	penalty = 'l2'
Stochastic Gradient Descent (SGD)	loss = 'hinge', penalty = 'l2'
LSTM, RNN	optimizer = 'Adam' loss = 'categorical cross_entropy' learning rate = 0.001 batch_size = 128, epoch = 100

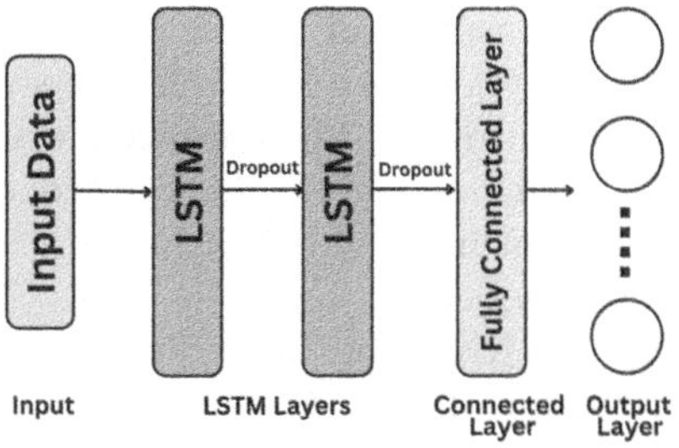

Fig. 6. Architecture of LSTM model.

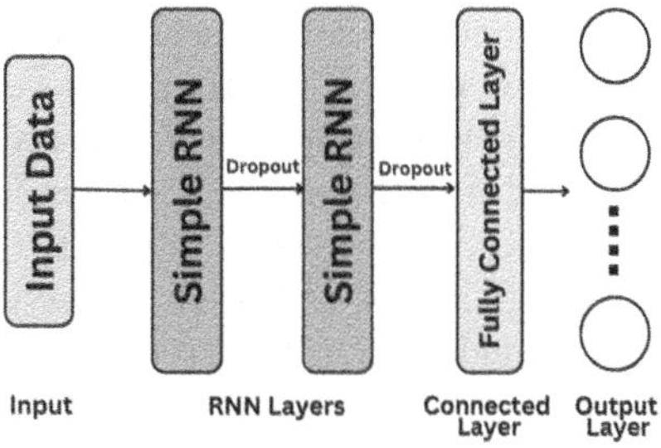

Fig. 7. Architecture of RNN model.

3 Experimental Results

The performance evaluation of ML (RF, SVM, LR, SGD) and DL (LSTM, RNN) models on the dataset for COD detection revealed promising outcomes. The dataset was divided into 80:20 ratios for training and testing (COD: 61 samples, Running: 88 samples).

The Confusion Matrices (CM) of the LSTM and RNN models are displayed in Figs. 8 and 9, respectively. Figures 10, 11, 12, and 13 on the other hand, compare the confusion matrix of RF, SVM, SGD, and LR machine learning models, respectively. The comparison of accuracy, precision, recall, and F1-score metrics for each model is shown in Table 2.

LSTM and RNN have achieved the highest accuracy of 0.7651, followed by SVM and RF at 0.7583. LR and SGD had lower accuracies of 0.7449 and 0.7248, respectively. Comparing the performance of the LSTM with the highest F1-score of 0.7059 and an accuracy of 0.7651 shows the balance in identifying True Positives (TP) while minimizing False Positives (FP), respectively. The SVM model also performs similarly to LSTM, with a strong recall and precision of 0.7049. The RNN and RF models perform with higher precision at 0.7600 and 0.7451, respectively; however, both have a similar lower recall of 0.6230 compared

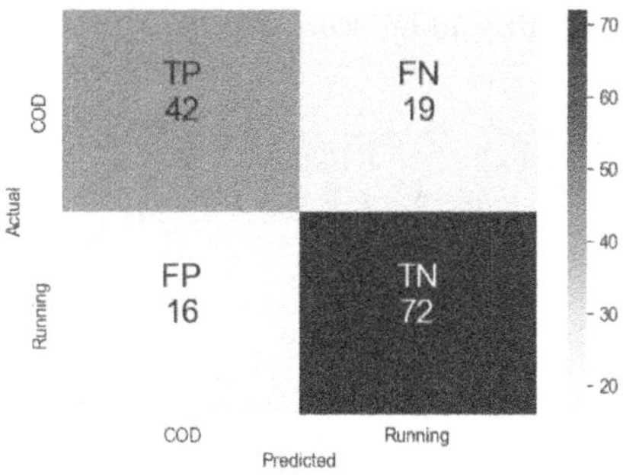

Fig. 8. CM of LSTM model.

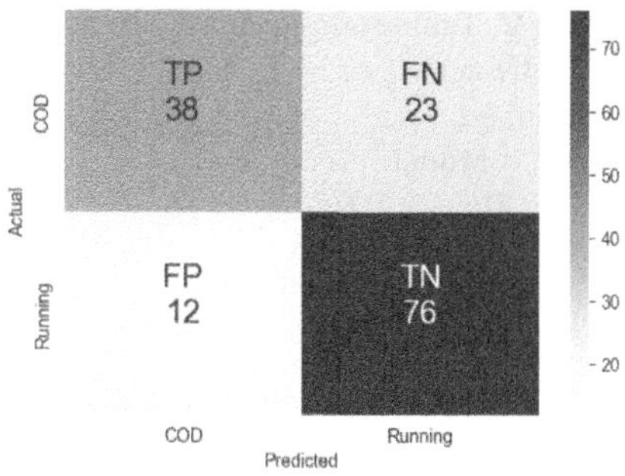

Fig. 9. CM of RNN model.

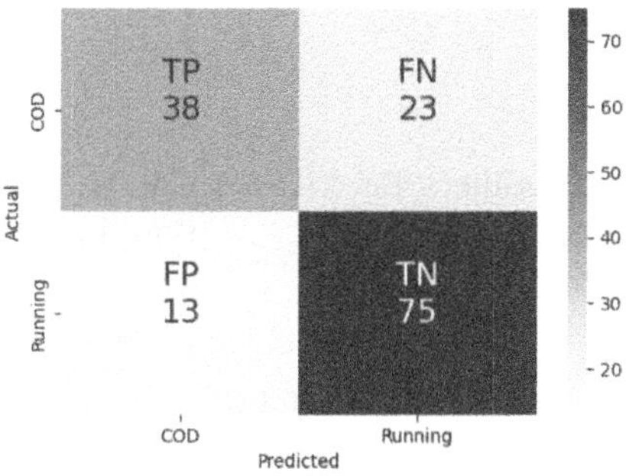

Fig. 10. CM of RF model.

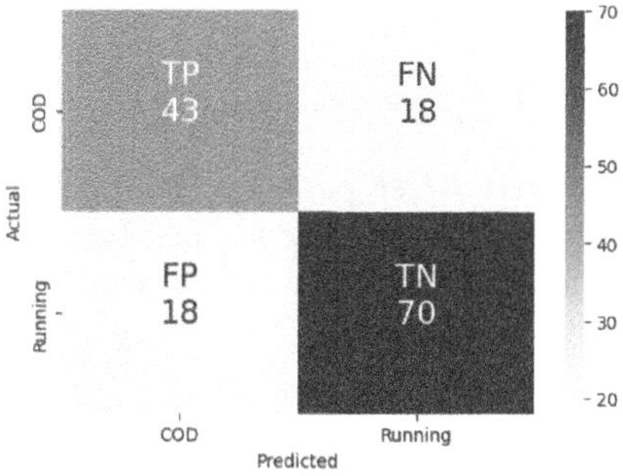

Fig. 11. CM of SVM model.

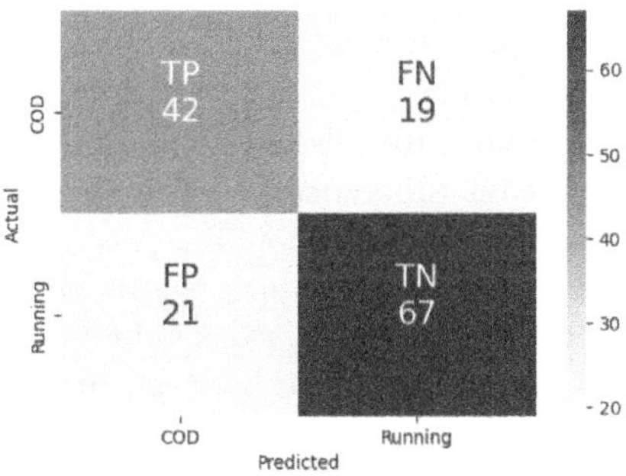

Fig. 12. CM of SGD model.

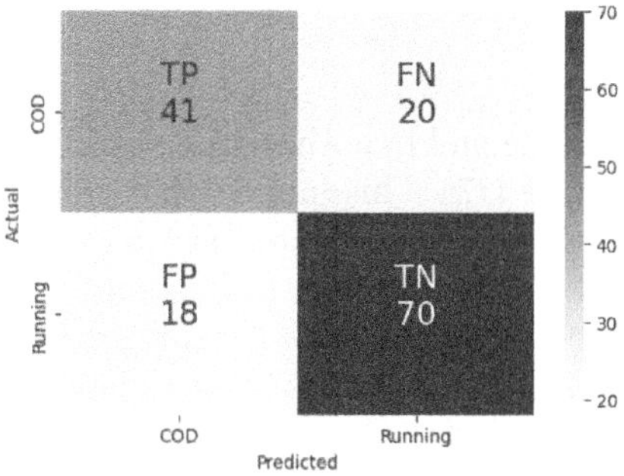

Fig. 13. CM of LR model.

with other models, meaning they detect COD with fewer FP but more False Negative (FN) detection of COD events. LR and SGD models also perform similarly with balanced evaluation metrics. These results conclude that LSTM is the best-performing DL model, and SVM is the best ML model.

Table 2. Table comparing Accuracy, Precision, Recall, and F1-score of different models for COD activity.

Model	Accuracy $\left(\frac{TP+TN}{TP+TN+FP+FN}\right)$	Precision $\left(\frac{TP}{TP+FP}\right)$	Recall $\left(\frac{TP}{TP+FN}\right)$	F1-score $\left(\frac{2TP}{2TP+FP+FN}\right)$
RF	0.7583	0.7451	0.6230	0.6786
SVM	0.7583	0.7049	0.7049	0.7049
LR	0.7449	0.6949	0.6721	0.6833
SGD	0.7248	0.6667	0.6885	0.6774
LSTM	0.7651	0.7241	0.6885	0.7059
RNN	0.7651	0.7600	0.6230	0.6847

3.1 Hypothesis Testing

We used *t-test* paired to compare the best classifiers for ML and DL models based on accuracy score, i.e., LSTM and SVM [7]. The accuracy of 10-fold cross-validation results on 80% training data is shown in Table 3.

Table 3. Table comparing Accuracy of 10 different folds.

Model	1st Fold	2nd Fold	3rd Fold	4th Fold	5th Fold	6th Fold	7th Fold	8th Fold	9^{h} Fold	10th Fold
LSTM	77.08	77.08	77.08	77.08	79.16	81.25	79.16	70.21	80.85	78.72
SVM	83.33	75.00	66.67	83.33	73.33	76.67	74.58	69.49	72.88	77.97

The null hypothesis for a paired *t-test* is that both models have equal performance [27]. This implies that there is no significant difference between the two models. The calculated *t-statistic* was -1.4047, with a corresponding *p-value* of 0.1936. We fail to reject the null hypothesis because the *p-value* is significantly higher than the chosen significance level of $\alpha = 0.05$. Therefore, we conclude that no statistically significant difference exists in the performance between models LSTM and SVM.

4 Discussion

Our research hypothesis proposed that DL algorithms outperform ML classification algorithms. We can now address our research questions and assess the viability of our hypothesis based on the results.

RQ1: How do ML and DL models perform on the COD dataset based on various evaluation metrics? The LSTM and RNN achieved the highest accuracy of 0.7651 with balanced precision, recall, and F1-score. SVM and RF also achieved almost similar accuracy of 0.7588, but RF showed lower recall values and higher precision, whereas SVM showed a balanced performance. LR and SGD models also showed comparable performance but slightly lower than others. There are differences in evaluation metrics, but not so much that it can be concluded that DL models outperform ML models.

RQ2: How do the best-performing ML and best-performing DL models compare in terms of accuracy when applied to COD? The statistical significance of the performance differences between the ML and DL models was assessed using a paired t-test. The results showed that the observed differences in performance metrics between the ML and DL models are not statistically significant.

Based on the discussion of RQs, the results do not strongly support our hypothesis. The LSTM model performs comparably with the RF model in average precision, F1 score, and accuracy. Moreover, the statistical test confirms the insignificant variations in performance. Therefore, although DL models like LSTM and RNN perform competitively, our study's results do not demonstrate that they outperform conventional ML models for COD classification. In contrast with previous HAR studies, ML and DL models have also performed similarly with datasets from the accelerometer and gyroscope sensors [12,21]. However, Abbaspour et al. [1] have presented that hybrid DL models have outperformed ML models on the PAMAP2 dataset for HAR.

5 Conclusion and Future Work

This study aims to address the gap in automated COD classification literature. In this article, we have preliminarily investigated the performance of DL and ML algorithms in classifying COD movement when running. Our findings revealed that top-performing DL-LSTM and ML-SVM exhibited comparable performance across key metrics. The statistical analysis showed no significant differences at the significance level of 0.05. This suggests that deep learning and machine learning algorithms are viable options for classification. Further work with larger datasets, different model configurations such as hybrid DL models, or a more comprehensive range of evaluation metrics may be necessary to comprehensively compare ML and DL models for COD classification. Testing the reliability of these models in an uncontrolled environment could also be explored in the future.

Acknowledgments. This work was supported with the financial support of the Science Foundation Ireland grant 13/RC/2094_P2 and co-funded under the European Regional Development Fund through the Southern and Eastern Regional Operational Programme to **Lero - the Science Foundation Ireland Research Centre for Software** (www.lero.ie). The authors thank STATSports for continuous support. The authors also thank the Regulated Software Research Centre (RSRC), where this research was conducted.

References

1. Abbaspour, S., Fotouhi, F., Sedaghatbaf, A., Fotouhi, H., Vahabi, M., Linden, M.: A comparative analysis of hybrid deep learning models for human activity recognition. Sensors **20**(19), 5707 (2020)

2. Akan, A., Cura, O.K.: Time-frequency signal processing: today and future. Digit. Signal Process. **119**, 103216 (2021)
3. Alanen, A., Räisänen, A., Benson, L., Pasanen, K.: The use of inertial measurement units for analyzing change of direction movement in sports: a scoping review. Int. J. Sports Sci. Coach. **16**(6), 1332–1353 (2021)
4. Ann, O.C., Theng, L.B.: Human activity recognition: a review. In: 2014 IEEE International Conference on Control System, Computing and Engineering (ICC-SCE 2014), pp. 389–393. IEEE (2014)
5. Bloomfield, J., Polman, R., O'Donoghue, P.: Physical demands of different positions in fa premier league soccer. J. Sports Sci. Med. **6**(1), 63 (2007)
6. Clarke, R., Mundy, P., Aspe, R., Sargent, D., Hughes, J.: Change of direction technical models: biomechanical principles (2018)
7. Demšar, J.: Statistical comparisons of classifiers over multiple data sets. J. Mach. Learn. Res. **7**, 1–30 (2006)
8. Dos' Santos, T., McBurnie, A., Comfort, P., Jones, P.A.: The effects of six-weeks change of direction speed and technique modification training on cutting performance and movement quality in male youth soccer players. Sports **7**(9), 205 (2019)
9. Faude, O., Koch, T., Meyer, T.: Straight sprinting is the most frequent action in goal situations in professional football. J. Sports Sci. **30**(7), 625–631 (2012)
10. Ferraz, A., Duarte-Mendes, P., Sarmento, H., Valente-Dos-Santos, J., Travassos, B.: Tracking devices and physical performance analysis in team sports: a comprehensive framework for research—trends and future directions. Front. Sports Active Living **5** (2023)
11. George, A.H., Shahul, A., George, A.S.: Wearable sensors: a new way to track health and wellness. Partners Univ. Int. Innov. J. **1**(4), 15–34 (2023)
12. Gjoreski, H., Bizjak, J., Gjoreski, M., Gams, M.: Comparing deep and classical machine learning methods for human activity recognition using wrist accelerometer. In: Proceedings of the IJCAI 2016 Workshop on Deep Learning for Artificial Intelligence, New York, NY, USA, vol. 10, p. 970 (2016)
13. Hewamalage, H., Bergmeir, C., Bandara, K.: Recurrent neural networks for time series forecasting: current status and future directions. Int. J. Forecast. **37**(1), 388–427 (2021)
14. Khalid, S., Khalil, T., Nasreen, S.: A survey of feature selection and feature extraction techniques in machine learning. In: 2014 Science and Information Conference, pp. 372–378. IEEE (2014)
15. Khana, H., Qasima, S.Z., Waqarb, A.: COD detection for injury prevention using sports analytics
16. Marques, J.B., Paul, D.J., Graham-Smith, P., Read, P.J.: Change of direction assessment following anterior cruciate ligament reconstruction: a review of current practice and considerations to enhance practical application. Sports Med. **50**, 55–72 (2020)
17. Meghji, M., et al.: An algorithm for the automatic detection and quantification of athletes' change of direction incidents using IMU sensor data. IEEE Sens. J. **19**(12), 4518–4527 (2019)
18. Mekruksavanich, S., Jitpattanakul, A.: Lstm networks using smartphone data for sensor-based human activity recognition in smart homes. Sensors **21**(5), 1636 (2021)
19. Müller, P.N., Müller, A.J., Achenbach, P., Göbel, S.: IMU-based fitness activity recognition using CNNs for time series classification. Sensors **24**(3), 742 (2024)

20. Munguia Tapia, E.: Using machine learning for real-time activity recognition and estimation of energy expenditure. Ph.D. thesis, Massachusetts Institute of Technology (2008)
21. Muralidharan, K., Ramesh, A., Rithvik, G., Prem, S., Reghunaath, A., Gopinath, M.: 1d convolution approach to human activity recognition using sensor data and comparison with machine learning algorithms. International Journal of Cognitive Computing in Engineering **2**, 130–143 (2021)
22. Nayak, S., Panigrahi, C.R., Pati, B., Nanda, S., Hsieh, M.Y.: Comparative analysis of HAR datasets using classification algorithms. Comput. Sci. Inf. Syst. **19**(1), 47–63 (2022)
23. Nedelec, M., McCall, A., Carling, C., Legall, F., Berthoin, S., Dupont, G.: The influence of soccer playing actions on the recovery kinetics after a soccer match. J. Strength Cond. Res. **28**(6), 1517–1523 (2014)
24. Nimphius, S., Callaghan, S.J., Bezodis, N.E., Lockie, R.G.: Change of direction and agility tests: challenging our current measures of performance. Strength Cond. J. **40**(1), 26–38 (2018)
25. Pienaar, S.W., Malekian, R.: Human activity recognition using LSTM-RNN deep neural network architecture. In: 2019 IEEE 2nd Wireless Africa Conference (WAC), pp. 1–5. IEEE (2019)
26. Ramanujam, E., Perumal, T., Padmavathi, S.: Human activity recognition with smartphone and wearable sensors using deep learning techniques: a review. IEEE Sens. J. **21**(12), 13029–13040 (2021)
27. Raschka, S.: Model evaluation, model selection, and algorithm selection in machine learning. arXiv preprint arXiv:1811.12808 (2018)
28. Reilly, B., Morgan, O., Czanner, G., Robinson, M.A.: Automated classification of changes of direction in soccer using inertial measurement units. Sensors **21**(14), 4625 (2021)
29. Robinson, G., O'Donoghue, P.: A movement classification for the investigation of agility demands and injury risk in sport. Int. J. Perform. Anal. Sport **8**(1), 127–144 (2008)
30. Saez, Y., Baldominos, A., Isasi, P.: A comparison study of classifier algorithms for cross-person physical activity recognition. Sensors **17**(1), 66 (2016)
31. Shen, Z., Zhang, Y., Lu, J., Xu, J., Xiao, G.: A novel time series forecasting model with deep learning. Neurocomputing **396**, 302–313 (2020)
32. Singh, D., et al.: Human activity recognition using recurrent neural networks. In: Holzinger, A., Kieseberg, P., Tjoa, A.M., Weippl, E. (eds.) CD-MAKE 2017. LNCS, vol. 10410, pp. 267–274. Springer, Cham (2017). https://doi.org/10.1007/978-3-319-66808-6_18
33. Tariq, M.U.: Advanced wearable medical devices and their role in transformative remote health monitoring. In: Transformative Approaches to Patient Literacy and Healthcare Innovation, pp. 308–326. IGI Global (2024)
34. Tous-Fajardo, J., Gonzalo-Skok, O., Arjol-Serrano, J.L., Tesch, P.: Enhancing change-of-direction speed in soccer players by functional inertial eccentric overload and vibration training. Int. J. Sports Physiol. Perform. **11**(1), 66–73 (2016)
35. Wundersitz, D.W., Josman, C., Gupta, R., Netto, K.J., Gastin, P.B., Robertson, S.: Classification of team sport activities using a single wearable tracking device. J. Biomech. **48**(15), 3975–3981 (2015)

Efficient Radar Scheduling Using Genetic Algorithms and Stochastic Heuristic Initialization

Tien Minh Dam, Long Viet Truong(✉), Hung Viet Bui, Tuan Anh Nguyen, and Tiem Manh Nguyen

Viettel High Technology Industries Corporation, Hanoi, Vietnam
{tiendm9,longtv28,hungbv,tuanna63,tiemnm}@viettel.com.vn

Abstract. Optimal radar scheduling is essential for ensuring the safety and security of critical areas. It requires a balance between maximizing coverage and managing operational constraints. This study addresses the complex Radar System Scheduling Problem, which involves efficiently allocating various radars across multiple sessions to maximize coverage while minimize the number of radars across all sessions. We propose a specialized Genetic Algorithm with custom operators to handle these constraints effectively. Additionally, we introduce a Stochastic Heuristic Initialization method to dynamically prioritize radar assignments, enhancing the flexibility and robustness of the scheduling process. By combining heuristic-based initialization, a constraint-preserving recombination operator, and a customized mutation method, our approach generates high-quality solutions that respect the problem's many constraints. Our research demonstrates the adaptability of genetic algorithms to real-world problems, showing significant improvements in scheduling efficiency and operational flexibility.

Keywords: Radar System Scheduling Problem · Genetic Algorithm · Heuristic Initialization · Stochastic Method · Industrial Decision Making

1 Introduction

Radar systems have become indispensable tools across various domains including air traffic control, weather forecasting, defense surveillance, and remote sensing. The efficient operation of these radar systems relies heavily on effective scheduling of radar tasks, which involves allocating scarce radar resources and operational requirements. The Radar System Scheduling Problem (RSSP) emerges as a critical challenge in optimizing the utilization and performance of defined radar systems over a set of time blocks. The problem is typically formulated as a multi-objective optimization task, focusing on minimizing the number of active radars while maximizing the coverage area.

In recent decades, considerable research in radar systems optimization has centered on the Radar System Placement Problem (RSPP), which addresses the

V. Julian et al. (Eds.): IDEAL 2024, LNCS 15346, pp. 192–201, 2025.
https://doi.org/10.1007/978-3-031-77731-8_18

optimal positioning of radar stations. Although similar in objective to RSSP, RSPP does not impose strict operational scheduling constraints. Existing literature has explored various metaheuristic approaches, including Genetic Algorithms (Kurdzo & Palmer [1]), Particle Swarm Optimizations (Boudjemaa R. [2]; Li et al. [3]), to tackle these challenges effectively. Tema et al. [4] compare between Simulated Annealing and Evolutionary Algorithms (NSGA-II and MOEA/D), showing that the latter achieve better results in RSPP.

Another type of Radar optimization is Radar Task Scheduling Problem aimed at maximizing the number of tasks completed by a single radar over a specified time period. However, this is not the case of Radar System optimization. The RSSP is closely related to Employee Scheduling Problem (Glover & McMillan [5]). This scheduling problem focuses on minimizing costs by assigning staff members to shifts across different days of the week while meeting manpower efficiency demands for each shift, with each employee restricted to a limited number of work hours. Evolutionary Algorithms (EAs) have been frequently employed to tackle these challenges (Tanomaru J. [6]; Aickelin & Dowsland [7]), by leveraging their ability to explore diverse solution spaces and aligning well with this type of the problem's encoding.

Kazimipour et al. [8] stated that population initialization plays a critical role in EAs since the population size is limited while the search spaces increase with the scales of the problem, which makes harder to find optimal regions. Two types of mentioned initializers are stochastic, where every individual is diverse between seeds, and deterministic, which always generate the same population. In constraint optimization, validating violations is highly considered before optimizing fitness functions. Undetermined initialization will result in infeasible individuals in population. Consequently, the crossover or mutation processes are more prone to failure in generating feasible offspring, thereby impeding the search process by either stagnating with an invalid population or causing slower convergence. In initializing for job-shop scheduling problem, Vlašić et al. [9] uses dispatching rules, a deterministic method that prioritises the job with earliest released time at each decision point of assigning next job to a machine. On other hand, deterministic approaches can easily lead to lack of diversity, potentially causing premature convergence in evolutionary algorithms.

Inspired by practical challenges, we observed that naive or fair random initialization methods often struggle to meet the stringent constraints of the problem. As a result, early generations of solutions frequently fail to satisfy these constraints. This difficulty arises because the constraints are particularly hard to meet through random approaches. To address this issue, we introduce a specialized initialization method. This method uses heuristics to generate initial solutions that are both high in quality and minimally violate constraints. By guiding the randomization process with heuristic rules, our approach ensures that the initial population of solutions is more feasible and closer to optimal, thereby improving the overall effectiveness and efficiency of the genetic algorithm from the start.

Given these motivations, we propose methods based on Genetic Algorithms tailored to solve the Radar System Scheduling Problem. Our approach incorporates specific modifications to adapt to the unique characteristics of RSSP. We employs our proposed Stochastic Heuristic Initialization, a mixture of stochastic and derministic initializer, to enhance initial population's quality and feasibility.

2 Problem Definition

Consider $\mathcal{R} = \{R_1, R_2, ..., R_n\}$ as the set of n radars. A coverage area A_i, with respect to radar $R_i \in \mathcal{R}$, is the area surrounding R_i and has a specific geometrical shape and location. The union coverage of two areas A_i and A_j is denoted as $A_i \cup A_j$. The operation $|A'|$ of area A' represents the size measurement of A' (in m^2 or km^2).

Each radar in $\mathcal{R}$ can be classified into one of p categories $\mathcal{C} = \{C_1, C_2, ..., C_p\}$. The set of radars falling under a category $C_h \in \mathcal{C}$ is a subset of $\mathcal{R}$, denoted as $\mathcal{R}(C_h) \subset \mathcal{R}$. As a result, if $C_l \in \mathcal{C}$ and $C_l \neq C_h$ then $\mathcal{R}(C_h) \cap \mathcal{R}(C_l) = \emptyset$.

The set of m sessions $\mathcal{S} = \{S_1, S_2, ..., S_m\}$ contains consecutive time blocks in order. In each sessions, a radar can only be in one of two states: operating or closed. The coverage area of a session is the union of all coverage areas of operating radar in that session.

Decision Variables: The decision variables of the problem is represented as $X = \{x_{ik} \mid i \in \overline{1,n}; k \in \overline{1,m}\}$, where x_{ik} binary variable indicating the operational status of radar R_i during session S_k ($x_{ik} = 1$ if R_i activated in S_k, otherwise 0).

Objective Functions: The problem is characterized by two objectives:

1. *Maximize total coverage area of all sessions* (to make the objective to be minimal optimized, we use the negative equivalent);
2. *Minimize the use of radars in all sessions.*

The overall objective functions of the problem is defined as:

$$\text{Min } F(X) = [F_1(X), F_2(X)]^T \text{ s.t. } \begin{cases} F_1(X) = -\sum_{k=1}^{m} \left| \bigcup_{i=1}^{n} x_{ik} \cdot A_i \right| \\ F_2(X) = \sum_{k=1}^{m} \sum_{i=1}^{n} x_{ik} \end{cases}$$

Constraints: A feasible solution must satisfy three constraints:

1. *Resource-Specific Minimum and Maximum*: The number of operating radars of each category in a session is limited between minimum and maximum bounds of the category:

$$\forall C_h \in \mathcal{C}, \forall S_k \in \mathcal{S} : lb(C_h) \leq \sum_{R_i \in \mathcal{R}(C_h)} x_{ik} \leq ub(C_h)$$

2. *Non-consecutive operation*: A radar cannot operate in any two consecutive sessions:

$$\forall R_i \in \mathcal{R}, \forall S_k \in \mathcal{S} : x_{ik} + x_{i(k+1)} \leq 1$$

3. *Multi-tasking Operation Limit*: A radar can be operate for a limited number of sessions in a day:

$$\forall R_i \in \mathcal{R} : \sum_{k=1}^{m} x_{ik} \leq UB$$

3 Proposed Algorithm

In this section, we opted to use Genetic Algorithm (GA) as the foundational algorithm owing to its capacity to effectively optimize solutions. We describe our proposed adjustments to address the radar scheduling problem as below:

Individual Representation: An individual's chromosome is represented by a matrix of dimensions $m \times n$. A gene at position $[k, i]$ $(k \in \overline{1, m}, i \in \overline{1, n})$ is a binary value equal to x_{ik}, depicts whether radar R_i is activated in session S_k.

Initial Population: We employ a method termed Stochastic Heuristic Initialization. This involves assigning radar units to sessions based on computed priorities, considering feasibility and potential improvements over the current session state. These priorities guide the selection of suitable radars for session assignment.

The process of constructing an initial individual is described in Algorithm 1. It starts with an empty radar schedule and assigns radars to sessions sequentially. Each session assignment considers a priority list P, which estimates the probability of activating an unallocated radar for that session. If a radar assignment violates constraints or the radar is already assigned, its priority is set to zero. After assigning a radar, priorities of other unallocated radars are updated based on their potential contributions. Supposing a session S_k with already assigned radars and an unassigned feasible radar R_i, the priority function is defined as follows:

$$Prio(R_i, S_k) = \frac{\left|A_i \cup \left(\bigcup_{j=1}^{n} x_{jk} \cdot A_j\right)\right|}{\left|\bigcup_{j=1}^{n} x_{jk} \cdot A_j\right|} - 1 + \epsilon$$

Here, ϵ denotes a small positive quantity ensuring $Prio(R_i, S_k) > 0$, thereby R_i will have slight probability of allocation even if it does not contribute to the coverage area. The priority list P serves as weights in random function for radar selection in line 13. The probability of assigning radar R_i to session S_k is expressed as:

$$Prob(R_i, S_k) = \frac{P_i}{\sum_{j=1}^{n} P_j}$$

Algorithm 1. Generate new Individual in Initial Phase

Ensure: A new feasible Individual

Initialize $Individual \leftarrow [0]_{m \times n}$
for session S_k **in** set $\mathcal{S}$ **do**
  Initialize Priority list $P \leftarrow \emptyset$
  for radar R_i **in** set $\mathcal{R}$ **do**
    $P_i \leftarrow +\infty$ **if** assign R_i to S_k is feasible **else** $P_i \leftarrow 0$
    Add P_i to list P
  for category C_h **in** set $\mathcal{C}$ **do**
    Randomize an integer q in range $[lb(C_h), ub(C_h)]$
    while $q > 0$ **do**
      Randomly choose radar $R_i \in \mathcal{R}(C_h)$ by using list P as weights
      $Individual[k, i] \leftarrow 1$
      $P_i \leftarrow 0$
      for radar R_j **in** set $\mathcal{R}$ **do**
        if $Individual[k, j] = 1$ **then continue**
        $p'_j \leftarrow Prio(R_j, S_k)$ **if** assign R_j to S_k is feasible **else** $p'_j \leftarrow 0$
        $P_j \leftarrow Min(P_j, p'_j)$
      $q \leftarrow q - 1$
return *Individual*

The outcome of this initial phase yeilds a feasible population, in which each individual has favorable objective values, thus providing a foundation for the main algorithm to be rapidly convergent.

Crossover: The crossover technique employed in this algorithm adheres to the principle of uniform crossover. In this method, each column within the offspring's chromosome, which is a radar setting across sessions, is stochastically selected from either of the two parent individuals. This ensures the preservation of each radar assignment across all sessions while promoting diversification.

Mutation: For mutation operations, we randomly exchange the values between an active radar and an inactive radar belonging to the same category within a session.

4 Experimental Results

In this experiment, we use a sample datasets of a radar system with coverage area extending over Vietnam territory. The radar system has $n = 160$ radars and operates in $m = 12$ sessions. The radars is divided into 2 categories: 150 short-range radars (C_1) and 10 long-range radars (C_2). Firstly, we test in single case where number of radar usage in a session is fixed. Secondly, we increase the radar usage rate to see the change in total percentage of radar coverage.

In the first experiment, the constraint parameters is fixed as follows: The usage bounds of short-range radars are $(lb(C_1), ub(C_1)) = (15, 25)$, while those of long-range radars are $(lb(C_2), ub(C_2)) = (0, 1)$. The maximum number of sessions a radar is activated in a day is $UB = 6$. In initial phase, we consider radars in long-range category before short-range radars. The GA runs with population of 100 individuals over 100 generations.

Figure 1 depicts the solution yielding the largest coverage area post-algorithm execution. The radar schedules are evenly distributed across the area within each operated session. Within the region which covered by long-range radar, minimal intersect with short-range radars is observed. Due to the large number of short-range radars required, regions appear where their coverage areas overlap each other.

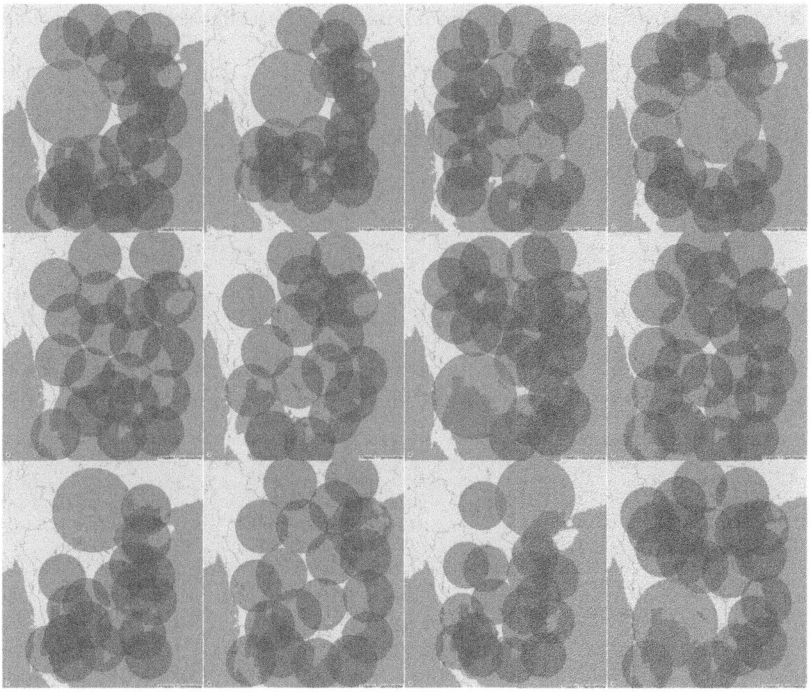

Fig. 1. Visualization of a radar system schedule over a single day, consisting of 12 sessions. The sessions are arranged sequentially from session 1 to session 12, displayed from left to right and top to bottom. The coverage areas are color-coded based on the category of the radars.

Figure 2 shows the best fitness values of each objective with respect to the generations. Both the maximal coverage area and the minimal number of radars converge uniformly through the process. However, the graphs highlight that the improvement of the fitness values from the initial to the final value is relatively modest, with 0.8% in coverage area and 2.2% in usage number. This finding

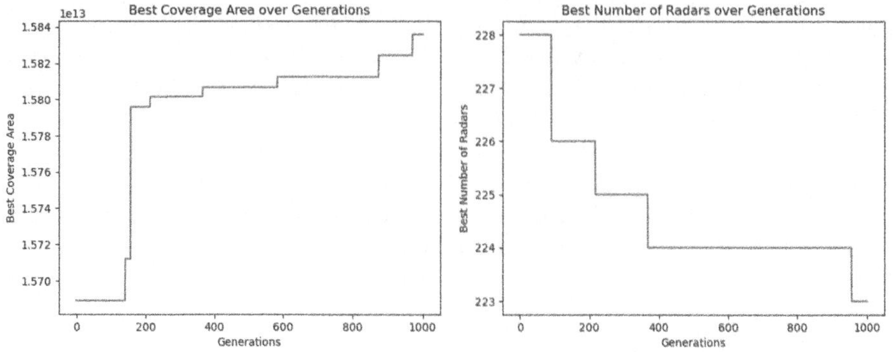

Fig. 2. Maximal coverage area and Minimal number of usage radars of population over generations

underscores the efficiency of the heuristic initialization approach employed in the optimization process.

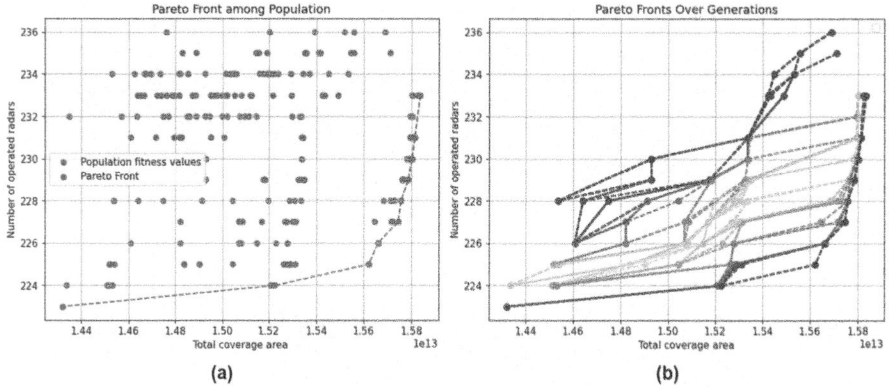

Fig. 3. Pareto Front founded by our Genetic Algorithm. (a) compares the Pareto Front with other fitness values of final population. (b) shows the changes of Pareto Front through generations (color change from blue to red where blues denote earlier iterations, reds denote final iterations) (Color figure online)

Figure 3(a) illustrates the Pareto Front, indicated by the red markers, compared to fitness values of the population. The transition from using 223 to 225 in total radars across sessions exhibits a significant increase in coverage area, whereas those above 225 show a consistent and gradual rise. This phenomenon potentially resulted from strategic deployment of long-range radars in individual. The remain fitness values are dispersed across the graph, highlighting the diversity of the population. These individuals shows the stochastic nature of the algorithm's initialization and its capacity to discover unconventional yet potentially valuable solutions. Figure 3(b) illustrates the progression of the Pareto Front across generations. It is apparent that the coverage area values do not show significant improvement over generations, which underscores the effectiveness of the heuristic initialization approach. The optimization of radar usages, initially lacking focus, is subsequently refined through the application of GA.

In the second experiment, we retained the same general settings as in the first experiment but modified the minimum and maximum number of radars allowed in each category per session. The purpose of this adjustment was to investigate whether increasing the number of radars would result in greater coverage areas consistently. To ensure a fair comparison, we focused on short-range radar category (C_1), and ignored long-range radar (C_2) in the dataset. This decision was made because the two radar categories differ significantly in their characteristics, and comparing them directly could skew the results. We conducted several test cases, each with different upper/lower-bound of C_1 radar usage settings. We use the metric of Coverage Percentage in each session:

$$\text{Coverage Percentage}(S_k) = \frac{|\bigcup_{i=1}^{n} x_{ik} \cdot A_i|}{|\bigcup_{i=1}^{n} A_i|} \cdot 100\%$$

where the numerator is the coverage area in session $S_k \in \mathcal{S}$ of a solution; the denominator is the maximum coverage area that a radar system can achieve in a session (an ideal solution allowing all radars to be activated). Note that, in very best case, this number can approach to 1, however, due to the constraints of consecutive radars, so it is can not to be 1 in most case. For each test case, we measured the maximum, minimum, average and median percentage of coverage areas achieved by the solution with the best coverage area in the final population.

Table 1. Maximum, Minimum, Mean and Median Coverage Percentage of each radar us age bounds

Usage bounds		Coverage Percentage (%)			
$lb(C_1)$	$ub(C_1)$	Max	Min	Mean	Median
5	10	78.13	59.69	69.89	70.91
10	15	84.59	76.32	82.00	82.85
15	20	91.20	85.84	88.64	88.05
20	25	95.25	89.47	92.65	92.80
25	30	98.67	91.66	95.60	95.70

The metrics are summarized in Table 1, providing a comprehensive view of how varying the number of radars affects overall coverage. The mean values exhibit a non-linear relationship with increasing usage bounds, showing a sharper escalation from the [6, 12] range (70% to 89%) to the [18, 24] range compared to the subsequent transition to the [30, 36] range (89% to 95%). This disparity arises from that lower usage bounds entail radar coverage areas in the system that are less likely to overlap significantly. Consequently, each additional radar deployed within such bounds contributes more to expanding the overall coverage area. Furthermore, the result reveals that the minimum and maximum values symmetrically around the median values, which may denotes a uniform distribution of the coverage areas of all sessions in each individual.

5 Conclusion

This study has presented an effective solution to the RSSP by leveraging a tailored Genetic Algorithm with the specialized Stochastic Heuristic Initialization method and custom operators. Our approach addresses the dual objectives of maximizing radar coverage while minimizing the number of operational radars across multiple sessions, all while adhering to strict operational constraints. The Stochastic Heuristic Initialization method, in particular, enhances the initial population's quality and feasibility, promoting rapid convergence and robust solution generation.

Experimental results demonstrate the effectiveness of our GA-based method in significantly improving scheduling efficiency and operational flexibility. The radar schedules produced are well-distributed, with minimal overlap in coverage areas, thus maximizing the system's overall coverage. Furthermore, the diversity in the population and the successful identification of the Pareto front underscore the algorithm's capability to navigate the solution space efficiently.

Disclosure of Interests. The authors have no competing interests to declare that are relevant to the content of this article.

References

1. Kurdzo, J.M., Palmer, R.D.: Objective optimization of weather radar networks for low-level coverage using a genetic algorithm. J. Atmos. Oceanic Technol. **29**(6), 807–821 (2012). ISSN 0739-0572, 1520-0426. https://journals.ametsoc.org/view/journals/atot/29/6/jtech-d-11-00076_1.xml
2. Boudjemaa, R.: Maximisation of weather radar network coverage based on accelerated particle swarm optimisation. Int. J. Metaheuristics **4**(3), 205–219 (2015). ISSN 1755-2176. https://doi.org/10.1504/IJMHEUR.2015.074420
3. Li, X., et al.: Fast antenna deployment method for multistatic radar with multiple dynamic surveillance regions. Signal Process. **170**, 107419 (2020). ISSN 01651684. https://linkinghub.elsevier.com/retrieve/pii/S0165168419304712
4. Tema, E.Y., Sahmoud, S., Kiraz, B.: Radar placement optimization based on adaptive multi-objective meta-heuristics. Expert Syst. Appl. **239**, 122568 (2024). ISSN 0957-4174. https://www.sciencedirect.com/science/article/pii/S0957417423030701
5. Glover, F., McMillan, C.: The general employee scheduling problem. An integration of MS and AI. Comput. Oper. Res. Appl. Integer Program. **13**(5), 563–573 (1986). ISSN 0305-0548. https://www.sciencedirect.com/science/article/pii/030505488690050X

6. Tanomaru, J.: Staff scheduling by a genetic algorithm with heuristic operators. In: 1995 IEEE International Conference on Systems, Man and Cybernetics. Intelligent Systems for the 21st Century, vol. 3, pp. 1951–1956 (1995). https://ieeexplore.ieee.org/document/538063
7. Aickelin, U., Dowsland, K.A.: An indirect genetic algorithm for a nurse-scheduling problem. Comput. Oper. Res. **31**(5), 761–778 (2004). ISSN 0305-0548. https://www.sciencedirect.com/science/article/pii/S0305054803000340
8. Kazimipour, B., Li, X., Qin, K.: A review of population initialization techniques for evolutionary algorithms. In: Proceedings of the 2014 IEEE Congress on Evolutionary Computation, CEC 2014 (2014)
9. Vlašić, I., Đurasević, M., Jakobović, D.: Improving genetic algorithm performance by population initialisation with dispatching rules. Comput. Ind. Eng. **137**, 106030 (2019). ISSN 03608352. https://linkinghub.elsevier.com/retrieve/pii/S0360835219304899

Towards a Communication Specification Language for Heterogeneous Service Orchestration Based on Process Calculus and Holonic Multi-agent Systems

Markus Schatten, Tomislav Peharda(✉), and Bogdan Okreša Đurić

Artificial Intelligence Laboratory, Faculty of Organization and Informatics, University of Zagreb, Zagreb, Croatia
{markus.schatten,tomislav.peharda,dokresa}@foi.unizg.hr

Abstract. In this paper an initial formalization for a novel communication specification language in multi-agent systems (MASs) based on Π-calculus and holonic MASs is introduced and analyzed. The novel language called Awkward Π-nguin (or APi for short) is aimed towards modeling and execution of complex service oriented architectures, heterogeneous cloud systems, containerized microservice architectures and similar complex systems by specifying the communication flows between processed (agents) and organizing them hierarchically into holons. We provide a literature review and initial motivation for introducing a new language as well as give preliminary results of the language development process.

Keywords: Multi-agent systems · Agent orchestration · Communication flows specification · process calculus

1 Introduction

MASs are valuable in performing complex tasks that require autonomy. The higher the number of agents in such a system, the more levels of complexities have to be addressed. Specification of communication flows is one of them. Agents exchange messages via communication channels to inform each other about their requests or status. Thus it is required to specify the communication flows. Since agents are autonomous and independent units, in traditional design approaches, for each agent, communication flows are implemented individually, which makes the code redundant but also unstable, less readable, and less extensible. A potential solution to this problem could be a new programming language for communication flows specification based on process calculus, a declarative engine that is capable of processing the specification, and a platform

Supported by the Croatian Science Foundation under the project number IP-2019-04-5824.

V. Julian et al. (Eds.): IDEAL 2024, LNCS 15346, pp. 202–213, 2025.
https://doi.org/10.1007/978-3-031-77731-8_19

to handle the coordination of agents. This enables consistent communication flows specification and results in an ability to orchestrate agents in MASs.

In computer science, a task executed repeatedly to deliver resources or impact a requesting party is often handled by an agent [5,15]. Agents are autonomous programs that perform tasks independently without human intervention [33]. They operate on a schedule, executing tasks periodically based on predefined rules or external stimuli, such as incoming messages [25].

In multi-agent system architectures, agents, where each has a different purpose, work together to achieve a common goal [14,33]. An example of a multi-agent system use case could be a scenario where we have 3 agents that communicate sequentially. The first agent collects weather data (temperature, humidity, etc.) from an external source and sends it to the second agent, which is responsible for forecasting weather in the next 6–12 h. Once the second agent completes computing the forecast, it sends the result to the third agent, which then informs the audience interested in such information [12,14,33].

A MAS can be described as a network of intelligent agents that collaborate as a unit to achieve a higher goal that might otherwise be too complex for a single agent to achieve [3,33]. Agents do not necessarily need to be in the same working environment but may rather be distributed across many different ones [6].

The complexity of MASs could be examined from multiple perspectives, one of which is the agent itself. Even though it is encouraged for agent logic not to be verbose but rather simple and specific, sometimes that may not be feasible [7,11,12]. As a consequence, an overly complex agent might have to communicate with multiple other agents, which implies it has to properly shape input and output depending on the agent it communicates with. Due to that, such an agent might not be as predictable in its behavior and also harder to maintain [6,7]. That may present an issue for the multi-agent system as a whole, especially in designing communication flows [14].

Another complexity arises from the possibility that two or more MASs (which means they are not in the same cluster) have to collaborate, however, they potentially employ different communication channels and communication protocols they use, such as Extensible Messaging and Presence Protocol (XMPP), message headers (i.e. authentication), message payload formats, e.g. JavaScript Object Notation (JSON)/eXtensible Markup Language (XML), etc., and general behavior that one has no control over, but can only adapt to [11,18,22].

Due to these difficulties, designing and implementing large and complex MASs may not be an easy task, which is why proper agent orchestration and coordination have a high impact on the scalability and maintenance of such systems [14–16].

In the following sections we will provide an overview of how a language that can address the above issues might be implemented. We will provide an initial formalization based on Π-calculus and holonic MASs. Firstly in Sect. 2 we provide an overview of related work. In Sect. 3 we shortly introduce the Π-calculus in order to be able to use it to specify a novel programming language. In Sect. 5 we provide an initial systematization of an orchestration language and in Sect. 6 we draw our conclusions and give guidelines for future work.

2 Related Work

Previously stated challenges in designing communication flows in MASs are the primary motivation for conducting this research. To support the research, two literature domains have been analyzed: (1) scientific papers and (2) practices from industry. These domains are recognized as the most relevant, as they cover approaches addressed from a scientific standpoint but also approaches from real-world use cases.

2.1 Communication Specification and Agent Orchestration in MASs

Communication specification in MASs deals with defining a protocol to be used to transport messages, message headers, the format of messages, communication flows, and other related characteristics within the traditional approaches to specifying communication in MAS [11,18,22].

Speaking of message transport, Corkill [4] proposes a broadcasting concept, where the idea is that each agent sends a message to all agents in a MAS so that they implement a filtering mechanism for irrelevant messages. This approach brings flexibility in the design and implementation of an agent but does not take into account performance concerns, as message traffic will grow high. Goldman and Ziberstein [9] build an idea on top of the broadcasting concept, however, it takes into account that external agents (outside the observed MAS) are likely not adjusted to filter messages that they do not have an interest in [4].

In [30] we propose an orchestration platform for hybrid artificial intelligence. The conceptual model of the platform is proposed to be implemented as a multi-agent system that takes care of load balancing, service allocation, and service orchestration via Docker Swarm or Kubernetes. This would also allow for support of the holonic organization concept, which indicates that MAS may consist of multiple agents where each agent might contain multiple other agents as part of it [10]. In [28] the described orchestration platform has been laid out in more detail, by giving a description of the semantics for the service orchestration, and supported communication characteristics, on a use case of digital twinning in autonomous vehicles. On a similar note, in [29] it has been explored how APIs could be used in computer games, especially massively multi-player online (MMO) games, for service orchestration. In MMOs, players usually enter a server that was started up on demand, by a controlling unit. Thus having an API that is capable of starting up the game server instances is analyzed more in-depth by the authors.

Whilst process calculus has been shown to be equivalent to λ-calculus in terms of expresivity, there have not been lots of attempts to implement programming languages based on it (as opposed to λ-calculus). A rare example is given in [19] where the benefits of using π-calculus as a fundamental for a strongly-typed concurrent programming language is explored. This is of high relevance to this paper, as the proposal is as well to use an approach of the process calculus family to formal modeling.

2.2 Service Orchestration Providers

An agent can be viewed as a service since it performs tasks to add value, with each agent providing at least one service. However, not all services are agents due to a lack of autonomy [8]. Communication flows in service-oriented architectures, especially microservices, are managed through orchestration [16]. Few service orchestration platforms exist, typically focusing on high-level orchestration without regard to service type. They primarily ensure services start in the correct order and within the same cluster, facilitating communication and load balancing [26,27].

Kubernetes [20] is an open-source platform that manages services through a declarative configuration, enabling the scaling and automated deployment of service clusters. It defines and monitors service co-existence, cluster membership, and server placement. However, Kubernetes does not manage in-depth communication between services. In its ecosystem, a service is a containerized application with access to its environment's resources, but specific features must be designed and implemented by the service itself [20,26,27].

Docker Swarm [21] is a work mode of the Docker containerization platform that enables orchestration tools similar to Kubernetes. The differences between the two arise from the definition of eco-system, wherein Docker Swarm is more lightweight on infrastructure, and the way how multiple clusters, so-called swarms, are combined. On the same note, Docker Swarm neither deals with communication between the actual services [21].

There is a good amount of other service orchestration providers that could be used as an alternative to the described two providers. Some of these are Azure Container Instances, Google Cloud Run, etc. Nonetheless, none of these examined providers emphasize the service communication or how services collaborate, but rather the infrastructure to enable cohabitation [32]. That makes sense since communication for a particular type of service differs from one another, thus it would be challenging to create a universal solution for all service types.

3 *Π* Calculus

Distributed programming is fundamentally based on the concept of process calculus [35]. Process calculus is a formalism that allows us to model concurrent systems by describing processes and communication channels, i.e., messages that are transmitted between processes. There are many implementations of process calculus, such as CCS, CSP, ACP, π-calculus, etc.

The π-calculus is a formal tool used for designing algorithms in various problem domains or application areas [17]. It provides a theoretical framework for describing interactions and communications between a multitude of concurrent systems. The π-calculus allows us to formally specify and reason about the behaviors of concurrent systems, which is particularly useful in the design and analysis of distributed systems and communication protocols.

It allows for describing and reasoning about concurrent processes and inter-process communication. It contains a small set of fundamental operators for describing the structure of processes. Algebraic laws are used that govern the behavior of processes. There are techniques by which we can check the correctness of processes.

Whilst the process calculus is fundamentally based on processes, in the following we shall refer to them as agents, since it is our aim to model MASs as systems of concurrent communication processes.

Definition 1. *Let* $\mathbb{N} = \{a, b, c, ..., z, a_1, b_1, ...\}$ *denote a potentially unlimited set of names that we will use to denote channels, variables, and variable values, and let* $\mathbb{A}$ *be a set of identifiers (processes or agents) defined by the following expressions:*

$$
\begin{aligned}
\textit{Prefixes:}\quad & \alpha ::= \bar{a}x \quad \textit{(output)}\\
& \alpha ::= a(x) \quad \textit{(input)}\\
& \alpha ::= \tau \quad \textit{(inactive)}\\
\textit{Agents:}\quad & P ::= 0 \quad \textit{(nil)}\\
& P ::= \alpha.P \quad \textit{(prefix)}\\
& P ::= P + P \quad \textit{(sum)}\\
& P ::= P \mid P \quad \textit{(parallel)}\\
& P ::= (x = y)?P \quad \textit{(match)}\\
& P ::= (x \neq y)?P \quad \textit{(mismatch)}\\
& P ::= (\nu x)P \quad \textit{(restriction)}\\
& P ::= A(y_1, ..., y_n) \quad \textit{(identifier)}\\
\textit{Definitions:}\quad & A(x_1, ..., x_n) \stackrel{def}{=} P \quad \textit{(where } i \neq j \Rightarrow x_i \neq x_j\textit{)}
\end{aligned}
$$

The semantics of these expressions are given bellow:

- The empty agent 0 cannot perform any actions.
- The output prefix, which is of the form $\bar{a}x.P$, means that the name x (data) is sent over the name a (channel) after which the agent continues as P.
- The input prefix, which is of the form $a(x).P$, means that some name (data) is received over the name a (channel) and is stored in the name x (variable) after which the agent continues as P.
- The inactive prefix, which is of the form $\tau.P$, means that the agent continues as P without the need for interaction with the environment (it proactively starts).
- The sum $P + Q$ indicates that the agent can either take on the role of P or the role of Q.
- The parallel composition $P \mid Q$ means that the processes P i Q run in parallel. This also means that all actions of P and all actions of Q will be executed. Also, both processes can communicate in parallel through the same channel.
- The parallel composition $P \mid Q$ means that the processes P and Q are executed in parallel. This means that all actions of P and all actions of Q will

be executed. Also, both processes can communicate in parallel through the same channel.
- The match of the form $(x = y)?P$ means that the agent will continue as P if and only if the names x and y are equal.
- The mismatch of the form $(x \neq y)?P$ means that the agent will continue as P if and only if the names x and y are different.
- The restriction of the form $(\nu x)P$ limits the name (variable) x to the current process, i.e., it cannot be used for communication with the environment (the variable is local to the current process). However, x can be used for communication between components within P. It can be interpreted as new x, i.e., creating a new variable x in the current context.
- Each identifier $A(y_1, ..., y_n)$ where n is the arity of A has a definition $A(x_1, ..., x_n) \stackrel{def}{=} P$ in which $x_i, i \in \{1, ..., n\}$ are mutually distinct names and behaves as P in which x_i are replaced with the corresponding y_i. We can interpret definitions as formal specifications, and identifiers as invocation (call) with actual arguments.

4 Holonic MASs

In multi-agent systems, a holon [24] is a hybrid entity that represents both an autonomous agent and a part of a larger system. This concept is derived from the idea of "holons" in systems theory, which suggests that everything is both a whole and a part.

In this context, a holon is an agent that operates independently, making its own decisions and interacting with other agents. At the same time, it also serves as a part of a larger, collective system where it collaborates and communicates with other agents. This allows for higher levels of organization and complexity, enabling multi-agent systems to perform tasks that are too large or complex for individual agents.

Holons can be arranged hierarchically in a multi-agent system, with holons at higher levels composed of multiple holons at lower levels. This kind of hierarchical arrangement can help to manage complexity in large systems and also help in scaling up the system. Each holon has its own goals, but it also contributes to the goals of the larger system of which it is a part. The balance between autonomy and cooperation is a key feature of holonic multi-agent systems.

5 Agents Orchestration in MAS Architecture

From the diligent literature analysis as well as the capabilities of the existing service orchestration providers, it is evident that there is room for improvement present in the aspect of specifying communication, which is especially important for multi-agent systems [7].

The main concern is that there is currently no unified approach examined and implemented on specifying communication flows between agents. That being said, the communication logic is encapsulated within the agent itself, which is not only hard to keep track of but also adds overhead to the agent logic that does not directly impact the purpose of an agent [15]. To simplify the design process of communication flows, the potential opportunity lies in constructing a programming language based on a process calculus approach for formal modeling that abstracts away the agent logic used to define communication flows [31]. Process calculus is a family of approaches with common features to describe interaction, communication, and synchronization between entities, which is why it may serve very well as a theoretical foundation for constructing a programming language to specify communication between agents [2]. Additionally, we propose a declarative engine that is capable of consuming communication flows' specification and using it for adequate agent orchestration.

Furthermore, each agent might be communicating the output in different formats (XML, JSON, etc.). That means each agent might need to implement various parsers, which is redundant [11,22]. An opportunity here would be to create a service orchestration platform that preprocesses all the messages that agents exchange (for example by converting messages from one format to another). By doing so, only the platform would implement various parsers, and each agent would just specify in what format they would like to receive the input. The platform could also be responsible for forwarding messages between the agents based on the specified communication flows.

Developing a programming language for specifying communication flows as well as the orchestration platform for an agent code would have the effect of making specifications more readable and modularly implemented. Additionally, an agent would then only have to worry about the task business logic it performs and not so much about communication. By introducing an orchestration platform, an agent would become much more lightweight. Additionally, the introduction of the platform would create an opportunity to encapsulate business logic that may be redundant if it lived inside an agent codebase, such as a message parser.

Having the basic requirements defined, we provide an initial grammar specified in Antlr4 [23] that describes the programming language syntax:

```
grammar APi ;
```

Every grammar constructed in the said technology starts by specifying the grammar name, which is APi (short for Awkward $\Pi - nquin$) in our case.

```
s_environment : ENVIRONMENT WS ':' NEWLINE ( iflow | oflow )+ ;
iflow : TAB INPUT WS INPUT_FORMAT WS s_input NEWLINE ;
oflow : TAB OUTPUT WS OUTPUT_FORMAT WS s_output NEWLINE ;
s_import : IMPORT IDENT NEWLINE ;

s_channel : CHANNEL WS IDENT WS ':' NEWLINE s_channel_spec ;
s_channel_forward : CHANNEL WS IDENT WS '.' NEWLINE ;
s_channel_spec : TAB s_input WS C_SENDS WS s_output NEWLINE;
```

```
s_agent : AGENT WS IDENT ( arglist WS )? ':' NEWLINE aflow+ ;
arglist : '(' IDENT (WS IDENT)* ')';
aflow : TAB valid_channel WS A_SENDS WS valid_channel NEWLINE ;
valid_channel : IDENT | SELF | NIL | STDIN | STDOUT | STDERR | VOID ;

pi_expr : '(' pi_expr ')'
     | pi_expr RESTART
     | pi_expr WS PARALLEL WS pi_expr
     | pi_expr WS ONSUCCESS WS pi_expr
     | pi_expr WS ONFAIL WS pi_expr
     | pi_expr WS pi_expr
     | IDENT ( arglist )?;
s_start : START WS pi_expr ;

api_program : ( s_import | s_environment | s_channel |
s_channel_forward | s_agent | s_start | COMMENT | NEWLINE )*? ;
```

Following are descriptions of different types of supported expressions. That being said, the language provides support for an environment (*s_environment*) where the idea is that it is an interface to communicate with a holon (usually used for communication between multiple holons). *s_import* expression is used to specify with what holon should the current holon communicate. That being said, after the import statement, the imported holon may be used in the specification of communications flows.

There is also a description of the channel expression which is a tying unit for communication between agents. Essentially, two agents communicate with each other by sending a message to a channel, which will be then responsible for transforming and forwarding the message to the interested agents.

Each agent could listen and send messages to multiple channels. To specify that, it is required to provide a pair in the following format $A \Rightarrow B$, where A specifies the sender, and B specifies the receiver. A & B could either be the current agent identificator (*self*), or a channel identificator.

The language also supports execution plans, which start with the keyword *start*. Execution plans are used to define dependencies between the execution of agent communication flows.

```
s_input : s_json | s_xml | s_regex ;
s_output : s_json | s_xml ;
IMPORT : 'import' ;
ENVIRONMENT : 'environment' ;
ONSUCCESS : '&' ;
RESTART : '+' ;
PARALLEL : '|' ;
START   : 'start' ;
SELF    : 'self' ;
AGENT   : 'agent' ;
```

```
CHANNEL : 'channel' ;
JSON : 'json' ;
XML : 'xml' ;
INPUT_FORMAT : '=>' ;
OUTPUT_FORMAT : '<=' ;
A_SENDS  : '->' ;
C_SENDS : TCP | UDP ;
TCP : '-->' ;
UDP : '*->' ;
```

Identifiers provided above are the subset of the ones that are available for constructing an expression. For example, they specify different types of communication protocols (UDP or TCP), dependencies in the execution plans, as well as input and output formats (JSON or XML).

Below is a basic example of communications flows specification using the described language which defines two agents, *agent_a* and *agent_b*, and a channel *channel_c*. Channel *channel_c* does not do any message transformation, thus only forwards messages it receives. The agent *agent_a* is defined to send a message to channel *agent_c*, while agent *agent_b* awaits the message. There is also an execution plan defined which indicates that both agents start at the same time.

```
channel channel_c .

agent agent_a :
    self -> c

agent agent_b :
    c -> self

start agent_a | agent_b
```

In our case, agents can be any program that can communicate, and should be wrapped inside an agent description which specifies how an agent is started and over which types of channels it communicates (for example the protocol used, the message format, type of communication - i.e. streamed or not etc.).

From this contemplation, it becomes evident that we have chosen to include some of the important aspects of Π-calculus for the declarative programming language, namely: (1) basic concepts including processes (agents), channels and data, (2) interprocess (interagent) communication, (3) possibly concurrent execution of processes. Additionally, we have included the idea of holons, which in our case are used similar to modules in contemporary programming languages like Python: a file represents a holon which can be imported to other holons in order to allow for communication between them.

6 Conclusion and Future Work

Multi-agent systems are complex in their nature mainly due to the fact that each individual agent may be designed and implemented differently. This is especially significant in terms of how it communicates with other agents of interest [7,14, 33]. Due to that, there are various aspects of communication that may differ from agent to agent, some of which are the transport protocol, message format, communication flows, etc. [18,22].

Multiple examined papers deal with communication flows mainly from the perspective of performance and information integrity [1,9,34], with few addressing a universal approach to defining communication to relieve the agent from implementing all aspects. Strategies such as broadcasting messages between all agents in MASs allow interested parties to decide if they wish to react [4], while others focus on specifying message formats to simplify agent design and interpretation [13]. Service orchestration providers emphasize stable and flexible infrastructure for interconnected services but allow agents to define communication, recognizing that it varies across services. These platforms, not specific to any particular type of service like MASs or databases, focus on abstracting the role of service in their ecosystems [8,16,26,27].

Based on the analyzed literature, there is room for improvement by constructing a programming language to specify communication flows in MASs. This would lighten agent logic and simplify communication maintenance and extension. Additionally, since current orchestration platforms do not emphasize communication, there is an opportunity to develop a middleware orchestration platform for MASs to handle message forwarding and preprocessing [20,21,32].

Herein we have provided a basic formalization of a declarative programming language that should be able to address some of the concerns outlined above. We have used ideas from Π-calculus and holonic MASs to design a novel programming language and provided an initial syntax grammar using Antlr4. Our future endeavors will be focused on implementing a working prototype of a language interpreter, the proposed declarative engine[1] as well as support for various use-cases that have been identified within the larger scope of the O_HAI ④ Games project.

Acknowledgement. This work has been fully supported by the Croatian Science Foundation under the project number IP-2019-04-5824.

References

1. Althnian, A., Agah, A.: Evolutionary learning of goal-oriented communication strategies in multi-agent systems. J. Autom. Mob. Robot. Intell. Syst. 52–64 (2015)
2. Baeten, J.C.: A brief history of process algebra. Theoret. Comput. Sci. **335**(2–3), 131–146 (2005)
3. Bansall, S.: Intelligent agent (2023). https://www.geeksforgeeks.org/agents-artificial-intelligence/. Accessed 20 June 2024
4. Corkill, D.D.: Design alternatives for parallel and distributed blackboard systems. University of Massachusetts at Amherst. Computer and Information Science [COINS] (1988)

[1] The current state of the implementation is available here: https://github.com/AILab-FOI/APi

5. d'Inverno, M., Luck, M.M., D'Inverno, M., Luck, M.: Understanding Agent Systems. Springer, Heidelberg (2004). https://doi.org/10.1007/978-3-662-10702-7
6. Fortino, G., Russo, W.: Eldameth: an agent-oriented methodology for simulation-based prototyping of distributed agent systems. Inf. Softw. Technol. **54**(6), 608–624 (2012)
7. Giles, C.L., Jim, K.-C.: Learning communication for multi-agent systems. In: Truszkowski, W., Hinchey, M., Rouff, C. (eds.) WRAC 2002. LNCS (LNAI), vol. 2564, pp. 377–390. Springer, Heidelberg (2003). https://doi.org/10.1007/978-3-540-45173-0_29
8. Gillis, A.S.: Microservices (2021). https://www.techtarget.com/searchapparchitecture/definition/microservices. Accessed 20 June 2024
9. Goldman, C.V., Zilberstein, S.: Optimizing information exchange in cooperative multi-agent systems. In: Proceedings of the Second International Joint Conference on Autonomous Agents and Multiagent Systems, pp. 137–144 (2003)
10. Horling, B., Lesser, V.: A survey of multi-agent organizational paradigms. Knowl. Eng. Rev. **19**(4), 281–316 (2004)
11. Janssen, M.: Complexity and Ecosystem Management: The Theory and Practice of Multi-Agent Systems. Edward Elgar Publishing (2002)
12. Luck, M., Marik, V., Stepankova, O., Trappl, R.: Multi-Agent Systems and Applications, vol. 2086. Springer, Heidelberg (2001). https://doi.org/10.1007/3-540-47745-4
13. Luncean, L., Becheru, A.: Communication and interaction in a multi-agent system devised for transport brokering. In: Proceedings of the 2015 Balkan Conference on Informatics: Advances in ICT, Craiova, Romania, pp. 2–4 (2015)
14. McArthur, S.D., et al.: Multi-agent systems for power engineering applications–part ii: technologies, standards, and tools for building multi-agent systems. IEEE Trans. Power Syst. **22**(4), 1753–1759 (2007)
15. Padgham, L., Winikoff, M.: Developing Intelligent Agent Systems: A Practical Guide. Wiley, Hoboken (2005)
16. Pankaj, R.: Service orchestration: what it is and why it matters (2023). https://www.awlindia.com/us-en/blog-details/service-orchestration-what-it-is-and-why-it-matters. Accessed 20 June 2024
17. Parrow, J.: An introduction to the π-calculus. In: Handbook of Process Algebra, pp. 479–543. Elsevier (2001)
18. Peterson, L.L., Davie, B.S.: Computer Networks: A Systems Approach. Elsevier (2007)
19. Pierce, B.C., Turner, D.N.: Pict: a programming language based on the pi-calculus. In: Proof, Language, and Interaction, pp. 455–494 (2000)
20. Powell, R.: Do i really need kubernetes? (2020). https://circleci.com/blog/do-i-really-need-kubernetes. Accessed 20 June 2024
21. Powell, R.: Docker swarm vs kubernetes: how to choose a container orchestration tool (2023). https://circleci.com/blog/docker-swarm-vs-kubernetes. Accessed 20 June 2024
22. Priya, B.: What are the components of the data communication system? (2022). https://www.tutorialspoint.com/what-are-the-components-of-the-data-communication-system. Accessed 20 June 2024
23. Rajan, H.: An Experiential Introduction to Principles of Programming Languages. MIT Press, Cambridge (2022)
24. Rodriguez, S., Hilaire, V., Gaud, N., Galland, S., Koukam, A.: Holonic multi-agent systems. Self-Organising Software: From Natural to Artificial Adaptation, pp. 251–279 (2011)

25. Rouse, M.: Agent (2017). https://www.techopedia.com/definition/1292/agent. Accessed 20 June 2024
26. Rudrabhatla, C.K.: Comparison of event choreography and orchestration techniques in microservice architecture. Int. J. Adv. Comput. Sci. Appl. **9**(8) (2018)
27. Rufino, J., Alam, M., Ferreira, J., Rehman, A., Tsang, K.F.: Orchestration of containerized microservices for IIoT using docker. In: 2017 IEEE International Conference on Industrial Technology (ICIT), pp. 1532–1536. IEEE (2017)
28. Schatten, M., Peharda, T., Tomicic, I.: Towards an orchestrated game development approach to digital twinning in autonomous vehicles. In: Central European Conference on Information and Intelligent Systems, pp. 3–8. Faculty of Organization and Informatics Varazdin (2022)
29. Schatten, M., Tomicic, I., Duric, B.O.: Towards application programming interfaces for cloud services orchestration platforms in computer games. In: Central European Conference on Information and Intelligent Systems, pp. 9–14. Faculty of Organization and Informatics Varazdin (2020)
30. Schatten, M., Tomičić, I., Đurić, B.O.: Orchestration platforms for hybrid artificial intelligence in computer games-a conceptual model. In: Central European Conference on Information and Intelligent Systems, pp. 3–8. Faculty of Organization and Informatics Varazdin (2020)
31. Singh, A., Ramakrishnan, C.R., Smolka, S.A.: A process calculus for mobile ad hoc networks. In: Lea, D., Zavattaro, G. (eds.) COORDINATION 2008. LNCS, vol. 5052, pp. 296–314. Springer, Heidelberg (2008). https://doi.org/10.1007/978-3-540-68265-3_19
32. Valdés, A.: The 5 best kubernetes alternatives (2021). https://hackernoon.com/the-5-best-kubernetes-alternatives-ks3f344k. Accessed 20 June 2024
33. Van Harmelen, F., Lifschitz, V., Porter, B.: Handbook of Knowledge Representation. Elsevier (2008)
34. Wei, C., Hindriks, K.V., Jonker, C.M.: The role of communication in coordination protocols for cooperative robot teams. In: ICAART (2), pp. 28–39 (2014)
35. Yoshida, N.: Graph notation for concurrent combinators. In: Ito, T., Yonezawa, A. (eds.) TPPP 1994. LNCS, vol. 907, pp. 393–412. Springer, Heidelberg (1995). https://doi.org/10.1007/BFb0026581

Counterfactual Explanations for Sustainable Tourism Indicators

Javier Saugar[1(✉)], Carmen Lancho[1], Marina Cuesta[1], Emilio L. Cano[1], Isaac Martín de Diego[1], and Antonio Amado[2]

[1] Rey Juan Carlos University, Data Science Laboratory, C/ Tulipán, s/n, 28933 Móstoles, Spain
{javier.saugar,carmen.lancho,marina.cuesta,emilio.lopez, isaac.martin}@urjc.es

[2] Dephimática, S.L. c/ de Ríos Rosas, 44. 3°C, 28002 Madrid, Spain
antonio.amado@dephimatica.es

Abstract. Counterfactual explanations are a well-known technique in *Explainable Machine Learning* (*XML*) to provide simple explanations on complex *Machine Learning* (*ML*) models. Through understandable "what if" scenarios, counterfactuals explore how changes in the input data affect the results of a model. This article leverages counterfactual explanations for sustainable tourism, an emerging approach within the tourism industry to mitigate the negative impacts of mass tourism on ecological systems and local communities. The proposed method analyzes the relationships between several *Sustainable Tourism Indicators* (*STI*s) defined for a specific tourist destination and its general sustainability assessment. It identifies the key changes needed in the *STI*s to achieve an improved global sustainability score. As a result, a decision-making system is offered for sustainable tourism management, which domain experts can use to make more informed decisions. The effectiveness of the proposed method is illustrated through its application to Mallorca, a popular Spanish tourist destination.

Keywords: Counterfactuals · Explainable Machine Learning · Explainable Artificial Intelligence · Sustainable Tourism

1 Introduction

Tourism is a major global industry, significantly influencing the world economy and experiencing rapid and steady growth. However, its rapid growth has led to uncontrolled tourism developments that negatively impact both natural environments and local communities, raising concerns among experts. As a response, sustainable tourism has emerged as a new approach and paradigm to mitigate the undesirable effects of mass tourism [1]. The *United Nations World Tourism Organization* (*UNWTO*), the leading tourism authority in the world, defines sustainable tourism as a "tourism that takes full account of its current and future

V. Julian et al. (Eds.): IDEAL 2024, LNCS 15346, pp. 214–220, 2025.
https://doi.org/10.1007/978-3-031-77731-8_20

economic, social and environmental impacts, addressing the needs of visitors, the industry, the environment, and host communities" [7].

Several tourism observatories around the world have focused efforts on defining key *Sustainable Tourism Indicators* (*STI*s) to measure the sustainability of different destinations, which are pivotal to tourism management [5]. However, the wide range of different *STI*s defined for each destination complicates a comprehensive analysis, making it difficult to draw global conclusions.

To address the challenge of a global analysis of all the *STI*s established for a destination, this work introduces a method to explore the impact of each indicator on the overall sustainability assessment in a specific destination. First, a global *STI*s indicator is computed encompassing all the available *STI*s. Then, counterfactual explanations [2] are employed to examine how individual changes in the *STI*s affect the final value of the global indicator.

Counterfactuals are among the most outstanding techniques in *Explainable Machine Learning* (*XML*) [4], a branch within the *Machine Learning* (*ML*) field that focuses on studying how and why a model makes its predictions to enhance transparency in the process. In particular, counterfactual analysis provides insight into the reasons behind a model decision in the form of easily understandable "what if" scenarios that explore how changes in the input variables of a data observation change its prediction.

Given a fitted model, a counterfactual explanation $\mathbf{x}$ of an instance $\mathbf{x}^*$ is the minimal modification required so that the altered instance is predicted with a different desired value, fulfilling the following properties: (1) $\mathbf{x}$ leads to the desired prediction (validity); (2) $\mathbf{x}$ is close to $\mathbf{x}^*$ in the input space (proximity); (3) $\mathbf{x}$ and $\mathbf{x}^*$ differ in a few variables (sparsity); (4) $\mathbf{x}$ is plausible according to the input distribution (plausibility); (5) $\mathbf{x}$ makes feasible changes, distinguishing between mutable and immutable features (actionability) and (6) $\mathbf{x}$ accounts for causal relations between features (causality).

As in many other domains, tourism can greatly benefit from *ML* [3]. Specifically, the use of counterfactuals in sustainable tourism decision-making is explored in [8], highlighting a causal effect-driven approach to adjust practices to meet sustainability goals. In this paper, counterfactual analysis is employed to achieve a comprehensive understanding of the effects of different *STI*s to reach a desired global sustainability score. As a result, a decision-making system is proposed to support domain experts in making more informed decisions toward sustainability goals. The present paper is structured as follows. The proposed methodology is presented in Sect. 2. Experiments are depicted in Sect. 3. Finally, Sect. 4 concludes.

2 Method

In this section, the proposed method for counterfactual reasoning within the *STI*s of a specific tourist destination is presented. First, a global indicator is built as a general sustainability measure by aggregating all *STI*s (each accounting for a specific concern). Then, the theoretical formulation of the optimization problem to obtain counterfactual explanations for the global indicator is tackled.

For simplicity, the global indicator is defined in this work as a weighted average of all specific *STI*s. However, different aggregations could be made according to domain expertise. Then, given a set of input variables $x_1, \ldots, x_p$ (each representing a *STI*), the global indicator is defined as its weighted average:

$$y = \sum_{j=1}^{p} w_j x_j, \tag{1}$$

where $\sum_{j=1}^{p} w_j = 1$ and $x_j \in [0,1], \forall j = 1, \ldots, p$. It can be assumed that the larger the input variables, the better. Therefore, the larger the global indicator y, the better. The stakeholder may not be satisfied with the value of the global indicator y, or may be interested in how changes in the input variables affect the outcome. Thus, the goal of the proposed methodology, based on counterfactual explanations, is to find the minimum and plausible changes that can be made to the input variables to achieve a desired change in the global indicator. For that specific goal, the following optimization problem is formulated:

$$\min \quad \sum_{j=1}^{p} \beta_j + \lambda \sum_{j=1}^{p} \delta_j \tag{2}$$

$$\text{s.t.} \quad y' - y = \sum_{j=1}^{p} w_j(x_j + \beta_j) - \sum_{j=1}^{p} w_j x_j \geq \epsilon \tag{3}$$

$$\beta_j \leq f_{std}(x_j), \qquad \forall j = 1, \ldots, p \tag{4}$$

$$x_j + \beta_j \leq 1, \qquad \forall j = 1, \ldots, p \tag{5}$$

$$\mu_l \leq \sum_{j=1}^{p} \delta_j \leq \mu_u \tag{6}$$

$$-m\delta_j \leq \beta_j \leq m\delta_j \qquad \forall j = 1, \ldots, p \tag{7}$$

$$\delta_j \in \{0,1\} \qquad \forall j = 1, \ldots, p \tag{8}$$

$$\beta_0, \ldots, \beta_p \geq 0, \ \beta_0, \ldots, \beta_p \in \mathbb{R} \tag{9}$$

$$m \gg 0, m \in \mathbb{R} \tag{10}$$

$$\lambda \geq 0, \lambda \in \mathbb{R} \tag{11}$$

The objective function (2) minimizes the changes made over the input variables (β_j), and the number of variables where a change is made (δ_j). This accounts, respectively, for proximity and sparsity properties. Constraint (3) ensures that the new value y' of the indicator has increased in at least ϵ, addressing the validity property. To guarantee the plausibility property, constraint (4) establishes the maximum change admitted in every variable x_j through a function $f_{std}(\cdot)$ of its corresponding standard deviation. The actionability property is ensured with constraint (5), which maintains each variable x_j within $[0,1]$. Constraint (6) contributes to the sparsity property by controlling the minimum and maximum number of x_j variables where a change is admitted. Finally, constraints (7) to (11) introduce a set of binary variables necessary for the problem. For example,

Table 1. Parameters for the three experiments. y is the value of the global indicator in 2024 for the first experiment and in 2023 for the second and the third.

Experiment	ϵ	f_{std}	μ_l	μ_u	λ
1	$0.01 \cdot y$	$0.3 \cdot std$	1	1	0
2	$0.05 \cdot y$	$0.55 \cdot std$	1	4	10
3	$0.25 \cdot y$	$0.9 \cdot std$	1	6	10

m (constraints (7) and (11)) is a constant large enough to let β_j achieve the optimal value when its associated δ_j equals 1.

The formulation of the problem involves different user-fixed parameters (λ, ϵ, $f_{std}(\cdot)$, μ_l and μ_u) that allow its adaptation to different scenarios, depending on the domain requirements.

3 Experiments

In this section, the proposed method is applied in a real scenario in both predictive and descriptive approaches[1]. The selected tourist destination to test the proposal is the popular Spanish city of Mallorca, which belongs to the *International Network of Sustainable Tourism Observatories* (*INSTO*)[2]. The following set of *STI*s, defined by the *Statistical Institute of the Balearic Islands* (*IBESTAT*), have been considered: $x_1 \propto$ total net 5-year schooling rate in both genders, $x_2 \propto$ population at risk of relative poverty (income $< 60\%$ median), $x_3 \propto$ unaccounted for water, $x_4 \propto$ estimated open establishments by period, island and category, $x_5 \propto$ percentage of annual amount of energy consumed from renewable resources compared to global energy consumption at target level per year, $x_6 \propto$ passengers arriving to the island by sea and air, $x_7 \propto$ maritime passenger traffic on regular lines by type of movement, port, month and year, and $x_8 \propto$ vehicle registrations in the Balearic Islands, by period, island, vehicle type and propulsion type. All of these variables are measured from the year 2000 to 2023. The sustainability of tourism in Mallorca is studied through this dataset, searching for the best changes in the *STI*s that increase the global sustainability score.

A global *STI* is computed following Eq. 1 with $w_j = 1/8$, $\forall j$. Each individual *STI* has been previously transformed through a max-min normalization to range in $[0, 1]$, and to ensure a positive interpretation, that is, the higher the better. Next, the proposed method is applied to achieve the counterfactual explanations in three different scenarios to illustrate different capabilities. The first is a predictive approach to obtain a counterfactual explanation for the year 2024, with the objective of highlighting which is the most important factor to focus on regarding sustainable tourism next year (recall that the dataset ends in 2023). The next two experiments are addressed from a descriptive perspective to better understand what changes in the *STI*s would have been necessary to

[1] The code can be found at https://github.com/URJCDSLab/CounterfactualsSTIs.
[2] https://insto.unwto.org/.

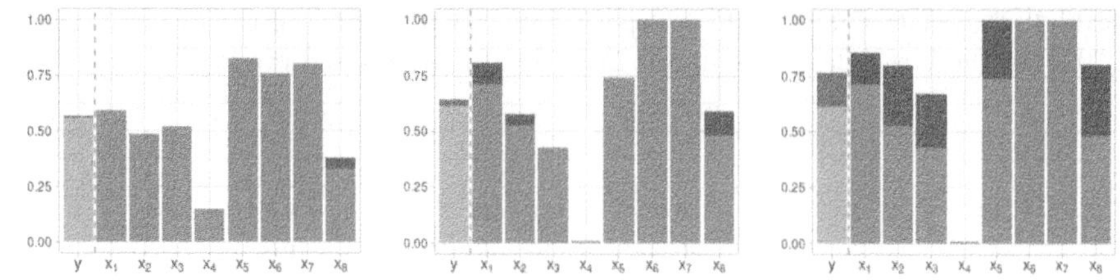

(a) Experiment 1 (b) Experiment 2 (c) Experiment 3

Fig. 1. Counterfactual changes (darker colors), and real (or predicted in case (a)) values (lighter colors), for the three experiments. The global indicator is the first green bar of each graph. The variables are presented in the transformed scale. (Color figure online)

increase the global *STI* in 2023 in a certain percentage. The parameters of the three experiments have been set by a domain expert and f_{std} has been set to be the standard deviation of the historical values of each variable (see Table 1).

The first experiment presents a predictive approach to estimate how to increase the prediction of the global *STI* for the next year. If this study was conducted at the beginning of the year, pertinent decisions could be made to improve the global tourism sustainability in Mallorca. The eight different *STI*s have been predicted using a moving average model of the last three years, and the prediction for the global sustainability indicator in 2024 is computed by averaging the eight individual predictions as in Eq. 1. After that, the proposed method is applied to increase at least 1% the predicted global score with a single change within the eight *STI*s, as an illustrative example of a simple counterfactual explanation. The results, presented in Fig. 1a, reveal that by just increasing the vehicles registered (x_8) by 0.045 points (which implies 1203 more registered cars, that is, a 3.053% of the predicted value), the global *STI* would increase from 0.558 to 0.564. Stakeholders can leverage this information to anticipate their actions for a more positive impact on sustainable tourism in Mallorca.

The second experiment here exposed takes a descriptive approach to gain a deeper understanding of global sustainability in tourism in Mallorca in the recent past. The proposed method is now applied to answer the following question: which values in the input variables would have made the global indicator a 5% higher in 2023? As Table 1 shows, the maximum change allowed in every *STI*s is higher than in the previous experiment to enable a greater increase in the global indicator. Besides, a maximum of four changes are allowed. The results obtained by the proposed method, shown in Fig. 1b, suggest changing a total of three variables: an enhancement of 0.091 in x_1, of 0.049 in x_2 and of 0.105 in x_8. This is translated as: a raise of 0.320% in the school enrollment, a decrease of 0.493% in the population at risk of relative poverty and an increase of 2836 cars registered (equivalent to 6.51% of the value of 2023), respectively. If these changes had occurred, the global indicator would be 0.644 instead of 0.613.

The third experiment also follows a descriptive approach but with a more demanding requirement: what changes would have been necessary in 2023 to obtain a global *STI* 25% higher? The results are shown in Fig. 1c. As can be seen, five changes need to be made, which is clearly more challenging than the results in the second scenario. Variables x_1, x_2 and x_8 also appear as necessary, but, in this case, with an increase of 0.141, 0.271 and 0.241 points, respectively. These gains are translated to an increase of 0.492%, a decrease of 2.74% and increase of 8508 vehicles (equivalent to 20.89%), respectively. The results obtained also suggest a change in x_3 and x_5: x_3 moves up 0.241, reducing a 0.015% the unaccounted for water, and x_5 experiences an increment of 0.258, which implies a gain of 0.781% renewable resources. Thus, the global indicator would have been a 25% higher, from 0.613 to 0.766. Although these changes are unrealistic, this third experiment is an illustrative example of how domain experts can interact with the proposed method to get deeper insights on the global *STI* functioning.

4 Conclusions and Future Work

In this paper, the potential of *XML* for sustainable tourism has been demonstrated through a counterfactual analysis within the *STI*s of a tourist destination to reach a better global sustainable score. An optimization problem for finding counterfactual explanations relating which changes of the individual *STI*s reach a desired growth on the global *STI*s combination has been proposed. The method has been assessed with a real dataset in three experiments showing its adaptability to distinct scenarios, its interpretability and its applicability independently of the *ML* expertise of the user. The achieved counterfactual explanations have proofed to be easily understandable and to offer valuable information, thus, supporting and empowering the stakeholders with decision-making.

Future work will focus on applying this methodology to other tourist destinations with larger datasets to address the scalability. Also on applying counterfactual explanations for time series [6] which will enrich the proposed method by considering the evolution and the trend of the different *STI*s.

Acknowledgements. This research is supported by grants from Rey Juan Carlos University (Ref: C1PREDOC2020) and the Spanish Ministry of Science and Innovation, under the Knowledge Generation Projects program: XMIDAS (Ref: PID2021-122640OB-100).

References

1. Amerta, I.M.S., Sara, I.M., Bagiada, K.: Sustainable tourism development. Int. Res. J. Manag. IT Soc. Sci. **5**(2), 248–254 (2018)
2. Guidotti, R.: Counterfactual explanations and how to find them: literature review and benchmarking. Data Mining Knowl. Discov. 1–55 (2022)
3. Madzík, P., Falát, L., Copuš, L., Valeri, M.: Digital transformation in tourism: bibliometric literature review based on machine learning approach. Eur. J. Innov. Manag. **26**(7), 177–205 (2023)
4. Molnar, C.: Interpretable machine learning. Lulu.com (2020)

5. Rasoolimanesh, S.M., Ramakrishna, S., Hall, C.M., Esfandiar, K., Seyfi, S.: A systematic scoping review of sustainable tourism indicators in relation to the sustainable development goals. J. Sustain. Tour. **31**(7), 1497–1517 (2023)
6. Theissler, A., Spinnato, F., Schlegel, U., Guidotti, R.: Explainable AI for time series classification: a review, taxonomy and research directions. IEEE Access **10**, 100700–100724 (2022)
7. WTO: Indicators of Sustainable Development for Tourism Destinations A Guidebook (English version). World Tourism Organization (2004)
8. Xia, H., Muskat, B., Li, G., Prayag, G.: AI-based counterfactual reasoning for tourism research. Ann. Tour. Res. **101**(C) (2023)

Tracking Healthy Organs in Medical Scans to Improve Cancer Treatment by Using UW-Madison GI Tract Image Segmentation

Bimal Kumar Sah and Doina Logofătu(✉)

Frankfurt University of Applied Sciences,
Nibelungenplatz 1, 60318 Frankfurt am Main, Germany
bimal.sah@stud.fra-uas.de, logofatu@fb2.fra-uas.de

Abstract. This research presents our approaches to the Kaggle UW-Madison GI Tract Image Segmentation Challenge. Radiation oncologists aim to emit X-ray beams targeted at the tumor while simultaneously avoiding the stomach and intestines. Oncologists can see where the tumor is located and administer a precise dose following the presence of tumor cells, which can change daily, thanks to more recent technologies like MR-Linacs. Currently, the task is to outline the location of the gastrointestinal tract (intestines and stomach) to change the direction of the X-ray beam such that the dose is sent to the tumor without touching the organs. This task can take from a few minutes to hours of treatment time unless there is an automated segmentation method that can automate the segmentation process. This research investigates an automated segmentation method based on deep learning that will quicken the segmentation process and eventually help more patients to benefit from effective treatments.

Keywords: U-NET · Inception-ResnetV2 · GI Tract segmentation

1 Introduction

Deep learning techniques can help in the segmentation task of tracing out the stomach and intestines. Building on the baseline UNet model, an investigation is done using Inception-ResnetV2 to improve performance on unseen test data in similar medical applications where patient data is scarce.

In 2019, Radiation therapy was used to treat gastrointestinal tract cancer in almost 5 million patients worldwide [1]. Radiation oncologists try to provide high radiation doses through X-ray beams targeted at tumors for 10–15 min each day for a period of 1–6 weeks. They are usually required to spend approximately 1 h per day manually tracing out the stomach and intestines in MRI scans to ensure that they direct x-ray beams to avoid those important organs before directing high doses of radiation to treat gastro-intestinal tract cancer [2].

V. Julian et al. (Eds.): IDEAL 2024, LNCS 15346, pp. 221–232, 2025.
https://doi.org/10.1007/978-3-031-77731-8_21

Figure 1 shows an example input MRI scan, the tumor (thick pink line), which is close to the stomach (thick red line), is targeted with various doses of radiation highlighted by the different colored thin lines (with a larger area representing a larger dosage). The position of the tumor and intestines, and stomach varies day to day, which eventually makes it hard for Radiation oncologists to trace such scans properly. This leads to prolonging the treatment from 15 min per day to about an hour, reducing the ability of radiologists to treat more patients, and can be difficult for the patients to tolerate.

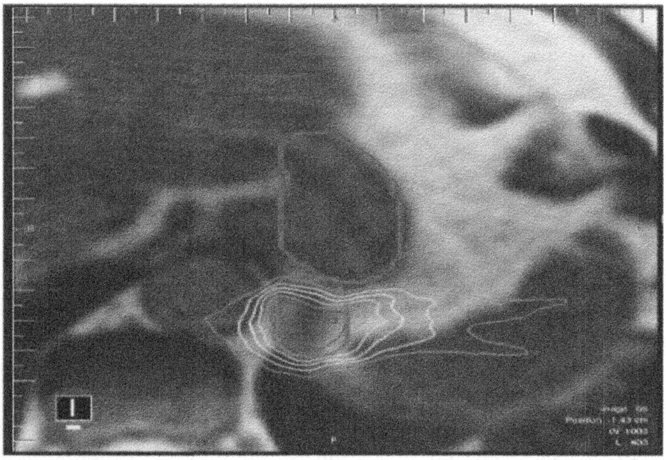

Fig. 1. An illustration of a GI tract with an intestinal and stomach mask [3]

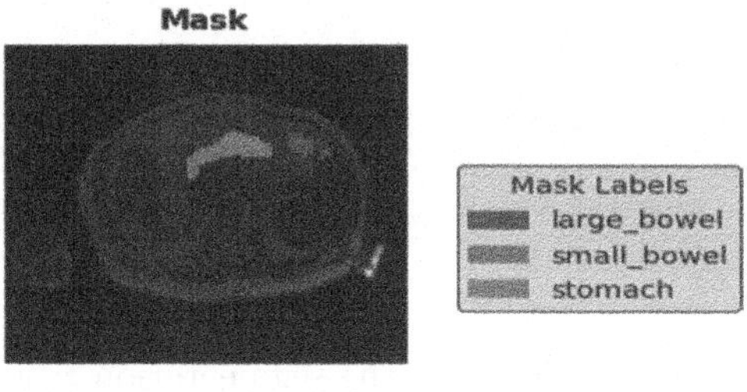

Fig. 2. An illustration of an Input with target masks of organs

In this research, a deep learning model that can automate the segmentation process by segmenting the intestines and stomach on MRI scans is developed. As seen in Fig. 2, our inputs are cross-sectional 2D images of an MRI scan from actual cancer patients who had 1–5 MRI scans on separate days during their radiation treatment provided by the UW-Madison Carbone Cancer Center.

With the help of UNet model prediction of masks for each pixel is made, indicating whether the pixel is an (i) large intestine, (ii) small intestine, or (iii) stomach; thus, indicating a multi-label segmentation problem.

2 Problem Statement and Related Work

Researchers have successfully created semantic and instance segmentation using various forms of U-Net architecture in medical imaging [4–6]. The authors [16] provided a deep learning-based technique that they used to classify images into the stomach, large intestine, and small intestine categories on their own. This transformer-based architecture, which integrates Le Vit units with CNN, was given the name LeVit-UNet. Their results showed a 0.77 Dice coefficient and a 0.69 Jaccard coefficient [16]. Chou A et al. [15] segmented intestine and stomach areas in MRI scans for the Kaggle GI Tract Image Segmentation dataset, in 2022 using Mask R-CNN and U-Net methods. Their top Mask R-CNN model receives a Dice score of 0.73 on the validation set, but their suggested technique has a Dice value of 0.51 [15].

This research investigates U-Net architectures to a new challenge of GI Tract Image segmentation in order to separate organs like the stomach, large bowel, and small bowel (Fig. 3). The goal of this research is to determine the ideal architectural design for masks that will highlight the large bowel, small bowel, and stomach during a scan. Applying deep learning to these activities typically begins with gathering the data and effectively preparing it for feeding into the model to produce helpful predictions.

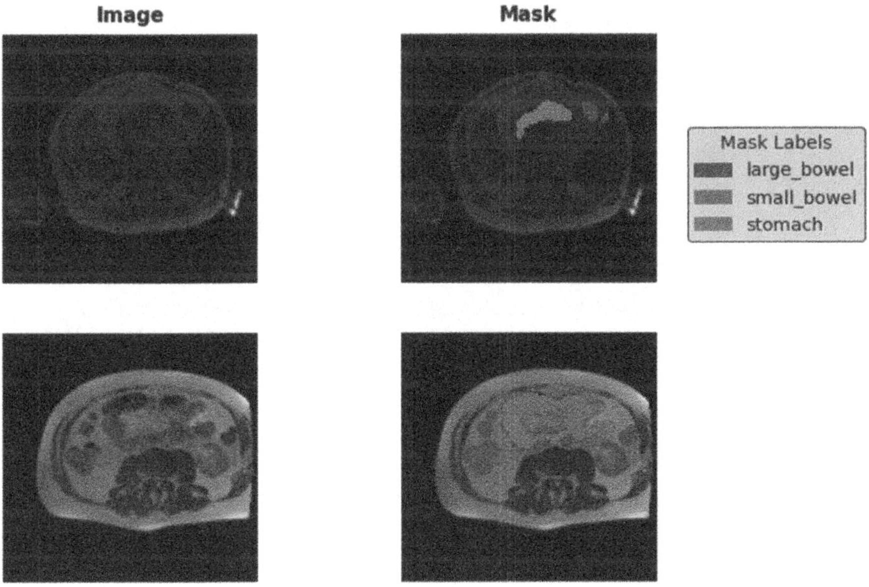

Fig. 3. Dataset (Scans with masks of organs)

3 Data Analysis and Pre-processing

The dataset used in this research is freely available on Kaggle in their latest competition named "UW-Madison GI Tract Image Segmentation". Using this competition, researchers are being encouraged to create an efficient deep learning model that can automate the segmentation process of the GI Tract Images. The scans are collected over a six-day period, and in order to conserve memory and bandwidth, the training masks are provided as RLE-encoded masks.

More than 115488 scans from the original dataset "train.csv" are included, together with IDs and masks for each training image. Each case/day folder in the train folder contains slice photos for a certain case on a specific day. The image filenames contain four integers, for example 276_276_1.63_1.63.png. The slice height/width (integers in pixels) and height/width pixel spacing make up these four numbers (floating points in mm). The resolution of the slide is determined by the first two. The final two measure the actual dimensions of each pixel.

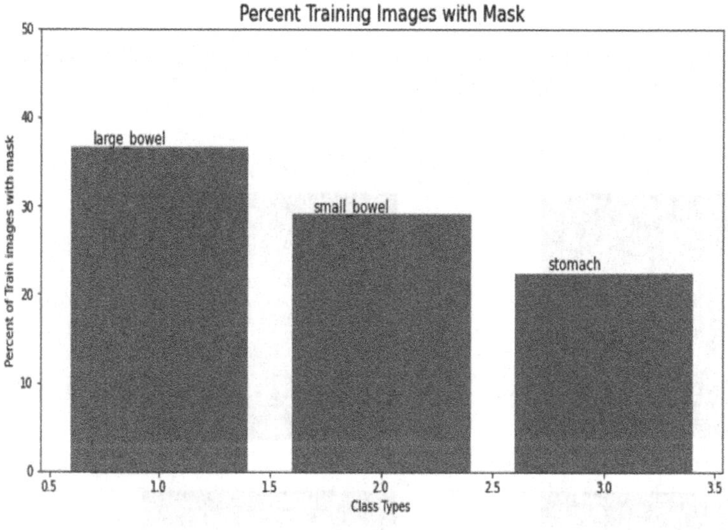

Fig. 4. Percent Training Images with Masks

The test df variable contains the first 300 rows of the original train dataset. A new dataset containing targets and labels is created and feature extraction on the original and test datasets is completed using a helper function.

Pre-processing of the dataset is done by rearranging and erasing the incorrectly categorized data from the dataset, and then the number of images taken with masks is determined. Plotting and masking the images with color patches for large bowel, small bowel, and stomach (plot of images of train dataset with the mask side by side) is shown in (Fig. 4).

On the training dataset, run StratifiedGroupKFold Cross-Validation [7]. Stratification will be used to handle/take care of the imbalance of classes in the training dataset. Using StratifiedGroupKFold Cross-Validation, divide dataset into training and validation datasets. Then, using training and validation data, create training and validation generators. Generators will be input into the model through training and validation data.

4 Model Structure

The convolutional layer, fully connected layer, and pooling layer are the three primary neural layers that make up a convolutional neural network, or CNN [8,9]. U-NET is a U-shaped encoder-decoder network architecture deep learning convolutional network made up of four encoder blocks and four decoder blocks linked by a bridge. The encoder is all about the covenant layers followed by pooling operation. It is used to extract the factors in the image. The second part decoder uses transposed convolution to permit localization [5] (Fig. 5).

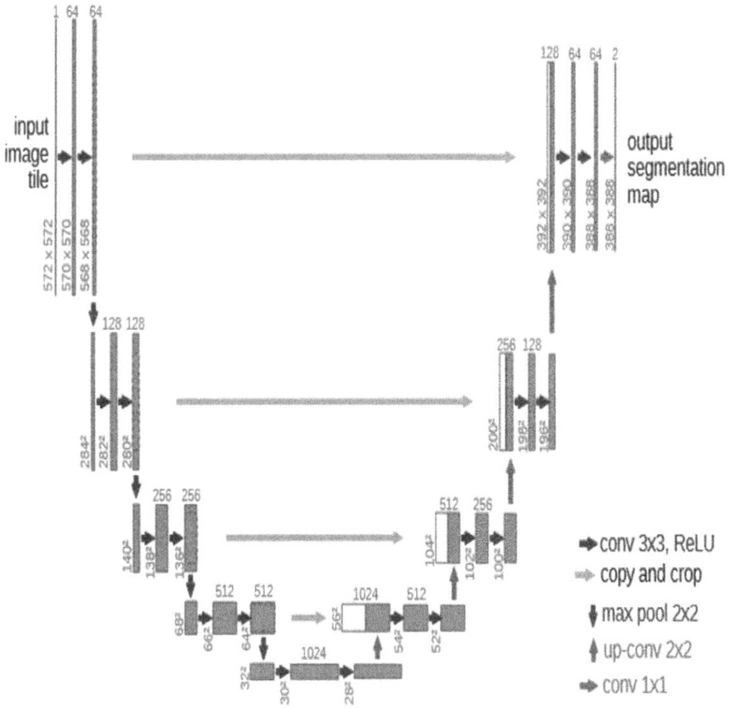

Fig. 5. U-Net architecture, original publication [5]

In this research, Inception-ResNet-v2 [10] encoder, a widely-used convolutional neural network architecture trained on an Imagenet database with more than 1 million images were used due to its proven performance in image segmentation tasks. The application of this model to MRI scan segmentation in this study provides new insights and a valuable approach to enhancing segmentation accuracy in medical imaging. The Inception-ResNet-v2 model was selected for this research due to its ability to handle large-scale image data and its incorporation of residual connections, which help mitigate the vanishing gradient problem common in deep architectures. These features make it particularly well-suited for MRI image segmentation tasks, where precision and computational efficiency are critical. Inception-ResNet-v2 has one input channel and three output channels. A separate segmentation mask corresponds to each of the three output channels. The output channels serve as masks for one of the three output channels.

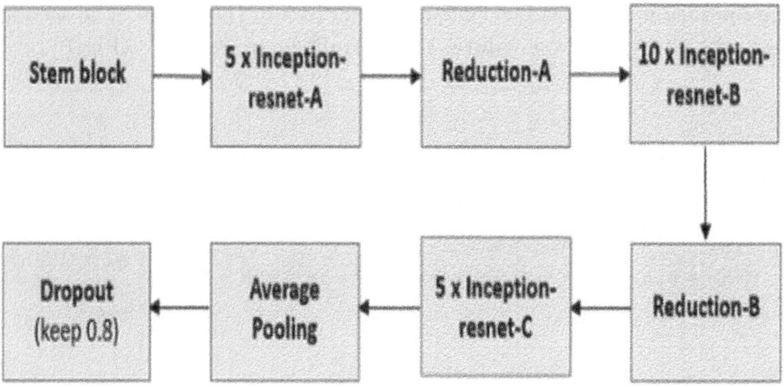

Fig. 6. The basic architecture of Inception-Resnet-v2

Inception-ResNet-v2 classifies each pixel in the MRI scans images into a particular class. The data is condensed using a down-sampling (max-pooling) method, and the segmentation masks that result are then reconstructed using up-sampling (up convolution/convolution transpose) rather than pooling layers to increase the output's resolution. A successive convolution neural layer in the decoder part can then learn to create a more precise output based on this data [5].

The Inception structure and the Residual connection were combined to create Inception-ResNet-v2. Multiple convolutional filters of various sizes were mixed with residual connections in the Inception-Resnet block. The incorporation of residual connections reduces training time while also avoiding the degradation problem caused by deep structures. The Inception-Resnet-v2 network architecture is shown in the figure below (Fig. 6).

5 Performance Measurement

In this research, "Intersection over Union" [12] and "Dice Coefficient" [11] were used to measure the model performance. And the combination of "Binary Cross Entropy Loss" [13] and "Dice Loss" [11] as a loss function for optimizing the model.

Performance Metrics: A typical metric for pixel segmentation that can also be changed to behave as a loss function is the Dice coefficient, also known as the Dice-Sorensen coefficient:

$$DSC = 2|X \cap Y|/|X| + |Y| \tag{1}$$

Similar to the Dice metric, the IoU metric, also known as the Jaccard Index, is determined by dividing the overlap of positive instances between two sets by the sum of their individual values:

$$J(A,B) = |A \cap B|/|A \cup B| = |A \cap B|/|A| + |B| - |A \cap B| \tag{2}$$

Used Loss Function. The usual binary cross-entropy (BCE) loss, which is typically the default for segmentation models, is combined with dice loss in this loss. Combining the two approaches permits some loss diversity while gaining from BCE's consistency. Anyone that has studied logistic regression will be familiar with the equation for multi-class BCE alone:

$$L = -\sum_{i=1}^{N}\sum_{c=1}^{C}(y_{i,c}\log(\hat{y}_{i,c}) + (1 - y_{i,c})\log(1 - \hat{y}_{i,c})) \tag{3}$$

where:

- N is the number of samples.
- C is the number of classes.
- $y_{i,c}$ is the true label for the i-th sample for class c (0 or 1).
- $\hat{y}_{i,c}$ is the predicted probability that the i-th sample belongs to class c.

6 Training

To train the model, we feed the training dataset to the model with the Adam optimizer [14] implementation of Keras with an initial learning rate of 1e-3. This learning rate is chosen based on empirical studies and preliminary experiments, which showed that it offers a good balance between convergence speed and stability. The training datasets contain the input images with their segmentation masks. The batch size of 30 is used which is informed by standard practices in the field and previous work on similar datasets. The model is trained for 20 epochs, which is determined through cross-validation experiments to ensure that the model achieves good performance without overfitting. The dataset is divided into a training dataset and a validation dataset using 80/20 split using StratifiedGroupKFold Cross-Validation to mitigate overfitting and assess model performance. The models are trained using a training dataset and the performance is evaluated using a validation dataset. For classification tasks with neural networks, the logistic sigmoid function is used as the activation function due to its effectiveness in producing probabilities for binary classification. The hyperparameters, including the learning rate, batch size, and number of epochs, are selected based on a combination of empirical evidence from related work and systematic experimentation. After the training, the model is saved in the 'UNET-model' folder. Then, a model.predict function is used to predict the tested results from the trained model. The performance of the model is described in the next section.

7 Experimentatal Results

The execution of the project was carried out on the Kaggle environment. The model was trained for 20 epochs tracking loss, Dice coefficient, and IoU. From the graph below (Fig. 7), we can see that both the training loss and validation loss are decreasing over time, with the validation loss fluctuating more than the training loss. This could indicate some generalization challenges, but overall the model is improving.

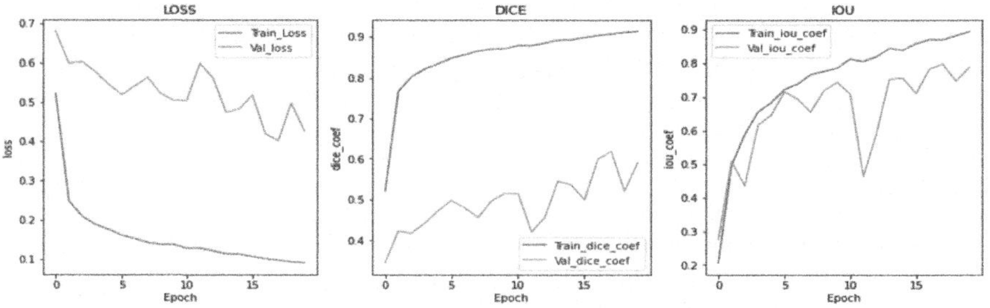

Fig. 7. Loss and Accuracy achieved using IoU, Dice and (BCE + Dice) Loss function

Increasing the number of records in the training dataset improves the performance of the model. Similarly, both training Dice coefficient and training IoU are steadily improving, but the validation Dice and validation IoU show variability, which suggests that the model might be overfitting slightly to the training data.

The improvement in both metrics for the training set indicates that the model is learning effectively, but the fluctuations in the validation set need attention. These values can range from 0 (worst) to 1 (best), and the higher these values, the better the segmentation accuracy.

Loss: The decreasing loss is a positive sign, but the difference between the training and validation loss may suggest a need for more data, better regularization, or more fine-tuning of the model to improve generalization.

Dice Coefficient: A score closer to 1 indicates better overlap between the predicted and actual segmentation. In this case, while the training Dice coefficient improves, validation fluctuates, which could mean that further improvements can be achieved with more epochs or adjustments in the learning rate.

IoU: Similar to Dice, IoU measures the overlap between the predicted segmentation mask and the ground truth. With IoU, a value near 1 shows strong model performance, and while the training set shows clear improvement, the validation IoU suggests that more tuning is needed for better generalization.

The Table 1 shows the experimental results of using the Inception-ResnetV2 encoder over 20 epochs. A loss of 9% on the training dataset indicates that the model is learning effectively, minimizing the difference between predicted and actual segmentation. An Intersection over Union score of 89% is quite high, showing that the predicted segmentation overlaps significantly with the ground truth. A Dice coefficient of 91% also confirms that the model is performing well in terms of overlap between predictions and ground truth, which is in line with the IoU score.

Table 1. Result achieved using Inception-ResnetV2 encoder

Training and Validation Accuracy over 20 epochs						
Encoder	Loss	IoU Coeff	Dice Coeff	Val Loss	Val IoU Coeff	Val Dice Coeff
Inception-ResnetV2	9%	89%	91%	42%	78%	59%

Finally, the trained model was tested with the test dataset and the predicted result is shown in the figure below. It can be comprehended from the result (Fig. 8) that the segmentation has been accurate for most of the predictions.

The regions corresponding to the Large Bowel, Small Bowel, and Stomach are well-defined in the majority of the samples, indicating that the model has learned to distinguish between the different parts effectively.

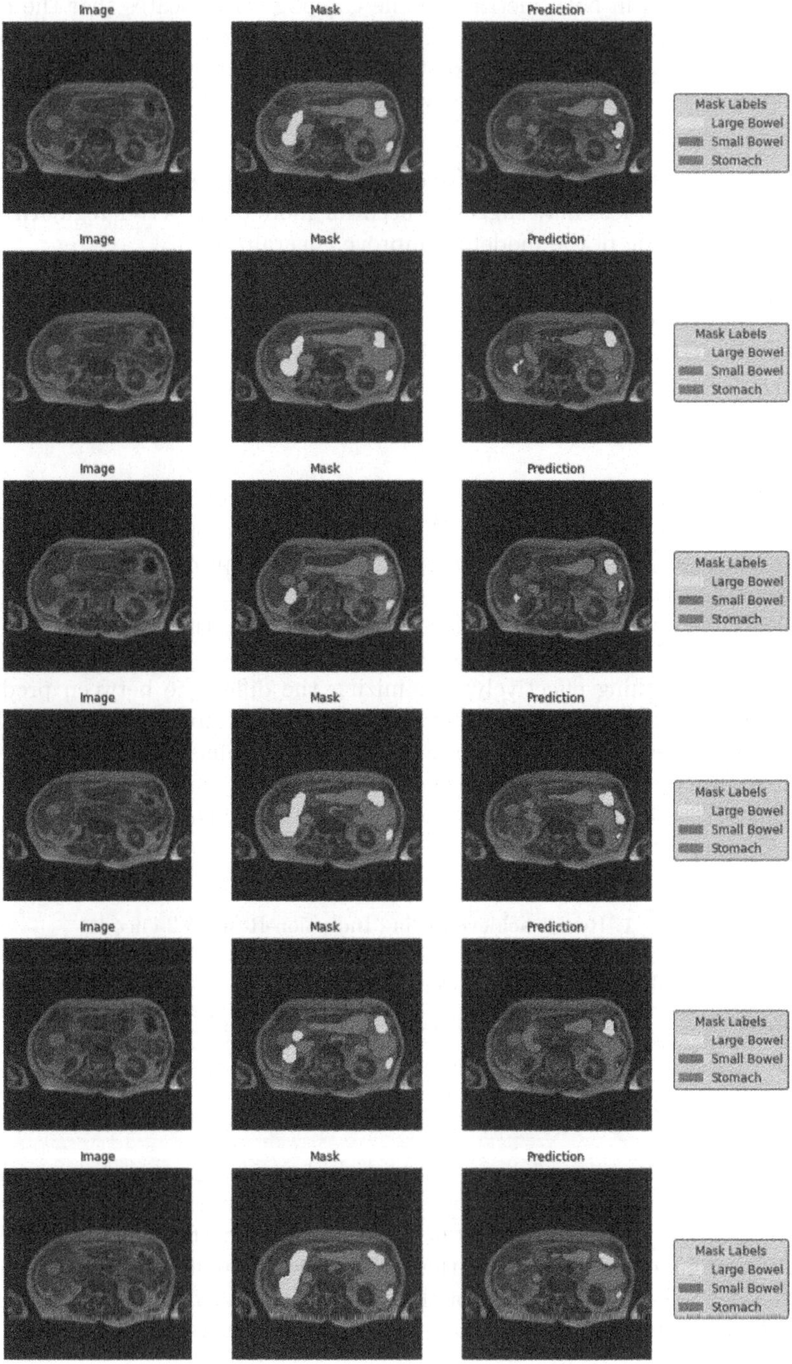

Fig. 8. Final Predictions after training

8 Conclusion and Future Work

In this research, we implemented and modified the U-Net model with an InceptionResNetV2 encoder and applied them to the GI Tract Image Segmentation Problem. Our Model achieved a Dice Loss Score of 91% on the training dataset and 62% on the test dataset. Similarly, the IoU Loss Score of 89% on the training dataset and 73% on the test dataset. The application of well-established architectures like U-Net and InceptionResNetV2 to this specific problem domain, combined with in-depth data analysis and preprocessing techniques, yields highly accurate and effective segmentation mask predictions. This innovative integration of existing models with tailored data handling represents a fresh approach to addressing the challenges in this field. U-Net has proven to be ideal for complex segmentation tasks, but building, scaling, and tuning it to achieve the desired output was a challenging and time-intensive process. Efficient data preprocessing followed by hyperparameter tuning of the InceptionResnetV2 Model helps to perform productive and accurate prediction of the segmentation masks in this research.

Though the segmentation is generally accurate, there may be slight misalignments or uncertainties in some images. These can be highlighted as areas for potential improvement, either through model refinement or additional data preprocessing. It's essential to emphasize that while the model performs well on many predictions, continued tuning could enhance performance for edge cases. The model demonstrated strong prediction accuracy for the small bowel and stomach, but struggled with the large bowel (particularly left part on the images Fig. 8), which, despite comprising 38% of the dataset, still faced performance issues compared to the smaller proportions of the other organs. This discrepancy may be attributed to the relative distribution of images, which could have led to less effective training for the large bowel. Future work should explore methods such as data augmentation or class re-weighting to address this imbalance and enhance prediction accuracy.

References

1. Rawla, P., Barsouk, A.: Epidemiology of gastric cancer: global trends, risk factors and prevention. Gastroenterol. Rev. **14**(1), 26–38 (2019). https://doi.org/10.5114/pg.2018.80001
2. Jaffray, D.A., Gospodarowicz, M.K.: Radiation therapy for cancer, Disease Control Priorities, Third Edition (Volume 3): Cancer, pp. 239–247, 2015
3. "UW-Madison GI tract image segmentation," Kaggle. https://www.kaggle.com/competitions/uw-madison-gi-tract-image-segmentation/overview. Accessed 13 Aug 2022
4. Jha, D., et al.: Resunet++: an advanced architecture for medical image segmentation. In: IEEE International Symposium on Multimedia (ISM). IEEE 2019, pp. 225–2255 (2019)

5. Ronneberger, O., Fischer, P., Brox, T.: U-Net: convolutional networks for biomedical image segmentation. In: Navab, N., Hornegger, J., Wells, W.M., Frangi, A.F. (eds.) MICCAI 2015. LNCS, vol. 9351, pp. 234–241. Springer, Cham (2015). https://doi.org/10.1007/978-3-319-24574-4_28
6. Ghosh, S., Chaki, A., Santosh, K.: Improved u-net architecture with vgg-16 for brain tumor segmentation. Phys. Eng. Sci. Med. **44**(3), 703–712 (2021)
7. "sklearn.model_selection.StratifiedGroupKFold," scikit-learn. https://scikit-learn.org/stable/modules/generated/sklearn.model_selection.StratifiedGroupKFold.html. Accessed 06 Aug 2022
8. Malhotra, P., Gupta, S., Koundal, D.: Computer aided diagnosis of pneumonia from chest radiographs. J. Comput. Theor. Nanosci. **16**(10), 4202–4213 (2019)
9. Dargan, S., Kumar, M., Ayyagari, M.R., Kumar, G.: A survey of deep learning and its applications: a new paradigm to machine learning. Arch. Comput. Methods Eng. **27**(4), 1071–1092 (2020)
10. Szegedy, C., Ioffe, S., Vanhoucke, V., Alemi, A.: Inception-v4, inception-ResNet and the impact of residual connections on learning. In: Proceedings of the AAAI Conference on Artificial Intelligence, vol. 31, no. 1, February 2017. https://doi.org/10.1609/aaai.v31i1.11231.
11. Zhang, Y., Liu, S., Li, C., Wang, J.: Rethinking the dice loss for deep learning lesion segmentation in medical images. J. Shanghai Jiaotong Univ. (Science) **26**(1), 93–102 (2021). https://doi.org/10.1007/s12204-021-2264-x
12. Zhou, D., et al.: IoU loss for 2D/3D object detection, 2019. https://doi.org/10.48550/ARXIV.1908.03851, https://arxiv.org/abs/1908.03851
13. Shafiei Neyestanak, M., Jahani, H., Khodarahmi, M., Zahiri, J., Yekaninejad, M.S.: A Quantitative Comparison between Focal Loss and Binary Cross-Entropy Loss in Brain Tumor Auto-Segmentation Using U-Net, SSRN Electronic Journal, 2022. https://doi.org/10.2139/ssrn.4142314
14. Kingma, D.P., Ba, J.: Adam: a method for stochastic optimization, 2017. arXiv: 1412.6980 [cs.LG]
15. Chou, A., Li, W., Roman, E.: GI Tract Image Segmentation with UNet and Mask R-CNN
16. Alam, M.J., Zaman, S., Shill, P.C., Kar, S., Hakim, M.A.: Automated gastrointestinal tract image segmentation of cancer patient using LeViT-UNet to automate radiotherapy. In: 2023 International Conference on Electrical, Computer and Communication Engineering (ECCE), pp. 1–5 (2023)

Low Consumption Models for Disease Diagnosis in Isolated Farms

I. X. Vázquez[1(✉)], A. M. García-Vico[2], H. Seker[3], and J. Sedano[1]

[1] ITCL Technology Center, 09001 Burgos, Spain
iago.vazquez@itcl.es
[2] Andalusian Research Institute in Data Science and Computational Intelligence (DaSCI), University of Jaén, 23071 Jaén, Spain
[3] Department of Information Systems, College of Computing and Informatics, The University of Sharjah, Sharjah, UAE

Abstract. The detection of bacterial and viral microbes is pivotal for both human and animal well-being in public health services and veterinary care, but it traditionally requires time-consuming procedures and expert technicians. However, the rise of Machine Learning and Deep Learning has led to a surge in the application of new techniques that can perform bacterial and viral detection faster and at a lower cost. Yet, despite that success, Deep Learning approaches tend to have high energy demands, which can in some contexts limit their application, increasing both costs and environmental concerns. In this study, a new hybrid methodology, in which an Artificial Neural Network was combined with a more energy efficient Spiking Neural Network, was employed to develop a model able to classify 18 species of Eimeria parasites, affecting both rabbits and chickens, from microscope images. We show how significant energy savings can be obtained from SNN layers, while their use in the model can improve its performance.

Keywords: Eimeria · Microorganisms Detection · Green AI · Spiking Neural Network

1 Introduction

The detection of bacterial and viral microbes is of paramount importance to the public health services and for veterinary health care. Historically, the identification of those microscopic entities has relied on labor-intensive techniques that require skilled personnel [20]. The rise of Machine Learning facilitates rapid recognition of pathogens for the diagnosis of various illnesses, reducing reliance on skilled personnel and minimizing the potential spread of infection, which can be especially useful on remote farms with limited access to both veterinary and computer services.

The difficulty of executing Deep Learning models on isolated farms in countries such as India and Pakistan, with limited access to various services, is the

V. Julian et al. (Eds.): IDEAL 2024, LNCS 15346, pp. 233–243, 2025.
https://doi.org/10.1007/978-3-031-77731-8_22

main motivation behind this research. Despite the scarce resources of isolated farms, farmers may well have access to a cellphone with a high-resolution camera in outlying villages, in such a way that a Deep Learning model could be run on a cellphone to recognize the infectious pathogen from microscopic images. But, the use of such models could impact the battery duration, so a path to reduce energy consumption is required. So, our efforts have focused on the development of a more energy-efficient deep learning model, in line with the concept of Green-AI [22], an emerging concept that is a call to action to develop energy-efficient and environmentally friendly [22] AI techniques.

Among the available energy-efficient Deep Learning techniques [16], an interesting approach is the Spiking Neural Network (SNN) [15], which process information in a sparse, asynchronous and binary signals referred as spikes [17]. One key advantage of SNNs over a traditional ANN is that their connections only consume energy when a spike crosses them. As there tend to be more connections than neurons, especially within large neural networks, the energy consumption levels of an ANN are more critical and higher than the processing produced in spiking neurons. Therefore, if the number of spikes is kept low, an SNN can produce more energetically efficient models than an ANN.

In this work, we present a hybrid model that combines SNN and traditional ANN layers [11,13] to perform Eimeria Parasite classification from microscopic images. That model can differentiate among 18 different species, 7 of them affecting to chickens and the other 13 affecting to rabbits. The main objective is to calibrate the balance between the energy efficiency of an SNN and the performance of an ANN, in order to be able to run the model on a cellphone with a minimal performance loss, if it is the case.

The main contribution of this paper is the development of the abovementioned model. The use of a single model for both animal species, decreases the computational resources required to perform the diagnosis. This fact, along with the higher energy efficiency achieved by it, eases its implementation on devices such as a cellphone, facilitating its use in remote and isolated farms with limited access to energy and computational resources.

On the other hand, microscopic images taken at different times can show illumination changes. Moreover, overlapping is expected to happen on them, as it can be difficult to take a picture of just one parasite. For those reasons, we have applied illumination changes and overlapping of different pictures to simulate real word conditions, and make our model more reliable in practical applications.

The paper is organized in the following way: in Sect. 2, similar research lines to this study are presented. In Sect. 3, the technologies used in our study are exposed. In Sect. 4, our proposed model is introduced. In Sect. 5, the experimental design is detailed. In Sect. 6, the results are presented and, finally, the key points of our proposal are summed up in Sect. 7.

2 Related Work

Microorganism detection has traditionally relied on microscopic examination [18] and culture techniques [19]. Features such as measurement size, shape, symmetry, and texture have been used by biologists [1] to classify microscopic images of Eimeria species. However, several shortcomings are associated with such time-consuming methods that require specialized laboratory technicians [14]. Additionally, microorganism classification can be challenging, due to the similarities between different species [20].

In recent years, researchers have proposed methods based on Machine Learning for automatic microorganism classification [26]. Probabilistic neural networks [24], bilateral filtering along with morphological image processing [6], Bayesian Classifiers [3] were proposed. Notably, a common thread among these methods is their reliance on threshold algorithms for image segmentation and feature extraction, such as shape, area, and size. While some of those methodologies achieve commendable accuracy, they often entail specific pre-processing steps that might not be suitable for other related problems.

However, in more recent times, Deep Learning techniques have been applied to this challenge. Convolutional Neural Networks were proposed [7]. Also, in combination with them, transformers [8] and transfer learning [4]. Models like Inception or U-Net [27] were also applied. Those advanced Deep Learning methodologies often obviate the need for explicit feature extraction, making the application of those methodologies more feasible for solving similar problems and for detecting other sorts of parasites with different shapes and features.

It is remarkable that other researchers have been focused on improving both performance and energy efficiency. For instance, [21] developed a model capable of detecting Covid-19 in chest X-ray images, combining CNNs with Quantum classifiers, and achieving high accuracy while reducing the computational cost compared to other Deep Learning-based approaches.

3 Background

In this section, the technologies used to develop our research are presented.

3.1 Spiking Neural Networks

An Spiking Neural Network (SNN) [15] is a kind of Artificial Neural Network that processes data through sparse, asynchronous binary signals referred to as spikes [17].

The advantage of the SNN in terms of energy efficiency is due to the fact that information is coded in the form of sparse spikes through time. In an SNN, Multiply-Accumulate (MAC) operations are performed with each voltage update, which is linearly dependent on the number of neurons, and when a spike is generated [12]. Meanwhile, the number of MAC operations in a traditional dense ANN mainly depends on the number of connections, which grows quadratically with regard to the number of neurons. For this reason, if the number of spikes that are generated remains low, it is clear that inference at SNNs can require less MAC operations than traditional ANNs, which carries to a higher energy efficiency.

Rate Coding. To feed an SNN with information, it is necessary to code input data somehow. The traditional input coding method in an SNN has been Rate coding [10]. This method is based on assigning higher spike frequency to higher input values. Numerical values are transformed into a sequence of spikes, which is represented as a vector whose elements can be either one (representing an spike) or zero (representing the absence of spikes). Spike positions are random, given that the information is encoded just in their frequency.

Neuron Types. Different models of spiking neurons have been developed over time. Among them, we have performed our research for two neuron models:

- *Current-Based Neuron (CUBA)* ([2]): Neuron that models a leaky resistor in a parallel combination with a capacitor. When the potential exceeds a fixed threshold, it resets to a predefined value and an spike is generated. Its inputs are defined using the expression in Eq. 1:

$$I_i(t) = -\sum_j \omega_{ji}(t)(\bar{V} - E_j) \tag{1}$$

  where i identifies the neuron that is being fed with current, ω_{ji}, the synaptic weights between the precedent neuron, j, and the current one, i, and E_j corresponds to the resting potential of the precedent neuron, j. $\bar{V}$ is the mean membrane potential, taken as a constant.
- *Resonate-and-fire Izhikevich neuron (RIF)*: Neuron model developed to show damped oscillations in the membrane potential. Its dynamics are described by the complex differential Eq. 2

$$\frac{dz}{dt} = z\omega + I(t) \tag{2}$$

  Here, $I(t)$ represents the input to the neuron at a given instant, t, while $\omega \in \mathbb{C}$ is an internal, complex constant related to the neuron, and $z \in \mathbb{C}$ is another complex number representing its internal state at instant t. As with the previous model, when a spike is generated, the internal state is reset to a predefined complex parameter.

4 Proposed Model

The core idea of our proposal is to fine-tune a pre-trained model (PTM) to perform a desired task, by using Network-based transfer learning [9], and gradually to substitute the previously trained final dense ANN layers for their SNN counterparts. The process of substituting ANN layers for SNN layers is continued and their best-quality configuration is selected according to a criterion that takes into account both performance and energy efficiency. That quality measurement guarantees that the most energy-efficient model with tolerable accuracy loss (if it is the case) is selected, making the final model more cost-effective and environmentally friendly than the original one.

4.1 Data Flow Through the Model

A vision transformer [5] was employed as the PTM. Concretely, the used model was the one presented in [25]. Images are scaled to 224×224 pixels, and the PTM is applied to them. After that, each of them is transformed into a set of 1000 features that are fed into a variable number of traditional ANN dense layers. Each single value of the last dense layer was coded into a sequence of spikes, measured in milliseconds. That sequence of spikes is generated by following rate coding, in such a way that its spike frequency corresponds to the numerical value that is being coded. The set of all sequences of spikes generated in that way is passed to the SNN layers. Finally, the last spiking layer will generate a sequence of output spikes for each class. The class with the highest frequency of spikes will be taken as the prediction for the input image.

In Fig. 1, an schematic representation of the presented flow can be seen.

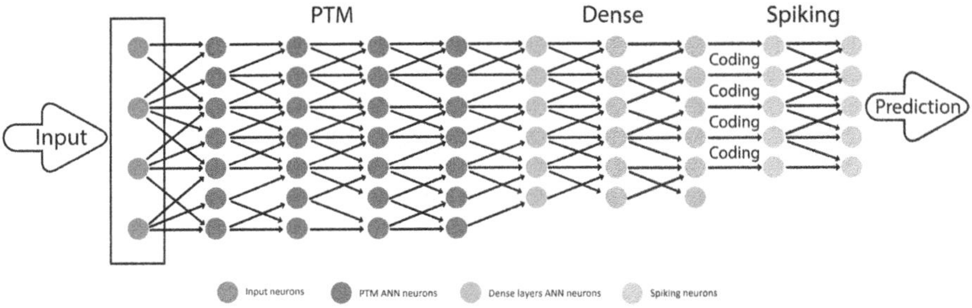

Fig. 1. Data flow through the proposed models.

4.2 Training Procedure

As processing in a Spiking Neural Network occurs through discrete events called spikes, outputs from spiking neurons are non-differentiable, which prevents the direct application of backpropagation. Therefore, SNN layers and ANN layers must be separately trained. Thus, the training process for a single model with SNN starts with the training of the model, M_{ANN}, which initially contains only traditional ANN layers. Once the model has been trained, the last n dense layers are replaced by their SNN equivalents, while the remaining traditional layers are kept frozen, giving rise to the model: M^n_{SNN}. Finally, the SLAYER [23] method is applied to train those newly added SNN layers, and a pre-fixed quality criterion is applied to rate the generated models, taking into account both the model performance and its energy efficiency.

As the level of tolerable performance loss can vary from one application to other, the best criterion to rate generated models might be to set a predefined minimum acceptable performance and to select the SNN model that achieves the lowest energy consumption, while maintaining a performance higher than the predefined threshold. This leads us to the definition of the quality metric Q, which can be defined as exposed in Eq. 3:

$$Q(M_{SNN}) = \begin{cases} 0 & \text{if } A_{M_{SNN}} < \alpha A_{M_{ANN}} \\ \frac{E_{M_{ANN}}}{E_{M_{SNN}}} & \text{other case} \end{cases} \tag{3}$$

where $E_{M_{SNN}}$ corresponds to the energy consumption associated to the M^n_{SNN} model being rated, and $E_{M_{ANN}}$, to the energy consumption associated to M_{ANN}. In the same way, $A_{M_{SNN}}$ corresponds to the accuracy achieved by M^n_{SNN}, while $A_{M_{ANN}}$, to the accuracy achieved by M_{ANN}. α is a parameter that defines the acceptable accuracy loss for SNN models.

Following this training procedure, six dense layers were added to the PTM to perform the first fine-tuning step using backpropagation. Afterwards, six candidates to be the optimal model were generated by recursively substituting the ANN traditional layers by their SNN counterparts, and trained by applying the Slayer method to those newly added layers.

5 Experimental Design

Two datasets acquired from [3] were used, each one contained images of oocysts belonging to microparasites of different species from the genus Eimeria. The first dataset consisted of 4460 images of 7 species of parasites that infect chickens, while the second one contained 7237 images of 11 species of parasites that infect rabbits. In the Fig. 2, some of the used images can be seen.

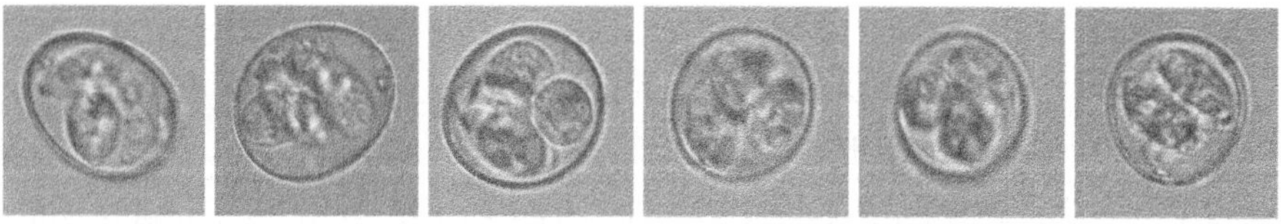

(a) acervulina (b) maxima (c) praecox (d) necatrix (e) mitis (f) tenella

Fig. 2. Images of the different parasites in the chicken dataset.

The previously presented PTM was applied. Then, as shown in Subsect. 4.2, six dense ANN layers were added and Network-based transfer learning was applied to fine-tune the M_{ANN} model with the two selected datasets combined. Finally, M^n_{SNN} models, with $n = 1, ..., 6$ final SNN layers each, were trained by following the procedure described in Sub-sect. 4.2.

This pipeline was applied to the combination of the chicken and rabbit datasets using the CUBA and RIF neurons, for exposure times of 2, 4, and 6 milliseconds, and for 800 and 1500 training epochs. Overlapping and brightness changes were applied to the input images, to simulate real-world microscopy images without any post-processing. Specifically, brightness changes were applied to one third of the images, overlapping of two random images was applied to other third (ensuring that the same image was not used in different folds), and both changes were applied to the last third. Translations were applied with overlapping to prevent the two cells from being perfectly aligned, which would not be realistic. Subsequently, for each exposure time, training and evaluation were conducted using a 5-fold cross-validation approach. Accuracy was determined by calculating the ratio of correctly classified examples to the total number of examples, while energy consumption was computed using the method described in Sub-sect. 5.1. Subsequently, a quality measure was assigned to each model following the approach outlined in Subsect. 4.2, and the model with the highest quality measurement was selected as the final M_{SNN} model. The minimum acceptable accuracy α for the SNN models was set at 97% of the accuracy achieved by the equivalent ANN model.

5.1 Energy Consumption Estimation

Following the methodology presented in [12], the energy consumption levels of a traditional ANN and an SNN were estimated.

Broadly, the energy consumption levels were estimated by counting the number of operations, N, required when inference was carried out on a single sample.

In the case of inference on ANNs, a MAC operation was performed each time information crossed a synapse from one layer to the following one. For ANNs composed by dense layers, the number of operations required for inference, N_{ann} could be estimated as:

$$N_{ann} = \sum_{i=1}^{l-1} n_i n_{i+1} \tag{4}$$

Here, l is the total number of layers, while n_i corresponds to the number of neurons at layer i.

On the other hand, the number of operations required to execute an SNN can be estimated as the sum of the total number of spikes generated plus the total number of voltage updates, following the Eq. 5.

$$N_{snn} = N_u + N_s = T(\sum_{l \in L} n_l + \sum_{i=1}^{l-1} \bar{s}_i n_i n_{i+1} + \bar{s}_l n_l) \tag{5}$$

where N_u is the number of operations due to voltage updates, N_s is the number of operations due to the generated spikes. T is the exposure time, L the set of the layers in the network, and n_i the number of neurons in the i-th layer. $\bar{s}_i$ is the mean number of spikes going from the i-th layer to the $i+1$-th layer each millisecond. The last term in the formula, $\bar{s}_l n_l$, represents the output spikes arising from the last layer of the model.

6 Results

The best models of the entire experimental program are listed in Table 1. The best model was selected for each exposure time and neuron model by following the quality metric Q as introduced in Eq. 3.

The neuron type used to run each experiment layers was identified in the table. Column T represents the exposure time set for the SNN layers, measured in milliseconds. The column Model indicates the selected architecture, following the notation introduced in Sect. 4.2, and the column Epochs shows the number of epochs used to train the corresponding model. The column N. Ops. represents the number of millions of operations required for a single inference step, which is proportional to the estimated energy consumption, as shown in Sub-Sect. 5.1. Additionally, Acc. showcases the accuracy of the selected models. Precision, recall, and F1 are listed in the following columns. Those values represent the mean of each of these metrics for each class in the dataset. Finally, the Q column shows the quality metric of the model. The best values for the columns N. Ops., Acc., Precision, Recall, F1, and Q are highlighted in bold for each dataset. ANN models with the same number of epochs as SNN models are also listed.

Table 1. Best models found for each exposure time and neuron type. T (ms) corresponds to the exposure time in milliseconds, while Q corresponds to the quality of the obtained model. The best results found for each metric are highlighted in bold.

Neuron	T (ms)	Model	Epochs	N. Ops. (M)	Acc.	Precision	Recall	F1	Q
ANN	N/A	M_{ANN}	1500	4.509 ± 0	0.749 ± 0.016	0.76 ± 0.02	0.749 ± 0.017	0.745 ± 0.021	1
	N/A	M_{ANN}	800	4.509 ± 0	0.764 ± 0.015	0.771 ± 0.016	0.765 ± 0.016	0.763 ± 0.016	1
CUBA	2	M^4_{SNN}	1500	$\mathbf{2.612 \pm 0.03}$	0.742 ± 0.028	0.752 ± 0.032	0.747 ± 0.027	0.742 ± 0.028	**1.726**
	4	M^3_{SNN}	800	3.637 ± 0.068	0.776 ± 0.018	$\mathbf{0.786 \pm 0.02}$	0.78 ± 0.017	0.778 ± 0.017	1.24
	6	M^4_{SNN}	1500	3.625 ± 0.218	0.745 ± 0.026	0.75 ± 0.031	0.749 ± 0.025	0.745 ± 0.027	1.244
RIF	2	M^4_{SNN}	1500	2.634 ± 0.035	0.745 ± 0.037	0.752 ± 0.04	0.749 ± 0.034	0.746 ± 0.037	1.712
	4	M^3_{SNN}	800	3.663 ± 0.077	$\mathbf{0.779 \pm 0.026}$	$\mathbf{0.786 \pm 0.031}$	$\mathbf{0.782 \pm 0.024}$	$\mathbf{0.78 \pm 0.026}$	1.230
	6	M^3_{SNN}	1500	4.003 ± 0.158	0.769 ± 0.021	0.777 ± 0.026	0.773 ± 0.018	0.77 ± 0.021	1.126

The most energy-efficient model found, in terms of the number of operations performed, was M^4_{SNN} with two milliseconds of exposure time and a CUBA neuron. The higher accuracy was achieved by the SNN model M^3_{SNN} with a RIF neuron. For both kinds of neurons, it is clear that M^4_{SNN} with 2 milliseconds of exposure time achieves a similar accuracy to the ANN, while their energy consumption is around 40% lower than the traditional ANN. On the other hand, the model M^3_{SNN} with 4 milliseconds of exposure time achieves a 19% of energy savings regarding the corresponding ANN, while its accuracy is a 3% higher than its ANN counterpart.

According to the obtained results, models with SNN layers have shown a remarkable robustness against illumination and overlapping changes. Moreover, the notable energy efficiency achieved by models with SNN layers paves the way to the development of highly efficient Deep Learning models which could be run on a cellphone. This would make Deep Learning models more accessible for farms with energy supply restrictions, contributing to the access of disadvantaged people, reducing the technological gap. Moreover, those high energy savings could help to develop more environmental-friendly models.

7 Conclusions and Future Work

In this study, a Deep Learning model that incorporates both ANN and SNN layers was developed to classify different microorganisms of various Eimeria species. Eimeria species are a common source of infection among farm animals, such as chickens and rabbits. Rapid detection is essential for minimizing both the impact of such illnesses on animal health and any consequent economic repercussions.

Our results show that models with Spiking layers are often able to get better results than their traditional counterparts, in addition to require a lower energy consumption at inference, saving battery on the device used to run them. Those results are promising regarding the development of more environmentally friendly models that can operate efficiently with fewer resources, without compromising performance, and set the stage for quick diagnosis in isolated and remote farms.

Acknowledgment. This research has been founded from the Recovery, Resilience and Transformation Plan under the project CICERO: CDTI Project CER-20231019 (CICERO).

References

1. Abdalla, M.A.E.: Developing pixel-based feature sets for intelligent identification of eimeria species from microscopic images (2018). https://api.semanticscholar.org/CorpusID:198334004
2. Brette, R., et al.: Simulation of networks of spiking neurons: a review of tools and strategies. J. Comput. Neurosci. **23**(3), 349–398 (2007). https://doi.org/10.1007/s10827-007-0038-6
3. Castañón, C.A., Fraga, J.S., Fernandez, S., Gruber, A., Costa, L.D.F.: Biological shape characterization for automatic image recognition and diagnosis of protozoan parasites of the genus eimeria. Pattern Recognit. **40**(7), 1899–1910 (2007). https://doi.org/10.1016/j.patcog.2006.12.006, https://www.sciencedirect.com/science/article/pii/S0031320306005206
4. Chen, W.C., Liu, P.Y., Lai, C.C., Lin, Y.H.: Identification of environmental microorganism using optimally fine-tuned convolutional neural network. Environ. Res. **206**, 112610 (2022) 10.1016/j.envres.2021.112610, https://www.sciencedirect.com/science/article/pii/S0013935121019113
5. Dosovitskiy, A., et al.: An image is worth 16 × 16 words: transformers for image recognition at scale. In: 9th International Conference on Learning Representations, ICLR 2021, Virtual Event, Austria, 3–7 May 2021. OpenReview.net (2021). https://openreview.net/forum?id=YicbFdNTTy

6. Fatima, T., Farid, M.S.: Automatic detection of plasmodium parasites from microscopic blood images. J. Parasit. Dis.**44**, 69–78 (2019). https://api.semanticscholar.org/CorpusID:202688228
7. Fuhad, K.M.F., Tuba, J.F., Sarker, M.R.A., Momen, S., Mohammed, N., Rahman, T.: Deep learning based automatic malaria parasite detection from blood smear and its smartphone based application. Diagnostics **10**(5) (2020).https://doi.org/10.3390/diagnostics10050329, https://www.mdpi.com/2075-4418/10/5/329
8. He, P., et al.: A reliable and low-cost deep learning model integrating convolutional neural network and transformer structure for fine-grained classification of chicken Eimeria species. Poult. Sci. **102**(3), 102459 (2023). https://doi.org/10.1016/j.psj.2022.102459, https://www.sciencedirect.com/science/article/pii/S0032579122007532
9. Iman, M., Arabnia, H.R., Rasheed, K.: A review of deep transfer learning and recent advancements. Technologies **11**(2) (2023).https://doi.org/10.3390/technologies11020040, https://www.mdpi.com/2227-7080/11/2/40
10. Kiselev, M.: Rate coding vs. temporal coding - is optimum between? In: 2016 International Joint Conference on Neural Networks (IJCNN), pp. 1355–1359 (2016). https://doi.org/10.1109/IJCNN.2016.7727355
11. Kosta, A.K., Apolinario, M.P.E., Roy, K.: Live demonstration: Ann vs SNN vs hybrid architectures for event-based real-time gesture recognition and optical flow estimation. In: 2023 IEEE/CVF Conference on Computer Vision and Pattern Recognition Workshops (CVPRW), pp. 4148–4149 (2023). https://doi.org/10.1109/CVPRW59228.2023.00436
12. Kucik, A.S., Meoni, G.: Investigating spiking neural networks for energy-efficient on-board AI applications. a case study in land cover and land use classification. In: 2021 IEEE/CVF Conference on Computer Vision and Pattern Recognition Workshops (CVPRW), pp. 2020–2030 (2021). https://doi.org/10.1109/CVPRW53098.2021.00230
13. Kugele, A., Pfeil, T., Pfeiffer, M., Chicca, E.: Hybrid SNN-ANN: energy-efficient classification and object detection for event-based vision. In: Bauckhage, C., Gall, J., Schwing, A. (eds.) DAGM GCPR 2021. LNCS, vol. 13024, pp. 297–312. Springer, Cham (2021). https://doi.org/10.1007/978-3-030-92659-5_19
14. Kumar, S., Arif, T., Alotaibi, A.S., Malik, M.B., Manhas, J.: Advances towards automatic detection and classification of parasites microscopic images using deep convolutional neural network: methods, models and research directions. Arch. Comput. Methods Eng. State Art Rev. **30**(3), 2013–2039 (2023). https://doi.org/10.1007/s11831-022-09858-w
15. Maass, W.: Networks of spiking neurons: the third generation of neural network models. Neural Netw. **10**(9), 1659–1671 (1997). https://doi.org/10.1016/S0893-6080(97)00011-7, https://www.sciencedirect.com/science/article/pii/S0893608097000117
16. Menghani, G.: Efficient deep learning: a survey on making deep learning models smaller, faster, and better. ACM Comput. Surv. **55**(12) (2023). https://doi.org/10.1145/3578938
17. Pfeiffer, M., Pfeil, T.: Deep learning with spiking neurons: opportunities and challenges. Front. Neurosci. **12** (2018). https://api.semanticscholar.org/CorpusID:53086859
18. Quinn, J.A., Nakasi, R., Mugagga, P.K.B., Byanyima, P., Lubega, W., Andama, A.: Deep convolutional neural networks for microscopy-based point of care diagnostics. In: Doshi-Velez, F., Fackler, J., Kale, D., Wallace, B., Wiens, J. (eds.) Proceedings

of the 1st Machine Learning for Healthcare Conference. Proceedings of Machine Learning Research, vol. 56, pp. 271–281. PMLR, Northeastern University, Boston, MA, USA, 18–19 August 2016. https://proceedings.mlr.press/v56/Quinn16.html
19. Rajapaksha, P., Elbourne, A.J., Gangadoo, S., Brown, R., Cozzolino, D., Chapman, J.: A review of methods for the detection of pathogenic microorganisms. Analyst **144 2**, 396–411 (2019). https://api.semanticscholar.org/CorpusID:53766416
20. Rani, P., Kotwal, S., Manhas, J., Sharma, V., Sharma, S.: Machine learning and deep learning based computational approaches in automatic microorganisms image recognition: methodologies, challenges, and developments. Arch. Comput. Methods Eng. **29**(3), 1801–1837 (2022). https://doi.org/10.1007/s11831-021-09639-x
21. Rao, G.E., B., R., Srinivasu, P.N., Ijaz, M.F., Woźniak, M.: Hybrid framework for respiratory lung diseases detection based on classical CNN and quantum classifiers from chest x-rays. Biomed. Signal Process. Control **88**, 105567 (2024). https://doi.org/10.1016/j.bspc.2023.105567, https://www.sciencedirect.com/science/article/pii/S1746809423010005
22. Schwartz, R., Dodge, J., Smith, N.A., Etzioni, O.: Green AI. Commun. ACM **63**(12), 54–63 (2020). https://doi.org/10.1145/3381831,
23. Shrestha, S., Orchard, G.: Slayer: spike layer error reassignment in time. ArXiv **abs/1810.08646** (2018)
24. Tchinda, B.S., Tchiotsop, D., Tchinda, R., Wolf, D., Noubom, M.: Automatic recognition of human parasite cysts on microscopic stools images using principal component analysis and probabilistic neural network. Int. J. Adv. Res. Artif. Intell. **4**, 26–33 (2015). https://api.semanticscholar.org/CorpusID:15677269
25. Touvron, H., Cord, M., Douze, M., Massa, F., Sablayrolles, A., Jégou, H.: Training data-efficient image transformers & distillation through attention (2021)
26. Zhang, J., et al.: A comprehensive review of image analysis methods for microorganism counting: from classical image processing to deep learning approaches. Artif. Intell. Rev. **55**(4), 2875–2944 (2022). https://doi.org/10.1007/s10462-021-10082-4
27. Zhang, J., et al.: LCU-Net: a novel low-cost u-net for environmental microorganism image segmentation. Pattern Recognit. **115**, 107885 (2021). https://doi.org/10.1016/j.patcog.2021.107885, https://www.sciencedirect.com/science/article/pii/S0031320321000728

Fast and Scalable Recommendation Retrieval Model with Mixed Attention and Knowledge Distillation

Dmytro Androsov(✉)

National Technical University of Ukraine "Igor Sikorsky Kyiv Polytechnic Institute", Beresteiskyi ave. 37, Kyiv 03056, Ukraine
androsovdmitry80@gmail.com

Abstract. Users' purchase decisions are highly influenced not only by their general preferences, but also most recent interactions with a given platform or marketplace. In the case of gathering insights on user behaviour patterns, sequential recommendations are essential for any customer-oriented business, since they enable the prediction and suggestion of the next best basket or action for a user based on their past interactions. The advent of deep neural networks has enhanced the precision of analyzing patterns in sequential recommendations. However, these models are limited by their restricted memory scope and a tendency to prioritize currently popular items, which can decrease recommendation diversity. To address this issue, it is proposed to perform "injecting" of external item-item interaction knowledge, by synthesizing knowledge graph from previously known interactions, into the multi-head attention mechanism during model training, followed by knowledge distillation afterwards for faster inference. Obtained results indicate that the proposed model achieves high inference speed and exceptional precision, effectively addressing the limitations of existing sequential recommendation models.

Keywords: Knowledge graph · Sequential recommender systems · Graph neural network · Attention network · Knowledge Distillation

1 Introduction

1.1 Recommender Systems Overview

Recommender systems (RS) are algorithms aimed at suggesting relevant items to users, tailored to their preferences and behaviour [1]. Relevancy in this case is a rather broad notion, but usually it is expressed in terms of rating, i.e. explicit score each user can assign to each item. On the other hand, if the rating system is not possible to implement, user feedback is considered implicit. This type of feedback is gathered from user interaction with items, such as clicks, views, purchases, and time spent on certain pages [2].

Recommender system taxonomies are diverse but one of the most popular ways to specify RS type is to assign it to one of the following categories:

V. Julian et al. (Eds.): IDEAL 2024, LNCS 15346, pp. 244–253, 2025.
https://doi.org/10.1007/978-3-031-77731-8_23

- Collaborative Filtering: This approach predicts the target users preferences by collecting such from users that behave similarly to them. Collaborative filtering can be item-to-item, or user-to-item [1–3].
- Content-Based Filtering: Recommendations are made by taking into account user profile features and past interactions to produce relevant items [1,2].
- Knowledge Based Systems: These systems leverage explicit knowledge about how particular item features meet user needs and preferences, often by applying some rule-based approach to match users with relevant items [2].
- Hybrid Systems: These RS combine approaches and methods stated above to achieve better precision of recommendations.

Each RS type has their strengths and weaknesses which defines the area of application. As an example, cllaborative filtering provides highly personalized recommendations by considering the collective preferences of users with similar tastes [1]. However, since it requires some information about customer, e.g. visits and purchases, CF can't be used to offer relevant items for new users, which is known as cold start problem [3].

On the other hand, cold start problem can be mitigated leveraging either content-based filtering or knowledge-based approaches. Moreover, content-based filtering does not rely on other users data, making it effective even with a small training set. [1], as well as knowledge-based systems are highly flexible, since popular behavioural patterns could be manually programmed [2]. However, since users are isolated and rules are manually defined, RS could become poorly scalable, since rules can be over-simplified or quantity of items in training data may be too little which may reduce RS exploration capabilities.

1.2 Machine Learning for Sequential Recommendations

Sequential recommender systems aim to predict next items a user will most likely interact with based on current interactions. In contrast with traditional recommendation systems that consider only user-item interactions at a single point in time, sequential RS can grasp the dynamics of user behaviour. This allows for more accurate and context-aware recommendations.

Consider the following multiset of user interactions $\{i_1^{(1)}, i_2^{(2)}, \ldots, i_k^{(t-1)}\}$, where $i_k^{(t)}$ represents the item interacted with at time t, the goal is to predict the next item $i_m^{(t)}$, $1 \leq m, k \leq |I|$. Formally, the task can be defined as:

$$\hat{i}^{(t)} = \arg\max_{i \in I} P(i | i_1^{(1)}, i_2^{(2)}, \ldots, i_k^{(t-1)}), \quad (1)$$

where I is the set of all items.

Various models have been developed to capture the sequential patterns in user behavior. These include Recurrent Neural Networks, such as Vanilla RNNs [4] and Long Short-Term Memory networks along with bi-directional LSTM [5,6], and more recently, Transformer models [7] and self-attention networks [8].

Despite both LSTMs and Transformers possess their remarkable ability to process sequences, attention-based networks are more preferrable to use over

LSTMs since of their ability to process sequences in parallel. This leads to significant speed improvements during both training and inference [7].

2 Proposed Approach

In order to solve the problem (1) it is proposed to build a recommendation retrieval model that utilizes both collaborative filtering, content-based and knowledge-based approaches. To leverage all the proposed approaches simultaneously most effectively it is proposed to use external knowledge about items, users and their relations to significantly enhance RS exploration capabilities and solve cold start problem.

One of the most natural ways to present such relations is to organize them in a special graph, called knowledge graph [9].

Knowledge graphs organize and utilize information about some concept, enabling a deeper understanding and more sophisticated interactions with data.

2.1 Synthesizing Knowledge Graph from Historical Data

To convert historical interactions into knowledge graph it is proposed to use Pointwise Mutual Information (PMI) measure. PMI compares the joint probability of two items occurring together to the product of their individual probabilities. Formally, the PMI between two items i and j is defined as:

$$\mathrm{PMI}(i, j) = \log \frac{P(i, j)}{P(i)P(j)}, \tag{2}$$

where $P(i, j)$ is the joint probability of items i and j co-occurring, and $P(i)$ and $P(j)$ are the individual probabilities of items i and j occurring independently [10].

To construct an item-item interaction graph using PMI, the following steps are proposed:

1. Data Collection: Gather user interaction data such as clicks, views, and purchases.
2. Computing Probabilities: Calculate the probabilities $P(i)$, $P(j)$, and $P(i, j)$ based on the frequency of interactions. For instance, $P(i)$ is computed as the ratio of users who interacted with item i to the total number of users.
3. PMI Calculation: For each type of interaction (e.g., clicks, purchases), compute the PMI score using the formula above. This results in individual PMI scores for each type of interaction [10].
4. Graph Construction: Construct the graph where nodes represent items, and edges determined by the PMI scores above the threshold $\alpha \in \mathbb{R}$.

Using PMI to create an item-item interaction graph leverages the statistical strength of item co-occurrences, capturing contextual relationships between items. This approach can improve recommendation quality by considering both individual item frequencies and their contextual associations.

2.2 Mixed Attention

Consider knowledge graph $G = (V, E)$, where V is a node set (e.g. items) and $E = (w_i, p_i)$ is an edge set with weights w_i properties p_i. Introducing the projection operator:

$$\mathrm{P} : G, p \longrightarrow \hat{G}, \tag{3}$$

where $\hat{G}$ is the subgraph G where only edges with property p are kept.

Consider user session $s = \{(i_1^{(1)}, p, i_2^{(2)}, \ldots, i_k^{(t-1)}\}$ along with action user action properties. Also consider matrices E_I, E_P for item and associated property embedding, respectively.

Mixed attention in this scope is defined as follows:

$$\begin{aligned} &\mathrm{MixedAttention}(Q, K, V) = \\ &= \mathrm{MHA}\left(Q\left(E_I\right), K\left(E_I\right), V\left(E_I\right)\right) || \\ &||\mathrm{MHA}\left(Q\left(E_P\right), K\left(E_P\right), V\left(E_P\right)\right) || \\ &||\mathrm{MHA}\left(Q\left(\hat{G}\right), K\left(\hat{G}\right), V\left(\hat{G}\right)\right) \end{aligned} \tag{4}$$

where MHA is Multi-Head Attention, || is a concatenation operation, and MHA of $\hat{G}$ is calculated as concatenated MHA for first order neighbors of $i_k^{(t)}$ in graph $\hat{G}$.

2.3 Knowledge Distillation

Since the proposed approach utilizes multiple traverses through session data and knowledge graph, it could cause issues with response time latency and constant need to update such graph. To overcome this, it is proposed to use a "simple" Multi-Head Attention network trained with knowledge distillation methods, particularly temperature scaling approach [11].

Temperature scaling in the context of knowledge distillation, is a technique used in machine learning to improve the transfer of knowledge from a larger, more complex model (the teacher) to a smaller, simpler model (the student) [11]. The process involves modifying the softmax output of the teacher model by a temperature parameter, which smooths the probability distribution and provides richer information to the student model.

The temperature scaling modifies raw output scores from the teacher model before applying the softmax function. The formula for temperature-scaled softmax is:

$$Q(i, T) = q_i(T) = \frac{\exp(z_i/T)}{\sum_j \exp(z_j/T)},$$

where z_i are the logits, T is the temperature parameter, and q_i are the softmax probabilities. When $T > 1$, the distribution becomes more uniform, and when $T = 1$, it corresponds to the standard softmax function.

The key steps in temperatured distillation are as follows:

1. Training the Teacher Model. The teacher model is trained on the original dataset to achieve high accuracy.
2. Temperature Scaling. The logits from the teacher model are divided by a temperature $T > 1$ before applying the softmax function. This results in a softened probability distribution.
3. Training the Student Model. The student model is trained to match the softened output of the teacher model.

The Kullback-Leibler divergence with the soft targets is often chosen as the loss function for the student model during distillation:

$$L = \mathrm{KL}(Q_t \parallel Q_s) = \sum_i Q_t(i, T) \log \frac{Q_s(i, T)}{Q_t(i, T)},$$

where Q_s and Q_t are the (scaled) softmax outputs of the student and teacher models, respectively.

2.4 Model Architecture

Overall, the architecture of model is depicted on the Fig. 1.

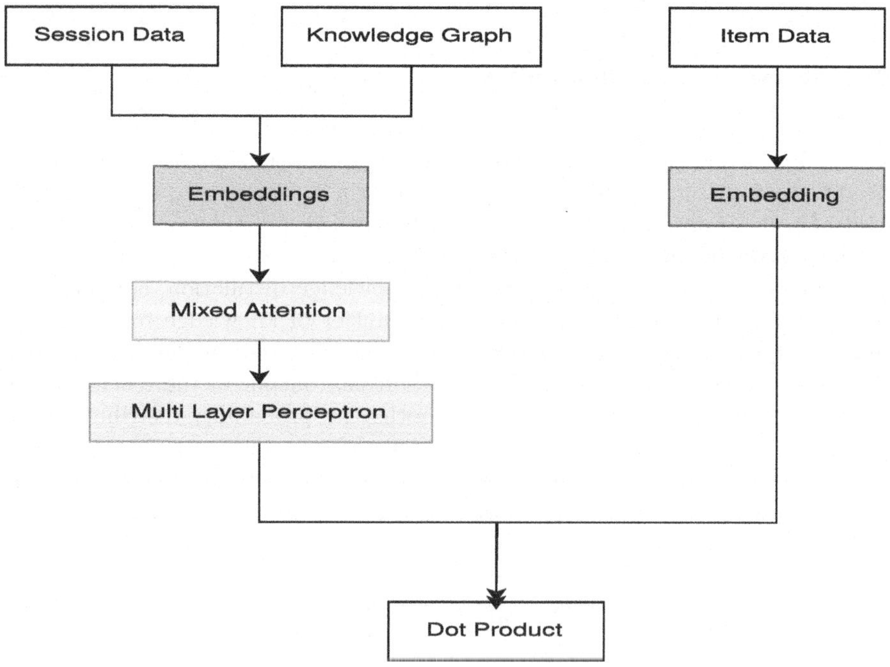

Fig. 1. Two-tower RS architecture.

On the other hand, for knowledge distillation process the model is visualized at Fig. 2.

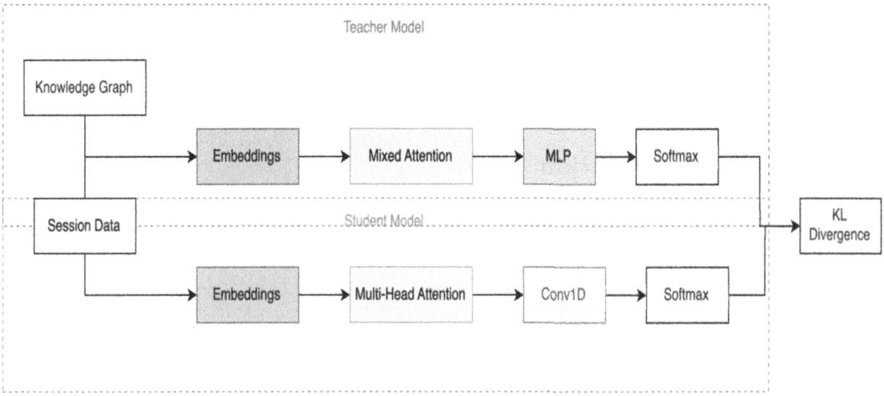

Fig. 2. Knowledge distillation process.

3 Experiment and Results

3.1 Dataset Description

To benchmark the proposed model, dataset of visits to an anonymous e-commerce store was chosen. The schema of the data is presented in the Table 1.

Table 1. Dataset fields description

Column name	Column type	Column description
user_id	unsigned bigint	user identifier
product_id	unsigned bigint	product identifier
category_id	unsigned bigint	item category identifier
event_type	string	interaction type (view, purchase)
user_session	base64	user session identifier

The dataset contains historical visit records, covering the period from October 2019 to November 2019. In total, it includes roughly 6.5 million user sessions, with the data encompassing around 69 million records.

This extensive dataset provides a detailed overview of user interaction along with the specified timestamp and associated action, such as viewing some item, purchasing it or adding it to basket, offering valuable insights for user behaviour analysis. In current work relations between user and items are proposed to be examined and potential insights to be leveraged by transforming this tabular dataset into graph, using PMI threshloding.

After multiple experiments it was decided to choose value $\alpha = 12$ as a PMI threshold, roughly corresponding to first quartile of PMI scores distribution. By applying this threshold it was obtained item-to-item relationship graph with 58618 nodes and 5230019 edges.

3.2 Models Configuration

Teacher Model. For the experiment purposes the following teacher model configuration was selected after performing hyperparameters tuning:

- Embedding dimension size - 64
- Number of multi-layer perceptron layers - 2

After training teacher model for 20 epochs, overfitting for NDCG@10 measure was achieved after the fifth epoch, thus the best model was trained on 5 epochs (see Fig. 3).

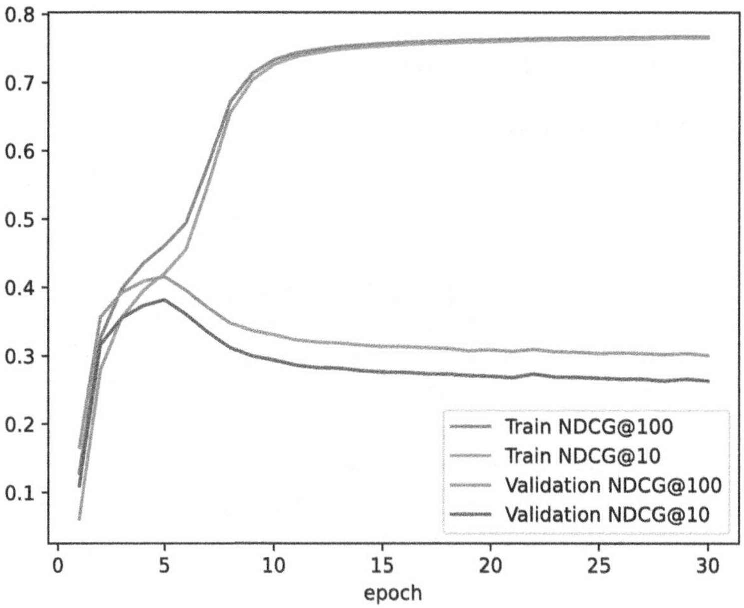

Fig. 3. NDCG metric evolution

Baseline Model. Since, one of the most popular frameworks for graph representation learning is the Node2Vec algorithm [12], it is proposed to utilize MHA model, similar to the teacher one, but with pre-computed Node2Vec embedding for each item. Node2vec is a prominent algorithm in graph representation learning that extends the concept of word2vec to graph data. It performs random walks on the graph to generate sequences of nodes, treating these sequences similarly to sentences in natural language processing. Node2vec employs a biased random walk strategy, balancing between breadth-first (BFS) and depth-first search (DFS) strategies to capture both local and global structural features [12].

For experiment purposes the following Node2Vec parameters were used:

- Random walk length - 10
- Number of random walks from each node - 5
- Parameter p (exploration) - 1
- Parameter q (exploitation) - 1
- Embedding dimension size - 64

Student Model. After the same grid search procedure the following student model configuration was achieved:

- Embedding dimension size - 64
- Number of convolution layers - 3
- Dropout factor - 0.45

As the loss measure, KL divergence was selected. The minimal value of such metric was achieved around the 10th epoch, thus this model snapshot was selected for further recommendation retrieval evaluation (see Fig. 4).

3.3 Results

For models performance evaluation next criteria were considered:

- Mean Average Precision (MAP@10, MAP@100).
- Normalized Discounted Cumulative Gain (NDCG@10, NDCG@100).
- Average time per retrieving top-100 candidates (ATI@100)

In order to guarantee robustness of results, multiple runs were executed on every model, and average of each metric on the test dataset was calculated.

These scores are shown in the Table 2

Table 2. Models evaluation results

Model	ATI@100, s	MAP@10	MAP@100	NDCG@10	NDCG@100
Non-distilled model w. Mixed Attention	2.670	0.3247	0.3310	0.3742	0.4093
Non-distilled model w. N2V	1.220	0.2551	0.2892	0.2988	0.3432
Distilled model	0.13	0.2672	0.2728	0.2952	0.3240

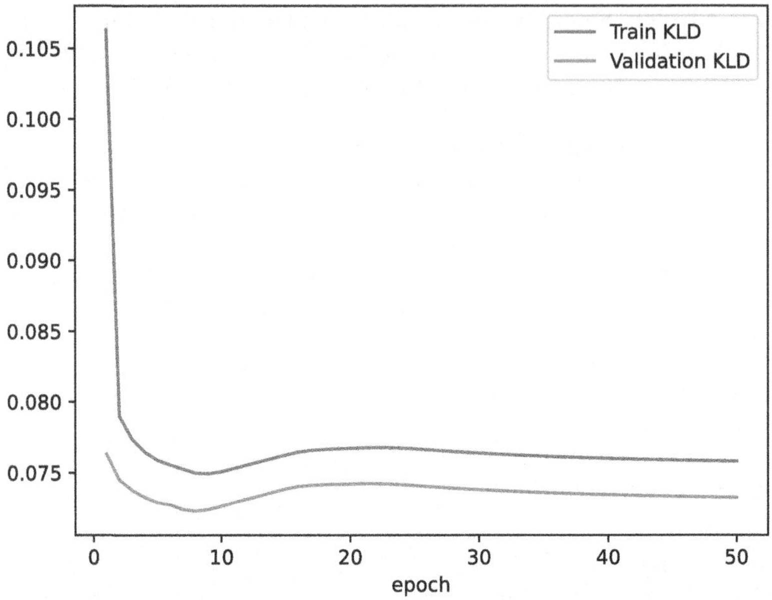

Fig. 4. KL divergence metric evolution

As it was observed, obtained distilled model have similar, albeit mostly lower results, compared to the one trained with Node2Vec embeddings. However, the offered inference speed makes the proposed model much more suitable for usage in production environments.

4 Conclusions

The precision of analyzing patterns in sequential recommendations has been enhanced recently by leveraging capabilities various deep learning models, such as RNNs and Transformers. However, these models are constrained by their restricted memory scope, since they are applied session-wise and thus consider only currently used items, which can decrease recommendation diversity. To mitigate this issue, a novel approach involving the "injection" of external item-item interaction knowledge has been proposed. This method synthesizes a knowledge graph from previously known interactions and integrates it into the multi-head attention mechanism during model training. Following this, knowledge distillation is applied for faster inference.

Initially, the knowledge graph is constructed using the Pointwise Mutual Information (PMI) metric. Subsequently, a Mixed Attention model is introduced, and lastly temperature scaling knowledge distillation is utilized to achieve a fast and scalable model for inference. The results indicate that the proposed model achieves high inference speed and still offers high precision, effectively addressing the limitations of existing sequential recommendation models.

The distilled model, though showing lower results than original model, and similar ones to model trained with Node2Vec embeddings, propose superior inference speed, which makes it more suitable for use in production environments.

References

1. Falk, K.: Practical Recommender Systems. Manning Publications, Shelter Island (2019)
2. Jannach, D., Zanker, M., Felfernig, A., Friedrich, G.: Recommender Systems. An Introduction. Cambridge University Press, Cambridge (2010)
3. Aggarwal, C.: Recommender Systems, 2nd edn. Springer, Cham (2016). https://doi.org/10.1007/978-3-319-29659-3
4. Hidasi, B., Karatzoglou, A., Baltrunas, L., Tikk, D.: Session-based recommendations with recurrent neural networks. In: Proceedings of the International Conference on Learning Representations (ICLR) (2016)
5. Xia Q., Jiang P., Sun F., Zhang Y., Wang X., Sui Z.: Modeling consumer buying decision for recommendation based on multi-task deep learning. In: Proceedings of the 27th ACM International Conference on Information and Knowledge Management (CIKM '18). Association for Computing Machinery, New York, NY, USA, pp. 1703—1706 (2018). https://doi.org/10.1145/3269206.3269285
6. Zhao, C., You, J., Wen, X., Li, X: Deep Bi-LSTM networks for sequential recommendation. Entropy(Basel) **22**(8), 870 (2022)
7. Vaswani, A., et al.: Attention is all you need. Adv. Neural Inf. Process. Syst. (NIPS) (2017)
8. Kang, W.C., McAuley, J.: Self-attentive sequential recommendation. In: Proceedings of the IEEE International Conference on Data Mining (ICDM) (2018)
9. Ji, S., Pan, S., Cambria, E., Marttinen, P., Yu, P.S.: A Survey on Knowledge Graphs: Representation, Acquisition, and Applications. arXiv preprint. https://arxiv.org/abs/2303.13948 (2022). Accessed 30 June 2024
10. Church, K.W., Hanks, P.: Word association norms, mutual information, and lexicography. Comput. Linguist. **16**(1), 22–29 (1990)
11. Hinton, G., Vinyals, O., Dean, J.: Distilling the Knowledge in a Neural Network. arXiv preprint. https://arxiv.org/abs/1503.02531. Accessed 6 July 2024
12. Grover, A., Leskovec, J.: Node2vec: scalable feature learning for networks. In: Proceedings of the 22nd ACM SIGKDD International Conference on Knowledge Discovery and Data Mining (2016)

Federated Learning for Vietnamese SMS Spam Detection Using Pre-trained PhoBERT

Hoang Quang Anh, Pham Tuan Anh, Pham Son Nguyen, and Phan Duy Hung(✉)

FPT University, Hanoi, Vietnam
{anhhqhe163092,anhpthe163897,nguyenpshe163872}@fpt.edu.vn,
hungpd2@fe.edu.vn

Abstract. SMS Spam Detection has increasingly garnered attention due to the widespread use of mobile devices. Currently, most SMS spam detection model training methods rely on centralized data collection, which poses numerous privacy threats and creates security vulnerabilities that expose sensitive information. This study aims to propose a training method that does not require data sharing between parties, based on a federated learning system. In this paper, we experiment with FedAvg, FedAvgM, and FedAdam algorithms using a fine-tuned PhoBERT model tailored for the SMS spam classification task. The results show that the FedAvg algorithm achieves high performance with an accuracy of 99.38% in the IID setting, while the FedAdam algorithm proves more effective in the Non-IID setting, yielding a model with an accuracy of up to 98.5%. This study demonstrates that models like PhoBERT trained with FL algorithms can achieve classification capabilities comparable to centralized data training methods, highlighting the significant potential of FL for natural language processing models without the need for centralized data collection.

Keywords: Vietnamese SMS spam detection · PhoBERT · Federated learning

1 Introduction

1.1 Problem and Motivation

In the digital age, content spam has become a prevalent issue, inundating platforms with irrelevant, misleading, or harmful information. It affects not only the user experience but also undermines the credibility of online platforms. Traditional methods such as keyword filtering, abnormal behavior detection, or training content filtering models can achieve some success; however, they still have significant limitations.

Federated Learning (FL) is a machine learning model that enables AI training in a distributed manner without sharing sensitive data. This opens up new opportunities to address the content spam problem more effectively and securely. By training the spam detection model directly on individual user devices, FL leverages the diversity of user data and contexts, thereby enhancing accuracy compared to centralized methods.

V. Julian et al. (Eds.): IDEAL 2024, LNCS 15346, pp. 254–264, 2025.
https://doi.org/10.1007/978-3-031-77731-8_24

1.2 Related Works

The email spam detection task has been well explored in the literature [1] and is commonly used in undergraduate machine learning courses as an introductory use-case for study and learning. Despite the familiarity of the email spam detection task, it continues to provide real challenges to practitioners due to data distribution drift, the adversarial nature of the environment in which the task is embedded, and class imbalance. Significant research e ort has been expended on this task, and sustained research is necessary to keep ahead of ever-evolving data and the changing landscape of spamming techniques.

The email spam detection task is framed as the development of an effective automated algorithm for differentiating spam versus ham (non-spam) emails. Spam detection methods have been classified in terms of rules-based, collaborative, and content-based approaches [2]. Rules-based approaches include checking incoming email addresses against white lists of allowed email addresses, black lists of commonly used spam email sources, and hand-crafted rules (e.g. that look for empty To: elds or a certain combination of keywords). Collaborative approaches compute, for example, a hash function on the content of example spam which is shared with the community and the hash function compared to new emails to detect spam [3–5]. These simple methods are unable to generalize well and are brittle, which is why content-based methods involving machine learning have been explored.

The field of machine learning has been revolutionized with the emergence of Large Language Models (LLMs), a powerful suite of deep learning methods that have exceptional abilities in understanding and learning from the structural, relational, and semantic patterns inherent in natural language datasets [6]. LLMs are extreme multi-taskers, able to replace most bespoke models for natural language tasks. For example they are very capable at text generation, sentiment detection, question answering, and text summarization. The release of ChatGPT and more recently GPT4 [7] to the public made waves around the world that continue to reverberate given the naturalness and human-like.

The concept of using federated learning to enhance spam detection models has been explored in various repositories. One study [8] focused on enabling privacy-preserving cyber threat detection with federated learning, highlighting the challenges of collecting individual-relevant and privacy-sensitive threat datasets due to enforcement of various privacy protection regulations. The authors systematically profiled the feasibility of federated learning for privacy-preserving cyber threat detection in terms of effectiveness, byzantine resilience, and efficiency.

Another study [9] used federated learning for SMS spam detection, collecting SMS spam messages reported by victims on Twitter and releasing a multilingual SMS spam dataset. The authors also combined newly released datasets to build and evaluate SMS spam detection models in federated learning settings. Federated learning has also been applied to phishing email detection [10], where it was used to train a joint deep learning model by harvesting rich distributed data held by each client in a default privacy mode. This approach enabled multiple organizations to participate in training deep email anti-phishing models without sharing email data. Additionally, researchers have explored the use of federated learning for keyword spotting [11] and mobile keyboard prediction [12], demonstrating its potential in various natural language processing tasks. Despite the popularity and importance of email spam detection, there is little prior work that explores the benefits of LLMs for the task. We address the gap in this paper.

1.3 Contribution

This research contributes to the development of a novel federated learning framework for Vietnamese spam SMS detection, which addresses the pressing concern of preserving private information between parties. Our system enables multiple parties to collaboratively train a robust spam SMS detection model without sharing sensitive data, thereby promoting privacy-preserving collaboration. Integrating the pre-trained Vietnamese language model PhoBERT and collecting Vietnamese data to expand the model's capabilities to detect spam SMS in both English and Vietnamese.

Evaluating the performance of our proposed model using a synthetic dataset, achieving an accuracy of 99.38% and demonstrating the effectiveness of our approach. Our research has significant implications for the development of privacy-preserving AI systems.

2 Theory Fundamentals

2.1 Federated Learning System

In federated learning (FL) systems, multiple clients collaboratively train deep learning (DL) models while maintaining the privacy and isolation of their respective local training datasets [14–16]. During the federated training process, a central server initializes a global model with random weights, which is then disseminated to K clients, each possessing their own private and sensitive training datasets [17].

Maintaining user privacy is a paramount concern in our system, which is why we employ Federated Learning (FL) to safeguard sensitive information. In contrast to traditional centralized model architectures, which can compromise user security by exposing personal data, FL ensures that data remains localized on individual devices. This approach enables the training of models on decentralized data sources, without the need for data sharing with external parties or servers.

In our FL system, models are trained on each device using local data, which comprises individual user messages. The model itself is not shared across parties; instead, only the model weights are aggregated and shared. This approach eliminates the need for centralized storage of individual message data, thereby protecting user privacy. The only information exchanged is the model weights, which are updated after training on each local dataset.

By keeping personal data securely on devices, FL provides an effective means of preserving user privacy. Our federated learning model architecture is illustrated in Fig. 1, which highlights the decentralized nature of our approach.

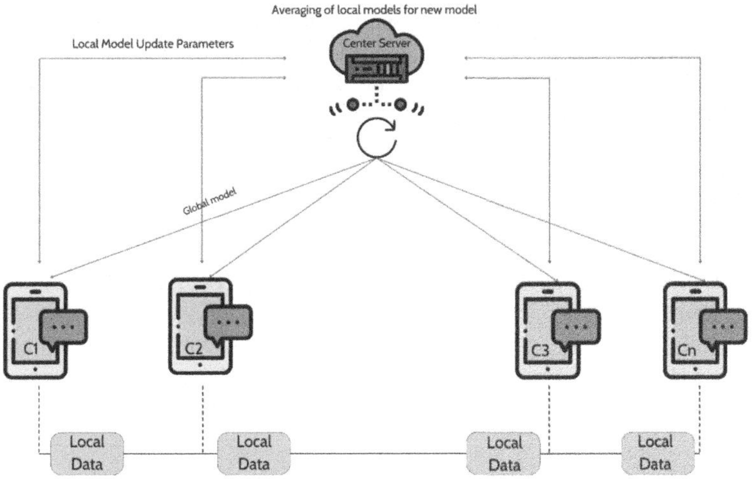

Fig. 1. FL system with clients holding sensitive training data.

2.2 The Pretrained Model PhoBERT

The PhoBERT [13] model is a pre-trained language model for Vietnamese, developed by VinAI. The model is a version of BERT (Bidirectional Encoder Representations from Transformers), including two versions: PhoBERT_Base and PhoBert_large, trained from 20 Gb of Vietnamese text data.

We chose PhoBERT for our SMS spam detection system due to its unique advantages:

- Capture contextual relationships: PhoBERT's transformational architecture enables modeling of complex relationships between words and phrases in Vietnamese SMS messages, allowing it to identify subtle patterns and unusual signs of spam.
- Handling noisy input: PhoBERT's resistance to input noise, including spelling errors, grammatical errors, and non-standard language, makes it well-suited for handling SMS messages, which often contain errors and inconsistencies.
- Generalize to unseen data: PhoBERT's pre-training on a large Vietnamese text corpus allows it to generalize well to new SMS messages, reducing the risk of overfitting and improving overall performance.
- PhoBERT's lightweight architecture and efficient computational requirements make it an ideal choice for deployment on mobile devices. Its smaller size and reduced latency enable fast and accurate SMS spam detection, even on resource-constrained mobile devices.
- PhoBERT's design and training make it especially suitable for Vietnamese text processing. The ability to handle Vietnamese-specific linguistic features such as accents and diacritics ensures that it can accurately analyze and classify Vietnamese SMS messages.

In conclusion, PhoBERT's efficiency, suitability for mobile applications and advantages in processing Vietnamese text make it the ideal choice for our SMS spam detection system. By leveraging PhoBERT as a pre-trained model, we can build a robust and accurate Vietnamese SMS spam detection system.

2.3 Federated Algorithms

a) **FedAvg**

Google introduced the seminal Federated Averaging (FedAvg) algorithm to tackle the optimization challenges in Federated Learning (FL) systems [17]. In this system, the central coordinator server aggregates the local model updates submitted by participating clients, generating a new global model through a weight averaging process. The FedAvg algorithm can be mathematically formalized as:

$$\Delta w^{t+1} = \sum_{k=1}^{k} \left(\frac{n_k}{n} \Delta w_k^t \right)$$

where K denotes the total number of clients participating in the federated learning process. n is the total number of training samples, n_k is number of data in client k. w_k^t are the model weights from client k for aggregation in federated round t. So, Δw^t represents the aggregated global model that is distributed from the server to the client. Then, the client k calculates the cumulative local gradient with $\Delta w_k^{t+1} \leftarrow w_k^{t+1} - \Delta w^{t+1}$, w_k^{t+1} are the updated weights the of new locally trained model [17, 18]. In this study, the reason FedAvg is used is that it is the base algorithm for FL optimization problems. In addition, there should also be a fair comparison with other federated learning algorithms.

b) **FedAvgM**

Vanilla FedAvg updates the weights via $w \leftarrow w - \Delta w$, where:

$$\Delta w = \sum_{k=1}^{k} \left(\frac{n_k}{n} \Delta w_k \right)$$

(n_k is the number of examples, Δw_k is the weight update from k'th client and $n = \sum_{k=1}^{k} n_k$, $v \leftarrow \beta v + \Delta w$ and update the model with $w \leftarrow w - v$ [19].

The FedAvgM algorithm, which has demonstrated outperformance to FedAvg in most Non-IID cases [19], is not included in our comparative study.

c) **FedAdam** [20]

FedAdam is a federated adaptive DL version of the popular Adam optimizer function. In FedAdam the EMA is used to calculate the gradients. The equation is:

$$e_t = \beta_2 e_{t-1}.(1 - \beta_2).\Delta w_t^2$$

3 Datasets

3.1 Data Description

In this study, the dataset is sourced from two different origins. The first is an 'English spam dataset' obtained from Kaggle, and the second is a 'Vietnamese dataset'. Due to the lack of readily available data online, our team collected data from individual devices. We gathered nearly 6000 messages (Table 1), comprising over 5000 non-spam messages and close to 1000 spam messages.

Table 1. Table of data description.

	Spam	Non-Spam	Total
English Data	9600	1800	11400
Vietnamese Data	5100	800	5900
Total	14700	2600	17300

3.2 Preprocessing

Since the Vietnamese dataset was self-collected, it often includes data types such as 'network notifications', 'deduction notifications', 'network advertisements', and 'personal messages'. Therefore, our team decided to divide the data into two groups: 'spam', which includes 'network notifications', 'deduction notifications', 'network advertisements', and 'personal messages'; and 'non-spam', which includes 'negative messages', 'meaningless messages', and 'annoying messages'.

3.3 Dataset Partitioning

In this study, the training and test datasets were created in a centralized learning environment by combining both languages. However, in practical Federated Learning (FL) settings, customer data predominantly exhibits Non-IID characteristics [1, 2]. Consequently, statistical assumptions for IID data cannot be applied to Non-IID data. Therefore, it is necessary to examine both data distribution types to ensure a fair comparison. The data will be divided into IID (Fig. 2) and Non-IID (Fig. 3).

IID Setting is to distribute the dataset equally among all three clients, ensuring that the number of English and Vietnamese entries is balanced for each client. And Non-IID Setting is to distribute different datasets to each client, ensuring that the datasets differ as much as possible.

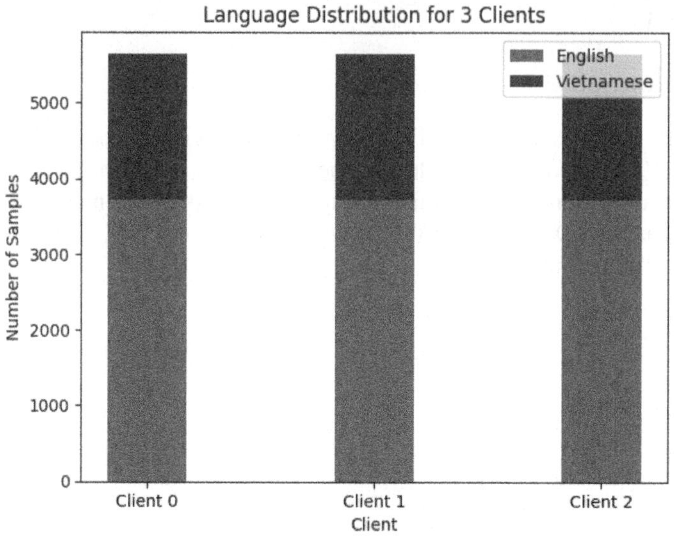

Fig. 2. IID setting

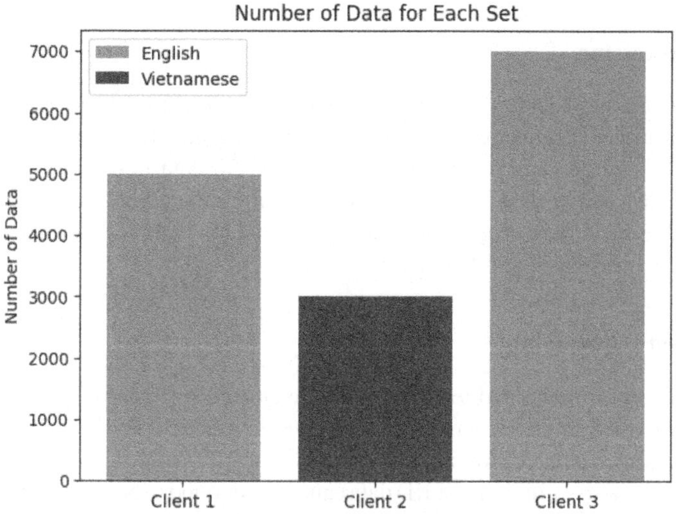

Fig. 3. Non-IID setting

4 Experiment and Results

Our research is implemented in Python and PyTorch frameworks. We perform experiments on Colab's GPU T4. An overview of the parameters and functions considered for training and evaluation is listed in Table 2.

Table 2. Parameters and functions used for model training

Parameters	Values
Number of clients	3
Federated of aggregation rounds	5
Epochs for local training	1
Learning rate	2e-5
Batch size	16
Local Gradient Optimizer	AdamW
Loss function	Cross-entropy

Our model is evaluated through 20% of the dataset based on the classification accuracy rate of the federated PhoBERT model trained with FedAvg, FedAvgM, FedAdam in IID and Non-IID settings on the dataset SMS spam. With centralized model performance we obtained is 98.73%.

With the IID setting, we obtained the results shown in Table 3. The results show very good performance with FedAvg and FedAdam.

Table 3. Accuracy comparison of Federated PhoBERT in IID settings.

Fed algorithms	Accuracy(%)
FedAvg	99.38
FedAvgM	87.36
FedAdam	99.32

With the Non-IID setting, we obtain the results in Table 4.

Table 4. Accuracy comparison of Federated PhoBERT in Non-IID settings.

Fed algorithms	Accuracy(%)
FedAvg	97.25
FedAvgM	87.03
FedAdam	98.5

The results presented in Tables 3 and 4 demonstrate the effectiveness of the proposed federated learning approach in SMS spam detection. We will analyze the results in more detail and discuss the implications of our findings:

- **Federated Learning Performance**
  The accuracy of the federated learning model in detecting SMS spam is significantly higher than that of the centralized model. In the IID setting, FedAvg and FedAdam achieve accuracy of 99.38% and 99.32%, respectively, outperforming the centralized model by 0.65%. In the Non-IID setting, FedAvg and FedAdam achieve 96.32% and 98.5% accuracy, respectively, still outperforming the centralized model. These results show that federated learning can improve SMS spam detection using non-centrally distributed data from many customers.
- **Comparison to Centralized Approach**
  Notably, the federated learning model achieved higher accuracy than the centralized model trained on the entire dataset with 5 epochs. This is an important finding because it suggests that federated learning can outperform traditional centralized approaches even when the centralized model has access to the entire dataset. This is likely due to the fact that federated learning can utilize the diversity of data distribution across clients, leading to a more power and general model.
- **Comparison of Federated Algorithms**
  The results also show that FedAvg and FedAdam perform better than FedAvgM in both IID and Non-IID settings. This shows that FedAvg and FedAdam are more effective in handling the heterogeneity of data distribution across customers. On the other hand, FedAvgM may require additional adjustments or modifications to adapt to Non-IID settings.
- **IID vs Non-IID Settings**
  The results highlight the importance of considering the data distribution across clients in federated learning. In the IID setting, the data distribution is identical across clients, and the federated learning models achieve high accuracy. However, in the Non-IID setting, the data distribution is non-identical across clients, and the accuracy of the federated learning models decreases. This suggests that federated learning models need to be designed to handle the heterogeneity of the data distribution across clients.

In conclusion, our analysis demonstrates the effectiveness of federated learning in SMS spam detection. The results suggest that federated learning can improve the detection of SMS spam by leveraging distributed data from multiple clients, and that FedAvg and FedAdam are effective algorithms for handling IID and Non-IID data distributions. Furthermore, the results highlight the potential of federated learning to outperform traditional centralized approaches, even when the centralized model has access to the entire dataset.

5 Conclusion and Future Works

In this paper, we propose a new method to detect SMS spam using PhoBERT, a language model for Vietnamese and federated learning methods. Our approach leverages PhoBERT's strengths in capturing semantic relationships in Vietnamese text data and federated learning in protecting user privacy. We demonstrated the effectiveness of our method in enhancing SMS spam detection models and compared its performance with traditional machine learning-based methods. Additionally, we also tested and compared different federated learning algorithms. The results show that FedAvg achieves stable performance on both IID setting and non-IID setting with an accuracy of 99.38% and 97.25%, respectively. Meanwhile, FedAdam is also more stable and effective than FedAvg with non-IID setting with accuracy of 99.32% and 98.5% respectively. As for FedAvgM, the efficiency is not good compared to the other two algorithms.

In future work, we plan to apply FL to other Vietnamese tasks, thereby coming up with reasonable optimization methods. Additionally, we are also interested in combining FL with other language models such as BERT and RoBERTa. This could explore applying our method to other natural language processing tasks, such as sentiment analysis and text classification [21–23].

References

1. Dada, E.G., Bassi, J.S., Chiroma, H., Abdulhamid, S.M., Adetunmbi, A.O., Ajibuwa, O.E.: Machine learning for email spam filtering: review, approaches and open research problems. Heliyon **5**(6), e01802 (2019)
2. Méndez, J.R., Fdez-Riverola, F., Díaz, F., Iglesias, E.L., Corchado, J.M.: A comparative performance study of feature selection methods for the anti-spam filtering domain. In: Perner, P. (eds.) Advances in Data Mining. Applications in Medicine, Web Mining, Marketing, Image and Signal Mining. ICDM 2006. LNCS, vol. 4065, p. 106120. Springer, Berlin, Heidelberg (2006). https://doi.org/10.1007/11790853_9
3. Attenberg, J., Weinberger, K., Dasgupta, A., Smola, A., Zinkevich, M.: Collaborative email-spam filtering with the hashing trick. In: Proceedings of the Sixth Conference on Email and Anti-Spam (CEAS 2009) (2009)
4. Shi, W., Xie, M., Huang, Y.: Collaborative spam filtering technique based on mime fingerprints. In: Proceedings of the 9th World Congress on Intelligent Control and Automation, p. 225230 (2011)
5. Wittel, G.L., Wu, S.F.: On attacking statistical spam filters. In: Proceedings of the Conference on Email and Anti-Spam (CEAS). Mountain View, CA, USA (2004)
6. Liu, Y., et al.: Roberta: a robustly optimized bert pretraining approach. arXiv:1907.11692 [cs.CL] (2019)
7. OpenAI: Gpt-4 technical report. arXiv:2303.08774 [cs.CL] (2023)
8. Rieke, N., et al.: The future of digital health with federated learning. arXiv:2003.08119 [cs.CY] (2021)
9. Kanaparthy, S., Padala, M., Damle, S.: Fair and federated face attribute classification with heterogeneous data. arXiv:2109.02351 [cs.LG] (2022)
10. Thapa, C., et al.: Evaluation of Federated Learning in Phishing Email Detection: arXiv:2007.13300 [cs.LG] (2021)
11. Leroy, D., et al.: Federated Learning for Keyword Spotting. arXiv:1810.05512 [eess.AS] (2019)
12. Hard, A., et al.: Federated Learning for Mobile Keyboard Prediction. arXiv:1811.03604 [cs.CL] (2019)
13. Dat, N.Q., Anh, N.T.: PhoBERT: Pre-trained language models for Vietnamese. arXiv:2003.00744 [cs.CL] (2020)
14. Gao, L., et al.: FedDC: Federated learning with non-IID data via local drift decoupling and correction. arXiv:2203.11751 [cs.LG] (2022)
15. Zhu, H., Xu, J., Liu, S., Jin, Y.: Federated learning on non-IID data: a survey. arXiv:2106.06843 [cs.LG] (2021)

16. Yang, Q., Liu, Y., Chen, T., Tong, Y.: Federated machine learning: concept and applications. ACM Trans. Intell. Syst. Technol. **10**(2), 1–19 (2019)
17. McMahan, H.B., Moore, E., Ramage, D., Hampson, S., y Arcas, B.A.: Communication-efficient learning of deep networks from decentralized data. arXiv:1602.05629 [cs.LG] (2016)
18. Bonawitz, K., Et al.: Towards federated learning at scale: system design. In: Talwalkar, A., Smith, V., Zaharia, M. (eds.) Proceedings of Machine Learning and Systems, vol. 1, pp. 374–388 (2019)
19. Hsu, T.-M.H., Qi, H., Brown, M.: Measuring the effects of nonidentical data distribution for federated visual classification. arXiv:1909.06335 [cs.LG] (2019)
20. Riedel, P., et al.: Performance analysis of federated learning algorithms for multilingual protest news detection using pre-trained DistilBERT and BERT. IEEE Access **11**, 134009–134022 (2023)
21. Hung, P.D., Hung, N.D., Diep, V.T.: URL classification using convolutional neural network for a new large dataset. In: Luo, Y. (eds.) Cooperative Design, Visualization, and Engineering. CDVE 2022. LNCS, vol. 13492, pp. 103–114. Springer, Cham (2022). https://doi.org/10.1007/978-3-031-16538-2_11
22. Minh, N.Q., Hung, P.D.: The system for detecting Vietnamese mispronunciation. In: Dang, T.K., Küng, J., Chung, T.M., Takizawa, M. (eds.) Future Data and Security Engineering. Big Data, Security and Privacy, Smart City and Industry 4.0 Applications. FDSE 2021. Communications in Computer and Information Science, vol. 1500, pp. 452–459 Springer, Singapore (2021). https://doi.org/10.1007/978-981-16-8062-5_32
23. Ngoc Tram, N., Duy Hung, P.: Analysing hot facebook users posts' sentiment using deep learning. In: Hassanien, A.E., Bhattacharyya, S., Chakrabati, S., Bhattacharya, A., Dutta, S. (eds.) Emerging Technologies in Data Mining and Information Security. AISC, vol. 1300, pp. 561–569. Springer, Singapore (2021). https://doi.org/10.1007/978-981-33-4367-2_53

Deep Learning Inference on Edge: A Preliminary Device Comparison

Manuel L. González[1](✉), Jorge Ruiz[1], Lidia Andrés[1], Randy Lozada[1], Erik S. Skibinsky[1], Jorge Fernández[1], Javier Sedano[1], and Ángel M. García-Vico[2]

[1] Instituto Tecnológico de Castilla y León, Burgos, Spain
{manuel.gonzalez,jorge.ruiz,lidia.andres,randy.lozada,erik.skibinsky, jorge.fernandez,javier.sedano}@itcl.es

[2] Andalusian Research Institute in Data Science and Computational Intelligence (DaSCI), University of Jaén, 23071 Jaén, Spain
agvico@ujaen.es

Abstract. The deployment and inference of Deep Learning models are investigated in this article on seven different Edge Computing (EC) devices. This study addresses the features, performance and limitations of the NVIDIA Jetson Orin NX and Nano, Google Coral DevBoard and USB, Intel Neural Compute Stick 2, NXP i.MX8 Plus and Xilinx Zynq UltraScale+ MPSoC ZCU104 on Deep Learning inference. Fully Connected, Convolutional and Long Short-Term Memory (LSTM) neural networks are implemented to test these EC devices. The benchmarking focuses on the performance metrics: inference latency, increase in error metric, and power consumption. The results show considerable variability among devices, with the ZCU104 and Jetson Orin achieving the lowest latencies across most models without any increase in the error metric. At the same time, Coral devices exhibit increased latency and error for complex convolutional models. NVIDIA Jetson devices, ZCU104 and Neural Compute Stick 2 are the only devices that support LSTM inference. The study also highlights differences in power consumption, with USB accelerators being the most energy-efficient.

Keywords: Edge Computing · Deep Learning · Real-Time Inference

1 Introduction

The proliferation of the Internet of Things (IoT) and its immense traffic generation is driving a paradigm shift toward computation. Deep learning (DL) which traditionally has been centralized in cloud computing infrastructures, is increasingly being deployed closer to data sources on different edge devices [1]. This transition will address the limitations in latency and bandwidth of cloud-centric models, and will also boost privacy and reliability because data is processed in

V. Julian et al. (Eds.): IDEAL 2024, LNCS 15346, pp. 265–276, 2025.
https://doi.org/10.1007/978-3-031-77731-8_25

the local system [2]. Therefore, the Edge Computing (EC) paradigm enables the real-time processing of large streams of data from sensors and devices but with a reduction in data transmission to the cloud [3].

EC applied concretely in DL models brings about the concept of Edge DL. Integrating DL algorithms directly onto edge devices facilitates real-time data processing, which is essential for applications that require instant decision-making, such as autonomous driving or surveillance [4]. However, deploying DL models at the edge is challenging due to constrained computational resources, limited power availability, and the necessity for models to work well under changing environmental conditions [5]. Thus, recent research has been mainly focused on optimizing the inference process of neural networks to make them fit within the edge hardware constraints. Model compression, lightweight neural architectures, dynamic adaptive computation, and other techniques have notably represented such approaches [6,7]. These approaches ensure that the performance remains robust from DL applications despite the small processing power of edge devices, paving the way for innovative applications that integrate AI into everyday devices and services.

EC devices specialized in DL are numerous and their applications are countless. From NVIDIA, the Jetson family has been applied in intelligent traffic management, with a YOLOv5-based network [8] or for real-time counting of people and bicycles [9]. These devices are also used in agricultural applications for real-time weed control systems [10]. EdgeDRNN [11], a recurrent neural network accelerator, has also been developed, demonstrating the ability of these devices to improve processing speed and energy efficiency for real-time data analysis. Although NVIDIA's Jetson devices are too popular, more EC devices have been applied in other fields. In healthcare, a real-time pose estimation system running on top of the Google Coral USB [12] has been optimized. In addition, it has been used as part of an IoT system for wildlife image classification [13]. For security applications, flame and smoke recognition has been developed using a system with the Intel Neural Compute Stick v2 [14]. Finally, Xilinx FPGAs have also been used to run YOLOv3 for road defect detection [15]. These examples highlight the versatility and efficiency of edge devices for deploying DL models in various fields.

As shown, different EC devices have been deployed in multiple applications. Understanding their specifications, limitations, and performance results is essential to develop and implement projects using these devices effectively. This study aims to establish a comparative benchmark between various EC devices over DL models. The devices under comparison are NVIDIA Jetson Orin NX, NVIDIA Jetson Nano, Google Coral DevBoard, Google Coral USB, Intel Neural Compute Stick 2, NXP i.MX8, and Xilinx Zynq UltraScale+ MPSoC ZCU104. The DL models employed will include Fully-Connected Networks (FCN), Convolutional Neural Networks (CNN), and Long Short-Term Memory (LSTM) Neural Networks with different complexity. Testing will be focused on evaluating the device constraints, inference latency, increase in error metrics and power consumption. The first part of this study will focus on explaining the EC devices and the frameworks that can be used to run DL models on them. Subsequently, the experimental setup will be described. Finally, a detailed analysis of the results and the conclusions will be presented.

2 Edge Computing Devices and Deep Learning Tools

2.1 NVIDIA Jetson and JetPack SDK

NVIDIA's Jetson family is designed to bring the power of DL from server GPU (Graphic Processing Unit) devices to the edge. These devices integrate NVIDIA's GPUs and ARM Cortex microprocessors. The Jetson family includes a wide range of products with different specifications. The most basic device is the Jetson Nano, the intermediate is the Jetson Orin NX and the device with the highest performance is the Jetson AGX Orin. The software stack, which includes NVIDIA's JetPack SDK, provides a comprehensive set of tools and libraries that enable developers to implement DL models. Jetson devices support various frameworks such as TensorFlow, PyTorch and ONNX.

DL models can be run using the integrated GPU with the JetPack SDK. The GPU can be leveraged using NVIDIA providers such as CUDA or TensorRT. However, data copies from RAM to GPU memory must be considered.

2.2 Google Coral and TensorFlow Lite

Google's Coral family includes several devices designed to enable efficient AI computing at the edge using TPU (Tensor Processing Unit). The Coral Devboard is a full-featured development board equipped with an ARM Cortex SoC, an Edge TPU and RAM. It has different connectivity options and can run on the Mendel Linux operating system. The Coral USB Accelerator, meanwhile, is a compact device that connects via USB-C to a host system and provides powerful AI capabilities using the Edge TPU. These devices are ideal for prototyping. Google's Coral family offers other versions with connectors such as M.2, PCIe, or System-on-Module (SoM) form factors for production.

TensorFlow Lite (TFLite) is a lightweight version of TensorFlow designed for mobile devices and peripherals. It enables the execution of DL models in resource-constrained environments. For inference on the Coral Edge TPU, models must be quantized to INT8 using this framework. This transformation reduces the size of the model but may result in a slight degradation of model accuracy.

2.3 Intel Neural Compute Stick 2 and OpenVINO

The Intel Neural Compute Stick 2 (INCS2) is a compact peripheral designed for efficient DL inference at the edge. It works with the Movidius Myriad X VPU (Vision Processing Unit). The VPU features a dedicated high-performance computation engine while maintaining low power consumption. The INCS2 connects to a host system via a USB-A interface, making it a plug-and-play solution compatible with Windows and GNU/Linux environments. The INCS2 can run multiple DL models simultaneously with 500 MB of onboard memory.

OpenVINO (Open Visual Inference and Neural Network Optimization) is an open-source toolkit developed by Intel to deploy DL models on hardware platforms, including INCS2. To use VPU acceleration, DL models must be converted to an intermediate representation using the OpenVINO model optimizer. This process involves transforming models from high-level frameworks such as TensorFlow or PyTorch to an FP16 format that the inference engine can process efficiently. Once the optimized model is loaded, the OpenVINO inference engine optimizes it for the target device, ensuring that the model takes advantage of the VPU acceleration.

2.4 NXP i.MX8 Plus and TensorFlow Lite

The NXP i.MX8 family is a series of processors based on ARM Cortex technology, designed for peripheral computing. These SoMs are designed to handle a wide range of computations and are compatible with numerous peripherals, making them adaptable for all types of applications. Among the different models, the i.MX8M Plus stands out for its integrated Neural Processing Unit (NPU) with a processing power of 2.3 TOPS, specifically for machine learning tasks. The NPU enhances the device's ability to perform DL inferences efficiently.

TFLite is also the library used to run deep learning models on the i.MX8M Plus. Models must be quantized to INT8 to use the NPU which might lead to a worsening of model evaluation metrics. The i.MX8M Plus employs OpenVX as a delegate to optimize and accelerate neural network models on the NPU. TFLite facilitates the conversion of TensorFlow models to a special format suitable for inference on edge devices, and the OpenVX delegate adapts these models to run efficiently on the NPU.

2.5 Xilinx Zynq UltraScale+ MPSoC ZCU104, Vitis AI and HLS Implementations

Field Programmable Gate Arrays (FPGAs) are integrated circuits designed to be configured by a programmer. Xilinx, a leader in FPGA technology, offers devices that allow custom hardware configurations by loading bitstreams. This flexibility makes FPGAs ideal for prototyping and debugging integrated circuits. Xilinx FPGAs, such as the Zynq UltraScale+ MPSoC (ZUS+) ZCU104, integrate ARM Cortex-A53 microprocessor and extensive programmable logic enabling powerful high-speed computation and data processing capabilities.

Xilinx offers the Vitis AI framework to implement DL inference on their FPGAs. Developers can deploy optimized DL models by implementing the IP-Core Deep Processing Unit (DPU) in FPGA devices. DPU needs the DL model to be quantized at INT8 precision. Finally, DL models must be compiled in a .xmodel file with all the inference instructions for the DPU.

Another way to execute DL models on FPGA is by implementing them using high-level synthesis (HLS). HLS offers a way to develop mathematical algorithms directly in hardware using high-level programming languages such as C++. Our research team from the ITCL developed custom DL models using HLS and the Xilinx library Basic Linear Algebra Subprograms (BLAS). Our previous work has demonstrated the successful implementation of FCN and LSTM models [16,17], exploiting the inherent parallelism of FPGA to achieve real-time in DL inference. In addition, we have tested the use of the Python framework HLS4ML [18], which is in charge of generating the necessary software to be integrated into HLS, from a DL model in .h5 format [19]. To carry out the synthesis it has been essential to use fixed data type FIX16,6 where 6 is the number of bits representing the integer part and 16 is the total number of bits.

Table 1. Specifications of the FCN models. The topology list represents the number of neurons on each layer. The input shape indicates the input data dimension, and the activation is the non-linear function applied at the end of each layer.

	FCN Model 1	FCN Model 2	FCN Model 3
Num. Layers	5	6	7
Topology	[32, 256, 64, 32, 8]	[32, 256, 128, 64, 32, 8]	[32, 128, 256, 128, 64, 32, 8]
Input Shape	1024	1024	1024
Activation	Tanh	Sigmoid	ReLU

3 Experimental Setup

3.1 Datasets and Model's Error Metrics

The selected datasets used to evaluate the DL models are not crucial for this work, as the focus is on measuring inference latency on EC devices and the impact on the DL error metric. However, the datasets used for evaluating the DL models are the following. For the FCN models, we use the 8 classes of the Fashion-MNIST dataset. The input size of the images is 28×28 pixels, so they are resized to 32×32 pixels and flattened into a vector of 1024 values. For the CNN models, two different datasets are used. CNN Model 1 uses the 8 classes of the CIFAR-10 dataset resized to 32×32 pixels. CNN Models 2 and 3 use 8 classes from the ImageNet dataset. We use this dataset to test CNN models on edge with 720p high-resolution images. For those classification tasks, category cross-entropy is the error metric. For the LSTM models, the dataset is a timeseries of sensor data from a wave energy converter with a power take-off system. This dataset is private so not too much detail could be explained. The input data are 4 sensor measurements recorded at 10 Hz frequency and the output data is the wave height with the same frequency. The task is to predict the wave height at the 1-minute horizon so the mean square error is used as the error metric.

3.2 Deep Learning Models

In this preliminary work, we evaluate three types of DL models: FCN, CNN, and LSTM. The architecture of each one is detailed below. Table 1 provides descriptions of the three FCNs, Table 2 details the three CNNs, and Table 3 outlines the three LSTMs. Note that the DL models have increasing complexity.

Table 2. Specifications of the CNN models. CNN includes convolutional, max-pooling, flatten, and fully connected layers. The topology's first list is the number of convolutional and max-pooling (in parentheses) kernels, the second list shows the number of neurons in fully connected layers. Kernel parameter k is the kernel shape and s is the stripe. Input shape is a tuple of the width, height, and channels. The activation function is applied to fully connected layers.

	CNN Model 1	CNN Model 2	CNN Model 3
Num. Conv Layers	3	3	6
Num. Max-P Layers	2	2	0
Num. Flatten Layers	1	1	1
Num. Fully-C Layers	2	2	2
Topology	[3, (3), 5, (5), 10] [16, 8]	[3, (3), 3, (3), 3] [16, 8]	[7, 10, 15, 10, 10, 10] [16, 8]
Kernel Params	5x[k:(3, 3), s:(1, 1)]	5x[k:(3, 3), s:(2, 2)]	6x[k:(3, 3), s:(2, 2)]
Input Shape	(32, 32, 3)	(1280, 720, 3)	(1280, 720, 3)
Fully-C Activation	Sigmoid	ReLU	ReLU

Table 3. Specifications of the LSTM models. The network includes LSTM and fully connected layers. The topology's first block lists the LSTM number of cells, while the second block shows the number of neurons in fully connected layers. The input shape indicates the input data dimension. The sequence length is the number of time steps. The activation function is applied to fully connected layers.

	LSTM Model 1	LSTM Model 2	LSTM Model 3
Num. LSTM Layers	2	3	4
Num. Dense Layers	2	2	2
Topology	[32, 64] [32, 1]	[32, 64, 128] [32, 1]	[32, 64, 128, 32] [32, 1]
Input Shape	4	4	4
Sequence Length	16	32	64
Fully-C Activation	Tanh	Sigmoid	ReLU

3.3 Selected Edge Computing Devices

We have selected seven EC devices to run the DL benchmark. Each device has different DL Accelerator technology and specifications. EC devices available are detailed in Table 4. The most powerful in terms of specification is the Jetson Orin NX. On the ZUS+ ZCU104, three different DL Accelerators are implemented. The first is the DPU accelerator from Xilinx, the second is the ITCL implementations of FCN [16], CNN and LSTM [17], and the third one is the implementations using HLS4ML.

3.4 Performance Metrics

Our analysis focuses on three key metrics that are crucial for evaluating the performance and efficiency of EC systems:

Table 4. Hardware specification of the edge computing devices. ID column will be used as identification in the results section. DL Accelerator is the technology which each device uses to execute DL models on edge. Notice that ZUS+ ZCU104 is used to test three implementations of DL Accelerators.

ID	Device	Company	ARM CPU	Cores	RAM	DL Acc.
JNANO	Jetson Nano	NVIDIA	Cortex-A57	4	4 GB	GPU
JORIN	Jetson Orin NX	NVIDIA	Cortex-A78AE	8	16 GB	GPU
CDEV	Coral DevBoard	Google	Cortex-A53	4	1 GB	TPU
CUSB	Coral USB	Google	-	-	-	TPU
NCS2	INCS2	Intel	-	-	-	VPU
IMX8	iMX8 Plus	NXP	Cortex-A53	4	4 GB	NPU
ZCUDPU	ZUS+ ZCU104	Xilinx	Cortex-A53	4	2 GB	DPU
ZCUITCL						ITCL
ZCUHLS						HLS4ML

1. Inference latency: We measure each device's time to perform single-data inferences for the different DL models. Low latency is sought to build real-time applications. We measure it in milliseconds. The measurements are performed for 1000 samples of the datasets and the mean is calculated.
2. Increase in error metric: This evaluates how DL models' precision is affected when running on EC devices. It helps to understand the trade-off between computational efficiency and model accuracy. It is measured as a percentage over the base value run on conventional hardware.
3. Device energy consumption: This metric allows us to evaluate the energy efficiency of each EC device. Is measured in idle and in inference mode. The measure is performed with a potentiometer and expressed in watts (W).

These metrics provide a comprehensive view of the performance of the seven EC devices evaluated in different DL models.

4 Results

This section presents the results from the benchmark of the selected EC devices using detailed DL models. Below, we present a thorough analysis of the metrics explained in the previous section.

Table 5 presents the latency results for FCN, CNN, and LSTM models on EC devices. For Jetson Devices, the best latency result between CUDA and

Table 5. Latency results, in milliseconds, for each DL model and DL accelerator. Latency columns report the average latency across the total number of iterations. Lower latency indicates better results. The best result for each model is shown in bold

	FCN			CNN			LSTM		
	1	2	3	1	2	3	1	2	3
JNANO	0.7972	0.8394	0.9372	0.7454	24.169	24.530	2.7236	6.9737	13.628
JORIN	0.2973	0.3305	0.2886	0.3598	5.2927	6.4111	0.8765	2.6557	3.9979
CDEV	**0.1991**	**0.2286**	0.2459	0.4223	76.134	120.31	-	-	-
CUSB	0.3434	0.4734	0.6263	0.4937	118.64	411.14	3.9445	8.6274	18.789
NCS2	1.9335	1.9310	1.8994	2.1093	60.792	64.427	-	-	-
IMX8	0.3478	0.3689	0.3352	0.3901	27.052	28.916	-	-	-
ZCUDPU	-	-	0.3975	-	**4.6044**	**5.1914**	-	-	-
ZCUITCL	0.2673	0.2967	0.3261	**0.1720**	16.100	-	0.5620	**1.4020**	**2.6410**
ZCUHLS	0.2277	0.2484	**0.2217**	0.1940	22.758	26.091	**0.2294**	2.7957	5.5043

Table 6. Percentage of increase in error metric for each DL model and DL accelerator. For FCN and CNN, the error metric is the binary cross-entropy, and for LSTM the error metric is the mean squared error. Finally, the data type of the DL model weights is shown for each DL accelerator

	Weight Type	FCN			CNN			LSTM		
		1	2	3	1	2	3	1	2	3
JNANO	FP32	0.0%	0.0%	0.0%	0.0%	0.0%	0.0%	0.0%	0.0%	0.0%
JORIN	FP32	0.0%	0.0%	0.0%	0.0%	0.0%	0.0%	0.0%	0.0%	0.0%
CDEV	INT8	14.3%	8.5%	24.4%	0.6%	43.8%	22.1%	-	-	-
CUSB	INT8	14.3%	8.5%	24.4%	0.6%	43.8%	22.1%	-	-	-
NCS2	FP16	0.6%	0.5%	3.7%	0.0%	0.4%	2.6%	0.0%	0.0%	0.1%
IMX8	INT8	14.3%	8.5%	24.4%	0.6%	43.8%	22.1%	-	-	-
ZCUDPU	INT8	-	-	3.3%	-	0.0%	0.0%	-	-	-
ZCUITCL	FP32	0.0%	0.0%	0.0%	0.0%	0.0%	-	0.0%	0.0%	0.0%
ZCUHLS	FIX16,6	7.9%	22.2%	44.1%	11.9%	27.1%	17.7%	0.0%	0.0%	0.0%

Tensor RT providers was chosen for all DL models. Not all models are executable on all devices. For instance, DPU executions are constrained due to Vitis AI's exclusive support for ReLU activations, which conflicts with the sigmoid or tanh activations used by FCN Model 1 and 2, and CNN Model 1. Additionally, CNN Model 3 could not be executed on the ZCU104 ITCL implementation due to the hardware requirements of the DL Accelerator. The Coral DevBoard, Coral USB, iMX8 Plus, and ZCU104 with DPU lack support for LSTM models.

The FCN latency results indicate that most devices achieve a mean inference latency below 0.5 ms, confirming that FCN models are well-suited for real-time execution on EC devices. However, exceptions include the Jetson Nano, where the latency peaks at 0.9 ms, and the NCS2, which shows a significant rise to

1.9 ms. The Coral DevBoard and ZCU104 with HLS4ML stand out by delivering the lowest latency across all three FCN models, largely due to their use of INT8 and FIX16,6 precision. While this reduction in precision helps lower latency, it comes with the trade-off of an increased error metric due to quantization (Table 6). Conversely, Jetson Orin and the ZCU104 ITCL implementations manage to maintain low latency without any error increase. Despite the varying complexity of the models, inference latencies remain fairly consistent across all EC devices, underscoring their capability to handle FCN models efficiently (Table 6).

CNN latency results exhibit significant variability across devices. A noticeable latency difference exists between model 1 and models 2 and 3, with minimal variation, except for the ZCU104 ITCL implementation. This suggests that Max-Pooling and Convolutional layers typically require similar computation times on most devices, with ZCU104 ITCL being a notable outlier. The FPGA ZUS+ ZCU104, leveraging DPU and ITCL implementations, achieves the lowest mean latency, consistently staying below 5.5 ms. Jetson Orin also performs well, maintaining latency under 6.5 ms for all models. Coral Devices perform efficiently for model 1 but exhibit a sharp latency increase for models 2 and 3, surpassing 100 ms, alongside significant error metric increases of 43.8% and 22.1% for models 2 and 3, respectively. Interestingly, ZCU104 DPU, Coral Devices, and i.MX8 Plus employs INT8 inference computations but shows a vast difference in error metric increases. Coral Devices and i.MX8 Plus relies on TFLite quantization, which incurs a high error increase, whereas ZCU104 DPU uses Vitis AI quantization with no increase in error metrics.

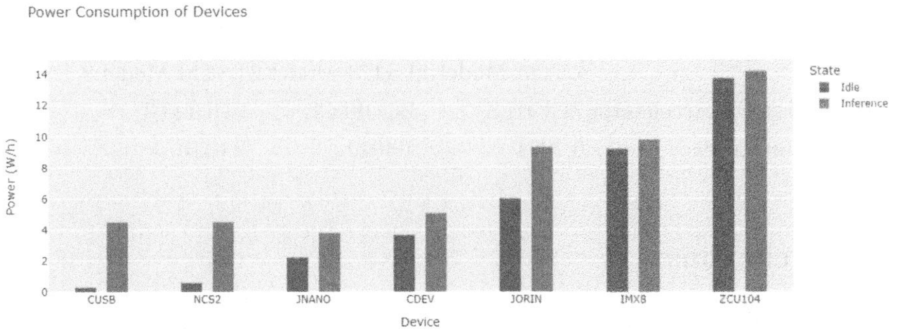

Fig. 1. Power consumption in watts for each device. Blue indicates idle, while red indicates inference power consumption. (Color figure online)

The LSTM latency results demonstrate that this architecture is only supported by EC devices capable of computing with FP16, FIX16,6, or FP32 precision. The ZCU104 HLS4ML and ZCU104 ITCL implementations achieve the lowest latencies, followed closely by Jetson Orin, with all models maintaining latencies under 5 ms. In contrast, Jetson Nano and NCS2 exhibit significantly higher latencies, nearing 20 ms, which can be attributed to the input data's sequence length. As sequence length increases, latency rises substantially across all EC devices due to LSTMs' inherently sequential processing. Unlike FCNs, LSTMs cannot leverage parallel computations, leading to latency increases that are independent of the number of layers, but instead driven by the complexity of processing time-dependent data.

Power consumption is also a critical metric (Fig. 1). All EC devices consume 14.2 W or less. USB accelerators like Coral USB and INCS2 have the lowest consumption, near 0 W at idle and 4.5 W during inference, due to minimal idle processes. Standalone devices have higher idle consumption due to operating system processes. The ZCU104 consumes the most power, peaking at 14.2 W, regardless of DL implementation (DPU, ITCL or HLS4ML). The iMX8 and Jetson Orin consume under 10 W, with iMX8 showing similar idle and inference consumption, and Jetson Orin showing a 2.5 W difference. Finally, Jetson Nano has the lowest inference consumption at 3.8 W, while the Coral DevBoard consumes 5.1 W during inference.

Table 7. The best DL accelerator for each model, based on latency performance without any increase in the error metric. The best latency is measured in milliseconds.

	FCN Model 1	FCN Model 2	FCN Model 3
Best DL Accelerator	ZCUITCL	ZCUITCL	JORIN
Best Latency	0.2673	0.2967	0.2886
	CNN Model 1	CNN Model 2	CNN Model 3
Best DL Accelerator	ZCUITCL	ZCUDPU	ZCUDPU
Best Latency	0.1720	4.6044	5.1914
	LSTM Model 1	LSTM Model 2	LSTM Model 3
Best DL Accelerator	ZCUHLS	ZCUITCL	ZCUITCL
Best Latency	0.2294	1.4020	2.6410

5 Conclusions

Edge DL enables real-time processing of data near the data source. This study has explored the deployment of DL models on various EC devices, highlighting their specifications, constraints, and performance. NVIDIA Jetson, Google Coral, INCS2, NXP i.MX8 and Xilinx ZCU104 devices are evaluated for FCN, CNN, and LSTM models.

The latency results highlight significant performance differences across the devices. For FCN models, most devices achieve latencies below 0.5 ms, with the ZCU104 HLS4ML and Coral DevBoard delivering the lowest latency, though the Coral DevBoard incurs an increased error due to INT8 quantization. Jetson Orin and ZCU104 with ITCL implementations emerge as the top DL accelerators, achieving low latency without sacrificing accuracy. In CNN models, ZCU104

ITCL and ZCU104 DPU stand out by offering the lowest latencies, with consistent performance across all models. For LSTM models, only devices that support FP16, FIX16,6, or FP32 precision can handle them effectively, with ZCU104 ITCL and HLS4ML implementations achieving latencies under 5 ms. However, LSTM latency is notably impacted by the sequence length of the input data, as EC devices struggle with the sequential processing demands of this architecture. Table 7 summarizes the best DL accelerator for each DL model with the lowest latency without any error metric increase. This table indicates that for the majority of DL models, the best-performing device is an implementation on the ZUS+ ZCU104.

FP32 inference on EC devices shows no increase in error metric, while FP16, FIX16,6 and INT8 suffer some degradation. Vitis AI INT8 quantization for ZCU104 DPU has negligible impact on measured error metrics, unlike TFLite INT8 quantization which worsens the error metric up to 13.3% in FCNs and 43.8% in CNNs on devices like Coral and iMX.8. In terms of power consumption all devices are under 14W being the ZCU104 the most power hungry and the USB devices the most efficient. That is not a fair comparison because USB devices are not standalone and need a powerful host like a standard PC or other DevBoard.

In conclusion, there is no single best device for all DL applications but FPGA implementations stand out. Each device has strengths tailored to specific model architectures and application needs. Jetson Orin and ZCU104 are ideal for applications requiring high performance and low latency across various models but with a high power consumption. In contrast, Coral USB and INCS2 are suited for low-power applications with less complex tasks. Effective deployment of DL models on EC devices necessitates a balanced approach, taking into account both performance and energy efficiency to select the most appropriate device for each specific use case.

Acknowledgments. This work has been founded from the Recovery, Resilience and Transformation Plan under the projects: Agrar.IA: Missions Program of R&D in Artificial Intelligence of the SEDIA from Spanish Ministry of Economic Affairs and Digital Transformation (MIA.2021.M01.0004) and CICERO: CDTI Project CER-20231019 (CICERO).

References

1. Shafique, M., et al.: TinyML: current progress, research challenges, and future roadmap. In: 2021 58th ACM/IEEE Design Automation Conference (DAC), pp. 1303–1306 (2021). https://doi.org/10.1109/DAC18074.2021.9586232
2. Liu, F., et al.: A survey on edge computing systems and tools. Proc. IEEE **107**(8), 1537–1562 (2019). https://doi.org/10.1109/JPROC.2019.2920341
3. Chang, Z., et al.: A survey of recent advances in edge-computing-powered artificial Intelligence of Things. IEEE Internet Things J. **8**(18), 13849–13875 (2021). https://doi.org/10.1109/JIOT.2021.3088875

4. Wang, X., et al.: Convergence of edge computing and deep learning: a comprehensive survey. IEEE Commun. Surv. Tutor. **22**(2), 869–904 (2020). https://doi.org/10.1109/COMST.2020.2970550
5. Genda, K.: Deep learning inference time guarantee in near future edge computing. In: 2023 International Conference on Computing, Networking and Communications (ICNC), pp. 220–225 (2023). https://doi.org/10.1109/ICNC57223.2023.10074015
6. Chen, J., Ran, X.: Deep learning with edge computing: a review. Proc. IEEE **107**(8), 1655–1674 (2019). https://doi.org/10.1109/JPROC.2019.2921977
7. Li, E., et al.: Edge AI: on-demand accelerating deep neural network inference via edge computing. IEEE Trans. Wirel. Commun. **19**(1), 447–457 (2020). https://doi.org/10.1109/TWC.2019.2946140
8. Zhang, Z.-D., et al.: CDNet: a real-time and robust crosswalk detection network on Jetson Nano based on YOLOv5. Neural Comput. Appl. **34**(13), 10719–10730 (2022). https://doi.org/10.1007/s00521-022-07007-9
9. Gomes, H., et al.: Counting people and bicycles in real time using YOLO on Jetson Nano. Energies **15**(23), 8816 (2022). https://doi.org/10.3390/en15238816
10. Assunção, E., et al.: Real-time weed control application using a Jetson Nano edge device and a spray mechanism. Remote Sens. **14**(17), 4217 (2022). https://doi.org/10.3390/rs14174217
11. Gao, C., et al.: EdgeDRNN: recurrent neural network accelerator for edge inference. IEEE J. Emerg. Sel. Top. Circuits Syst. **10**(4), 419–432 (2020). https://doi.org/10.1109/JETCAS.2020.3040300
12. Dos Santos Melício, B.C., et al.: DeepRehab: real time pose estimation on the edge for knee injury rehabilitation. In: Farkaš, I., Masulli, P., Otte, S., Wermter, S. (eds.) ICANN 2021. LNCS, vol. 12893, pp. 380–391. Springer, Cham (2021). https://doi.org/10.1007/978-3-030-86365-4_31
13. Zualkernan, I., et al.: An IoT system using deep learning to classify camera trap images on the edge. Computers **11**(1), 13 (2022). https://doi.org/10.3390/computers11010013
14. Kristiani, E., et al.: Flame and smoke recognition on smart edge using deep learning. J. Supercomput. **79**(5), 5552–5575 (2023). https://doi.org/10.1007/s11227-022-04884-8
15. Chi, T.-K., et al.: An edge computing system with AMD Xilinx FPGA AI customer platform for advanced driver assistance system. Sensors **24**(10), 3098 (2024). https://doi.org/10.3390/s24103098
16. Lozada, R., et al.: Performance/resources comparison of hardware implementations on fully connected network inference. In: Yin, H., Camacho, D., Tino, P. (eds.) IDEAL 2022. LNCS, vol. 13756, pp. 348–358. Springer, Cham (2022). https://doi.org/10.1007/978-3-031-21753-1_34
17. González, M.L., et al.: Exploring the implementation of LSTM inference on FPGA. In: 2023 3rd International Conference on Electrical, Computer, Communications and Mechatronics Engineering (ICECCME), pp. 1–5 (2023). https://doi.org/10.1109/ICECCME57830.2023.10252205
18. Duarte, J., et al.: Fast inference of deep neural networks in FPGAs for particle physics. J. Instrumentation **13**(07), P07027–P07027 (2018). https://doi.org/10.1088/1748-0221/13/07/P07027. ISSN 1748-022
19. FastML Team. fastmachinelearning/hls4ml. Version v0.8.1 (2023). https://www.dgithub.com/fastmachinelearning/hls4ml

Causal Explanation of Graph Neural Networks

Hichem Debbi(✉)

Department of Computer Science, Laboratory of Informatics and its Applications of M'sila (LIAM), University of M'sila, M'sila, Algeria
hichem.debbi@univ-msila.dz

Abstract. Graph Neural Networks (GNNs) are currently used in many real-world applications. With this notable spread, the development of sophisticated techniques for explaining their decisions becomes highly necessary. Although many works have been proposed in the aim of explaining their predictions, on different aspects, such as nodes, edges, and features, they all tend to generate the explanations as subgraphs. In this paper, we will show that relying only on explanatory subgraphs is not sufficient explanation tool, especially that these subgraphs could range from small graphs to untraceable ones within the same model. In this regard, we propose a causal explanation framework based on the rigorous structural model of causality. We show that our framework does not compete with existing explanation framework for GNNs, but rather acts as a complementary approach.

Keywords: Graph Neural Networks (GNNs) · Causality · Explainability · Counterfactuals

1 Introduction

Graphs represent a powerful data representation tool that can be used in many real-life applications such as social networks [4], biology and medicine [30,31], and molecule structures [20,23]. However, working with graphs presents a real challenge, since we need to deal with both node features and edge features in a very synergistic way. With the increasing performance of deep neural networks in many research tasks such as computer vision and natural language processing, there is motivation to use deep neural networks for graph data analysis. In this regard, Graph Neural Networks (GNNs) have been proposed as a state-of-the-art tool for deep learning on graphs. Thanks to robust mechanisms such as the recursive message passing scheme, which encodes and aggregates information from neighboring nodes and passe it along the edges of the graph, GNNs achieved state-of-the-art performance on many tasks such as graph classification [28], node classification [6,13] and link prediction [2]. Moreover, to achieve better accuracy, different operations on graphs have been proposed such as Graph Convolutional Neural Networks [7,10], graph attention [21], and graph pooling [24].

V. Julian et al. (Eds.): IDEAL 2024, LNCS 15346, pp. 277–288, 2025.
https://doi.org/10.1007/978-3-031-77731-8_26

Despite their performance, similar to all deep learning architectures, they fail to explain their decisions, and thus they still have black-box nature. Such an explanation ability becomes more necessary because it would make learning on graphs structures more efficient, robust and safe. The explanation of deep neural models was first and largely addressed in the context of computer vision [5,18] and natural language processing [15,26], mainly by analyzing the underlying relationships behind the predictions of these models. Other existing models have been proposed to explain the decisions of any Machine learning classifier such as LIME [17]. Due to its importance, explanation in the context of GNNs has been also widely addressed by many researchers.

For GNN explainability, many challenges can emerge, first GNNs represent combination of data structures, thus identifying critical substructures that could maximize a model's prediction must be achieved by performing many combinations. In addition, in GNNs, nodes are featured, and thus both node features and edges representing the connectivity must be evaluated in order to achieve better explanation. In the literature, we have many GNN explainers such as GNNExplainer [22], PGExplainer [14] and SubgraphX [27], with different taxonomies and classification [1,25]. Globally, explanation methods for GNNs can be classified into two categories: self-interpretable frameworks and post-hoc explanation frameworks. Self-interpretable frameworks try to incorporate the explainability methods directly inside the GNN model by finding an informative subgraph for the prediction. However, one challenge with this approach is that the quality of the explanation and the prediction accuracy are dependent. Post-hoc explanation methods, which encompasses the well-known existing explanation methods, try to generate an explanation E for any graph G, using the model's input D and prediction output Y. Some of these methods need access to the internal parameters of the model or embeddings; in this case, we call them white-box methods. Otherwise, they are called black-box methods.

Post-hoc explanation methods can use different techniques such as gradients [16], surrogate models [9,29], and perturbation methods [14,22,27]. Our approach has a strong relationship to the last one, where we typically try to provide important subgraphs as explanations by altering the input and checking the actual prediction score. GNNExplainer, which is one of the earliest works on GNN explainability, generates the explanation as a subgraph that maximizes the influence on the prediction, in addition to subsets of important node features. However, due to its focus on node features, sometimes the generated subgraph might be disconnected [27]. Moreover, edges have no sufficient consideration in this approach.

As the main evaluation metrics for any GNN explanation method, we consider usually two metrics: fidelity and sparsity. While fidelity helps to evaluate the explanation method in terms of prediction accuracy of the generated subgraph when compared to the input graph, sparsity is a conciseness measure of the generated subgraph, where the most desirable explanation is the one having the least number of important elements by ignoring non-relevant ones. However, sparsity and fidelity are highly correlated. Thus, fidelity scores should be compared under the same level of sparsity scores. We suggest that the level of sparsity should vary sometimes to allow the generated subgraph to carry more information. However, we may face here a tradeoff between the information we get and the conciseness of the subgraph. In addition, when we deal with large graphs having large ground-truth explanations, especially having large number of features, most GNN explanation methods could fail to generate reliable explanations [3]. Therefore, to deal with such issues, we propose the following: we further explain the explanations generated by existing explainers, especially when these explanations tend to be large. We propose a causal-based explanation framework that complements existing explainers by considering the subgraphs generated as a state-space search, in which we identify only the most relevant parts as pairs (node-feature). To our knowledge, this is the first approach that generates explanations as state-feature pairs and considers complementing existing explainers, not competing with them. We rely on the definition of structural equation modelling of Halpern and Pearl [8] for formally defining the causes and explanation.

The rest of this paper is organized as follows: Sect. 2 presents the main related works, including perturbation and causal-based methods. In Sect. 3, we present the formal background of our work, which is mainly related to GNNs and the structural equation definition of causality. Section 4 represents the main part of the paper, where we introduce the causal explanation method for GNNs by formally defining the actual cause in the context of GNNs. Section 5 includes a running practical example of our framework on the Cora dataset. Section 6 concludes the paper.

2 Related Work

Perturbation-Based Methods

GNNExplainer is among the first post-hoc explanation methods proposed in literature [22]. Given a trained model M and a prediction Y, it identifies a small indicative subgraph G_e with only subset of influential node features. It is applicable to any graph learning task including node classification, link prediction, and graph classification. It applies a simultaneous process that incorporates both graph and feature masks to identify the subgraph and ignores non-relevant features. Formally, GNNexplainer can be identified as an optimisation framework that tries to maximize the mutual information given a prediction Y, a subset of nodes V_{G_e} and a subset of features X_{G_e}. However, such a strategy could sometimes result in a disconnected explanatory subgraph [27]. In addition, the sub-set of features X_{G_e} are identified equivalently for all nodes of the sub-graph G_e. What we propose here is complement these works through explaining the explanatory subgraph G_e, by identifying node/node feature pairs as the actual causes for such a prediction.

Instead of relying on nodes importance and particularly node features, SubgraphX directly applies the Monte Carlo Tree Search (MCTS) algorithm [19] to explore the subgraph. The importance of the subgraph is measured then using

Shapley values. Although SubgraphX can be considered as a perturbation-based method similar to GNNExplainer, it has different strategy by not targeting directly nodes and features masks as GNNEXplainer does. It assigns Shapley values as scores to the generated explanatory subgraphs using Monte Carlo sampling by considering coalition sets in the game and their rewards. Their idea is more practical and more human-intelligible, since it targets the sub-graph as one block and thus it helps to generate a connected subgraph, in contrast to GNNexplainer, which could ignore node interaction. However, one drawback of this method is the computational cost, since it needs to explore multiple subgraphs using the MCTS algorithm.

PGExplainer [14] is another perturbation-based method that builds on a similar idea to GNNExplainer but targets edges and assigns them scores based on their probability distribution. The distribution of each edge is considered as a Bernoulli distribution. The main problem targeted by this work is generalizing GNNExplainer, which is instance-based and must be invoked for each individual node. However, PGExplainer tries to generate explanatory graphs by first generating a probabilistic model for the graph data, parameterized with a neural network (MLP). Since the MLP parameters are shared across a multiple set of explained instances, this would result in a global view of the GNN model when explaining individual instances.

Causal-Based Methods

While causality has been addressed by many researchers in text and image explainability with different approaches, the use of causality for GNN explainability is still limited. GEM [12] is the first approach that tries to resolve some issues and fill some gaps of existing GNN explainers. It is very similar to PGExplainer, since it is heavily based on estimating the causal effect to generate the subgraph. However, it comes with some differences. While PGExplainer needs to access the internal parameters of the GNN model, as it relies on node embeddings from the target GNN model, GEM directly targets the original graph space. Additionally, GEM does not need existing knowledge about certain motifs in the dataset, which can help with explanations as PGExplainer does, because it addresses some correlations through the causal learning process. GEM employs Granger causality, which is originally used for forecasting in time series data. Granger causality is based on the assumption that a variable X is considered a Granger cause for Y, if predicting Y based on all sufficient knowledge, including X, is better than predicting Y without considering the availability of X data. GEM applies this causality test to explain GNNs by considering the cause as the explanation subgraph, which is obtained by quantifying the edges' causal contributions. The resulting explanations can be considered graph rules, which are used to ensure that the explanations are human-intelligible.

In contrast to GEM, which identifies causal contribution in the graph space, OrphicX [11] identifies causal contribution in the embedding space. With a more complex architecture, OrphicX relies on two modules: an inference network and a generative network. The aim of the inference network is to learn latent causal features, which are then added to some spurious features and passed together to the generative network, which is responsible for generating the causal explanatory subgraph. They use an information flow measurement technique, which is similar to mutual information, to quantify the causal contribution of the latent features on model prediction, also considering the existence of confounders. Thus, the resulting explanatory subgraph should be as informative as possible, since OrphicX learns only to consider causal features independently of any spurious features. The aim is to provide a global view of the target GNNs by generating subgraphs that causally contribute to the predictions for any instance. While GEM relies on Granger causality, OrphicX relies on the concept of Structural Causal Model (SCM). Similar to GEM, OrphicX does not need any access to the model's inner parameters or provided subgraph patterns.

3 Background

3.1 GNNs

Let $G = (V, E)$ be a graph where $V = v_1, v_1, ..., v_n$ denotes the set of n nodes and $E = V \times V$ denotes the set of m edges. A graph can be represented using an adjacency matrix $A \in [0, 1]^{n \times m}$, where $a_{i,j} = 1$ if an edge connects two nodes i and j, and $a_{i,j} = 0$ otherwise. In addition, nodes in V are associated with d-denominational node features denoted X, such that $X \in \mathbb{R}^{n \times d}$. A graph Neural Network (GNN) model, which is used to learn powerful representations over graphs structures can be then formulated as a function $f : V \rightarrow Y$, which assigns to each node in V a label from a finite set Y, and this is called node classification; or can be formulated as $f : G \rightarrow Y$ in the case of graph classification.

The GNN model is trained with an objective function $\mathcal{L} : y \times \tilde{y} \rightarrow s$ that computes the loss $s \in \mathbb{R}$, by comparing the model's prediction $\tilde{y}$ to a ground truth label y. A GNN model embeds each node into a low-dimensional space by performing message passing.

Given a pretrained GNN model M, a graph G, and a prediction outcome $\tilde{y}$ of M, the objective of explanation in GNN is to generate a subgraph G_e of G. The subgraph G_e is a mask of nodes, nodes' features and edges that are most relevant for the prediction $\tilde{y}$. In node classification, G_e is typically generated around the target node, and k-hop neighborhoods are often considered.

3.2 Causality

All the definitions of causality in literature are mainly based on counterfactual reasoning, event A is a cause of event B if, had A not happened then B would not have happened. Unfortunately, this definition does not capture all the cases we face in real world.

Due to the importance of counterfactuals, Halpern and Pearl have extended the definition of counterfactuals by Lewis to build a rigorous mathematical model of causation, which they refer to as structural equations [8].

Halpern and Pearl have extended the Lewis counterfactual model [11] to what they refer to as structural equations for modelling the causal influence made by random variables. They introduce the causality model M as a tuple $M = (U, V, F)$, where random variables are partitioned into two sets, the exogenous variables U, whose values are determined by factors outside the model M, but they should be represented to encode the context, and the endogenous variables V whose values are determined by set of functions F, where each $f_{V_i} \in F$ is a mapping from $U \cup (V \setminus V_i)$ to V_i. Thus, each f_{V_i} tells us the value of V_i given the values of all other variables in $U \cup V$.

We call a Boolean combination of primitive events a basic causal formula. We call a Boolean combination of basic causal formulas a causal formula. A causal formula φ is true or false in causal model, given a context. We write $(M, \overrightarrow{u}) \models \varphi$ if φ is true in causality model M given a context $\overrightarrow{u}$, where $\overrightarrow{u}$ represents a setting for the variables in U. We write $(M, \overrightarrow{u}) \models [\overrightarrow{Y} \leftarrow \overrightarrow{y}](X = x)$, where $\overrightarrow{Y} \subset V$, if X has a value x in M given a context $\overrightarrow{u}$ and the assignment $\overrightarrow{y}$ to $\overrightarrow{Y}$. The formulas that are allowed to be causes for φ are ones of the form $X_1 = x_1 \wedge ... \wedge X_k = x_k$ which is abbreviated to the form $\overrightarrow{X} = \overrightarrow{x}$.

4 Causality for GNNs Explanation

Let us give the following formulation: X is the set of features, G is the input graph, G_e is its explanatory subgraph, and V_{G_e} is the set of nodes of G_e. Let us denote by $\tilde{y} = M(x)$ the output prediction of a model M given an input x, and p its prediction probability, where input x can be a graph in case of graph classification. The aim of explanation in GNNs is to explain the prediction $\tilde{y}$. To explain $\tilde{y}$ we need to ask the causal question: which features values in the subgraph G_e cause the prediction $\tilde{y}$. We denote these features by X_{G_e}, with their values they refer to the causal Boolean formula φ in the causality definition of Halpern and Pearl. Since, the definition of causality by Halpern and Peral deals with the notion of contexts, we should also ask in addition the following question: in which nodes these features values are considered causes. Here, we consider the set of nodes V_{G_e} as the possible contexts, thus, features values which are considered causal in some nodes may not considered causes in other nodes. Now, since $\tilde{y}$ is returned with a probability p, we would be able to reshape the causal question as follows: which features values in the subg-graph G_e cause the prediction $\tilde{y}$ to achieve a probability p.

4.1 Causality Model for GNNs

With respect to the definition of causality by Halpern and Pearl, a causality model consists of: a set of exogenous variables U, a set of endogenous variables V and set of functions F. We can adopt this definition to the explanatory subgraph G_e as follows.

Causality Model. A causality model for the explanatory subgraph G_e is a tuple $\mathcal{C}_{G_e} = (V_{G_e}, X_{G_e}, F_{G_e})$, where the set V_{G_e} refers to exogenous variables, where every context is a node $v \in V_{G_e}$, X_{G_e} is the set of endogenous variables, which refer to causal features, and F_{G_e} associates with every feature $X_i \in X_{G_e}$ a value function f_{X_i} that determines the value of X_i, given a node v and the values of other variables in X.

The causal influence in $\mathcal{C}_{G_e}$ is modeled by the edges in the explanatory subgraph.

As this definition stands for causality model, we need to introduce conterfactuals into our model since they serve as a basic definition for actual causes. As we mentioned earlier, a prediction $\tilde{y}$ is returned with prediction probability p. Therefore, we should define countefactuals in terms of the prediction probability. To do so, we should ask the following question: what would be the probability of $\tilde{y}$, when switching the value of a feature X_i at a specific node v, while keeping all the values of all features the same ?

In the aim of answering such question, let us introduce first the definition of criticality. Let us denote by $\widehat{G}_{v,X_i=x_i \to X_i=x_i'}$ the resulting graph by switching the value of a feature X_i from x_i to x_i' in a node v, such that $v \in G_e$.

Criticality. We say that a pair $(v, X_i = x_i)$ is critical for the prediction $\tilde{y}$ given a graph G, if the prediction probability p' of $\tilde{y}$ given $\widehat{G}_{v,X_i=x_i \to X_i=x_i'}$ is less than the original prediction probability p. We refer to the difference $p - p'$ the counterfactual score. In the case of $p - p' \leqslant 0$, there is no counterfactual causation.

That is, for a prediction $\tilde{y}$, we say that the value of a feature X_i is critical in a node v if changing its value in this node decreases the prediction probability. We rely on the prediction probability instead of switching the truth of the entire prediction $\tilde{y}$, because, in almost all situations, it is typically impossible to switch the decision only by changing the value of a single feature in one single node, especially when nodes are represented by multiple features. Now, given the definition of criticality, we can introduce the definition of actual cause.

Actual Cause. We say that a pair $(v, X_i = x_i)$ is an actual cause for the prediction of $\tilde{y}$, if $(v, X_i = x_i)$ is critical, or there exists a subset of features W of X, such that switching their values in v makes $(v, X_i = x_i)$ critical.

That is, in the context of GNN explanation, and given the explanatory subgraph G_e, we refer to the actual cause as a pair $(v, X_i = x_i)$. $(v, X_i = x_i)$ is said to be cause, if it is critical, or it can be made critical by switching the values of set of variables W in a given node v.

5 Examples and Experiments

We consider the Cora dataset, which is one of well-known datasets for graph learning. It is a citation network that consists of 2708 scientific publications, which range in seven (07) main classes. The network consists of 5429 edges that connect 2708 publications. Each publication is represented by a word vector of 1433 unique words, where 1/0 refers to the presence/absence of the corresponding word. These 1433 words represent the features of the dataset.

We consider here the task of node classification. We can generate an explanatory graph for any given node, where the prediction class is one of the seven (07) classes: Case Based, Genetic Algorithms, Neural Networks, Probabilistic Methods, Reinforcement Learning, Rule Learning and Theory. We use GNNExplainer as an en explanation tool. Although, our method can work with any GNN explainer, which returns the explanation as a subgraph, we choose GNNExplainer for its simplicity, and more importantly, because it returns the set of most important features. So, we may rely on some information for speeding up the process of finding most important causes, by considering only the most important features.

We try to generate an explanatory subgraph G_e for the node 10, this node is employed by many examples in literature. Node 10 is returned by the GNN model as class 0. We can define a causality model for this graph as follows: the endogenous variables refer to the 1433 features of CORA, every feature could be causal depending on the context variable, which encode a node $v \in V_{G_e}$, and the values function F_{G_e}, which will tel us whether a feature is present in a node or not. As we mentioned in definition, the causes are identified as pairs (node, feature), where removing a feature in a specific node, if decreases the prediction probability, it is considered causal; otherwise it is not causal.

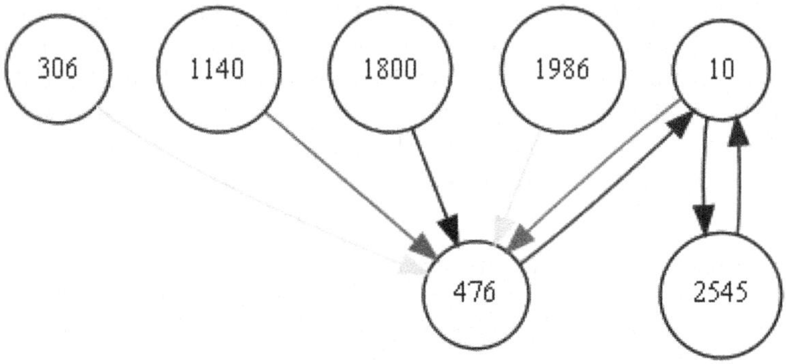

Fig. 1. A subgraph generated by GNNexplainer as an explanation for the node 10

The explanatory subgraph G_e as generated by GNNExplainer is depicted in Fig. 1, and the list of important features as generated by GNNExplainer are depicted in Fig. 2. We should mention that GNNExplainer returns the list of important features over the entire graph. Now, we try to identify the causes given the generated sub-graph G_e by GNNExplainer for the nodes 10, 476, 2545, 1800, 306, 1140, 1986 and the features 1263, 1336, 758, 581, 1177, 1301, 1292, 507, 1159.

The set of causes as generated by our algorithm for this example ordered with respect to the counterfactual score, form most important causes to least important are depicted in Table 1.

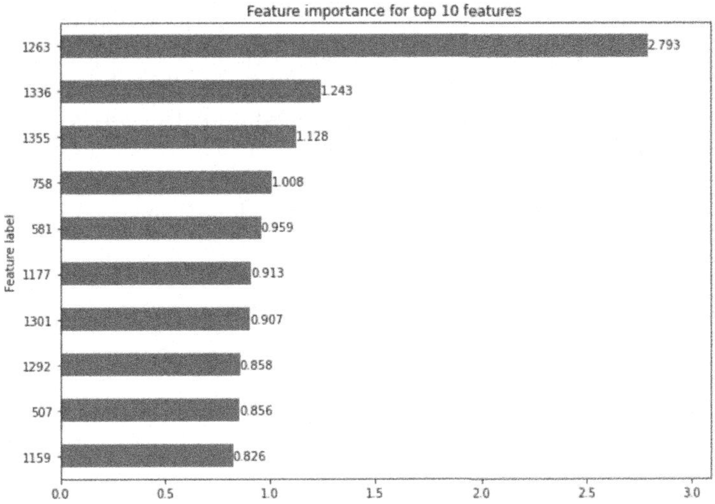

Fig. 2. Top-10 features generated by GNNExplainer as an explanation for the node 10

Table 1. The set of causes generated as nodes/node features pairs ordered with their importance.

Cause	Node	Feature	Counterfactual Score
1	10	1263	0.0063842
2	2545	1263	0.004440
3	10	1301	0.003748
4	476	1336	0.002854
5	2545	1301	0.0026049
6	2545	758	0.00182284
7	1140	1336	0.0017304

As we see from the results, it is very convenient to have the actual causes of the prediction as pairs (node/feature). For instance, it is not necessarily the case that the most important feature is always important in all nodes. For example, in the fourth most important cause, the most important feature in node 476 is 1336, although feature 1263, as returned by GNNExplainer, has a higher contribution compared to 1336. This sample example shows the importance of identifying the causes as pairs, in which the feature must be identified given the context, which is a node, and the causal influence is led by the edges of the graph. Moreover, it could also happen that a feature with lower importance, for instance, 1301 compared to 1336, is returned as a cause in node 10 as the third top cause, although feature 1301 (7th important feature) is less important than 1336 (2nd important feature) with respect to the results returned by GNNExplainer. These obtained results show again the importance of returning explanations in the context of GNN as pairs of nodes and features.

To show the challenges associated with existing explanation techniques, let us take the same model and modify the selected node to be 476. The resulting explanatory subgraph for this node is depicted in Fig. 3. This node has a large k-hop neighborhood, and therefore as we see in the figure, the visualization becomes nearly impossible. This example clearly shows the need for complementary explanation techniques for the generated explanatory subgraphs.

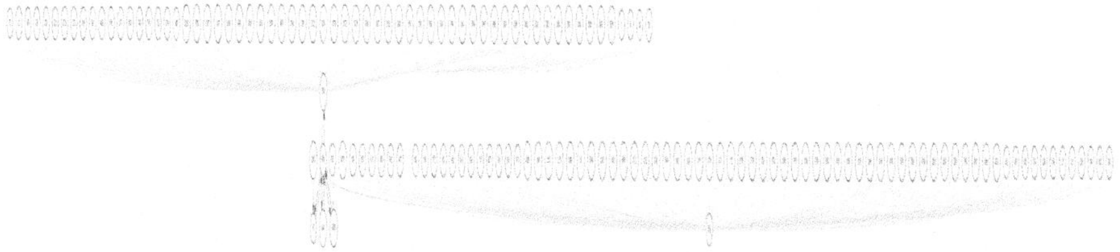

Fig. 3. A large subgraph generated by GNNexplainer as an explanation for the node 476

6 Conclusion

In this paper, we introduced a causal-based technique for explaining GNNs. Despite the task at hand, we showed that providing explanations as a set of actual causes is practical and helpful in many different ways. Although causality has been investigated before for explaining GNNs, relying on the definition of actual cause based on the Structural Equation Model (SEM) is introduced for the first time. Compared to existing explanation methods, this technique does not compete with them but rather acts as a complementary technique. We argued that even good explanation tools that provide small explanatory subgraphs can benefit from our technique by pointing out the relevant set of causes instead of the entire subgraph.

References

1. Amara, K., et al.: Graphframex: towards systematic evaluation of explainability methods for graph neural networks. arXiv preprint arXiv:2206.09677 (2022)
2. Cai, L., Ji, S.: A multi-scale approach for graph link prediction. **34**(04), 3308–3315 (2020)
3. Chirag, A., Owen, Q., Himabindu, L., Marinka, Z.: Evaluating explainability for graph neural networks. Nat. Sci. Data **10**(144) (2023)
4. Cho, E., Myers, S.A., Leskovec, J.: Friendship and mobility: user movement in location-based social networks. In: Proceedings of the 17th ACM SIGKDD International Conference on Knowledge Discovery and Data Mining, KDD 2011, pp. 1082–1090. Association for Computing Machinery (2011)

5. Dabkowski, P., Gal, Y.: Real time image saliency for black box classifiers. In: Proceedings of the 31st International Conference on Neural Information Processing Systems, NIPS 2017, pp. 6970–6979 (2017)
6. Gao, H., Ji, S.: Graph u-nets. In: Chaudhuri, K., Salakhutdinov, R. (eds.) Proceedings of the 36th International Conference on Machine Learning. Proceedings of Machine Learning Research, vol. 97, pp. 2083–2092 (2019)
7. Gao, H., Wang, Z., Ji, S.: Large-scale learnable graph convolutional networks. In: Proceedings of the 24th ACM SIGKDD International Conference on Knowledge Discovery & Data Mining, pp. 1416–1424. ACM (2018)
8. Halpern, J.Y., Pearl, J.: Causes and explanations: a structural-model approach: Part I: causes. In: Proceedings of the Seventeenth Conference on Uncertainty in Artificial Intelligence, UAI 2001, pp. 194–202 (2001)
9. Huang, Q., Yamada, M., Tian, Y., Singh, D., Chang, Y.: Graphlime: local interpretable model explanations for graph neural networks. IEEE Trans. Knowl. Data Eng. **35**(7), 6968–6972 (2023)
10. Kipf, T.N., Welling, M.: Semi-supervised classification with graph convolutional networks. In: ICLR (2017)
11. Lin, W., Lan, H., Wang, H., Li, B.: Orphicx: a causality-inspired latent variable model for interpreting graph neural networks. In: 2022 IEEE/CVF Conference on Computer Vision and Pattern Recognition (CVPR), pp. 13719–13728 (2022)
12. Lin, W., Lan, H., Li, B.: Generative causal explanations for graph neural networks. In: International Conference on Machine Learning (2021). https://arxiv.org/pdf/2104.06643.pdf
13. Liu, M., Gao, H., Ji, S.: Towards deeper graph neural networks. In: Proceedings of the 26th ACM SIGKDD International Conference on Knowledge Discovery & Data Mining, KDD 2020, pp. 338–348 (2020)
14. Luo, D., et al.: Parameterized explainer for graph neural network. In: NIPS 2020 (2020)
15. Madsen, A., Reddy, S., Chandar, S.: Post-hoc interpretability for neural NLP: a survey. ACM Comput. Surv. **55**(8) (2022)
16. Pope, P.E., Kolouri, S., Rostami, M., Martin, C.E., Hoffmann, H.: Explainability methods for graph convolutional neural networks. In: 2019 IEEE/CVF Conference on Computer Vision and Pattern Recognition (CVPR), pp. 10764–10773 (2019)
17. Ribeiro, M.T., Singh, S., Guestrin, C.: Why should I trust you?: explaining the predictions of any classifier. In: KDD 2016, pp. 1135–1144 (2016)
18. Selvaraju, R.R., Cogswell, M., Das, A., Vedantam, R., Parikh, D., Batra, D.: Grad-CAM: visual explanations from deep networks via gradient-based localization. In: 2017 IEEE International Conference on Computer Vision (ICCV), pp. 618–626 (2017)
19. Silver, D., et al.: Mastering the game of go without human knowledge. IEEE Trans. Knowl. Data Eng. **550**, 354–359 (2017)
20. Sun, M., Zhao, S., Gilvary, C., Elemento, O., Zhou, J., Wang, F.: Graph convolutional networks for computational drug development and discovery. Brief. Bioinform. **21**(3), 919–935 (2019)
21. Velickovic, P., Cucurull, G., Casanova, A., Romero, A., Lio, P., Bengio, Y.: Graph attention networks. In: ICLR (2018)
22. Ying, R., Bourgeois, D., You, J., Zitnik, M., Leskovec, J.: Gnnexplainer: generating explanations for graph neural networks. In: Advances in Neural Information Processing Systems, vol. 32, pp. 9240–9251 (2019)

23. You, J., Liu, B., Ying, R., Pande, V., Leskovec, J.: Graph convolutional policy network for goal-directed molecular graph generation. In: Proceedings of the 32nd International Conference on Neural Information Processing Systems, NIPS 2018, pp. 6412–6422 (2018)
24. Yuan, H., Ji, S.: Structpool: structured graph pooling via conditional random fields. In: ICLR (2020)
25. Yuan, H., Yu, H., Gui, S., Ji, S.: Explainability in graph neural networks: a taxonomic survey. IEEE Trans. Pattern Anal. Mach. Intell. **45**(05), 5782–5799 (2023)
26. Yuan, H., Chen, Y., Hu, X., Ji, S.: Interpreting deep models for text analysis via optimization and regularization methods. In: AAAI 2019/IAAI 2019/EAAI 2019 (2019)
27. Yuan, H., Yu, H., Wang, J., Li, K., Ji, S.: On explainability of graph neural networks via subgraph explorations. In: ICML, pp. 12241–12252 (2021)
28. Zhang, M., Cui, Z., Neumann, M., Chen, Y.: An end-to-end deep learning architecture for graph classification. In: AAAI 2018/IAAI 2018/EAAI 2018 (2018)
29. Zhang, Y., Defazio, D., Ramesh, A.: Relex: a model-agnostic relational model explainer. In: AIES 2021, pp. 1042–1049 (2021)
30. Zitnik, M., Agrawal, M., Leskovec, J.: Modeling polypharmacy side effects with graph convolutional networks. Bioinformatics **34**(13), i457–i466 (2018)
31. Zitnik, M., Nguyen, F., Wang, B., Leskovec, J., Goldenberg, A., Hoffman, M.M.: Machine learning for integrating data in biology and medicine: principles, practice, and opportunities. Inf. Fusion **50**, 71–91 (2019)

The Contribution of Social Sciences Driven User Studies to the Development of Human-Centered Artificial Intelligence

Magdalena Wójcik(✉)

Jagiellonian University, Kraków, Poland
magda.wojcik@uj.edu.pl

Abstract. The aim of the paper is to indicate the categories of results that can be obtained in the course of social sciences inspired user studies and to discuss their potential for further development of human-centered XAI. The paper reviews selected qualitative methods and techniques, including: in-depth interviews, focus groups, Delphi method, usability tests with the TAP protocol, UX techniques and observations with the guided tour module as examples of research focused on diagnosing users' opinions, behaviors or experiences. A case study of the implementation of selected methods and techniques in an interdisciplinary research project in the field of XAI was discussed, showing in practice the value of methods driven from social science in the context of AI research. As a result, preliminary conclusions were formulated regarding a new approach to the evaluation of XAI methods, as well as prospects for further research.

Keywords: Explainable Artificial Intelligence · Human-centered Artificial Intelligence · User Studies · Social Science Methods

1 Background and Motivation

The need to include users in the process of evaluating explainable artificial intelligence (XAI) methods is increasingly emphasized in the subject literature, especially in relation to critical areas such as medicine, economics and security. As Arrieta et al. stated, XAI: "(…) 1) produce more explainable models while maintaining a high level of learning performance (e.g., prediction accuracy), and 2) enable humans to understand, appropriately trust, and effectively manage the emerging generation of artificially intelligent partners" (2020, p. 83).

It has clearly been stated numerous times in highly cited papers in the field of computer science that users should not only be recipients of final AI products, but also an active participants in the process of designing, testing and evaluating AI solutions (Miller, 2019) (Ghassemi, Oakden-Rayner & Beam, 2021) (Longo et al. 2023). This is important not only in terms of providing people with user-friendly, understandable and positive experiences with AI products, but also in terms of building public trust in AI by increasing the degree of its transparency and understandability.

V. Julian et al. (Eds.): IDEAL 2024, LNCS 15346, pp. 289–300, 2025.
https://doi.org/10.1007/978-3-031-77731-8_27

Developing research on artificial intelligence created for and with the cooperation of humans requires a multilateral, interdisciplinary approach, which is noted by authors in the field of computer science. As L. Longo et al. emphasizes: "XAI has evolved into an exceedingly multidisciplinary, interdisciplinary, and transdisciplinary field" (2023, p. 2). In this context, the contribution of social sciences is pointed out as particularly valuable, as it is a domain that have a long, well-established tradition and methodology of conducting user studies. As T. Miller rightly stated, a large part of current studies on XAI is based on: "researchers' intuitions of what constitutes a 'good' explanation" (2019, p. 2), which may be misleading due to the fact that: "very experts who understand decision-making models the best are not in the right position to judge the usefulness of explanations to lay users" (Miller, 2019, p. 2). It is therefore necessary to conduct extensive research with users taking inspiration from well-established and proven social science methods.

Drawing direct inspiration from the postulates presented in papers by Ghassemi, Oakden-Rayner & Beam, L. Longo et al., T. Miller and other authors, who clearly stated the need to involve users in the evaluation of AI explanations, and thus the need to examine their needs and feelings and behaviors; the author of this paper systematically characterized the key methods and techniques of qualitative user research, proving their usefulness for computer science on specific examples and by discussing a case study.

2 Objective and Contribution

The aim of the paper is to indicate the categories of results that can be obtained in the course of social sciences inspired user studies and to discuss their potential for further development of human-centered artificial intelligence research. The specific objectives of the study include:

- identifying key methods and techniques of user studies in social sciences,
- developing a typology of user studies methods and techniques depending on the category of obtained results,
- synthetic characterization of identified methods and techniques in relation to their potential usefulness for the evaluation of XAI methods,
- presenting an example of practical use of the potential of social sciences for AI research,
- formulating perspectives for further research integrating the potential of social sciences and engineering.

The paper reviews selected methods and techniques for examining users' opinions, behaviors and experiences, including: in-depth interviews, focus groups, Delphi method, usability tests with the TAP protocol, participant observations with the guided tour module, as well as UX questionnaires and Kano approach. The attention was focused exclusively on qualitative methods and techniques as an approach typical for social sciences.

A case study of the implementation of selected methods and techniques in an interdisciplinary research project in the field of XAI was discussed, showing in practice the value of methods driven from social science in the context of AI research.

As a result, preliminary conclusions were formulated regarding a new approach to the evaluation of XAI methods and prospects for further research were outlined.

3 Methods

To establish the state of research and select key methods and techniques of user studies the method of literature review was used. With regard to the subject literature, a systematic search was carried out according to the adopted queries in database indexing journals with high IF - Scopus, and additionally, selected resources indexed by Google Scholar were also included. Methodological papers on user studies in the field of social science published in English in the years 2014–2024 were taken into account. The collected material was initially analyzed qualitatively based on abstracts and keywords to identify the most relevant contributions. A selected corpus of publications covering methodological works in the social sciences, consisting mainly of papers or monographs, was then further reviewed to identify the most frequently repeated references to user research methods and techniques. It is worth noting that at this stage many publications that related to social science research methodology, but did not necessarily directly concern user studies, were rejected. Publications concerning only quantitative methods were also deliberately omitted, with a conscious decision to focus only on qualitative methods as more useful from the point of view of the subject of considerations in the article.

Then, a case study of the implementation of selected methods and techniques in practice in an interdisciplinary project in the field of XAI was described. The purpose of the study, the used methods and the categories of results obtained were characterized, drawing conclusions regarding the usefulness of user research methods taken from social sciences for developing a new approach to the evaluation of XAI methods.

4 Relevant Subject Literature

In the social sciences, methodological works are published relatively frequently, both in the form of short papers dealing with narrow issues, e.g. relating to a specific method or technique or a specific area of their application, as well as in the form of monographs and collective publications dealing with methods and techniques of research in social sciences in a broad, review manner (Eriksson & Kovalainen, 2015) (McNabb, 2016). Many papers drew attention to the classic methodological dilemma related to using a quantitative vs. qualitative approach, as well as the pros and cons of both (Gerring, 2017) (Baškarada, S. & Koronios, 2018), often emphasizing the benefits of using the mix-method approach (Tashakkori & Teddlie, 2021). In relation to qualitative research, special attention was paid to the need to maintain strict methodological rigor, the need to thoroughly document the research process, and importance was attached to the personal and social skills of researchers (Johnson & Rasulova, 2017) (Johnson, Adkins & Chauvin, 2020). Many works also referred to difficulties related to the development of qualitative material and emphasized ethical and legal issues related to conducting research involving people (Jerolmack & Murphy, 2019) (Shaw, 2023).

5 Results

As a result of the literature review several of the most popular user studies methods and techniques taken from the field of social science research were selected and synthetically characterized, emphasizing the categories of results they can provide. Finally, a practical example of implementing some methods in an interdisciplinary project in the field of XAI was described.

5.1 Methods and Techniques of User Studies Aimed at Obtaining Opinions

One of the basic qualitative methods for obtaining information from users is an in-depth interview based on sourcing information during a conversation. It may take a structured form, in which the researcher strictly follows a predefined pattern of questions; unstructured form that gives the researcher greater flexibility in reacting to the situation and conducting the conversation in various directions within a defined area of interest; or semi-structured, mixed approach. The structured variant allows for greater comparability of research conducted by different scientists and is also easier to develop. On the other hand, it may limit the framework of the study and favor the omission of observations spontaneously reported by users and not assumed in the original concept of the study (Brinkmann, 2014). The selection of the interview variant, as well as the number and way of formulating questions, the duration of the interview and the circumstances of its conduct (in the research laboratory vs. in the natural environment of the respondents) depend on the objectives of the study. The advantages of the interview, regardless of the variant adopted, include its universal nature and the high flexibility, which allows it to be used as the main method of obtaining information from users or as an auxiliary tool. The scope of interview results depends not only on the adopted parameters, but also on the researcher's communication and psychological competences, as well as on the selection of respondents and their individual characteristics (Knott et al., 2022).

A variant of interviews are focus groups, often used in public opinion research and marketing studies. Focus, is a form of discussion on a specific, usually quite narrow topic, led by a moderator in a group of deliberately selected people (Nuttavuthisit, 2019). The advantage of focus groups over one-on-one interviews is the ability to observe how respondents' opinions, beliefs and preferences change in social situations under the influence of group dynamics. This allows for a more realistic representation of human decision-making processes, taking into account their social nature. On the other hand, focus groups may not be the most effective method in researching sensitive topics (Kruger et al., 2019).

An interesting method of collecting group opinions is Delphi method. It involves asking questions to a group of deliberately selected experts, often representing various fields of research and areas of practical activities, in order to collect, average and develop a consensus of their opinions on complex, multidimensional phenomena. Answers to questions are collected through individual contact between the researcher and particular experts. Then, the researcher prepares conclusions and sends them to all involved experts, asking them to comment on the results until opinions on a given topic are shared and averaged (Lund, 2020). This method is particularly useful in developing scenarios for the future, as well as in analyzing complex problems and controversial issues on which it is difficult to reach a consensus using other methods. The disadvantage is the time-consuming nature of this process, difficulty in obtaining the involvement of high-quality experts and the loss of the individual character of qualitative research in favor of averaging and reaching consensus. However, it is still a valuable method of obtaining expert opinions in relation to complex problems.

5.2 Methods and Techniques of User Studies Aimed at Diagnosing Behavior

Usability tests are one of the most popular methods of testing the usability of websites and applications, providing real insight into the actual way people interact with various information systems. This aspect distinguishes usability tests from methods and techniques aimed at obtaining opinions (declarations), such as previously discussed interviews or focus groups. It is important to note that the role of usability tests is to validate systems, not people, however, it is still possible to gain insight into their behavior in relation to the system. During tests conducted by a moderator according to a research scenario, users watch the tested websites or applications (sometimes also prototypes) and perform practical tasks typical for using a specific type of system. Users' voices and images are recorded, and sometimes the time required to complete tasks is also measured. The entire process is observed by an additional person or group of people invisible to the test participant through using a one-way mirror or sharing the desktop (Geisen & Bergstrom, 2017). Think-aloud protocol (TAP), which involves encouraging the test participant to constantly verbalize the actions taken, emotions felt or difficulties experienced, is a popular module added to usability tests which allows to understand better users' motivations to engage in specific behaviors and provides insight into their thought processes (Wolcott & Lobczowski, 2021).

Another classic way of researching users' actions and behaviors is observation, which can be either participatory - when researchers are simultaneously participants of the situation under study, or non-participatory, when they remain fully neutral and separate from the research context. Observation is based on the process of knowing the characteristics of people, objects or environments in a purposeful way using researchers senses (Ciesielska, Boström & Öhlander, 2018). Observation, regardless of the variant, involves the perception, collection, description and interpretation of data obtaining from the environment, without changing it, which is important and distinguishes this method from experiment. The advantages of observation as a research method include, first of all, the ability to directly, undisturbedly learn about human behavior as it occurs in natural conditions. The disadvantages include the possibility of distortion of the research due to the imperfection of human senses, the subjective nature of the research due to the personal characteristics of the observer and sometimes also the dynamics of interaction between the researcher and the participant.

A research technique aimed at diagnosing human behavior, that is currently gaining popularity in social sciences, is so-called guided tour. It is a way of learning about human activities in which the subject (participant) of research acts as a guide in his or her space (physical or virtual) and presents, tells and shows it from his or her subjective perspective (Thomson, 2018). This is particularly useful in examining behaviors and routines practiced by people intuitively and often difficult to detect in other types of research due to their obvious nature to the research participant but interesting and unknown from the researcher's point of view. The disadvantage of using this technique is that control over the course of the study is largely given to the participants and it is difficult to predict the categories of study results. Also noteworthy is the high degree of subjectivity of the data obtained, which, depending on the context, may be considered as an advantage or a disadvantage.

5.3 Methods and Techniques of User Studies Aimed at Examination of Experiences

Nowadays, one of the most important categories of user studies is research focused on understanding user experiences. It is assumed that it is no longer enough to design a system that will be functional, but it is necessary to design it in such a way that using the system is a positive, unique and memorable experience for the user. It fits into the broader context of the assumptions of the so-called experience economy, defined as an economic model in which the highest value for customers, for which they are willing to pay, is not the product or service itself, but the experience related to the commercial process (Wójcik, 2019). Researching experiences is not an easy task due to their highly subjective nature, but many methods and techniques in this area have already been developed.

One of the most popular and longest-standing technique is the user experience questionnaire. It is a technique that uses a research tool based on a questionnaire divided into sections related to specific features of the tested system in relation to users' feelings. The individual sections contain terms reflecting the user's emotions on a scale from terms related to negative experiences to positive ones (Wöbbekind, Mandl & Womser-Hacker, 2021). The advantages of this technique include the simplicity of its use and the ease of processing the results by the researcher. The disadvantage is the rather general nature of the results obtained, giving an overall idea of the user's feelings, but leaving no room for explanations or specification of their source. In this context, the Kano approach may be helpful as a complementary technique. In this type of study, users are asked to indicate the features of the tested system that are considered by them as attractive, expected, indifferent, necessary or are considered a failure. This is a technique often used in researching people for marketing purposes because it allows to translate users' subjective feelings into specialized language and organize people's often chaotic reflections about products and services into clear categories of results (Park, Lee & Back, 2020). The disadvantages of this technique include the relatively high difficulty for users in completing Kano questionnaires, especially on their own without the assistance of a researcher. The study requires a thorough explanation to participants of how to define the individual features of the system under study about which they are asked. However, if used correctly, it can provide valuable insight into the features of the system, taking into account users' experiences related to them, and also allow for comparison of the tested system with other, similar ones.

5.4 Implementations of User Studies in the Field of Artificial Intelligence Research – Literature Review

In the subject literature, descriptions of different practical application of user studies in the field of artificial intelligence research can be found, although not that often.

Different meta-analyses of the literature on XAI available so far, show that there are several dozen relevant papers that refer, at least to some extent, to user studies, however in the individual papers it is often difficult to find detailed descriptions of the applied user research methodology with a clear indication categories of results obtained. For instance, a literature review by Rong et al. done on a group of 97 papers on human-based XAI evaluations, clearly showed that: "user evaluations are still rather sparse and incorporate hardly any insights from cognitive or social sciences" (2024). Similarly, a literature review conducted on a group of 58 papers by Haque, Islam and Mikalef showed that: "limited number of studies exist that measure the user perceptions of the transparency, understandability, and usability of an XAI system" (2023).

In a relatively few papers reporting specific examples of XAI user research, the most common were those referring to diagnosing uses' opinions; rarely papers show the practical use of social sciences' methods and techniques to study behaviors and experiences.

Interesting example of using interview to diagnose opinions is reported by Liao, Gruen and Miller. Authors interviewed 20 UX designers working on various AI products to identify gaps between the current approach to XAI and contemporary design practices focused on building user experiences. It is worth noting, however, that this study did not take into account the opinions of end users (Liao, Gruen & Miller, 2020). The literature review showed that focus groups were a particularly popular method of researching opinions of AI users, especially in sensitive areas such as safety, medicine or finance. Example can be paper by Aitken et al. in which the results were reported from a series of focus research conducted with various target groups in the banking sector. The results of the study allowed to determine users' the needs, preferences, but also concerns related to AI (Aitken et al. 2020). Rarely, the Delphi method was also used to diagnose opinions on AI. An example can be the study by Boesl, Prassler and Bode, which indicates the potential of the Delphi method as a non-biased alternative for other types of predictive research on technology development (2017).

Overall it can be concluded that rather unsatisfactory number of examples of the implementation of user studies in the practice of AI research, especially in the layer of diagnosing behaviors and experiences - despite numerous calls for such activities in the subject literature - shows a certain gap between theory and practice and reveals scope for further research. To fill this gap and show the practical application of selected social science methods and techniques in the field of XAI research, a case study related an interdisciplinary research project was described.

5.5 Practical Implementation of Selected Types of User Studies in the Area of XAI - Case Study

The aim of the research described as a case study was to examine the extent to which explanations generated by artificial intelligence are understandable to various groups of users, especially to people without a background in computer science and engineering.

A tabular dataset on the edibility of wild mushrooms was selected for the study. A group of domain experts - mycologists and a group of students representing various levels of knowledge about fungi (usually low to medium) and various degrees of experience in data analytics and visualization (students of humanities and social sciences, as well as students of computer science) were invited to participate in the study.

A mix-method research approach was used based on a combination of various qualitative elements. An experiment based on an adapted usability tests scheme combined with the TAP protocol was conducted with all participants to observe in practice the way they interacted with AI explanations and the actual level of their understanding of XAI (behavior research level). Additionally, to determine users' feelings and experiences related to the use of XAI, simplified UX questionnaires and elements of the Kano approach were used (experience research level). Finally, unstructured interviews were woven into the main core of the experiment which was particularly useful during the study with a group of domain experts who, due to their knowledge, had many comments to share, especially about the quality of the dataset and the consequences of using AI explanations in practice (opinion research level).

Overall 39 research individual sessions with users were conducted which is considered to be a high number of participants in qualitative research. Each session lasted on average 2 h witch gave about 80 h of recording converted into a text version so it could be analyzed further. Every hour of recording is about 20 pages of transcription which adds up to a sizable text corpus. Transcriptions provided by the software as a result of automatic processes were often not satisfactory in terms of quality for the purposes of the study so most transcriptions had to be either extensively edited or made manually from scratches. Finally, the transcriptions were subjected to the tagging process in accordance with the assumptions of content analysis procedure using MaxQDA software (Lindgren, Lundman & Graneheim, 2020).

The use of an original combination of various qualitative methods and techniques, oriented on obtaining different categories of results, provided unique insight into the problems of actual usefulness of AI explanations for users, opening up the possibility of a new way of evaluating XAI methods.

At the level of obtaining opinions, it was found that comments of non-expert students gave abundance of practical tips concerning improvements of the usability of AI explanations, especially in terms of aesthetics and information organization, while the comments of experts were almost solely focused on the quality of the dataset itself, especially in terms of its consistency with reality. Non-experts pointed out problems with the readability and comprehensibility of explanations, while clearly indicating their preferences for the types of explanations they consider to be more understandable and user-friendly. Domain experts, on the other hand, were focused on giving opinions, most often critical, about the quality of the dataset itself, highlighting the observed discrepancies between the model's performance and the biological reality of fungi.

At the level of diagnosing behaviors, it was found that AI explanations are treated by users as a tool enabling them to get cognitive access to the content generated by AI models. However, AI explanations clearly played different roles for different user groups. For non-specialists XAI was treated as s a potential source of knowledge that provide sole, yet usually insufficient access to AI model's output. For experts, on the other hand, viewing different types of XAI was considered to be an opportunity to compare their own knowledge with AI, which often ended with a critical assessment of the compliance of AI models with real life scenarios. With that note, conducted research allowed to conclude that, still for the group of domain experts XAI have a potential to serve proper mediation function as a gateway giving insight into the dataset itself and AI model's operation rules.

At the level of examining experiences, the methods and techniques used allowed to determine users overall emotions and feelings related to XAI, especially in terms of subjectively felt degree of difficulty, level of readability and aesthetics, and level of attention stimulation. It was also possible to determine the degree of difficulty of XAI methods compared to other types of data visualization previously seen by users in their everyday lives.

6 Conclusions

The described case study showed that user studies methods and techniques driven from social sciences can be successfully used in practice in interdisciplinary projects as a support for broadly understood computer science. It has been proven, that applying correctly selected combination of social science inspired methods and techniques to study the reception of AI explanations provided unique and otherwise difficult to obtain insight into their usefulness, understandability and user-friendliness. This input can be further used as a starting point for discussing new approaches to assessing the quality of AI explanation; one that takes into account the human-centered artificial intelligence perspective. This, in turn, at the application level, may lead to overall better adaptation of XAI to the needs of different groups of users.

It is worth noting, that user studies methods known from social sciences can be used - as in the case study described - to research users' opinions, behaviors and experiences connected with already existing solutions, or to diagnose users' needs, skills, preferences or fears before starting the design stage. The selection of an appropriate combination of methods and techniques from the wide methodological catalog of social sciences depends on the objectives of the study and the group of users being studied. However, it is worth triangulating different methods, taking into account the categories of results that are needed to be obtained.

Social sciences, with a long tradition of studying people, can offer methodological background useful for creating more user-friendly AI solutions. Providing people with clear, understandable and approachable explanations can serve not only to increase usability of XAI, but also raise the overall transparency of AI solutions, which is an important factor in the process of building social trust in AI. Including users in the loop of creating technical solutions addressed to them, at various stages, by examining their opinions, behaviors and experiences, is necessary to create proper human-centered AI.

7 Limitations and Prospect for Further Research

It should be noted that due to the short format of the paper, only selected, qualitative methods and techniques were discussed, which of course does not constitute the full catalog of user studies methods known in social sciences. In this context, it would be worth considering in the future the usefulness of other, especially quantitative, methods and techniques of social sciences for the development of XAI, which were not the subject of considerations in this paper, however, they also carry considerable potential for the interdisciplinary cooperation.

The division into methods and techniques for examining opinions, behaviors and experiences adopted in this paper is conventional and serves to organize the discussion - due to the large diversity of variants of the described methods and techniques, there may be some elements falling into several different categories depending on the circumstances of the study. Nevertheless, thinking about the potential of methods and techniques taken from social sciences in terms of the categories of produced results, although burdened with certain limitations, can be used as a good, general direction for further considerations. In this context, it would be worth in the future to also develop other approaches to categorizing user studies results, than those presented in this study.

The paper constitutes a starting point towards a broader reflection on the methodological framework for developing a new approach to the evaluation of XAI methods. This challenging, multidimensional topic could undoubtedly benefit from closer cooperation between representatives of technical and social sciences. The foundations for the use of social science methods and techniques in AI research outlined in the paper will be continued and developed in subsequent publications.

Acknowledgments. The research discussed as a case study was conducted by the team: Magdalena Wójcik, Monika Krakowska, Magdalena Zych, Dorota Rak and Paloma Korycińska with the support of Szymon Bobek, Maciej Mozolewski and Grzegorz J. Nalepa as a part of XPM project funded by the National Science Centre, Poland under the CHIST-ERA programme grant agreement Np. 857925 (NCN UMO-2020/02/Y/ST6/00070). This paper is part of a project that has received funding from the European Union's Horizon Europe Research and Innovation Programme, under Grant Agreement number 101120406. The paper reflects only the authors' view and the EC is not responsible for any use that may be made of the information it contains.

Disclosure of Interests. Author was a paid participant in the XPM project funded by the National Science Centre, Poland under the CHIST-ERA programme grant agreement Np. 857925 (NCN UMO-2020/02/Y/ST6/00070). Author is a participant in the the European Union's Horizon Europe Research and Innovation Programme, under Grant Agreement number 101120406.

References

Aitken, M., et al.: Keeping it human: a focus group study of public attitudes towards AI in banking. In: ESORICS 2020. LNCS, vol. 12580, pp. 21–38. Springer, Cham (2020). https://doi.org/10.1007/978-3-030-66504-3_2

Arrieta, A.B., et al.: Explainable artificial intelligence (XAI): concepts, taxonomies, opportunities and challenges toward responsible AI. Inf. Fusion **58**, 82–115 (2020)

Baškarada, S., Koronios, A.: A philosophical discussion of qualitative, quantitative, and mixed methods research in social science. Qual. Res. J. **18**(1), 2–21 (2018)

Boesl, D.B., Prassler, E., Bode, M.: Proposing a delphi-study as non-biased alternative to studies and roadmaps predicting the future of robotics, automation A.I. In: 2017 International Conference on Current Trends in Computer, Electrical, Electronics and Communication (CTCEEC), Mysore, India, pp. 47–49 (2017)

Brinkmann, S.: Unstructured and semi-structured interviewing. In: The Oxford Handbook of Qualitative Research, pp. 277–299. Oxford University Press, Oxford (2014)

Ciesielska, M., Boström, K.W., Öhlander, M.: Observation methods. In: Qualitative Methodologies in Organization Studies: Volume II: Methods and Possibilities, pp. 33–52. Palgrave Macmillan, Cham (2018)

Eriksson, P., Kovalainen, A.: Qualitative Methods in Business Research: A Practical Guide to Social Research. Sage, London (2015)

Geisen, E., Bergstrom, J.R.: Usability Testing for Survey Research. Morgan Kaufmann, Cambridge (2017)

Gerring, J.: Qualitative methods. Annu. Rev. Polit. Sci. **20**, 15–36 (2017)

Ghassemi, M., Oakden-Rayner, L., Beam, A.L.: The false hope of current approaches to explainable artificial intelligence in health care. Lancet Digit. Health **3**(11), e745–e750 (2021)

Haque, A.B., Islam, A.N., Mikalef, P.: Explainable artificial intelligence (XAI) from a user perspective: a synthesis of prior literature and problematizing avenues for future research. Technol. Forecast. Soc. Change **186** (2023), https://www.sciencedirect.com/science/article/pii/S0040162522006412. Accessed 10 July 2024

Jerolmack, C., Murphy, A.K.: The ethical dilemmas and social scientific trade-offs of masking in ethnography. Sociol. Methods Res. **48**(4), 801–827 (2019)

Johnson, S., Rasulova, S.: Qualitative research and the evaluation of development impact: incorporating authenticity into the assessment of rigour. J. Dev. Effectiveness **9**(2), 263–276 (2017)

Johnson, J.L., Adkins, D., Chauvin, S.: A review of the quality indicators of rigor in qualitative research. Am. J. Pharmac. Educ. **84**(1) (2020). https://doi.org/10.5688/ajpe7120. Accessed 03 July 2024

Knott, E., Rao, A.H., Summers, K., Teeger, C.: Interviews in the social sciences. Nat. Rev. Methods Primers **2**(1) (2022). https://doi.org/10.1038/s43586-022-00150-6. Accessed 02 July 2024

Kruger, L.J., Rodgers, R.F., Long, S.J., Lowy, A.S.: Individual interviews or focus groups? Interview format and women's self-disclosure. Int. J. Soc. Res. Methodol. **22**(3), 245–255 (2019)

Liao, Q.V., Gruen, D., Miller, S.: Questioning the AI: informing design practices for explainable AI user experiences. In: Proceedings of the 2020 CHI Conference on Human Factors in Computing Systems, pp. 1–15. Association for Computing Machinery, New York (2020)

Lindgren, B.-M., Lundman, B., Graneheim, U.H.: Abstraction and interpretation during the qualitative content analysis process. Int. J. Nurs. Stud. **108** (2020). https://doi.org/10.1016/j.ijnurstu.2020.103632. Accessed 03 July 2024

Lund, B.D.: Review of the Delphi method in library and information science research. J. Doc. **76**(4), 929–960 (2020)

Longo, L., et al.: Explainable artificial intelligence (XAI) 2.0: a manifesto of open challenges and interdisciplinary research directions (2023), arXiv:2310.19775. Accessed 09 July 2024

McNabb, D.E.: Research Methods for Political Science: Quantitative and Qualitative Methods. Routledge, New York (2016)

Miller, T.: Explanation in artificial intelligence: insights from the social sciences. Artif. Intell. **267**, 1–38 (2019)

Nuttavuthisit, K.: Focus group interview. In: Nuttavuthisit, K. (ed.) Qualitative Consumer and Marketing Research, pp. 141–164. Springer, Singapore (2019). https://doi.org/10.1007/978-981-13-6142-5_6

Park, H., Lee, M., Back, K.J.: Exploring the roles of hotel wellness attributes in customer satisfaction and dissatisfaction: application of Kano model through mixed methods. Int. J. Contemp. Hosp. Manag. **33**(1), 263–285 (2020)

Rong, Y., et al.: Towards human-centered explainable AI: a survey of user studies for model explanations. IEEE Trans. Pattern Anal. Mach. Intell. **46**(4), 2104–2122 (2024), https://ieeexplore.ieee.org/abstract/document/10316181. Accessed 10 July 2024

Shaw, I.: Ethics and the Practice of Qualitative Research. Routledge, New York (2023)

Tashakkori, A., Teddlie, C.: SAGE Handbook of Mixed Methods in Social & Behavioral Research. Sage, London (2021)

Thomson, L.: The guided tour: a research technique for the study of situated, embodied information. Libr. Trends **66**(4), 511–534 (2018)

Wolcott, M.D., Lobczowski, N.G.: Using cognitive interviews and think-aloud protocols to understand thought processes. Curr. Pharm. Teach. Learn. **13**(2), 181–188 (2021)

Wöbbekind, L., Mandl, T., Womser-Hacker, C.: Construction and first testing of the UX kids questionnaire (UXKQ): a tool for measuring pupil's user experience in interactive learning apps using semantic differentials. In: Menschund Computer 2021 – Tagungsband, pp. 473–484. ACM, New York (2021)

Wójcik, M.: How to design innovative information services at the library? Libr. Hi Tech **37**(2), 138–154 (2019)

Towards Reliable Drift Detection and Explanation in Text Data

Robert Feldhans(✉) and Barbara Hammer

Bielefeld University, Bielefeld, Germany
{rfeldhans,bhammer}@techfak.uni-bielefeld.de

Abstract. When delivered to the market, machine learning models face new data which are possibly subject to novel characteristics – a phenomenon known as concept drift. As this might lead to performance degradation, it is necessary to detect such drift and, if required, adapt the model accordingly. While a variety of drift detection and adaptation methods exists for standard vectorial data, a suitable treatment of text data is less researched. In this work we present a novel approach which detects and explains drift in text data based on their representation via transformer embeddings.

In a nutshell, the method generates suitable statistical features from the original distribution and the possibly shifted variation. Based on these representations, drift scores can be assigned to individual data points, allowing a visualization and human-readable characterization of the type of drift.

We demonstrate the approach's effectiveness in reliably detecting drift in several experiments.

Keywords: Drift Explanation · Text Data · Transformer · Visualization

1 Introduction

In an open environment, machine learning models face the challenge of new and possibly evolving data after their deployment. In this context, one of the fundamental assumptions underlying classical machine learning techniques might be violated, the fact that training data draws from the same distribution as the evolving new data set. This phenomenon is called concept drift. Concept drift encompasses two different kinds of drift: Real drift, which refers to changes in the target concept and virtual drift, which refers to changes in the data only affecting the input distribution, without affecting the target concept [8]. While instances of pure real and virtual drift in text data are rare in the real world, one example for real drift may be a sudden or gradual change in the meaning of a word, while an example for virtual drift may be the relative increase of published novels of a specific genre when considering all published novels. Both kinds of drift can cause a decline in model performance [17], which makes retraining of

V. Julian et al. (Eds.): IDEAL 2024, LNCS 15346, pp. 301–312, 2025.
https://doi.org/10.1007/978-3-031-77731-8_28

the model necessary. As it is often impractical to retrain and redeploy a machine learning model, it is important to detect and inspect such drift, as its cause might be only temporary and will revert given time, or it might not affect the downstream model. For instance, a temporary surge in public transport usage during a major event may not justify the cost and effort of retraining the control model. Identifying such circumstances is not always straightforward and in practice, it is often unclear why and how new situations and data differ from the known [12].

We focus our work on text data, as texts pose special challenges with regard to explanations, i.e., their lack of objective and easily humanly readable comparability in contrast to scalar data. In recent work, text embeddings generated with the transformer architecture [26] have become the state-of-the-art way to work with texts, e.g., [1,2,4,24]. Our goal is to provide efficient drift detection and explanation methods that are generally applicable. We consider the scenario of two independent distributions of text, for which the quantification of mutual drift and, if present, its explanation is targeted. In the following, we refer to these distributions as *original* and possibly *drifted* distributions, although their order can be arbitrary. The drift event or setting itself can be anything that causes or correlates with the drift, e.g. a restaurant changing a cook, and thus causing their reviews to change. This event may even be unknown and subject to investigation of via machine learning techniques.

This broad scenario subsumes relevant cases, such as time series data, which can be subdivided into subsequent time windows. Another example are two sets of user reviews divided into positive and negative (as shown later in Fig. 2a). In this case, the investigation of drift targets whether there is a difference in attitude; an explanation of such drift highlights why users reflected on their experience in the way they did. Thus, our main objective is to devise an algorithm which is capable of (i) detecting drift in two data sets characterizing texts, and (ii) explaining or visualizing important changes in the underlying data distributions if such drift is present.

The remainder of this article is structured as follows: In Sect. 2, we briefly outline related work. Section 3 describes our novel approach. Thereafter, Sect. 4 presents experiments and results. Finally, Sect. 5 contains our final conclusions.

2 Related Work

Explainable AI has become a widely researched field in the last years, and excellent introductions and overviews such as [10] exist, mostly targeted towards vectorial or tabular data. Our approach is tailored to text-data, for which transformer models constitute current state-of-the-art models. Research in that direction subsumes popular approaches centering around its attention heads for transformer model explanations; yet, it is debated in the literature whether this is a valid approach [14,21,28]. Liu et al. [15] use the attention weights of the model for explanations in the classification task, hinting this could be another possible approach to explain the behavior of transformer models.

While many approaches -our among them- use post-hoc explanations which are fundamentally independent of the used model, [25] propose an interactive explanation process where the model itself is exalted with an explanation component.

Counterfactual explanations, as described by [20,22,27], are often described as being easily understandable by humans. Lately, there has been some work on bringing this quality to transformer models [30]. Counterfactual explanations have also been used to explain concept drift [12].

Concept drift itself and its detection is widely researched [9,13,16,23]. Drift in itself has also been analyzed and categorized into distinct patterns e.g. in the time domain [6]. Yet, most technologies focus on vectorial data, and only first extensions of drift detection and explanation to textual data have been proposed [7,19], struggling not only with the high dimensionality of text embeddings but also its intrinsic huge variety. In the following, we target drift detection (and visualization) for text data, focusing on robustness and simple parameterization of the technology.

3 Method

Our proposed approach, in its basic form, aims to detect drift, and it can be augmented for drift explanation. The proposed algorithm assumes that two textual data sets are given in the form of vector embeddings, such as BERT embeddings [5]. The challenge is to robustly characterize whether drift is present and, if so, which type of drift. Unlike the approach of Madaan et al. [19], which relies on complex density models to describe the original and drifted data set, we propose efficient yet sufficiently powerful robust statistical features to describe the data sets and its potential difference.

3.1 Drift Detection

In essence, the proposed drift detector creates a number of features which robustly characterize the original and drifted distribution, respectively, by means of pairwise similarities of their mean vectors. These are computed for different splits of the two data sets of different sizes. These distributions can be compared using a test statistics, leading to a robust drift measure via averaging.

In detail, the first step generates features for both distributions in a slightly different way: Features for the original distribution are generated in a fitting step; features for the drifted distribution are generated in a testing step. In both cases, the features are independently generated for a number of chunk sizes. The set of chunk sizes C which we choose is

$$C = \{F_6, F_7, ... F_{20}\} \tag{1}$$

where F_n denotes the nth Fibonacci number. While the choice of this particular set is not important, repeating the feature generation with several hyperparameters in the form of $c \in C$ aides in creating a robust result.

Algorithm 1. Fitting step of the proposed drift detector

original data : $\mathbb{X}$
for $c \in C$ **do**
    $r \leftarrow \emptyset$
    for 1...512 **do**
        $X_i \xleftarrow{\$} \mathbb{X}, ||X_i|| = c$
        $X_j \leftarrow (x \in \mathbb{X} | x \notin X_j)$
        $\hat{\imath} \leftarrow \sum X_i$
        $\hat{\jmath} \leftarrow \sum X_j$
        $R \leftarrow R \cup cosinesim(\hat{\imath}, \hat{\jmath})$
    end for
    $\bar{r}_c \leftarrow mean(R)$
    $\sigma(r_c) \leftarrow stdev(R)$
end for
return $\bar{r}_c, \sigma(r_c) \forall c \in C$

Algorithm 2. Testing step of the proposed drift detector

original data : $\mathbb{X}$
drifted data : $\mathbb{Y}$
for $c \in C$ **do**
    $S \leftarrow \emptyset$
    for 1...512 **do**
        $Y_i \xleftarrow{\$} \mathbb{Y}, ||Y_i|| = c$
        $X_j \xleftarrow{\$} \mathbb{X}, ||X_j|| = n - c$
        $\hat{\imath} \leftarrow \sum Y_i$
        $\hat{\jmath} \leftarrow \sum X_j$
        $S \leftarrow S \cup cosinesim(\hat{\imath}, \hat{\jmath})$
    end for
    $S_c \leftarrow S$
end for
return $S_c \forall c \in C$

Pseudo-code for the fitting step of the algorithm can be seen in Algorithm 1. For each chunk size c, the n data points of the original distribution are randomly split into two groups of size c and $n - c$. The cosine similarity of the sum of all vectors of both groups is then calculated, estimating the similarity of the two means of the split. These steps are repeated 512 times. The average cosine similarity $\bar{r}_c$ as well as standard deviation $\sigma(r_c)$ is then chosen as a feature to characterize the original distribution.

Pseudo-code for the testing step of the algorithm can be seen in Algorithm 2. The testing step mirrors the fitting step in that c random samples from the drifted distribution and $n - c$ samples of the original distribution are chosen at random. The cosine similarity of the sum of all vectors of both groups is then calculated just as before. These steps are repeated 512 times again. However, this is not summarized as average cosine similarity $\bar{s}_c$ and standard deviation $\sigma(s_c)$, rather, the full set of cosine similarities S_c are returned.

Algorithm 3. Probability estimation step of the proposed drift detector

$features\ original\ data : \bar{r}_c, \sigma(r_c)$
$features\ drifted\ data : S_c$
for $c \in C$ **do**
  $P \leftarrow \emptyset$
  for $s \in S_c$ **do**
    $P \leftarrow P \cup pdf(\frac{r_c - s}{2 \cdot \sigma(r_c)})$
  end for
  $\bar{p}_c \leftarrow mean(P)$
  $\sigma(p_c) \leftarrow stdev(P)$
end for
return $\bar{p}_c$, $\sigma(p_c) \forall c \in C$

Using these full sets of cosine similarities, probability estimations for each chunk size c are calculated using the characteristics of the original data as follows (see Algorithm 3): For each chunk size c, probability estimations P are calculated for each cosine similarity $s \in S_c$ calculated in Algorithm 2 using a probability density function characterized by $\bar{r}_c$ and $\sigma(r_c)$. This step could generally also be handled via any other two sample test, but we found the PDF to yield better results during drift explanation.

Lastly, a final score q is calculated as a weighted average of all preliminary probability estimations $\bar{p}_c$ according to weight factors w_c with

$$w_c = \frac{\bar{p}_c - \sigma(p_c)}{\bar{p}_c + \sigma(p_c)} \tag{2}$$

where p_c are the probability estimations of the proposed algorithm for each c, $\bar{p}_c$ the arithmetic mean of p_c and σ denotes the standard deviation. Note that the resulting score does not constitute a p-value.

3.2 Drift Explanation

We will now briefly introduce changes to the previously described approach to allow for drift explanation. In essence, during the testing step a drift score is assigned to each sample of the drifted distribution: the higher the score, the more similar it is to the original distribution. The testing step is then repeated with data of the original distribution. The result are two score distributions, which can then be compared in a histogram of the scores, for example. As each score is indicative of the data points similarity to the original distribution, similar scores indicate similar data points. For instance, if the drifted distribution is the result of a novel class of samples in an otherwise relatively uniform original distribution -i.e. virtual drift-, this group of samples would have a collectively lower score and would register as a new cluster in the score distribution of the drifted data. On the contrary, a general shift in score of all points in the distribution may indicate real drift.

Algorithm 4. Testing step of the drift explanation component

$\mathit{original\ data} : \mathbb{X}$
$\mathit{drifted\ data} : \mathbb{Y}$
$T_{y,c} \leftarrow \emptyset \ \ \forall c \in C, \forall y \in \mathbb{Y}$
for $c \in C$ **do**
  for 1...5120 **do**
    $Y_i \xleftarrow{\$} \mathbb{Y}, ||Y_i|| = c$
    $X_j \xleftarrow{\$} \mathbb{X}, ||X_j|| = n - c$
    $\hat{\imath} \leftarrow \sum Y_i$
    $\hat{\jmath} \leftarrow \sum X_j$
    for $y \in Y_i$ **do**
      $T_{y,c} \leftarrow T_{y,c} \cup cosinesim(\hat{\imath}, \hat{\jmath})$
    end for
  end for
  $\bar{T}_{y,c} \leftarrow mean(T_y, c)$
end for
return $\bar{T}_{y,c} \forall c \in C, \forall y \in \mathbb{Y}$

The testing step in Algorithm 2 is adjusted as seen in Algorithm 4. The set of scores $T_{y,c}$ for each point y of the drifted distribution is just the set of cosine similarities it is involved in calculating. To ensure each data point is taken into account a sufficient number of times, we increase the repetition of the random sampling step tenfold to 5120. This set of scores $T_{y,c}$ for each point y is then averaged as $\bar{T}_{y,c}$. To calculate the singular, final score T_y for each data point y of the drifted distribution, the scores lastly weighted by the weight factors w_c of Eq. 2.

4 Experiment Setup

We evaluate our approach in several experiments. First, we test the drift detection capabilities of our approach in two scenarios with real-world data sets. We then test the soundness of the drift explanations our approach generates. For the Same Distribution and Different Distribution experiments, we compare our approach with the two general drift detectors, i.e., the kernel two sample test (KTS) [11] and the least squares density difference (LSDD) [3]. In contrast to our approach, both of these detectors provide a p-value, which would indicate a drift event given a result under 0.05.

4.1 Datasets

Sentiment Data Set. The Large movie review data set [18] contains 50k movie reviews, equally divided in highly polarizing positive and negative sentiment.

Twitter Election. The Twitter Election data set is composed of tweets that refer to the two candidates of the 2020 US presidential election (i.e., these tweets use the hashtag #Biden and/or #Trump). The data set contains ca. 521k #Biden tweets and 680k #Trump tweets, divided according to their publishing time into four subsets. Two of these subsets are from random times a few days before the election and thus relatively similar. The third and forth subsets were published during the last TV debate and the election day respectively and thus differ from the first two subsets and each other. A more in-depth description of this data set is also available [7].

Both data sets are pre-processed; i.e., all texts embedded into 768 dimensional vector data using a pre-trained model provided by the Hugging Face library [29] called 'bert-base-multilingual-cased'.

4.2 Drift Detection Experiments

Same Distribution. In this scenario we test the drift detector with two sets of data from the same distribution of data. As such, the drift detector should not detect drift. We conduct this experiment to evaluate whether the proposed approach does indiscriminately detect drift when there is none. For the Sentiment data set, we compare two distributions made up of an equal number of positive and negative data points. We use the Twitter data set in two different ways for this experiment. First, we draw data randomly from the complete set in the Twitter overall scenario, then we compare the two similar sets of data in the Twitter similar scenario. Each scenario is repeated on unique data a number of times depending on available data, i.e. Sentiment 20, Twitter overall 20, Twitter similar 10 times. Our approach generates high scores in all three scenarios (see Table 1). This is especially true for the Twitter similar scenario, in which KTS and LSDD deliver a score indicating a drift. Furthermore, it should be noted that both KTS and LSDD struggle to generate consistent results with the provided data as shown by their relatively high standard deviation across all scenarios. Looking at individual results, this resulted in them generating both sub- and above-0.05 p-values in the Sentiment and Twitter similar scenario, making them unreliable in these real world scenarios.

Table 1. Drift detector scores of Same Distribution exp. (mean result of all runs, higher ∅ is better)

Scenario	Our		KTS		LSDD	
	∅	stdev	∅	stdev	∅	stdev
Sentiment	0.9113	0.0272	0.629	0.2289	0.6000	0.2337
Twitter overall	0.8946	0.0180	0.3878	0.1940	0.4910	0.2048
Twitter similar	0.7313	0.0819	0.0225	0.0186	0.0575	0.0674

Different Distribution. In this scenario we test the drift detector on datasets which contain known drift, i.e. the sentiment data set and the twitter data set.

This provides insight into drift detector performance under real-world conditions. For the Sentiment data set, we compare two distributions drawn exclusively from positive and negative data points respectively. We again utilize the Twitter data set in two different ways: In the first scenario, we compare its' third and first subsets, i.e. Tweets from the time of the TV debate and a random point in time a few days earlier respectively. Secondly, we compare its' first and fourth subsets, i.e. Tweets from the time of election day. KTS and LSDD successfully detect a difference in distribution in every case. Our approach generates comparable scores in the Twitter scenarios, but delivers a slightly higher score for the sentiment scenario. However, when comparing this score to the Same Distribution experiments score, a clear difference can be seen and a low similarity is indicated (Table 2).

Table 2. Results of the Different Distribution experiments (mean result of all runs, lower ∅ is better)

Dataset	Our		KTS		LSDD	
	∅	stdev	∅	stdev	∅	stdev
Sentiment	0.0899	0.0270	0.0020	0.0000	0.0000	0.0000
Twitter TV Debate	0.0022	0.0039	0.0020	0.0000	0.0000	0.0000
Twitter Election Day	0.0063	0.0103	0.0020	0.0000	0.0000	0.0000

4.3 Drift Explanation Experiment

In this experiment, we evaluate the drift explanation part of our approach by applying the drift explanation to the data sets of the Different Distributions experiment. Due to the nature of our data sets, we can infer - at least partially - the reasons or kind of drift in them. In all cases, we show the resulting score distributions as histograms.

From a drift classification point of view, the twitter data set presents three distinct drift events. The first lies between the first two subsets and should show no detectable drift and therefore no significant change in score distribution, similar to the Same Distribution experiment. The second event lies between the first two subsets and the third subset containing the tweets during the last TV debate. Considering the focus on specific topics during the TV debate, we suspect a moderate amount of virtual drift between those subsets, and thus expect a score distribution with a moderate number of relatively low-score data points. The third event lies between the first two subsets and the fourth subset containing the tweets during the election day. We suspect a considerable amount of virtual drift in this event due to the influx of tweets with new, election-day related concepts such as the hashtag #IVoted-, and thus expect a score distribution with a high number of relatively low-score data points. We show the resulting score distributions in a combined histogram (see Fig. 1). Both of the drift events' score distributions can be seen clearly differing from the original distribution.

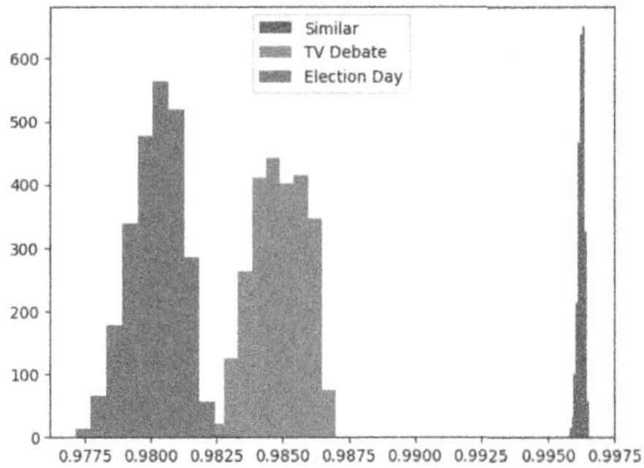

Fig. 1. Results of Twitter drift explanation experiment

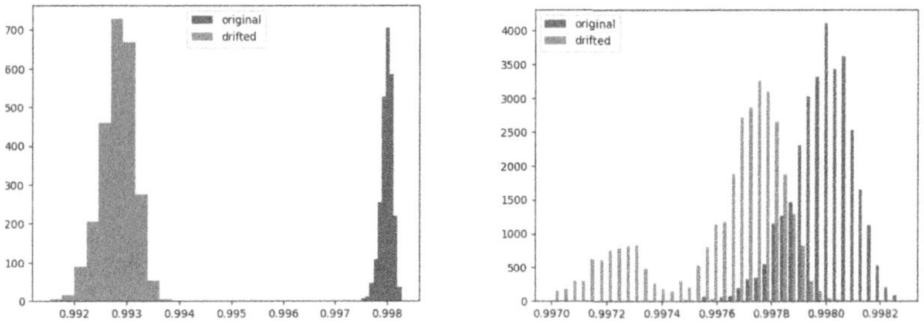

(a) Positive-negative drift (b) Artificial drift

Fig. 2. Sentiment Data Drift explanation experiment

Considering the sentiment data set as a drift from one sentiment to another, it presents as a fundamental concept drift where all data points of the post-drift subset differ from the data points of the pre-drift subset. We would thus expect our approach to show just that: A clear distinction between pre- and post-drift score distributions.

We also consider a third, artificial scenario in the form of the emergence of a new class of data in the post-drift subset. To achieve this setup, we consider the positive sentiment subset of the sentiment data set as the pre-drift subset and a mixture of 90% positive and 10% negative data points as the pre-drift subset. We would thus expect a post-drift score distribution made up of a great amount of high-score data points and a small amount of relatively low-score data points. In a real world scenario this would indicate the emergence of a new class of data or an occurrence of concept drift.

The results of both experiments can be seen in Fig. 2. For the drift event from positive to negative sentiment, a clear distinction between positive and negative distributions can be observed in Fig. 2a. While the difference in score distributions is not as great as in Fig. 1, the scope of the data set is more focused. In the results of the artificial drift scenario (see Fig. 2b), a part of the drifted data can be seen forming distribution with a lower score. A lower score among the original part of the drifted data can be seen and is expected, as combined score results with the drifted points of the data lower their similarity score.

5 Conclusion

Regarding our main objective, we devised an algorithm to detect and explain drift, and showed its feasibility in both real world and artificial scenarios. We demonstrated the drift detection part of our approach to perform more reliable than general drift detectors in the text domain. Additionally, the drift explanation part of our approach was shown to produce distinguishable results between different variants of drift. In future work, we would like to further explore our approach on real world and automatically generated data. Given enough data sets, it might be possible to train a classifier on the scores of our drift characterizer and as such classify drift automatically.

Acknowledgments. The authors were supported by SAIL. SAIL is funded by the Ministry of Culture and Science of the State of North Rhine-Westphalia under the grant no NW21-059A.

References

1. Acheampong, F.A., Nunoo-Mensah, H., Chen, W.: Transformer models for text-based emotion detection: a review of BERT-based approaches. Artif. Intell. Rev. **54**(8), 5789–5829 (2021)
2. Berabi, B., He, J., Raychev, V., Vechev, M.: TFix: learning to fix coding errors with a text-to-text transformer. In: International Conference on Machine Learning, pp. 780–791. PMLR (2021)
3. Bu, L., Alippi, C., Zhao, D.: A PDF-free change detection test based on density difference estimation. IEEE Trans. Neural Netw. Learn. Syst. **29**(2), 324–334 (2018). https://doi.org/10.1109/TNNLS.2016.2619909
4. Chi, Z., Dong, L., Ma, S., Mao, S.H.X.L., Huang, H., Wei, F.: mT6: multilingual pretrained text-to-text transformer with translation pairs. arXiv preprint arXiv:2104.08692 (2021)
5. Devlin, J., Chang, M., Lee, K., Toutanova, K.: BERT: pre-training of deep bidirectional transformers for language understanding. CoRR abs/1810.04805 (2018). http://arxiv.org/abs/1810.04805
6. Dongre, P.B., Malik, L.G.: A review on real time data stream classification and adapting to various concept drift scenarios. In: 2014 IEEE International Advance Computing Conference (IACC), pp. 533–537 (2014). https://doi.org/10.1109/IAdCC.2014.6779381
7. Feldhans, R., et al.: Drift detection in text data with document embeddings. In: Yin, H., et al. (eds.) IDEAL 2021. LNCS, vol. 13113, pp. 107–118. Springer, Cham (2021). https://doi.org/10.1007/978-3-030-91608-4_11. ISBN 978-3-030-91608-4

8. Gama, J., Žliobaitė, I., Bifet, A., Pechenizkiy, M., Bouchachia, H.: A survey on concept drift adaptation. ACM Comput. Surv. (CSUR) **46** (2014). https://doi.org/10.1145/2523813
9. Gama, J.a., Žliobaitundefined, I., Bifet, A., Pechenizkiy, M., Bouchachia, A.: A survey on concept drift adaptation. ACM Comput. Surv. **46**(4) (2014). https://doi.org/10.1145/2523813. ISSN 0360-030
10. Gilpin, L.H., Bau, D., Yuan, B.Z., Bajwa, A., Specter, M., Kagal, L.: Explaining explanations: an overview of interpretability of machine learning (2018). https://doi.org/10.1109/DSAA.2018.00018
11. Gretton, A., Borgwardt, K.M., Rasch, M.J., Schölkopf, B., Smola, A.: A kernel two-sample test. J. Mach. Learn. Res. **13**(25), 723–773 (2012). http://jmlr.org/papers/v13/gretton12a.html
12. Hinder, F., Vaquet, V., Brinkrolf, J., Hammer, B.: Model-based explanations of concept drift. Neurocomputing **555**, 126640 (2023). https://doi.org/10.1016/J.NEUCOM.2023.126640
13. Hu, H., Kantardzic, M., Sethi, T.S.: No free lunch theorem for concept drift detection in streaming data classification: a review. WIREs Data Min. Knowl. Discovery **10**(2), e1327 (2020). https://doi.org/10.1002/widm.1327, https://wires.onlinelibrary.wiley.com/doi/abs/10.1002/widm.1327
14. Jain, S., Wallace, B.C.: Attention is not explanation. In: NAACL (2019)
15. Liu, S., Le, F., Chakraborty, S., Abdelzaher, T.: On exploring attention-based explanation for transformer models in text classification. In: 2021 IEEE International Conference on Big Data (Big Data), pp. 1193–1203 (2021). https://doi.org/10.1109/BigData52589.2021.9671639
16. Lu, J., Liu, A., Dong, F., Gu, F., Gama, J., Zhang, G.: Learning under concept drift: a review. IEEE Trans. Knowl. Data Eng. **31**(12), 2346–2363 (2019). https://doi.org/10.1109/TKDE.2018.2876857
17. Lu, J., Liu, A., Dong, F., Gu, F., Gama, J., Zhang, G.: Learning under concept drift: a review. CoRR abs/2004.05785 (2020). https://arxiv.org/abs/2004.05785
18. Maas, A.L., Daly, R.E., Pham, P.T., Huang, D., Ng, A.Y., Potts, C.: Learning word vectors for sentiment analysis. In: Proceedings of the 49th Annual Meeting of the Association for Computational Linguistics: Human Language Technologies, pp. 142–150, Association for Computational Linguistics, Portland (2011). http://www.aclweb.org/anthology/P11-1015
19. Madaan, N., Manjunatha, A., Nambiar, H., Goel, A., Saha, D., Bedathur, S.: Detail: a tool to automatically detect and analyze drift in language. In: Proceedings of the 35th Conference on Innovative Applications of Artificial Intelligence (AAAI/IAAI) (2023)
20. Mahajan, D., Tan, C., Sharma, A.: Preserving causal constraints in counterfactual explanations for machine learning classifiers. CoRR abs/1912.03277 (2019). http://arxiv.org/abs/1912.03277
21. Michel, P., Levy, O., Neubig, G.: Are sixteen heads really better than one? In: Wallach, H., Larochelle, H., Beygelzimer, A., d'Alché-Buc, F., Fox, E., Garnett, R. (eds.) Advances in Neural Information Processing Systems, vol. 32, Curran Associates, Inc. (2019). https://proceedings.neurips.cc/paper/2019/file/2c601ad9d2ff9bc8b282670cdd54f69f-Paper.pdf
22. Mothilal, R.K., Sharma, A., Tan, C.: Explaining machine learning classifiers through diverse counterfactual explanations. In: Proceedings of the 2020 Conference on Fairness, Accountability, and Transparency, FAT* 2020, pp. 607–617. Association for Computing Machinery, New York (2020). https://doi.org/10.1145/3351095.3372850. ISBN 978145036936

23. Nishida, K., Yamauchi, K.: Detecting concept drift using statistical testing. In: Corruble, V., Takeda, M., Suzuki, E. (eds.) DS 2007. LNCS (LNAI), vol. 4755, pp. 264–269. Springer, Heidelberg (2007). https://doi.org/10.1007/978-3-540-75488-6_27
24. Phan, L.N., et al.: SciFive: a text-to-text transformer model for biomedical literature. arXiv preprint arXiv:2106.03598 (2021)
25. Schramowski, P., Friedrich, F., Tauchmann, C., Kersting, K.: Interactively generating explanations for transformer language models. CoRR abs/2110.02058 (2021). https://arxiv.org/abs/2110.02058
26. Vaswani, A., et al.: Attention is all you need (2017). https://arxiv.org/pdf/1706.03762.pdf
27. Verma, S., Dickerson, J.P., Hines, K.: Counterfactual explanations for machine learning: a review. CoRR abs/2010.10596 (2020). https://arxiv.org/abs/2010.10596
28. Wiegreffe, S., Pinter, Y.: Attention is not not explanation. CoRR abs/1908.04626 (2019). http://arxiv.org/abs/1908.04626
29. Wolf, T., et al.: Transformers: state-of-the-art natural language processing. In: Proceedings of the 2020 Conference on Empirical Methods in Natural Language Processing: System Demonstrations, pp. 38–45, Association for Computational Linguistics (2020). https://www.aclweb.org/anthology/2020.emnlp-demos.6
30. Yang, L., Kenny, E.M., Ng, T.L.J., Yang, Y., Smyth, B., Dong, R.: Generating plausible counterfactual explanations for deep transformers in financial text classification. CoRR abs/2010.12512 (2020). https://arxiv.org/abs/2010.12512

Using Diffusion Models for Data Augmentation on Limited Rodent OCT Datasets

Fernando García-Torres[1], Rocío del Amor[1], Sandra Morales-Martínez[1], Álvaro Barroso[2], Björn Kemper[2], Jürgen Schnekenburger[2], and Valery Naranjo[1](✉)

[1] Instituto Universitario de Investigación en Tecnología Centrada en el Ser Humano, Universitat Politècnica de València, Valencia, Spain
vnaranjo@dcom.upv.es

[2] Biomedical Technology Center of the Medical Faculty, University of Muenster, Mendelstr. 17, 48149 Muenster, Germany

Abstract. Optical coherence tomography (OCT) has transformed ophthalmology by enabling accurate diagnosis of retinal diseases. OCT in rodents is used in research studies of ophthalmic diseases and treatment evaluation. However, the scarcity of data limits the effectiveness of deep learning models for this purpose. This study explores the use of synthetic data generation to improve deep learning models for OCT image analysis from rodent retinas. To address this, the research leverages advanced class-conditional generative models to create high-quality synthetic images of ex-vivo rodent retinas embedded in two different mediums (solid resin and liquid media) with a reduced initial database. The conditional Denoising Diffusion Probabilistic Model (DDPM) proved to be the best model, as it produces more realistic images that closely mimic the characteristics of real rodent retinal scans. By augmenting the training datasets with these synthetic images, the performance of deep learning models in segmenting retinal layers improved by 7%. This approach reduces reliance on scarce biological samples and animal testing, promoting ethical research practices while improving the accuracy and efficiency of AI-driven medical diagnostics. The integration of synthetic data generation represents a significant advance in medical imaging, offering a scalable solution to data scarcity, generating reliable samples in two mediums and promoting the development of more robust and accurate diagnostic tools.

Keywords: Synthetic data generation · Deep learning · Optical coherence tomography · Image analysis · Retinal diseases · Denoising Diffusion Probabilistic Models · Medical imaging

1 Introduction

Optical coherence tomography (OCT) has revolutionised the field of ophthalmology by providing high-resolution cross-sectional images of the retina, enabling

V. Julian et al. (Eds.): IDEAL 2024, LNCS 15346, pp. 313–324, 2025.
https://doi.org/10.1007/978-3-031-77731-8_29

early and accurate diagnosis of various retinal diseases [1]. Conditions such as diabetic retinopathy, age-related macular degeneration and glaucoma can be detected and monitored with unprecedented accuracy thanks to OCT technology [2,3]. The images produced by OCT make it possible to identify subtle pathological changes in the retinal layers, which is crucial for timely intervention and effective treatment planning [4].

The integration of deep learning into OCT image analysis has further improved diagnostic capabilities [5,6]. Deep learning models, in particular convolutional neural networks (CNNs), have demonstrated remarkable success in automatically identifying and classifying retinal abnormalities from OCT scans [7,8]. These models can learn complex patterns and features from large datasets, outperforming traditional image analysis methods in speed and accuracy. The synergy between OCT and deep learning represents a significant advance in the early detection and management of retinal diseases, paving the way for more personalised and effective healthcare solutions [9,10].

To ensure consistent performance and comparability between different OCT systems, researchers have developed advanced calibration standards using tissue-mimicking phantoms [11]. While these phantoms help standardise measurements, they only partially replicate the complex properties of actual retinal tissue [12]. Consequently, ex-vivo retina studies have become vital, providing accurate simulations for validating OCT systems. However, maintaining the durability and proper preservation of ex-vivo samples is challenging because they have traditionally been preserved in fluid and perishable media. In recent studies [13,14], methodologies have been developed for preserving rodent retinas in resin, effectively maintaining their biological properties. These studies enhance reproducibility and reliability in OCT-based diagnostics, ensuring more consistent and accurate results across different systems and clinical applications. By contrast, obtaining ex-vivo retinal samples often requires the sacrifice of animals, which raises ethical issues and limits the availability of these samples, further restricting their use for studying retinal pathologies or developing deep learning algorithms to aid diagnosis.

To overcome the problem of limited data availability, synthetic image generation algorithms have emerged as a powerful solution. By augmenting datasets with synthetic images, researchers can enhance the training and performance of deep learning models, ensuring robust and accurate diagnostic tools without the need for extensive biological samples [15]. One of the most commonly used approaches is Generative Adversarial Networks (GANs), which have achieved promising results in generating retinal images [16]. GANs are being increasingly utilized for the generation of retinal disorder images for various applications, such as data augmentation in segmentation models, denoising, and post-intervention prediction, among others [17]. However, their primary limitation is the need for extensive datasets to prevent model collapse and ensure that the generated images closely resemble real ones [18]. Another novel approach is the introduction of Denoising Diffusion Probabilistic Models (DDPMs). These generative models have been used in the field of ophthalmology for applications such as noise reduction or segmentation of retinal layers in OCT images of human patients [19].

To the best of the authors' knowledge, although there is work in the scientific literature on the generation of retinal images in other conditions, there is no framework for the generation of images of ex-vivo rodent retinas in different mediums, used to improve the performance of deep learning algorithms. Note that OCT images are susceptible to speckle noise, which degrades their quality. When generating synthetic OCT data, it is crucial to replicate this effect without compromising the definition of the retinal layers, as this would reduce the accuracy of retinal layer segmentation algorithms. Through this research, we propose a class-conditioned generative algorithm for the generation of high-quality retinal samples embedded in two mediums (solid resin and fluid medium) starting from a limited database. Thus, we seek to provide a holistic solution that minimises the use of animals in medical research while optimising the accuracy and efficiency of deep learning models. This not only promotes ethical research practices but also advances the field of AI-assisted medical diagnostics. By leveraging synthetic data and advanced deep learning techniques, we aim to improve the performance of supervised segmentation models, thereby facilitating more accurate and timely diagnosis of retinal conditions.

2 Methodology

2.1 Class-Conditioned Generative Adversarial Networks

In recent years, GANs have attracted considerable interest due to their capability to produce high-quality synthetic data, especially in the domain of image generation. In the following, we will present the generative adversarial networks we use in a class-conditioned approach to generate retinal images in two different mediums.

Progressive GAN (PGAN) introduced a phased approach to training, starting from low resolutions and gradually increasing to the final image resolution [20]. This method enabled the generation of higher quality and more stable images as the network architecture was enhanced. On the other hand, StyleGAN improved the quality of generated images by introducing a style-based generation model [21]. This model separates the image content from the style, allowing for more precise control over synthetic images' specific characteristics such as texture, colour, and shape. StyleGAN2, an extension of the original StyleGAN, introduced several advancements that improved the resolution and diversity of generated images [22]. One of the notable advancements is the use of adaptive instance normalization (AdaIN), which allows for more fine-grained control over the style of generated images [23].

2.2 Class-Conditioned Denoising Diffusion Probabilistic Model

Denoising Diffusion Probabilistic Models are a class of generative models that have gained significant attention for their ability to produce high-quality and diverse samples [24]. The fundamental idea behind diffusion models is to gradually add noise to a data sample and then learn to reverse this noising process, ultimately generating new samples from random noise. The forward process, also known as the noising process, is a Markov chain that progressively adds Gaussian noise to the data. Starting from a data sample x_0, the process iteratively transforms it into a noise vector $\mathbf{x_T}$ through a series of steps. Mathematically, this is defined as:

$$q(x_t \mid x_{t-1}) = \mathcal{N}(x_t; \sqrt{\alpha_t} x_{t-1}, (1 - \alpha_t)I), \quad (1)$$

where x_t and x_{t-1} represent the states of the system at two consecutive time steps, α_t is a schedule parameter that controls the noise level at each step t, and I is the identity matrix.

The reverse process, also known as the denoising process, aims to learn the data distribution by reversing the forward process. Starting from the noise vector $\mathbf{x_T}$, the model learns to denoise it step by step, recovering the original data distribution $p_\theta(x_0)$. The reverse process is defined by

$$p_\theta(x_{t-1} \mid x_t) = \mathcal{N}(x_{t-1}; \mu_\theta(x_t, t), \Sigma_\theta(x_t, t)), \quad (2)$$

where μ_θ and Σ_θ are parameters learned by the model, representing the mean and variance of the denoising distribution. Figure 1 presents a schematic of the conditional DDPM algorithm, illustrating its main components and the data flow during the synthetic image generation process.

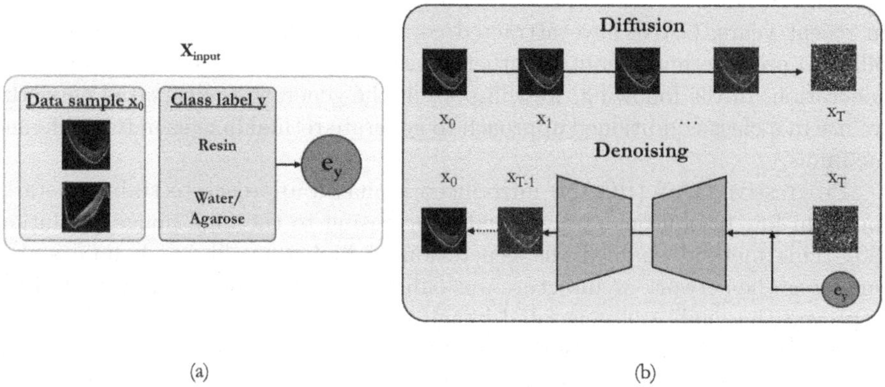

Fig. 1. Class-Conditioned DDPM Diagram. (a) Input data: Data sample x_0 and Class embedding e_y; (b) Diffusion and Denoising processes.

Class-conditioned diffusion models extend the basic framework of diffusion models by incorporating class labels into the generation process [25]. This modification allows the model to generate samples conditioned on specific class information, enhancing the capability to produce class-specific outputs. In a class-conditioned diffusion model, the class label is integrated into the denoising process. Class labels $\mathbf{y}$ are embedded into a continuous space using an embedding layer, transforming discrete class indices into vectors of real numbers (Fig. 1a). The input to the diffusion model is augmented to include both the data sample and the class embedding (Fig. 1b). This can be represented as

$$x_{\text{input}} = x \| e_y, \tag{3}$$

where e_y is the class embedding. This concatenated input is fed into the model, allowing the class information to influence the denoising process.

3 Experimental Setting

3.1 Dataset

The dataset used in this study consists of OCT grayscale scans of rodent retinas. The images were acquired using a high-resolution OCT system (Thorlabs Ganymede Series, Thorlabs GmbH, Luebeck, Germany) and the procedure described in [13]. The database used for synthetic image generation is subdivided into two classes according to the medium in which the retinas were embedded: solid resin or fluid medium (water or agarose). The main motivation for subdividing the database in this way lies in the differences in morphology, resolution and contrast of the retinal layers depending on the medium. The dataset comprises 12 samples of retina sections embedded in resin [13], yielding a total of 1,226 images, and 5 samples of retina sections embedded in the fluid medium, resulting in a total of 1,437 images. The models were trained using the entire set of images (Table 1). Due to the variety of image sizes, they were scaled to a size of 256×256 pixels.

Table 1. Description of the dataset of the two mediums.

Dataset	Samples	Avg. images per sample	Total images
Resin	12	102.17 ± 89.95	1226
Water/Agarose	5	287.40 ± 74.15	1437

From the 12 samples embedded in resin, 137 images were manually segmented to develop supervised deep learning models, with 9 of these samples (95 images) allocated for training and 3 for testing purposes (42 images). The images used in this model were resized to 512×256 pixels as the manually segmented layer did not cover the full width of the retina.

3.2 Evaluation

To evaluate the performance of the proposed approaches we compute the Fréchet Inception Distance (FID), a measure of the similarity between the distribution of real images and generated images [26]. A lower FID score indicates that the generated images are closer to the distribution of real images, implying higher quality and diversity.

As a further evaluation method, a deep-learning model was used to assess the quality of the resolution of the synthetically generated images and to validate the use of these images as data augmentation for the development of deep-learning models with OCT rodent retinal images. In particular, the model aimed to segment the retinal nerve fiber layer (RNFL) of rodent retinas embedded in resin in OCT images. The segmentation model consists of a U-Net based on the one proposed by Ronneberger et al. [27]. The U-Net architecture consists of a downsampling and an upsampling path, with 5 convolutional blocks in each branch. Batch normalisation is used to stabilise and speed up training by normalising activations within each minibatch.

3.3 Implementation Details

All GANs were trained using the Adam optimiser, with a learning rate of 0.001, a batch size of 32. We trained the PGAN and StyleGAN over 200 epochs per resolution (from a size of 4 to 256). The StyleGAN2 was trained during 300 epochs, with a latent space dimension z and tan intermediate latent space dimension w of 256. Additional training stability is provided by a gradient penalty of 10. The software used was Pytorch for building and training the models.

The DDPM was trained using the dataset described over 200 epochs with the Adam optimiser, with a learning rate of 0.001 and a batch size of 16. We defined 5 downsampling and upsampling blocks in each branch, with 2 of them with spatial self-attention respectively. The loss function utilised was Mean Square Error (MSE). The model's architecture is adapted to handle the class-conditioned inputs. In our implementation, we use the `UNet2DModel` from the Diffusers library [28], configured to accept both the image and class embedding, with size 4. The architecture is tailored to process these inputs through a series of convolutional and attention layers, ensuring that the class information is effectively integrated at each stage of the network.

The U-Net used as the evaluation method was trained over 300 epochs with the Adam optimiser, a learning rate of 0.001 and a batch size of 8. We used the Categorical Cross Entropy loss function. This model was built in Tensorflow. We use the Dice score to assess the semantic segmentation predicted by the U-Net.

Regarding the hardware, we used NVIDIA RTX 3090 24 GB x 1, 525.60.11 drivers & CUDA 12.0, MSI Z270 Gaming PRO Carbon (MS-7A63); 32 GB and Intel i7-7700K (4.2 GHz).

4 Results

4.1 Synthetic Retinal Image Generation

We attempted to train the conditional PGAN and StyleGAN, however these models did not converge, thus preventing us from presenting any results. A potential reason for this non-convergence could be the reduced number of training images available in this study.

Once the StyleGAN2 and DDPM were trained, we randomly inferred the exact number of images per class as the original dataset to measure the FID of both models, comparing the original versus the synthetically generated images. The FID results of each model are shown in Table 2.

Table 2. FID metrics of the models in the two categories of mediums.

Model	Resin	Water/Agarose
StyleGAN2	201,64	187,18
DDPM	**69,22**	**70,40**

The DDPM model shows significantly lower FID scores than StyleGAN2, indicating that the diffusion model produces higher-quality images, as lower FID scores suggest a closer alignment with the distribution of real images.

This can be proved qualitatively in Fig. 2, which shows some examples of images generated in the two types of medium by the proposed models. The images generated with the DDPM model make it possible to distinguish the different layers of the retina. In addition, it reliably reproduces the noise characteristic of the original OCT images, as well as the retinal shapes of the specimens and textures. In contrast, the StyleGAN2 model produces images of lower quality and definition, resembling the shape and curvature of the retinas but not the resolution of the layers.

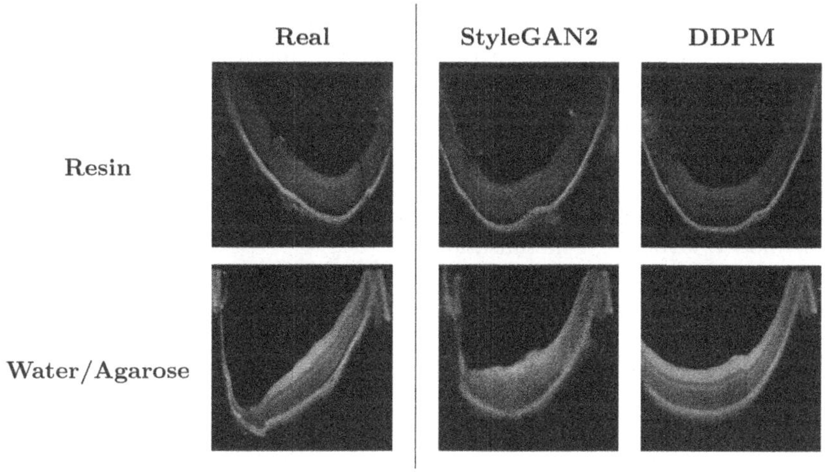

Fig. 2. Comparison of examples of real images of the two mediums with images generated by the proposed models.

4.2 Synthetic Data Augmentation for Segmentation Model

As previously discussed, the aim of generating new rodent retinal images in these mediums is to alleviate the shortage of samples available for training and improving supervised deep-learning models. As a method of validating the quality and resolution of the images obtained by synthetic image generation algorithms, we propose to introduce a series of images from the inference of the best generative model to the training of a segmentation model. The objective of this experiment is to prove that taking a sample of these inference images can serve as augmentation of the database to enhance the performance of the model. To overcome this challenge we use a segmentation model to automatically segment the RNFL of retinas embedded in resin.

Regarding the inference images, 100 generated images were taken, half from each medium, using DDPM due to its superior FID score. These images were manually segmented following a qualitative analysis of their resolution to obtain their masks. The next step involved integrating the DDPM inference images and masks into the segmentation model training alongside the real resin dataset used for training, maintaining the same initial test set reserved for this purpose.

Table 3. Segmentation Model Comparison: Training Images and Dice Scores.

Model	Training images	Dice
Initial	95	0.78
Augmented	193	**0.85**

Table 3 presents the Dice metric results on the test set for the segmentation models. The initial model trained solely on real images achieved a Dice score of 0.78, whereas incorporating the segmented inference images during training raised the Dice score to 0.85. Therefore, feeding the initial segmentation model with images generated by a DDPM algorithm improves performance. It should be noted that, although the validation is conducted only on resin-embedded samples, the augmentation of data in two mediums provides greater robustness to the model, favouring generalisation.

An example of a resin-embedded retinal image together with manual segmentation of the RNFL as ground truth (GT) is shown in Fig. 3. The mask predicted by the initial model has some irregularities in the RNFL and contains some areas below this layer that do not correspond to the area of interest. The Augmented model predicts the area of interest more accurately and although the mask is not perfect, it is smoother than the initial model.

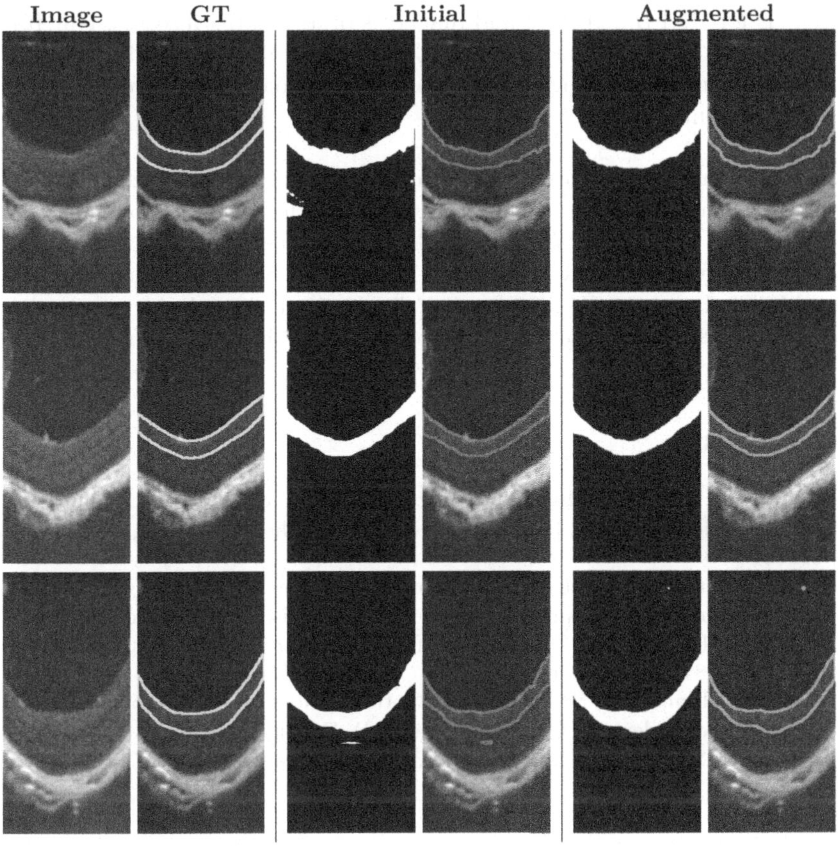

Fig. 3. Comparison between the GT and the prediction (mask and contour) of the Segmentation Models: Initial and Augmented.

5 Conclusion

Developing deep learning models for diagnostic and research purposes for retinal diseases requires retinal samples, which often involve using animals for medical studies. To address the problems associated with this, we propose a framework for generating synthetic images in two different mediums using two class-conditioned generative models for data augmentation of an automatic segmentation model. With a limited database of OCT retinal images, the most reliable generative model among those used in this study proved to be the DDPM. Some of its images were manually segmented to retrain an initial segmentation model trained with real images, yielding better quantitative and qualitative results. Our approach aims to reduce the number of retinal samples and thus the number of specimens needed for experimentation, without compromising the development of models for RNFL detection and diagnosis of associated diseases. Our framework represents a significant step towards more ethical and cost-effective medical research, ensuring that progress in diagnostic technologies does not come at the expense of animal welfare.

Funding Information. This work has received funding from Horizon Europe, the European Union's Framework Programme for Research and Innovation, under Grant Agreement No. 101070062 (SEQUOIA). Generalitat Valenciana partially funded this work through project CIPROM/2022/20. The work of Sandra Morales has been co-funded by the Universitat Politècnica de València, Spain through the program PAID-10-20. The work of Rocío del Amor has been supported by the Spanish Ministry of Universities under an FPU Grant (FPU20/05263).

References

1. Tălu, S-D.:. Optical coherence tomography in the diagnosis and monitoring of retinal diseases. Int. Sch. Res. Not. **2013**(1) (2013). https://doi.org/10.1155/2013/910641
2. Qin, J., An, L.: Optical coherence tomography for ophthalmology imaging. In: Wei, X., Gu, B. (eds.) Optical Imaging in Human Disease and Biological Research. AEMB, vol. 3233, pp. 197–216. Springer, Singapore (2021). https://doi.org/10.1007/978-981-15-7627-0_10
3. Tan, C.S., Lim, L.W., Sadda, S.V.R.: Optical coherence tomography angiography in macular disorders. In: Chang, A., Mieler, W.F., Ohji, M. (eds.) Macular Surgery, pp. 45–64. Springer, Singapore (2020). https://doi.org/10.1007/978-981-15-7644-7_4
4. Müller, P.L., Wolf, S., Dolz-Marco, R., Tafreshi, A., Schmitz-Valckenberg, S., Holz, F.G.: Ophthalmic diagnostic imaging: retina. In: Bille, J.F. (ed.) High Resolution Imaging in Microscopy and Ophthalmology, pp. 87–106. Springer, Cham (2019). https://doi.org/10.1007/978-3-030-16638-0_4
5. Leandro, I., et al.: OCT-based deep-learning models for the identification of retinal key signs. Sci. Rep. **13**(1) (2023). https://doi.org/10.1038/s41598-023-41362-4
6. Ho, P.-G.: Image Segmentation. BoD - Books on Demand (2011). Google-Books-ID: vDiaDwAAQBAJ
7. Akil, M., Elloumi, Y., Kachouri, R.: Chapter 2 - Detection of retinal abnormalities in fundus image using CNN deep learning networks. In: El-Baz, A.S., Suri, J.S. (eds.) State of the Art in Neural Networks and their Applications. Academic Press (2021). https://doi.org/10.1016/B978-0-12-819740-0.00002-4
8. Morales, S., et al.: Retinal layer segmentation in rodent OCT images: local intensity profiles & fully convolutional neural networks. Comput. Methods Program. Biomed. **198** (2021). https://doi.org/10.1016/j.cmpb.2020.105788
9. Diaz-Pinto, A., Morales, S., Naranjo, V., Köhler, T., Mossi, J.M., Navea, A.: CNNs for automatic glaucoma assessment using fundus images: an extensive validation. Biomed. Eng. Online **18**(1), 29 (2019). https://doi.org/10.1186/s12938-019-0649-y
10. Del Amor, R., et al.: Towards automatic glaucoma assessment: an encoder-decoder CNN for retinal layer segmentation in rodent OCT images. In: 2019 27th European Signal Processing Conference (EUSIPCO), A Coruna, Spain, pp. 1–5. IEEE (2019). https://doi.org/10.23919/EUSIPCO.2019.8902794

11. Wang, H., Liu, W., Hu, Z., Li, X., Li, F., Duan, L.: Model eye tool for retinal optical coherence tomography instrument calibration. J. Innov. Opt. Health Sci. (2021). https://doi.org/10.1142/S1793545821500103
12. Strupler, M., et al.: Toward an automated method for optical coherence tomography characterization. J. Biomed. Opt. **20**(12), 126007 (2015). https://doi.org/10.1117/1.JBO.20.12.126007
13. Barroso, A., et al.: Durable 3D murine ex vivo retina glaucoma models for optical coherence tomography. Biomed. Opt. Express **14**(9), 4421–4438 (2023). https://doi.org/10.1364/BOE.494271
14. Barroso, A., et al.: Durable ex vivo mouse retina 3D tissue models for optical coherence tomography. In: Label-free Biomedical Imaging and Sensing (LBIS) 2024, vol. 12854, pp. 21–23. SPIE (2024). https://doi.org/10.1117/12.3002538
15. Zha, X., Shi, F., Ma, Y., Zhu, W., Chen, X.: Generation of retinal OCT images with diseases based on cGAN. In: Medical Imaging 2019: Image Processing, vol. 10949, pp. 544–549. SPIE (2019). https://doi.org/10.1117/12.2510967
16. Zheng, C., et al.: Assessment of generative adversarial networks model for synthetic optical coherence tomography images of retinal disorders. Transl. Vision Sci. Technol. **9**(2), 29 (2020). https://doi.org/10.1167/tvst.9.2.29
17. Kugelman, J., Alonso-Caneiro, D., Read, S.A., Vincent, S.J., Chen, F.K., Collins, M.J.: Data augmentation for patch-based OCT chorio-retinal segmentation using generative adversarial networks. Neural Comput. Appl. **33**(13), 7393–7408 (2021). https://doi.org/10.1007/s00521-021-05826-w
18. You, A., Kim, J.K., Ryu, I.H., Yoo, T.K.: Application of generative adversarial networks (GAN) for ophthalmology image domains: a survey. Eye Vision **9**(1), 6 (2022). https://doi.org/10.1186/s40662-022-00277-3
19. Hu, D., Tao, Y.K., Oguz, I.: Unsupervised denoising of retinal OCT with diffusion probabilistic model. In: Medical Imaging 2022: Image Processing, vol. 12032, pp. 25–34. SPIE (2022). https://doi.org/10.1117/12.2612235
20. Karras, T., Aila, T., Laine, S., Lehtinen, J.: Progressive growing of GANs for improved quality, stability, and variation (2018)
21. Karras, T., Laine, S., Aila, T.: A style-based generator architecture for generative adversarial networks. IEEE Trans. Pattern Anal. Mach. Intell. **43**(12), 4217–4228 (2021). https://doi.org/10.1109/TPAMI.2020.2970919. Publisher: IEEE Computer Society
22. Karras, T., Laine, S., Aittala, M., Hellsten, J., Lehtinen, J., Aila, T.: Analyzing and improving the image quality of StyleGAN. In: 2020 IEEE/CVF Conference on Computer Vision and Pattern Recognition (CVPR), Seattle, WA, USA. IEEE (2020). https://doi.org/10.1109/CVPR42600.2020.00813. ISBN 978-1-72817-168-5
23. Huang, X., Belongie, S.: Arbitrary style transfer in real-time with adaptive instance normalization. In: 2017 IEEE International Conference on Computer Vision (ICCV), Venice, pp. 1510–1519. IEEE (2017). https://doi.org/10.1109/ICCV.2017.167
24. Ho, J., Jain, A., Abbeel, P.: Denoising diffusion probabilistic models. In: Advances in Neural Information Processing Systems, vol. 33, pp. 6840–6851. Curran Associates, Inc. (2020)
25. Rombach, R., Blattmann, A., Lorenz, D., Esser, P., Ommer, B.: High-resolution image synthesis with latent diffusion models. 2022 IEEE/CVF Conference on Computer Vision and Pattern Recognition (CVPR), New Orleans, LA, USA, pp. 10674–10685. IEEE (2022). https://doi.org/10.1109/CVPR52688.2022.01042

26. Heusel, M., Ramsauer, H., Unterthiner, T., Nessler, B., Hochreiter, S.: GANs trained by a two time-scale update rule converge to a local nash equilibrium. In: Advances in Neural Information Processing Systems, vol. 30. Curran Associates, Inc. (2017)
27. Ronneberger, O., Fischer, P., Brox, T.: U-net: convolutional networks for biomedical image segmentation. In: Navab, N., Hornegger, J., Wells, W.M., Frangi, A.F. (eds.) MICCAI 2015. LNCS, vol. 9351, pp. 234–241. Springer, Cham (2015). https://doi.org/10.1007/978-3-319-24574-4_28
28. von Platen, P., et al.: Diffusers: state-of-the-art diffusion models (2024). https://github.com/huggingface/diffusers. Accessed 20 June 2024

Employing Explainable AI Techniques for Air Pollution: An Ante-Hoc and Post-Hoc Approach in Dioxide Nitrogen Forecasting

Pedro Oliveira(✉), Francisco Franco, Afonso Bessa, Dalila Durães, and Paulo Novais

ALGORITMI/LASI Centre, University of Minho, Braga, Portugal
pedro.jose.oliveira@algoritmi.uminho.pt,
{pg47187,pg53597}@alunos.uminho.pt, {dad,pjon}@di.uminho.pt

Abstract. With the advancement of Artificial Intelligence (AI) techniques in different areas of our society, namely with the use of Machine and Deep Learning models, some challenges must be faced. One of these challenges is responding to the lack of transparency in these models, which makes it difficult to explain the results they obtained. This research centres on predicting the concentration of nitrogen dioxide (NO_2), a critical air pollutant, using a Long Short-Term Memory model (LSTM) along with applying Explainable AI (XAI) techniques to understand the predictions made. Two explainability techniques were applied: an *ante-hoc* approach with an Attention layer and a *post-hoc* approach using Shapley Additive Explanations (SHAP). The Attention layer identified carbon monoxide (CO) and NO_2 as the most powerful features, while the SHAP analysis highlighted NO_2 as the predominant contributor to the predictions, followed by particulate matter ($PM_{2.5}$) and CO. The results demonstrated that the target value significantly impacts the model's forecast for both XAI techniques.

Keywords: Air Pollution · Explainable Artificial Intelligence · Long Short-Term Memory · Time Series

1 Introduction

In recent years, there has been a significant increase in the use of Artificial Intelligence (AI) and Machine Learning (ML) to solve problems in various areas of science and industry [7]. Specifically, developed Deep Learning (DL) models can perform many tasks with increasingly better performance times and results applicable to real-world scenarios [14]. However, the complexity and opacity of these models, often referred to as BlackBox models, raise concerns about the transparency of their decision-making processes and the reliability of their outcomes [3]. As a result, Explainable Artificial Intelligence (XAI) has gained prominence

V. Julian et al. (Eds.): IDEAL 2024, LNCS 15346, pp. 325–336, 2025.
https://doi.org/10.1007/978-3-031-77731-8_30

due to the inadequacy or lack of formalization in many AI applications, particularly concerning ethical issues [3]. Explainability refers to a model's ability to offer human-understandable insights into its decision-making process, which helps stakeholders trust its predictions [4].

For this, there are two alternatives to explainability. *Ante-hoc* explainability refers to methods incorporated into the design of a DL model from the beginning, aiming to make the model more interpretable and transparent. Alternatively, *post-hoc* explainability refers to methods applied after forming a DL model to explain the model's predictions [11].

One of the areas where the explainability of DL models can be applied is air pollution. Burning fossil fuels has increased air pollution over the last few decades, which has had dire effects on our world, including climate change [5]. Air pollutants come in many forms, including nitrogen dioxide (NO_2). This pollutant may harm vegetation or photochemical haze, among other environmental issues [6]. Moreover, it might result in cardiovascular disorders or pulmonary issues for people [8]. Since it is a precursor to the formation of various pollutants, including ozone (O_3) and fine particles ($PM_{2.5}$), monitoring this pollutant is crucial [17].

This study aims to design, tune and evaluate a DL model, namely Long Short-Term Memory (LSTM), due to its ability to deal with time series forecasting problems. This model makes the NO_2 forecast for the next 24 h. Two explainability approaches are applied to understand how the model behaves. The first uses an *ante-hoc* approach, an Attention layer, to understand each feature's weight at each timestep. Secondly, a *post-hoc* approach was used, namely the Shapley Additive Explanations (SHAP), to understand the importance of the features in each timestep after the trained model. The remainder of the manuscript is organized as follows: the next section covers related work. The third section details the case study, including data collection and processing and model conception. The fourth section describes the experimental setup and evaluation metrics. The fifth section presents the results and explainability techniques for the best model. Finally, the sixth section concludes with future work.

2 Related Work

Paper [10] presents a novel approach for predicting multivariate time series data using Temporal Attention Convolutional Neural Networks (TACNN). This model improves interpretability and performance by dynamically weighing time steps using temporal attention. Convolutional Neural Networks (CNNs) are integrated to capture intricate temporal patterns and dependencies. Numerous tests demonstrate that TACNN's interpretability does not degrade performance, underscoring its usefulness in producing clear and accurate predictions.

Paper [13] investigates the use of various ML techniques to predict $PM_{2.5}$ levels, a critical air pollutant, across Singapore. The study reveals the importance of transparent models and uses XAI to improve output understanding. The authors compare neural networks, decision trees, random forests, and linear regression. Historical $PM_{2.5}$ data, meteorological variables, and geographic data are used to explain predictions with XAI methods like SHAP scores and Local Interpretable Model-Agnostic Explanations (LIME). The findings indicate that simpler models, such as decision trees and linear regression, are easier to interpret, while more complex models like neural networks offer better performance. By identifying key factors influencing $PM_{2.5}$ levels, the XAI techniques enhanced understanding of Singapore's air quality dynamics.

The paper [2] explores an innovative approach to modelling and understanding the behaviour of toluene, a volatile organic compound, in the atmosphere. By combining hybridized metaheuristics, the XGBoost ML algorithm, and SHAP, the authors develop an interpretable model to predict toluene concentrations and identify key influencing factors. The hybridized metaheuristics optimize model parameters, XGBoost builds the predictive model, and SHAP explains the predictions. Using atmospheric data, including meteorological variables and pollutants concentrations, the model achieves accurate predictions. Furthermore, the SHAP values, which highlight important factors influencing toluene behavior, offer a clear window into the model's decision-making process.

The paper [16] investigates the application of SHAP to improve the transparency and interpretability of the predictions of the NO_2 concentration. The study aims to identify important influencing factors and predict NO_2 levels by integrating SHAP with sophisticated ML models. Historical NO_2 data, meteorological variables, and other pertinent data are used to train the models. Findings indicate that SHAP not only increases prediction accuracy but also makes each variable's contribution more clear, demonstrating the advantages of combining ML and XAI approaches for better air quality management.

3 Material and Methods

This section details the case study dataset, including its presentation, analysis, and processing. It also explains the model design and the two developed explainability techniques.

3.1 Case Study Dataset

This study's dataset included information on various air pollutants in Porto, gathered from the Open Weather Map API between November 25th, 2020 (1 a.m.) and December 31st, 2022 (11 p.m.). As shown in Table 1, the dataset under research comprises eight features and 18,215 records collected daily. Seven of these features correspond to pollutants like $PM_{2.5}$ and PM_{10}, which refer to particulate matter up to 2.5 and 10 μm in diameter, respectively. These particles can penetrate the respiratory tract, posing health risks and indicating air quality.

The first step was checking if the dataset contained missing values and timestamps. After carefully analysing the dataset, it was verified that there were no missing values. However, 30 missing timestamps were detected. Considering that in a time series problem, the temporal sequence of the data is crucial for applying the DL models, these missing timestamps need to be treated. The treatment of the missing timestamps will be explained in the following subsection.

Table 1. Features present in the dataset.

Feature	Description	Type	Unit
Date	Temporal Registration	Date & Time	Date and Hour
Carbon monoxide (CO)	CO measurement value	double	$\mu g/m^3$
Nitrogen Monoxide (NO)	NO measurement value	double	$\mu g/m^3$
Nitrogen Dioxide (NO_2)	NO_2 measurement value	double	$\mu g/m^3$
Ozone (O_3)	O_3 measurement value	double	$\mu g/m^3$
Sulfur Dioxide (SO_2)	SO_2 measurement value	double	$\mu g/m^3$
Particulate Matter 2.5 ($PM_{2.5}$)	$PM_{2.5}$ measurement value	double	$\mu g/m^3$
Particulate Matter 10 (PM_{10})	PM_{10} measurement value	double	$\mu g/m^3$
Ammonia (NH_3)	NH_3 measurement value	double	$\mu g/m^3$

Considering that NO_2 is the target for the study of explainability in this research, it becomes essential to understand how this feature changes over time in the collected dataset. Figure 1 presents this analysis.

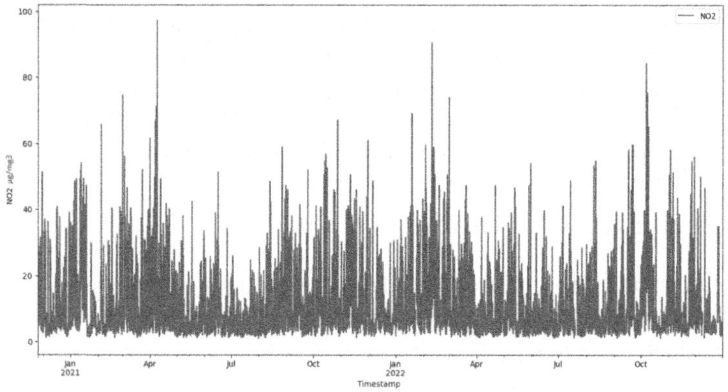

Fig. 1. NO_2 concentration over the years.

Three peaks in the measurements can be observed in April 2021 and February and October 2022. The minimum recorded value was 0.89 $\mu g/m^3$, while the maximum value reached 97.33 $\mu g/m^3$.

Next, it was also essential to understand whether or not the data followed a normal distribution, as this information will be necessary for choosing the methodology for analyzing data correlation. With this in mind, the Kolmogorov-Smirnov test was used with a p-value less than 0.05. Throughout this test, it was possible to verify that none of the features in the dataset presented a normal distribution.

3.2 Data Processing

One of the essential steps before developing a DL model is data processing. As indicated in the previous subsection, 30 missing records were identified to obtain a flawless temporal sequence. This way, once the missing temporal records were identified, they were inserted into the dataset. However, with this insertion of new observations, missing values emerged in each of them, which needed to be filled. In this sense, as the temporal observations entered were reasonably spaced in time, the average of the last 24 h was used in each feature to fill in the missing values.

Subsequently, it was necessary to analyze the features' correlation with the study's target (NO_2). Spearman's correlation test was applied because the features did not present a normal distribution, as seen previously. Figure 2 presents the generated correlation matrix.

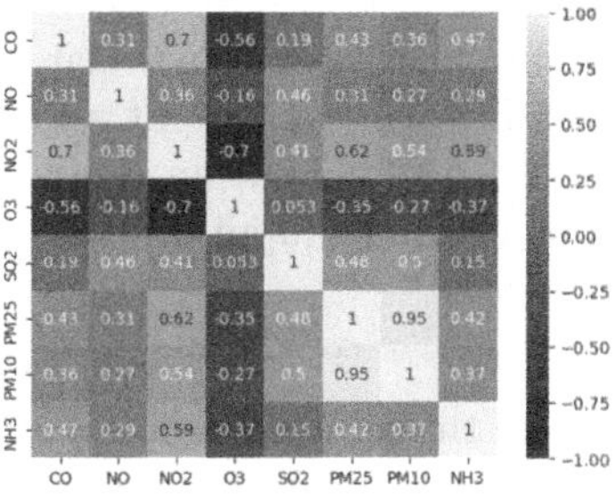

Fig. 2. Correlation matrix of dataset's features.

By analyzing this matrix, we can verify that the features significantly correlating with NO_2 are CO, O_3, $PM_{2.5}$, PM_{10} and NH_3. All have a positive correlation except O_3, which has a negative correlation of -0.7. Hence, considering the results, the features that did not present a correlation with the target were removed, with the final dataset having a total of 6 features (which included the target feature). Finally, it was necessary to normalize the data. Considering that LSTM works internally with the hyperbolic tangent, they were normalized between −1 and 1 using MinMaxScaler. Furthermore, the problem was transformed into a supervised problem by applying the sliding window technique.

3.3 Model Conception

Extensive research indicates that LSTM models effectively capture temporal dependencies within time series problems [1]. An LSTM architecture is specifically chosen for its ability to maintain long-term memory and alleviate issues such as vanishing gradients [12]. Essential to LSTM functionality are specialized gates—input, forget, and output gates—that regulate information retention [1].

Since this study focuses on explainability techniques to interpret the model's predictions, one of the strategies applied was an Attention-based strategy with an *ante-hoc* approach. This strategy's architecture is broken down into multiple phases, as Fig. 3 illustrates. The Attention layer processes the input first, breaking it down into three stages to determine the weights. Using a learned weight matrix W and a tanh activation applied to the linear transformation of the input sequence, attention scores S are calculated in the first phase:

$$S = \tanh(W \cdot \text{inputs})$$

In the subsequent stage, a softmax function is employed to transform the attention scores S into attention weights α, resulting in a probability distribution:

$$\alpha = \text{softmax}(V \cdot \text{ S})$$

Here, the learned weight matrix V ensures that the final attention weights α sum to 1 by converting attention scores into a scalar.

The last step involves multiplying the input sequence element-wise by the attention weights α to produce the final output vector C, which draws attention to the most important elements in the input sequence:

$$C = \alpha \odot \text{inputs}$$

After the Attention layer, the context vector is fed into the LSTM layer, which processes this information to capture complex temporal dependencies. Depending on the tuning process, the number of LSTM layers can vary, as presented in Sect. 4. After that, the output is passed through a dense layer followed by a dropout layer to prevent overfitting. Finally, the output layer, implemented as a dense layer with one unit, generates the final prediction.

The *ante-hoc* approach ensures the model is interpretable from the start, allowing for insights into which features are most influential during the forecasting process. This preventative transparency is crucial to building trust in the model, as it provides understandable feedback about its functioning and decision-making criteria.

The other technique applied was Shapley Values, implemented through the SHAP library, which is essential for interpreting the predictions made by our trained model [9]. After training the LSTM model with the Attention layer on the prepared time series data, we used a DeepExplainer to interpret its predictions. The DeepExplainer computes Shapley values using the reshaped data. This reshaped data consists of overlapping sequences of a fixed length obtained from the best candidate model presented in the results section. Each sequence captures the temporal dependencies in the time series data, representing the general characteristics of the input features. These values demonstrate how each feature affects the model's output, revealing its contribution to the overall prediction [15]. This *post-hoc* analysis not only helps validate the model's performance but also provides transparency. It increases the model's reliability by making

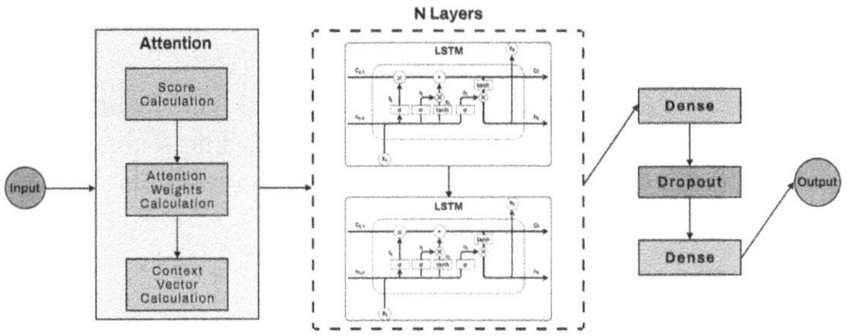

Fig. 3. Architecture of the Attention-based LSTM.

it more straightforward to identify the features that significantly influence the predictions.

4 Experiments

This section details the experimental configuration, focusing on tuning LSTM model hyperparameters to select the best model for explainability techniques. All models included an Attention layer. Table 2 summarizes the hyperparameters and their values.

Table 2. Set of hyperparameters' search space

Parameters	Values
Timesteps	[12, 24, 36]
Batch Size	[5, 10, 20]
Neurons	[32, 64, 128]
Layers	[3, 4, 5, 6]
Activation	[relu, tanh, sigmoid]
Dropout	[0.0, 0.25, 0.5]

It should be mentioned that the Grid Search function and two callbacks—Early Stopping and ReduceLROnPlateau—were used during the tuning process. The learning curve analysis was used to determine the number of epochs considered, resulting in 125 for this hyperparameter. Furthermore, a specific cross-validator for time series prevented the sequence from breaking. The cross-validator used was TimeSeriesSplit, using a k value of 4. Furthermore, to guarantee the reproducibility of the results, a seed value of 91195003 was used in all experiments. Finally, the error metric Root Mean Squared Error (RMSE) was used to evaluate all candidate models.

5 Results and Discussion

This section has two subsections. The first presents the result of the tuning process developed to find the best candidate model. The second section shows the outcomes of using the explainability techniques employed.

5.1 Selection of the Best Model Candidate

Considering the search process for the best hyperparameters described before, Table 3 summarizes the results of the best five candidate models.

Table 3. Top-five LSTM candidate models

Layers	Neurons	Dropout	Activation	Timesteps	Batch Size	RMSE
3	128	0.25	relu	36	10	1.87
3	64	0.25	tanh	24	10	1.94
4	32	0.0	tanh	24	10	1.97
3	64	0.5	tanh	24	5	1.98
3	128	0.25	relu	24	20	1.98

The table shows that the best candidate model presented a RMSE of 1.87, using 3 layers and 128 neurons. Furthermore, this model requires a day and a half of input in order to forecast the next 24 h of NO2. It does this by using 36 timesteps, or 36 h in the past. Overall, we found that only the best model needs these many time steps, with the remaining four only requiring one day to make the desired prediction. Additionally, it is evident that the models with 128 and 64 neurons, which are more significant numbers, require fewer layers (3) than the candidate model with 32 neurons, which requires 4 layers.

5.2 Explainability Results

Regarding the explainability techniques used in this study, this subsection presents the results obtained by evaluating the weights in the Attention layer and applying Shapley Values. Both techniques used were based on the best candidate model identified in the previous subsection.

Attention Layer. Concerning the first technique, the weights extracted from the Attention layer were first analyzed. Figure 4 illustrates a heatmap representing the intensity of those weights along the timesteps.

Before analyzing the figure itself, it must be noted that the input timesteps range from 0 to 35 on the horizontal axis, with 0 representing the oldest and 35 the most recent. Furthermore, on the vertical axis, input features in the range

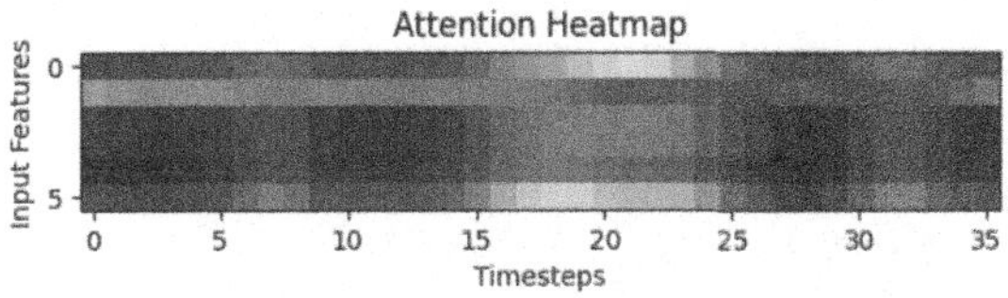

Fig. 4. Heatmap of Attention layer weights.

[0, 5] are identified with the following correspondence: 0 - CO, 1 - O_3, 2 - $PM_{2.5}$, 3 - PM_{10}, 4 - NH_3 and 5 - NO_2.

Image analysis allows us to extract the two features that Attention found most relevant: CO and NO_2, around timestep 20, which corresponds to 15 h ago. The same happens for $PM_{2.5}$, PM_{10} and NH_3. Regarding O_3, the highest values are recorded at the beginning, indicating that the closer to the present, the less relevance this variable holds. Despite this, and the high correlation of -0.7 that O_3 also presents with NO_2, no timestep is as relevant as with CO. This is a demonstration that correlation does not imply attention or causality, as correlation only indicates a relationship but does not point to any direction of causation. Figure 5 shows the evolution of Attention values throughout the timesteps to support this analysis and verify the abovementioned analysis.

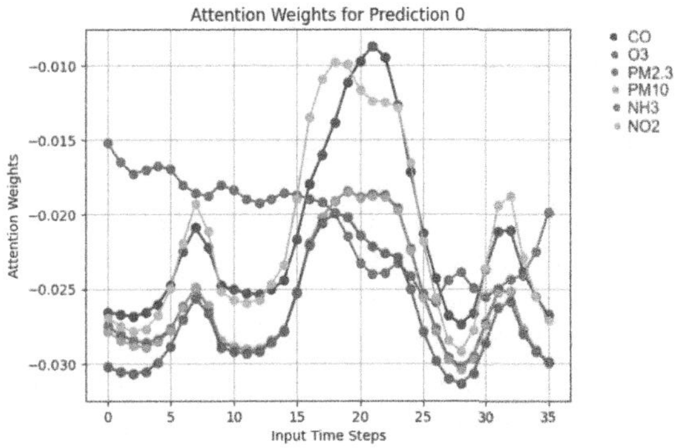

Fig. 5. Feature attention weights variation over timesteps.

Shapley Values. Using the SHAP technique, the focus was to assess each feature's effect to gain a greater understanding of the predictions made by the model. Due to computational limitations, the 7224 most recent observations from the training set were selected, which was the most computable regarding available hardware. Of these 7224, 7200 were used for the SHAP explainer to learn, with the remainder being used to compute the Shapley Values of the 24

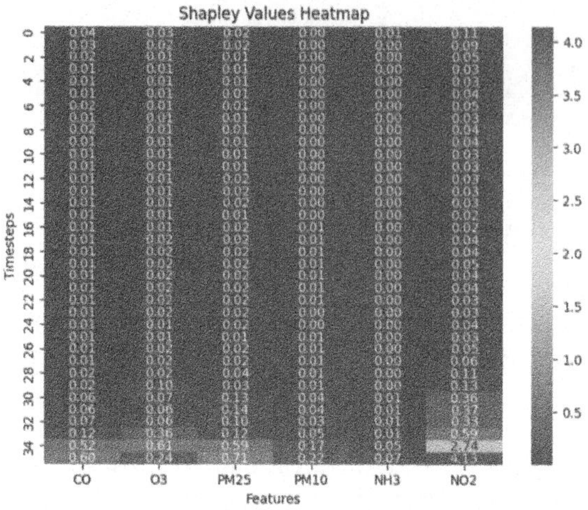

Fig. 6. Heatmap of Shapley Values for each feature per timesteps.

most recent timesteps. The values obtained for these timesteps are presented in Fig. 6.

The figure represents a heatmap of the Shapley Values of all features at each timestep. This 2D representation is achieved after adding all the values, in modules, of dimension 0 (samples) of the output. The timesteps are ordered from bottom to top, from newest to oldest. Importance values range from 0 to a maximum of 4.13. As seen from the matrix, the most relevant feature for the forecast is NO_2. This has the 3 highest importance values in timesteps 33, 34 and 35, these being 0.59, 2.74 and 4.13, respectively. These timesteps refer to the 3 h prior to the forecast. In the hour before the forecast, the second most relevant feature is CO, followed by O_3. At the initial timestep, O_3 decreases, while $PM_{2.5}$ and CO show higher values after NO_2. Throughout the entire time interval, NO_2 remains the most critical feature for the forecast, with NH3 and PM10 never proving relevant. As shown in Fig. 7, at the 24-hour mark, there is a decrease in all recorded values. However, even with this decrease, a change is observed with $PM_{2.5}$ being the second most influential, followed by O_3 and CO. Nevertheless, NO_2 remains the most important feature for the forecast.

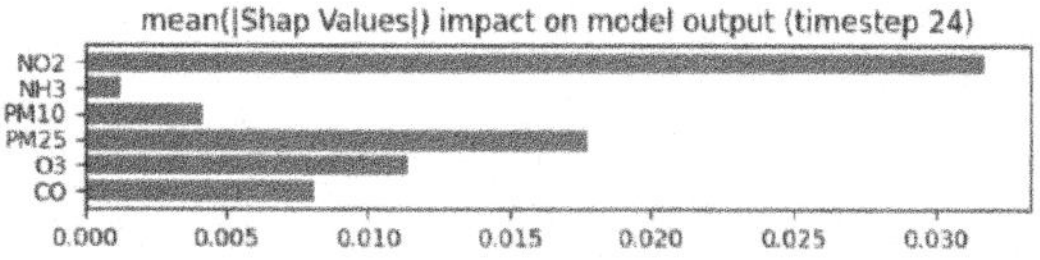

Fig. 7. Shapley Values at timestep 24

6 Conclusions

The evolution of ML, particularly DL, has transformed the AI landscape in recent years. As models become complex, understanding their decision-making, predicting behaviour, and ensuring alignment with human values become crucial. Consequently, transparency in these models is imperative. This study highlights the importance of accurate air pollution forecasting, especially for NO_2, due to its significant health and environmental impacts. We employed two explainability techniques for interpreting the target's predictions: an *ante-hoc* approach with the Attention layer and a *post-hoc* approach using Shapley Values for variable importance.

After selecting the best LSTM candidate model, the *ante-hoc* explainability allowed the model to be interpretable from the beginning. Our analysis revealed that CO and NO_2 were the most relevant features. Conversely, *post-hoc* explainability quantified how each feature contributes to the model's predictions after training. According to the results, NO_2 consistently had the highest importance, with $PM_{2.5}$ and CO coming in second and third, particularly in the final three hours before the forecast. In both strategies, verifying that the target value produces the most relevant impact on the model's performance was possible.

Regarding future work, several steps can be taken. Firstly, exploring additional explainability techniques, such as LIME or Feature Permutation, should be considered to provide more model interpretation. Since only the LSTM model was presented, developing additional models like Gated Recurrent Units and Convolutional Neural Networks will be a significant step forward since it will enable a comprehensive comparison of the features' impact in the NO_2 forecasting.

Acknowledgements. This work is financed by National Funds through the Portuguese funding agency, FCT - Fundação para a Ciência e a Tecnologia within project 2022.06822.PTDC (https://doi.org/10.54499/2022.06822.PTDC). The work of Pedro Oliveira was supported by the doctoral Grant PRT/BD/154311/2022 financed by the Portuguese Foundation for Science and Technology (FCT), and with funds from European Union, under MIT Portugal Program.

References

1. Ahmed, D.M., Hassan, M.M., Mstafa, R.J.: A review on deep sequential models for forecasting time series data. Appl. Comput. Intell. Soft Comput. **2022**(1), 6596397 (2022). https://doi.org/10.1155/2022/6596397
2. Bacanin, N., et al.: The explainable potential of coupling hybridized metaheuristics, XGBoost, and shap in revealing toluene behavior in the atmosphere. Sci. Total Environ. **929**, 172195 (2024)
3. Chamola, V., Hassija, V., Sulthana, A.R., Ghosh, D., Dhingra, D., Sikdar, B.: A review of trustworthy and explainable artificial intelligence (XAI). IEEE Access (2023). https://doi.org/10.1109/ACCESS.2023.3294569
4. Dwivedi, R., et al.: Explainable AI (XAI): core ideas, techniques, and solutions. ACM Comput. Surv. **55**(9), 1–33 (2023). https://doi.org/10.1145/356104

5. Hossain, M.R.: Killing billions to save millions? Analyzing the double jeopardy of fossil-fuel-led economic development in bangladesh. Environ. Dev. Sustain. 1–32 (2023). https://doi.org/10.1007/s10668-023-03497-2
6. Manisalidis, I., Stavropoulou, E., Stavropoulos, A., Bezirtzoglou, E.: Environmental and health impacts of air pollution: a review. Front. Public Health **8**, 14 (2020). https://doi.org/10.3389/fpubh.2020.00014
7. Mukhamediev, R.I., et al.: Review of artificial intelligence and machine learning technologies: classification, restrictions, opportunities and challenges. Mathematics **10**(15), 2552 (2022). https://doi.org/10.3390/math10152552
8. Orellano, P., Reynoso, J., Quaranta, N., Bardach, A., Ciapponi, A.: Short-term exposure to particulate matter (pm10 and pm2. 5), nitrogen dioxide (no2), and ozone (O3) and all-cause and cause-specific mortality: systematic review and meta-analysis. Environ. Int. **142**, 105876 (2020). https://doi.org/10.1016/j.envint.2020.105876
9. Ozyegen, O., Ilic, I., Cevik, M.: Evaluation of interpretability methods for multivariate time series forecasting. Appl. Intell. 1–17 (2022). https://doi.org/10.1007/s10489-021-02662-2
10. Pantiskas, L., Verstoep, K., Bal, H.: Interpretable multivariate time series forecasting with temporal attention convolutional neural networks. In: 2020 IEEE Symposium Series on Computational Intelligence (SSCI), pp. 1687–1694. IEEE (2020)
11. Retzlaff, C.O., et al.: Post-hoc vs ante-hoc explanations: XAI design guidelines for data scientists. Cogn. Syst. Res. **86**, 101243 (2024). https://doi.org/10.1016/j.cogsys.2024.101243
12. Sen, J., Mehtab, S.: Long-and-short-term memory (LSTM) networks architectures and applications in stock price prediction. Emerg. Comput. Paradigms: Principles Adv. Appl. 143–160 (2022). https://doi.org/10.1002/9781119813439.ch8
13. Sunder, M.S., Tikkiwal, V.A., Kumar, A., Tyagi, B.: Unveiling the transparency of prediction models for spatial PM2. 5 over Singapore: comparison of different machine learning approaches with explainable artificial intelligence. AI **4**(4), 787–811 (2023)
14. Taye, M.M.: Understanding of machine learning with deep learning: architectures, workflow, applications and future directions. Computers **12**(5), 91 (2023). https://doi.org/10.3390/computers12050091
15. Van Zyl, C., Ye, X., Naidoo, R.: Harnessing explainable artificial intelligence for feature selection in time series energy forecasting: a comparative analysis of grad-cam and shap. Appl. Energy **353**, 122079 (2024). https://doi.org/10.1016/j.apenergy.2023.122079
16. Vega, G.M., Aznarte José, L.: Shapley additive explanations for NO2 forecasting. Eco. Inform. **56**, 101039 (2020)
17. Wang, Y., et al.: Contrasting trends of PM2. 5 and surface-ozone concentrations in China from 2013 to 2017. Natl. Sci. Rev. **7**(8), 1331–1339 (2020). https://doi.org/10.1093/nsr/nwaa032

Predicting Employee Attrition in a Multi-company Setting

Adriano Gomes[1], Luís M. Silva[2(✉)], and João Pedro Cruz[2]

[1] Department of Economics, Management, Industrial Engineering and Tourism, University of Aveiro, 3810-193 Aveiro, Portugal
aog@ua.pt

[2] Center for Research and Development in Mathematics and Applications (CIDMA), Department of Mathematics, University of Aveiro, 3810-193 Aveiro, Portugal
{lmas,pedrocruz}@ua.pt

Abstract. This paper describes the creation of a database and a machine learning model to predict employee attrition. Our proposal deals with attrition by considering 3 classes (voluntary, involuntary and no attritors) giving a more complete view of the loss of qualified personnel to the Human Resources Management. Of the several machine learning models tested to solve the problem, XGBoost stood out as the best performing one on a dataset with more than four thousand employees and twenty-one features collected from three independent companies from different industrial sectors. The model, evaluated on a 20-run experiment, achieved an overall mean accuracy of 78.5%, corresponding to the correct classification of 52.6% of the voluntary attritors, 78.9% of the involuntary attritors and 81.6% of the non-attritors, showing that voluntary attritors are harder to discriminate.

Keywords: Employee attrition · Machine learning · Real data · Human resources management · Industry 4.0

1 Introduction

Employee attrition[1], that is, the leave of a collaborator from a company, has been a major concern as losing highly qualified personnel has tremendous impacts in several aspects of a company's daily life. In fact, company's expert collaborators carry valuable knowledge that is difficult to transfer to new collaborators [2,6,14] and loosing such expertise has high costs. Among others, waiting periods while the position is not filled or the training times of a new employee are factors that increase the value and uncertainty of the cost [6]. To reduce such uncertainty and cost, companies wish to anticipate the intention of an employee to leave its position and thus undertake a negotiation process to retain the collaborator or, alternatively, to adequately prepare a hiring process [6]. The study of employee attrition started decades ago [7,9,10] and since 1985 behavioural and organizational models for attrition included more complex concepts by considering

[1] Literature also use *employee churn* or *employee turnover*.

V. Julian et al. (Eds.): IDEAL 2024, LNCS 15346, pp. 337–348, 2025.
https://doi.org/10.1007/978-3-031-77731-8_31

contextual variables such as stress, satisfaction related variables or interpersonal relationships for example [6]. It was only in the past 10 to 15 years that the problem has been more consistently tackled from a data analysis perspective. Some studies focused on statistical methods (descriptive statistics, correlation, hypotheses testing) [2,13,16,19,20]. In [12], the International Business Machines Corporation (IBM) reported an "Artificial Intelligence" technology that could predict the intention of employees to leave and since then it attracted the attention of companies as well as academics from the Machine Learning (ML) perspective. We consider that predicting employee attrition is an Industry 4.0 (I4.0) problem because I4.0 refers to a dramatic update in the management of industrial processes [8] as well as in the management of Human Resources Management (HRM) [14].

Attrition has essentially being treated as a binary problem of non-attritors vs attritors (predicting if a person is working or not) or non-attritors vs voluntary attritors (by predicting the will of the employee to leave). Most papers focus on the prediction of overall attrition, and others focus only on predicting voluntary attrition, since companies usually define objectives of reducing costs with replacements and talent retention. In contrast, Saradhi & Palshikar [15] proposed a different way of approaching the problem by comparing it to the more common problem of customer churn, where non-churners, voluntary churners and involuntary churners are considered. Still, the authors studied three binary models which they called "Resigned + released vs Retained", "Resigned vs Retained" and "Released vs Retained". Involuntary attrition is commonly disregarded as it results from company decisions, but distinguishing voluntary from involuntary attrition will be fundamental for HRM to understand reasons for attrition. In fact, detecting employees that the company chooses to let go will help HRM to focus on understanding reasons for employees that choose to leave. In this sense, we propose a concise definition for each class of attrition, associating to each class the "leave motives" established in current legislation. Thus, non-attritors include those still working in the company or left by a natural cause such as death or retirement. On the other hand, voluntary attritors are those that choose to leave, encompassing any motive that comes from employee initiative. Finally, involuntary attritors are defined to be those whose exit is due to the company's initiative, such as any contract end anticipation or non-renewal. The "Leave Motive" record (and communication to specific public services) is a legal obligation and was used to define each one of the classes, as presented in Table 1. Note that "mutual agreement" was both considered as voluntary or involuntary attrition, depending on the company involved in this work. In fact, this leave motive comes from a mutual desire of the two parts hindering a clear choice of any of the attrition types. Given the residual number of employees in this situation, it was accepted the inability to detect it and "mutual agreement" was included in the no-attrition class.

While most studies devote their efforts to solve the attrition problem within a specific company data or even from several companies but within a common industrial sector, this work deals with data from multi-sector companies to ana-

Table 1. Classes of attrition

No attrition	Voluntary attrition	Involuntary attrition
Active	Worker's initiative	Company's initiative
Death	Cessation of No-Term Contract	Fixed-Term Contract Termination
Pre-Retirement	Advanced Cessation of Fixed-Term Contract	Uncertain-Term Contract Termination
Early Retirement Disability Retirement	Advanced Cessation of Uncertain-Term Contract	Collective Dismissal
Old-age Retirement	Job Abandonment	Fair Cause Dismissal
Other reasons	End of Trial Period Worker's Initiative	Extinction of Employment
Mutual Agreement		End of a Trial Period Company's Initiative
		Term of Mandate

lyze the feasibility of an unique attrition model. Being part of a broader project, developed in Portugal involving several companies that aims to build an integrated platform for HRM based on real data, our contribution comes in 3 dimensions by:

- providing a clear definition of attrition linking each facet to the legal motives of contract termination
- providing guidelines and standards for the creation of a multi-company database for the attrition problem
- developing a machine learning model to serve as a broad predictive tool for the attrition status in any company.

The rest of the paper is structured as follows: Sect. 2 provides a description of the methodology, by discussing guidelines for a multi-company data fusion and uniformization that took to the proposed database, as well as to describe how the problem is tackled from the machine learning perspective. Section 3 is devoted to the results, presenting model comparison through several performance measures. Finally, the conclusions and guidelines for future work are left for Sect. 4.

2 Methodology

2.1 Data Collection

The data was collected from three totally independent companies belonging to the "electronics" (Company 1), "furniture" (Company 2) and "household" (Company 3) sectors. Due to different collecting and storage procedures between companies, a pre-processing step of data unifying and standardization was conducted to reach an overall consistent dataset so that each variable is interpreted

Table 2. List of collected variables and corresponding coding standards.

Variable	Definition
Age	Age at present time or at date of leaving, when applicable
Sex	'male', 'female'
Marital Status	'Married', 'Single', 'Widowed', 'Divorced', 'NMP' (non-marital partnership)
Nationality	Unique country code (ISO CODE)
Number of holders	Numeric value - number of holders
Number of dependents	Numeric value - number of dependents
Number of dependents with disabilities	Numeric value - number of dependents with disabilities
Academic Qualifications	8 numeric values corresponding to the Portuguese National Qualification Framework: '1' - 2nd cycle of basic education, ..., '8' - Doctorate
Field of Study	From the Portuguese General Directorate of Higher Education: 'Teacher/Trainer Training and Educational Sciences', 'Mathematics and Statistics', 'Informatics', 'Engineering and Related Techniques', etc.
Work-home Time	Estimated work-home duration of travel, using the employee and company postal-codes
Contract Type	'No_Term', 'Fixed_Term', 'Uncertain_Term'
Schedule	'Fixed', 'Rotating', 'Concentrated'
Schedule Morning	'Yes', 'No'
Schedule Afternoon	'Yes', 'No'
Schedule Night	'Yes', 'No'
Schedule Weekend	'Yes', 'No'
Day Off	'Fixed_Week', 'Fixed_Weekend', 'Rotating'
Qualifications Level	Hierarchical level in the company: 'Superior Managers', 'Middle Managers', 'Managers, Foremen, Masters and Team Leaders', 'Highly Qualified Professional', 'Qualified Professional', 'Semi-qualified Professional', 'Unqualified Professional', 'Practitioner and Apprentice'
Area	Concerted working areas according to the employee's department and function: 'Administration/Management', 'Financial', 'Information Technologies', 'Human Resources', 'Logistics', 'Commercial/Marketing', 'Engineering and Research', 'Production', 'Quality', 'Maintenance'
Salary Rate	According to Eq. (1)
Years at Company	Years working at the company

and coded in the same way by each of the involved companies. Table 2 presents the complete set of variables used in this work as well as the coding standards used across the involved companies.

Personal information related variables such as Marital Status, Academic Qualifications and Field of Study were defined according to well established legal standards. Variables that reflect some relation between the employee and the respective company were collected and standardized across companies. Due to the highly specific organizational structures in each company (reflecting different productive sectors), the Area variable was created to represent the organizational position of the employee in what concerns department and function. Therefore, 10 broad levels were established that univocally cover each pair department/function across different companies. The Work-home Time, representing the duration (in seconds) of the travel between home and work locations, was estimated using free services [5,11] based on postal codes. These were transformed into geographical coordinates and Work-home Time is then computed between the two sets of coordinates (the company's and the employee's). Finally, salary information, which is considered an important variable for attrition prediction (e.g., [2,3]) was obtained indirectly by creating a salary rate variable computed as a percentage of wage difference compared to the median of the salaries of workers in the same group (peers) of the respective company:

$$SalaryRate = \frac{Salary - Median\ salary\ of\ the\ group}{Median\ salary\ of\ the\ group} \times 100 \qquad (1)$$

This expression for salary rate hides the employee salary value (data protection regulations prevented the collection of such data) but preserves the personal comparison mental factor that could affect attrition. The grouping variable can be different between companies so, the choice of how to group employees is left for the company as they better understand their own organizational structure. Also, the value of Salary was left for the company to choose whether to consider brute salary or to include other types of compensations such as allowances and bonuses.

The dataset used to train the models contains 21 features for 4162 employees, distributed across companies according to Table 3.

Table 3. Employee distribution regarding Company and attrition class. [%] represents percentage within each company; (%) represents percentage within total.

Attrition	Company 1	Company 2	Company 3	Total
No	1134 [93,9%]	1626 [90,5%]	644 [55,7%]	3404 (81,8%)
Voluntary	64 [5,3%]	120 [6,7%]	232 [20,0%]	416 (10,0%)
Involuntary	10 [0,8%]	51 [2,8%]	281 [24,3%]	342 (8,2%)
Total	1208 (29,0%)	1797 (43,2%)	1157 (27,8%)	4162

2.2 Machine Learning Methodology

This paper focuses in predicting attrition as a classification task. Multiple types of classification models can be found in the literature, from the most theoretically or technically simple, such as k-Nearest Neighbours (kNN), Naïve Bayes (NB), Linear Discriminant Analysis (LDA), to the most complex ones, such as support vector machines, neural networks or boosting methods [18]. The no free lunch theorems [17] advise us to test different models and evaluate the most appropriate for the data at hand, but as shown in the next section, most of the algorithms are tree based models as they are known to deal remarkably well with categorical (predictive) variables.

In this sense, our experimental procedure is divided in two phases. Phase 1 is devoted to the identification of the model that best suits the data using python's pycaret library [1]. This library can evaluate and auto-tune a large set of algorithms, providing a comparison according to several performance measures. The chosen model is then thoroughly evaluated in Phase 2. All results are obtained for a test set comprising 30% of the original dataset while 70% are left for training, stratifying by class and company to ensure proper learning of the problem. In Phase 2, the performance is evaluated in 20 different runs of train/test splitting to eliminate the influence of a particular split and to get statistical significance. In each run, the model is trained and tuned for AUC using 10-fold cross validation and the final model chosen with the F1-score. Mean and standard deviation of the performance measures of those 20 models are then obtained.

In any phase, we had to deal with class imbalance, the problem of high difference in class proportions. Among several strategies to deal with it, a combination of undersampling to the majority class (non-attritors) and oversampling of the other classes was used, which significantly improved the performance of the classification models. Oversampling is performed according to SMOTE-NC, a modified version of SMOTE (Synthetic Minority Over-Sampling Technique) that handles both numerical and categorical variables, by creating simulated observations from the existent ones using a nearest neighbours approach [4]. The result was an equilibrium between classes with the same number of employees in each class of attrition and a balance between companies, to avoid bias towards any company. Evaluation of the models is performed with several standard measures such as Accuracy (ACC), Recall (Sensitivity, True Positive Rate TPR), Precision (Positive Predictive Value PPV), F1-score (F1), Area Under the Curve (AUC) or the Balanced Error Rate (BER) [18], as looking to a set of measures provides a more thorough analysis of the performance of any model.

3 Results

3.1 Exploratory Data Analysis

An initial exploratory data analysis presented some interesting findings that we briefly discuss. The number of employees is quite balanced between companies but is highly imbalanced between non attritors and attritors (voluntary and involuntary), particularly in Company 1 and Company 2. The age of the employees within the 3 companies varies from 18 to 72 years old with Company 1 having older workers in general. The family structure is also quite similar between companies, consisting in 1 to 2 holders and 0 to 4 dependents (mostly 1 dependent) with rare cases of disabled dependents. Significant differences between companies can be found in Academic Qualifications, Salary Rate or Years at Company. In Company 1 higher education (levels 6, 7 and 8) are more frequent with at least 25% of the workforce having level 6 or more, while in the other companies, level 6 or higher is an exception considering that 75% of the workforce have level 3 or less. Salary Rate presents a higher mean value and variability within Company 3. The distribution of the workforce in terms of Years at Company is also very different among companies. In general, most of the workers are young, and consequently with less years at the respective company. Company 1 has also an important workforce having more than 10 years at the company, while Company 2 does not have workers with more than 15 years of work.

Production is the dominant area in all companies, with Engineering and Research at Company 1 and Logistics at Companies 2 and 3 being the second Area with most collaborators. This is an important aspect that differentiates the 3 companies and relates with the Qualification. Restricting to groups with more than 5% of all the workforce, in Company 1 around 38% of the workers are Superior Managers in different areas, while in Companies 2 and 3 most of the workers (around 70%) are less or equal to Semi-Qualified Professionals (mainly in Production and Logistics). This is clearly a consequence of the industrial sector each company belongs to. Contract Type is also an important aspect for the relation between a company and its collaborator. The data presented clear differences, being Company 3 the one where Fixed Term contracts dominate.

In relation to the attrition outcome, the most relevant aspects show that voluntary attritors are typically older and with more years of experience in the company. Attrition is more present in single male workers, with no dependents. In fact, the existence of dependents makes the overall proportion of attrition to decrease. Different behaviours can be observed between companies regarding the predictors, such as Qualification and Area: attrition is almost concentrated in the highest qualification level for Company 1, while for Company 3, the highest proportions of attrition are observed in extremal qualification levels; Company 3 shows an increased attrition at shop floor areas, such as Production or Logistics (see Fig. 1).

3.2 Predictive Model for Attrition

The first modelling step (Phase 1) was devoted to identify the model better suited to this data. Table 4 shows the corresponding results, obtained with python's pycaret package, using a 10-fold cross validation methodology applied to 15 algorithms.

It is quite clear that boosting models present overall best performances, particularly Light Gradient Boosting Machine (LightGBM), Categorical Boosting Classifier (CatBoost) and Extreme Gradient Boosting (XGBoost). Although LightGBM exhibits the best performance in all the scores of Table 4, we decided to look more deeply and analyse the performance in each class on the test set,

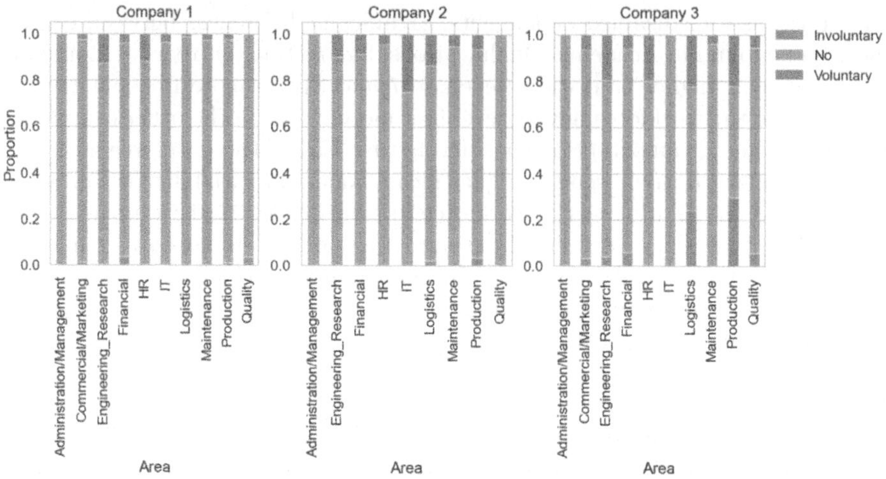

Fig. 1. Distribution of attrition according to Area and company.

Table 4. Performance measures for Phase 1.

Model	Accuracy	AUC	Recall	Precision	F1
Light Gradient Boosting Machine	0.829	0.951	0.829	0.831	0.829
Categorical Boosting	0.823	0.944	0.823	0.824	0.823
Extreme Gradient Boosting	0.820	0.946	0.820	0.821	0.820
Gradient Boosting Classifier	0.804	0.936	0.804	0.806	0.804
Random Forest	0.787	0.937	0.787	0.789	0.786
Decision Tree	0.755	0.816	0.755	0.756	0.755
Extra Trees	0.753	0.919	0.753	0.755	0.752
Adaptative Boosting	0.731	0.822	0.731	0.729	0.728
Logistic Regression	0.704	0.855	0.704	0.710	0.698
Ridge Classifier	0.700	0.000*	0.700	0.709	0.693
Linear Discriminant Analysis	0.695	0.869	0.695	0.704	0.691
Naive Bayes	0.687	0.823	0.687	0.711	0.681
K Nearest Neighbors	0.554	0.721	0.554	0.577	0.550
Quadratic Discriminant Analysis	0.450	0.650	0.450	0.535	0.356
Support Vector Machine (Linear Kernel)	0.403	0.000*	0.403	0.421	0.296

*AUC for models that do not support probability estimation is shown as 0.000

specifically at Recall and the F1-Score. As we can see in Table 5, when focusing on the interest class, voluntary attrition, XGBoost shows a higher Recall while CatBoost a higher F1-score. By averaging the class results, we conclude that XGBoost presents a slightly better performance, showing a more balanced performance.

Table 5. Recall and F1-score by class for each boosting model. Last columns are the arithmetic averages.

Model	Involuntary Attritors		Voluntary Attritors		Non-Attritors		Average	
	Recall	F1	Recall	F1	Recall	F1	Recall	F1
LightGBM	0.8333	0.6911	0.4720	0.3122	0.7959	0.8679	0.7004	0.6420
CatBoost	0.8431	0.6693	0.4640	0.3213	0.8027	0.8726	0.7033	0.6390
XGBoost	0.8235	0.6971	0.4960	0.3212	0.7939	0.8672	0.7045	0.6475

We may also look at an individual level, by analysing the estimated posterior probabilities for each class. Each point, in Fig. 2 represents an employee prediction, coloured by its true class. We can see that most employees are in the correct class with high probability, but some of the non-attritors are predicted to be voluntary attritors (green points in the Voluntary row). Would these employees be future attritors that are still working? Another interesting aspect is that the CatBoost algorithm has more difficulty to clearly recognize involuntary attritors, as we find the corresponding posterior probabilities spreads more when compared to the other algorithms. Given the above analysis, XGBoost was the chosen model for the next phase.

In Phase 2, we evaluate XGBoost thoroughly using a 20-run experimental procedure. To tune XGBoost models we use "multi:softprob" (to obtain probabilities at the output), AUC as the evaluation metric, and a grid of hyperparameters as follows (best choices are marked in bold):

- Booster: **gbtree**; dart
- eta: 0; 0.2; 0.4; 0.6; **0.8**; 1
- gamma: **0**; 1; 10; 100
- max_depth: 1; 5; **10**
- colsample_bytree: **0.5**; 1
- subsample: **0.7**; 1
- lambda: 0; **1**; 10
- alpha: 0; **1**; 10

For each run we compute the confusion matrix on the test set to obtain Recall and Precision in each class, Balanced Error Rate (BER) and Accuracy. Mean and standard deviations, over the 20 runs, are presented in Table 6 as well as 95% confidence intervals.

The model has an overall error of 21% and a BER of 29%, thus presenting some lack of equilibrium when learning the 3 classes. In fact, this is caused by the class of voluntary attritors that present lower values of Recall and Precision. Note, however, that, contrary to the training sets, the test set is not balanced by any under/oversampling technique. Error estimates in underrepresented classes are generally pessimistic. False positives on the voluntary class are particularly problematic, meaning that non-attritors and involuntary attritors are being predicted as voluntary attritors. In practice, human resources management would have to monitor employees signalled as potential attritors, excluding possible or known involuntary attritors and following the other cases with retention techniques if desired. The other 2 classes are quite well identified, particularly the non-attritors.

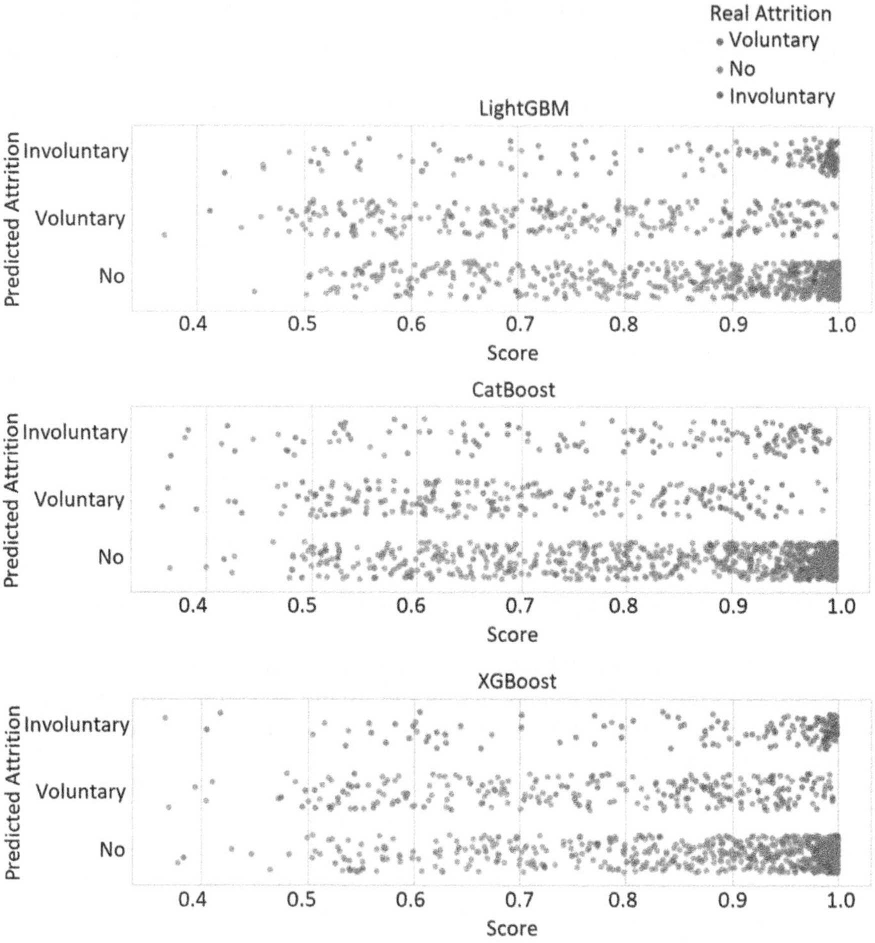

Fig. 2. Predicted probabilities for each class for the best 3 models.

Table 6. Results of the 20-run experiment for the tuned XGBoost model.

	Involuntary		Voluntary		No		Overall	
	Recall	Precision	Recall	Precision	Recall	Precision	BER	Accuracy
Mean	0.789	0.665	0.526	0.260	0.816	0.954	0.290	0.785
(Std)	(0.033)	(0.033)	(0.053)	(0.026)	(0.015)	(0.006)	(0.024)	(0.014)
CI 95%	[0.773; 0.804]	[0.649; 0.680]	[0.501; 0.550]	[0.247; 0.272]	[0.809; 0.823]	[0.951; 0.957]	[0.279; 0.301]	[0.778; 0.792]

4 Conclusion

The present paper deals with the problem of employee attrition, proposing a machine learning model able to make predictions in a multi-company setting. The first challenge was to devise a unique database with clear standards for data collection that could be implemented by the 3 independent companies that act in different industrial sectors. For that, an enormous effort to reach a consensus between different points of view and different data systems was made, resulting in the database used in this work. Nevertheless, improvements can be made regarding the type of collected information, namely variables that can help modelling the attrition problem. We argue that adding variables measuring performance or engagement, for example, would be very valuable but were not available at the time of this work. Boosting models based on decision trees are the most appropriate to this type of data, because of its easiness in dealing with categorical variables. The results have shown that voluntary attritors is the class with the lowest performance, greatly influenced by a considerable number of false positives. However, it is important to note that we are facing a dynamic problem in the sense that an employee that is currently working may be in an attrition process and willing to leave in a year or so. Thus, some of such false positives are eventually future true positives. In this sense, looking to posterior probabilities may be more valuable from the company point of view than a hard decision because it puts, for each employee, all the possibilities in perspective. Future work focusses on improving the database for attrition by including performance and engagement variables, in a consistent way, updating the predictive model with such information. Moreover, given the varying nature of the relation of the employees in the company, a dynamic model should also be considered over a temporal version of the database.

Acknowledgments. The present work was developed in the scope of the Project Augmented Humanity [POCI-01-0247-FEDER-046103 and LISBOA-01-0247-FEDER-046103], financed by Portugal 2020, under the Competitiveness and Internationalization Operational Program, the Lisbon Regional Operational Program, and by the European Regional Development Fund. Second and third authors were also supported by CIDMA and funded by the Fundação para a Ciência e a Tecnologia, I.P. (FCT, Funder ID = 50110000187) under Grants UIDB/04106/2020 and UIDP/04106/2020 (https://doi.org/10.54499/UIDB/04106/2020) (https://doi.org/10.54499/UIDP/04106/2020).

Disclosure of Interests. The authors have no competing interests to declare that are relevant to the content of this article.

References

1. Ali, M.: PyCaret: an Open Source, Low-Code Machine Learning Library in Python (2020). https://www.pycaret.org, pyCaret version 1.0.0
2. Basariya, R., Ahmed, R.R.: A study on attrition-turnover intentions of employees. Int. J. Civ. Eng. Technol. **10**(1), 2594–2601 (2019)
3. Brown, T., Green, B.: Is "The Great Resignation" Real? We Analyzed Salary Increases and Turnover to Make Sense of a Competitive Labor Market. https://humancapital.aon.com/insights/articles/2021/is-the-great-resignation-real
4. Chawla, N.V., Bowyer, K.W., Hall, L.O., Kegelmeyer, W.P.: Smote: synthetic minority over-sampling technique. J. Artif. Intell. Res. **16**, 321–357 (2002). https://doi.org/10.1613/jair.953

5. GeoPy: Geopy (2.2.0) (2021). https://geopy.readthedocs.io/en/stable/
6. Holtom, B.C., Mitchell, T.R., Lee, T.W., Eberly, M.B.: Turnover and retention research: a glance at the past, a closer review of the present, and a venture into the future. Acad. Manag. Ann. **2**(1), 231–274 (2008)
7. Hom, P.W., Griffeth, R.W., Sellaro, C.L.: The validity of mobley's (1977) model of employee turnover. Organ. Behav. Hum. Perform. **34**(2), 141–174 (1984)
8. Kavakli, E., Buenabad-Chavez, J., Tountopoulos, V., Loucopoulos, P., Sakellariou, R.: Specification of a software architecture for an industry 4.0 environment. In: 2018 Sixth International Conference on Enterprise Systems (ES), pp. 36–43. IEEE (2018)
9. March, J.G., Simon, H.A.: Organizations. John Wiley & Sons (1958)
10. Mobley, W.H., Griffeth, R.W., Hand, H.H., Meglino, B.M.: Review and conceptual analysis of the employee turnover process. Psychol. Bull. **86**(3), 493 (1979)
11. OpenRouteService: Openrouteservice (2.2.2) (2020). https://openrouteservice.org/
12. Rosenbaum, E.: IBM Artificial Intelligence can Predict with 95% Accuracy Which Workers are About to Quit Their Jobs. https://www.cnbc.com/2019/04/03/ibm-ai-can-predict-with-95-percent-accuracy-which-employees-will-quit.html
13. Rubenstein, A.L., Kammeyer-Mueller, J.D., Wang, M., Thundiyil, T.G.: "Embedded" at hire? Predicting the voluntary and involuntary turnover of new employees. J. Organ. Behav. **40**(3), 342–359 (2019)
14. Salvadorinho, J., Teixeira, L.: Organizational knowledge in the i4. 0 using BPMN: a case study. Procedia Comput. Sci. **181**, 981–988 (2021)
15. Saradhi, V.V., Palshikar, G.K.: Employee churn prediction. Expert Syst. Appl. **38**(3), 1999–2006 (2011)
16. Takeuchi, N., Takeuchi, T.: A longitudinal investigation on the factors affecting newcomers' adjustment: evidence from Japanese organizations. Int. J. Hum. Resour. Manag. **20**(4), 928–952 (2009)
17. Wolpert, D.H.: The lack of a priori distinctions between learning algorithms. Neural Comput. **8**(7), 1341–1390 (1996). https://doi.org/10.1162/neco.1996.8.7.1341
18. Zhou, Z.H.: Machine Learning, 1st edn. Springer Singapore (2021). https://doi.org/10.1007/978-981-15-1967-3
19. Zhu, X.: Forecasting Employee Turnover in Large Organizations (2016). https://trace.tennessee.edu/utk_graddiss/3985/
20. Zhu, X., et al.: Employee turnover forecasting for human resource management based on time series analysis. J. Appl. Stat. **44**(8), 1421–1440 (2017)

A Deep-Learning Approach for the Identification of New Subtypes of Lung Cancer

Tuhin Banerjee[1] and Andrea Corradini[2](✉)

[1] Canon EMEA, Amstelveen, The Netherlands
tuhin2024banerjee@gmail.com
[2] MCI, Innsbruck, Austria
andrea.corradini@mci.edu

Abstract. Lung cancer is the deadliest cancer in the world. It is caused by unchecked cell division of damaged cells in the lungs forming tumors that eventually prevent the lung from functioning properly. Identification of novel unsupervised subtypes of lung cancer is critical to reveal new insights into the underlying biology of cancer and ensure that patients receive specialized precision treatment based on the subtype of cancer they are suffering from. The ability of modern sequencing tools to produce patient-specific RNA sequencing (RNA-seq) gene expression data has transformed cancer research by offering in-depth understanding of the molecular landscape of cancer. This paper reports on a pipeline that comprise of a Deep Autoencoder (DAE) model coupled with hierarchical agglomerative clustering (H-Clust). It aims to identify new unsupervised lung cancer subtypes from RNA-seq expression samples collected from a publicly available dataset. Further, a deep learning (DL) model, Artificial Neural Network (ANN) is used to classify a patient's data into one of the newly identified subtypes.

Keywords: lung cancer subtype · TCGA · RNA-seq · deep learning · DAE · encoder-decoder · agglomerative clustering · ANN

1 Introduction

A study by the Institute for Health Metrics and Evaluation (IHME) highlights a concerning finding that lung cancer is the most common cause of cancer-related deaths worldwide accounting to approximately 2 million deaths [1]. Lung cancer is caused by uncontrolled cell division leading to the development of tumors that obstruct normal lung function. Two common histological lung cancer subtypes are lung adenocarcinoma (LUAD) and lung squamous cell carcinoma (LUSC). The investigation of further potential clinical distinctions among these subtypes is an active area of research [2].

The Cancer Genome Atlas (TCGA) is a valuable resource for researchers studying cancer biology and genomics. This database was developed as a collaborative effort between the National Cancer Institute (NCI) and the National Human Genome Research Institute (NHGRI) and it aims to provide a comprehensive understanding of the genomic

V. Julian et al. (Eds.): IDEAL 2024, LNCS 15346, pp. 349–360, 2025.
https://doi.org/10.1007/978-3-031-77731-8_32

changes that occur in various types of cancer. TCGA offers a large and diverse dataset that is openly accessible to researchers. Furthermore, it ensures consistency and quality standardized data in storing and reporting genomic sequence variations [3]. Most of the researchers who analyze and study genomic data have relied on the TCGA database to acquire their input data.

The fundamental principle of molecular biology is the central dogma. It states that genetic information flows in one direction only, meaning that DNA (Deoxyribonucleic acid) produces RNA (Ribonucleic Acid), and RNA produces protein. RNA is found in all living cells and has characteristics similar to the DNA [4]. RNA sequencing is a technique that utilizes next-generation sequencing technology to analyze the sequence and quantify RNA present in a biological sample [5]. Modern sequencing tools that analyze RNA-seq data, can offer a comprehensive and detailed view of the molecular landscape of cancer. This information is instrumental in unraveling the complexities of cancer biology and advancing personalized approaches to diagnosis and treatment. It is crucial for doctors to determine the particular subtype that a patient is experiencing in order to offer targeted treatment rather than a generic treatment that may have negative effects [6]. This paper proposes a solution pipeline based on a DL based approach as they are capable of handling high-dimensional data [7]. The pipeline utilizes RNA-seq gene expression data from lung cancer patients to identify cancer subtypes through an unsupervised method. The pipeline also includes a neural classifier to classify the samples using the newly discovered subtypes.

Researchers have used a wide range of Machine Learning (ML) techniques to address various classification-related [8, 9] and clustering research problems [10]. Many researchers have used images as input data such as MRI (Magnetic Resonance Imaging) [11] and biopsy slides [12] to categorize various types of cancer. Some studies used DL based models like Convolutional Neural Networks (CNN) [12, 15, 16], Long Short-Term Memory (LSTM) [13], Autoencoders [14] for classifying and clustering applications. However, we identified a dearth of research that employs DL techniques to categorize and identify new subtypes of lung cancer using RNA sequencing data.

In this study, an unsupervised DL-based approach is employed to identify new unsupervised subtypes of lung cancer from the RNA sequencing gene expression data that is rich in biological information about the patients who suffer from lung cancer. This clustering approach aims to leverage the strengths of both ML and DL algorithms. Additionally, a DL classifier is then used to classify the newly identified subtypes.

2 Related Work

A study presented in [8] discusses the benefits of using gene expression data for cancer subtype classification and highlights the challenge of handling very high dimensional data. In that study, the lung cancer cell line data are normalized using paper average normalization and kernel-based clustering feature selection is employed to choose the most important genes. The extracted features are fed to two ML models, notably Support Vector Machine (SVM), and Random Forest (RF) with both models achieving around 90% accuracy. The SVM model produced more stable results and trained faster than RF.

In another study presented in [9], the Wisconsin Breast Cancer dataset with pre-labeled data was used to build classifier models using two classical ML algorithms, notably Naïve Bayes (NB) and K-Nearest Neighbors (KNN). The KNN produced a slightly higher accuracy of around 97.51% compared to the NB's accuracy of 96.19%. However, it was noted that the KNN's time complexity increases more than that of NB with a larger dataset. That research did not aim to identify any new subtypes of cancer, nor any deep learning approaches were used. In [10], the authors proposed a novel approach to identify potential subtypes of cancer using gene expression data. They pointed out the limitations of the popular method of calculating the number of natural clusters and proposed the use of a Silhouette Index Hierarchical Agglomerative clustering (SilHAC) technique to overcome these limitations. They built a hierarchical clustering algorithm using Silhouette index-based criteria, compared their deterministic algorithm with other non-deterministic algorithms for clustering, and found that SilHAC produced more stable results. The authors highlighted the importance of identifying novel subtypes of cancer for developing targeted drugs to enhance patient survival. However, they did not use any deep learning techniques or created any models to classify the identified subtypes. In another study [13], the authors proposed a deep-learning approach for classifying the RNA data of five subtypes of kidney cancer from the TCGA database. They used Neighborhood Component Analysis (NCA) for feature extraction and LSTM for classification. They addressed the issue of imbalanced data using data augmentation and cross-validation techniques. They achieved an accuracy of approximately 95% and a Matthews Correlation Coefficient value of 0.92 in 10 runs. However, they did not identify any novel subtypes and used the already labeled data. Furthermore in [14], deep learning algorithms were used to classify RNA samples of hematopoietic cancer into five subtypes. A DAE was employed for feature extraction and its effectiveness was compared with traditional feature extraction algorithms like Principal Component Analysis (PCA) and Non-negative Matrix Factorization (NMF). The performance of various classifiers was also evaluated, finding that SVM with DAE produced the best results. Important genes contributing to each subtype using Shapley Additive exPlanations (SHAP) were identified but no new cancer subtypes discovered.

Alternatively, rather than using gene expression data, other researchers clustered and classified the subtypes of various cancers using picture data. The study in [12] makes use of CNNs to classify subtypes of lung carcinoma using biopsy slides. They employed four pretrained transfer learning CNNs and one CNN that was trained from scratch. They evaluated the performance of each model and found that both the Xception model and the CNN trained from scratch achieved the best performance with an AUC macro average of around 0.90. They suggested using a subset of data by excluding one subtype, which improved the performance of the CNN trained from scratch with an AUC macro average of approximately 0.96. The use of MRI data to generate breast cancer subtypes that can complement the already available molecular and histopathologic subtypes was proposed in [11]. The authors of that study argued that MRI provides a global and unbiased picture of the tumor and the peripheral areas of affected tissues, and can assist in stratifying breast cancer patients. They extracted quantitative features of MRI to represent the tumor morphology and used a ML unsupervised consensus clustering to identify three breast cancer subtypes. These subtypes were found to be highly correlated with varying survival rates and reveal a unique set of underlying biological pathways. The authors suggest that their work can aid in the direction of precision medicine as assessing the image subtypes provides a better understanding of the relationship with potential biological markers. In another work [15], the researchers proposed use of unsupervised learning techniques for uncovering histological patterns that can have predictive importance without the need for time-consuming tissue labeling and human inspection of numerous slides. They compared the benefits of unsupervised learning techniques to supervised techniques and highlighted the importance of finding novel subtypes using unsupervised techniques for better understanding and segregating cholangiocarcinoma, leading to improved treatment and patient survival rates. They suggested using a novel architecture comprising a convolutional autoencoder with a pre-trained ResNet18 on the ImageNet, updating the parameters in every layer of the encoder while training the entire model with the training dataset, and creating the decoder with five convolutional layers with a stride and padding value of 1. They used up-sampling before the convolution to enhance the feature map sizes. In [16], pan-cancer RNA gene expression data from the TCGA database was used to classify 33 different types of cancer using a CNN. The dataset was transformed into 2D images, reshaping each one-dimensional record into a normalized 106×106 image. An average accuracy of around 99.3% was achieved in 200 epochs using proper dropout layers and 15-fold cross-validation to handle overfitting. Similar research was also conducted in [17].

A noticeable gap in existing literature is the scarcity of research utilizing DL methods to categorize and distinguish subtypes of lung cancer through RNA sequencing data. In our study, we use the RNA gene expression data to discover new unsupervised subtypes of lung cancer using DAEs paired with unsupervised clustering ML algorithms like K-means and H-Clust. Unlike the existing histological subtypes from TCGA that need histological analysis to differentiate different subtypes of cancer, unsupervised techniques do not need prior knowledge about the samples. The proposed pipeline further makes use of the newly discovered subtypes to train an ANN classification model to categorize patients in one of the new subtypes of lung cancer.

3 System Overview

We developed a solution pipeline that can be broadly divided into 4 areas. First step is to gather the data and perform necessary preprocessing, followed by dimensionality reduction of the data to build an informative dataset which can be used by the third part of the pipeline to cluster the data for identifying the unsupervised subtypes. The reason for using a DL-based approach instead of standard ML algorithms is due to the very high-dimension of the data [7]. Given that the dimensionality of each sample is quite huge, we first reduced the sample dimensionality using DAEs and then clustered the data with H-Clust. Finally, we utilized an ANN model to categorize the samples using these newly discovered subtypes. Figure 1 shows the overall solution pipeline.

Data collection is the starting point of the solution pipeline, as necessary RNA-seq gene expression data for lung cancer is collected from the TCGA. The data collected has to be pre-processed for filtering out irrelevant 0 value data, low variance data, etc. The produced output acts as the input for the next sub-section where the dataset dimensionality is reduced using a DAE. The informative data with the reduced dimensionality as produced by the DAE is then fed into the clustering component. We tried out a few

ML algorithms such as H-Clust and K-Means, but eventually we chose H-Clust because of its better results. The newly created unsupervised subtypes are then labelled with the original dataset to create a labelled dataset with the newly identified unsupervised subtypes, which is used to train (and classify) the last pipeline component i.e., the neural classification module. The following section of the paper goes into further detail about the solution pipeline.

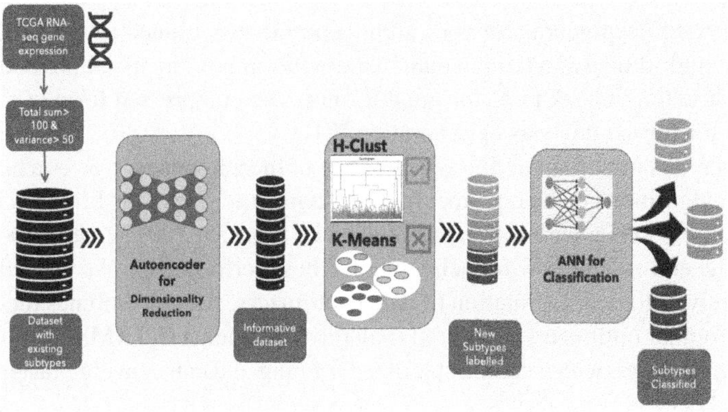

Fig. 1. Overall solution pipeline

3.1 Data Gathering and Preprocessing

We retrieved 1149 RNA-seq files from the TCGA GDC (Genomic data commons) portal for lung cancer that contains samples for four different histological subtypes of lung cancer notably LUAD (Lung Adenocarcinoma), LUSC (Lung Squamous Cell Carcinoma), LUAD-CN (Acinar Cell Neoplasms) and LUAD-SN (Serous Neoplasms). These files contain different types of strategies with which the gene expression is calculated, however, for this project, as the first step of data preprocessing, only the data marked as "tpm_unstranded" was selected. The "tpm_unstranded" is the RNA expression quantification that is normalized across samples and it makes the expression quantification of samples comparable [18]. Several other researchers have used the data marked as "tpm_unstranded" from the RNA-seq data for their ML applications [19].

3.2 Dimensionality Reduction

Every sample file has total 60661 genes (features), out of which many genes has value 0.0. As the gene expression samples have high dimensionality, it is crucial to gather only the informative data containing the important genes that can contribute to the subsequent prediction algorithms. For this, many researchers have used only those genes which have higher variance. For the purpose of our research, we used two filters. Firstly, the genes that have a sum total of the expression value less than 100 were removed. This way, the genes that have a good volume of non-zero readings were selected as they will actually contribute to the learning of the models in the subsequent steps. Also, the features that have high variance contribute more to the learning of the models when compared to features having very less variance [20]. So, as the second filter, the genes that have variance less than 50 were removed from the entire list of genes.

Even after applying the basic filters in the previous section, the gene expression samples still contain very high number of genes, and it is important to handle such high dimensionality of the data by further reducing the dimension of records. For this to occur, we decided to use DAEs which are a type of deep learning models. An autoencoder is an unsupervised deep neural network architecture that is trained to reduce the dimensionality of high-dimensional input data by transforming it into a compressed form, and then reconstructing it back to the original format. The compressed form of the data can be utilized for further analysis or generation [21].

However, it is important to find which shape of the autoencoder gives a better output for the next algorithms in the pipeline. In order to find the size of the bottleneck layer, we created multiple DAEs with different bottleneck sizes (75, 100, 125 and 200) and then we tested the end-to-end flow for achieving the best performance. We trained the DAEs using Adaptive Moment Estimation (Adam) optimizers. Adam combines the benefits of two other popular optimizers, AdaGrad (Adaptive Gradient) and RMSProp (Root Mean Square Propagation), which enable stabilized training and can converge faster than other optimizers [22].

3.3 Clustering

The compressed data i.e. the informative dataset with the reduced dimensionality that is produced by our DAE was then used by ML clustering techniques. To generate unsupervised subtypes of lung cancer, we focused on two techniques K-Means and H-Clust. After several experiments, we excluded K-means because we noticed that the final clustering solution provided by this approach could not produce clear-cut segregation as subtypes were overlapping with each other.

As already pointed out in Sect. 2, many researchers have used agglomerative clustering that works well with biological data. In order to evaluate the results of the clustering pipeline, we checked whether the unsupervised clusters (subtypes) are not overlapping with the existing subtypes that have already been classified by TCGA.

3.4 Classification

The input dataset's labels pertaining to the TCGA's pre-existing subtypes were substituted with the unsupervised subtypes generated by the clustering algorithms. Then, for classification, we used a feedforward ANN model that was trained using the gene expression data with the newly discovered unsupervised subtype labels.

Necessary preprocessing steps like scaling, and data balancing were carried out for training and classification purposed. The data was scaled with min-max scaling technique and then the number of records for each class (subtype) were balanced using Synthetic Minority Oversampling Technique (SMOTE) [23]. Once the data is balanced, we tuned the network's hyper-parameter by employing optimization techniques [24].

4 Results and Evaluation

It is essential to thoroughly evaluate the models against the research objectives because it helps in understanding how effective the models are to the problem at hand. We evaluated our system on two fronts i.e., the clustering and the classification results.

4.1 Dimensionality Reduction and Clustering

We developed several DAE models with varying bottleneck layer widths. The end-to-end solution was then evaluated using the datasets generated by each DAE; that is, we computed the ANN's F1 score using all of the DAEs' outputs (see Table 1). Based on the obtained F1 scores, we settled for a DAE with a bottleneck layer size of 200.

Table 1. F1 scores of DAEs with different bottleneck layer size

Autoencoder bottleneck layer size	F1 score
75	0.96
100	0.95
125	0.92
200	0.98

In the next step, the informative dataset with the reduced dimensionality was used to create clusters using two different unsupervised ML clustering techniques H-Clust and K-Means. Compared to K-means, H-clust has several advantages that make it well-suited for certain types of health-related applications. One of strengths of H-Clust is its ability to generate easily interpretable visualizations of the clustering outcomes. Furthermore, unlike k-means, H-Clust does not require advanced knowledge of the number of clusters, to learn complex data relationships for creating clusters [25].

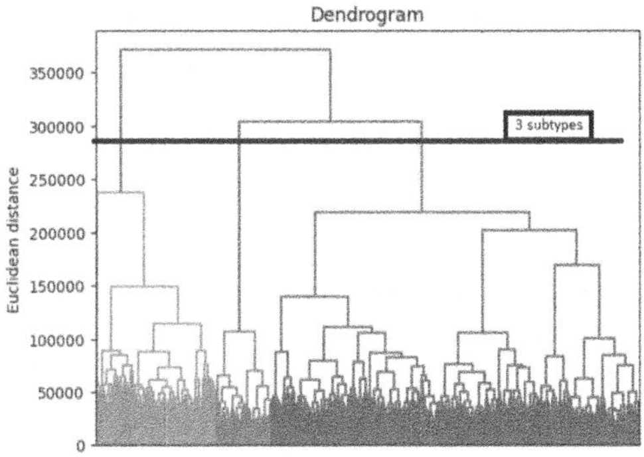

Fig. 2. Dendrogram Hierarchical Agglomerative Clustering output showing the three clusters (each cluster being in different color)

Figure 2 shows the dendrogram produced by H-Clust. It illustrates clear segregation of the samples resulting in three clusters. Whereas when the informative dataset was used to train the K-Means model, it failed to produce clear cut segregation of clusters as there were overlap among the clusters.

Several researchers have used explainable AI to establish a difference between subtypes to identify the contributing biomarkers [14]. It is a better way with which the number of groups and the uniqueness of the newly identified subtypes can be established when compared to the already existing subtypes of TCGA. For example, if different unique genes contribute dominantly to different subtypes, then all such subtypes can be considered as valid segregation of subtypes. Otherwise, if there is an overlap of such dominant genes for different subtypes then, it is beneficial to keep them together as a single subtype instead of forming multiple subtypes. However, the respective biomarkers are not studied as a part of this study after creating the dendrogram. Instead of that, we simply selected three subtypes to keep the number of samples in each subtype comparable with the TCGA subtypes.

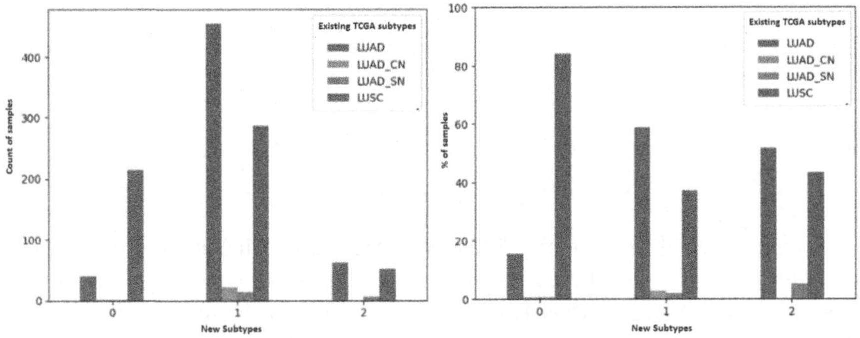

Fig. 3. Distribution of count of the newly identified subtypes against the existing subtypes from TCGA (left); Percentage of distribution of newly identified subtypes across existing subtypes (right)

Then, we carried out a simple check by comparing the newly identified unsupervised subtypes with the set of all existing subtypes in TCGA. Specifically, we verified how many samples for the three identified unsupervised subtypes of cancer are a match with that from the existing LUSC, LUAD, LUAS_SN, and LUAD_CN subtypes of cancer already provided by the TCGA. This is done to make sure that the clustering solution is producing novel subtypes and not reinventing the existing subtypes. Figure 3 (left) shows the distribution of the newly identified subtypes against the subtypes classified by TCGA. Moreover, Fig. 3 (right) also shows the percentage of existing subtypes present in each newly identified unsupervised subtype. It is clearly visible that unsupervised types 1 and 2 have less than 60% contribution from a single existing subtype. However, type 0 has more than 84% of samples matching with LUSC. Hence, two subtypes were unique as they did not match heavily with any one of the existing TCGA subtypes.

4.2 Classification of the Newly Identified Subtypes by the Clustering Solution

There are many performance metrics available to evaluate a classifier model. Choosing the appropriate metric is important to evaluate the model's performance and determine how well the model is meeting the research goals. We chose the F1 score as the evaluation metric in this study.

Both false positive and false negative cases can have a huge impact in the treatment of patients [26]. A false positive prediction can prompt needless and expensive testing whereas a false negative case can be even more fatal as it can delay the detection of the correct subtype the patient is suffering from and thereby delaying the precise treatment. F1 score takes account of both precision and recall; this way, we tried to balance these important factors. Also, accuracy is another performance metric that is commonly used however it can be misleading in use cases with imbalanced dataset [27]. For increasing the confidence of the clinical experts and get the best possible performance, we handled the class imbalance of the data and used the F1 score as the performance metric.

The results are shown by means of the confusion matrix in Fig. 4 (left). The confusion matrix shows the performance of the classification model that is classifying the subtypes of the lung cancer. It can be easily observed that the model has gone wrong in only 9 cases out of the 466 samples used to build the confusion matrix. The rows of this matrix represent the actual categories of the subtypes, while the columns represent the predicted categories. The diagonal cells of the matrix show the number of records that were correctly classified. For example, the top left cell shows that 162 records were correctly classified as belonging to subtype 0. The subtype 1 was instead misclassified 8 times out of 152 times. Whereas, as shown in bottom right corner, 151 records were correctly classified as subtype 2.

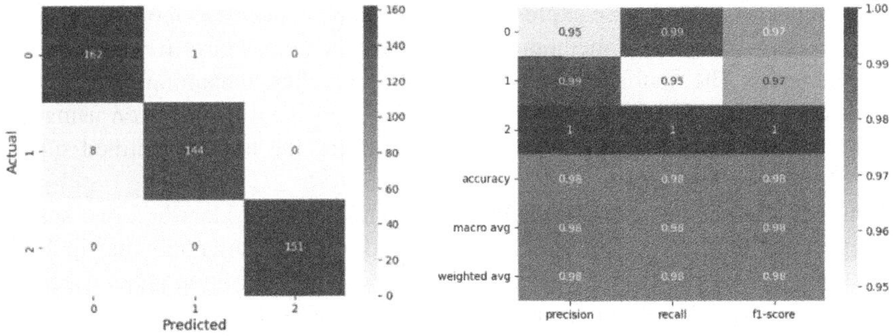

Fig. 4. Confusion matrix results (left); Classification report results of the model (right)

The ANN produced a very high performance of 0.98 F1 weighted average score as shown in Fig. 4 (right). This result outperforms the results of another similar study reported in Sect. 2 [14], where researchers in the area of hematopoietic cancer where they used an autoencoder to reduce the dimensionality and used SVM to classify the samples achieving a 0.93 F1 measure.

The model rarely made mistakes for the types 1 and 2 with a rather high precision score. Even for type 0, 162 out of 170 predicted positives were actually positives and hence have 0.95 precision. This results in an overall high precision of 0.98. The model also performed very well at finding actual instances of each type, as indicated by its high recall score for types 0 and 2. By contrast, there were a few missed cases for type 1, with 8 instances going undetected which can be fatal for the patients.

Evaluating a model performance comprehensively requires metrics that consider both precision and recall. The F1 score, which is the harmonic mean of these two metrics, provides a more holistic perspective. In this study, the F1 scores for all three classes were exceedingly 0.97 for each class and overall 0.98 F1 score. This high performance across all classes demonstrates the model's efficiency in correctly identifying instances of each type. Furthermore, the macro and weighted F1 scores exhibited minimal disparity, indicating that the model did not exhibit significant bias towards any particular class. This balanced performance across all classes further reinforces the model's generalizability and reliability.

5 Conclusions

We are living in an age where rich biological data like RNA gene expression is readily available in platforms like TCGA. It is possible because of the advanced sequencing tools that can find details of the genetic makeup of a patient in such detail. At the same time, we also have a variety of state-of-the-art DL algorithms that can analyze such biologically rich and large datasets. In this research, we proposed a combination of deep autoencoders and ML hierarchical clustering using the RNA-seq gene expression data to find new unsupervised subtypes of lung cancer. Further, we checked whether any novel lung cancer subtypes were identified or not.

This study shows how the power of DL techniques along with the gene expression RNA data can be used to identify the unsupervised subtypes of lung cancer. As the dataset used to collect the RNA gene expression data for lung cancer was the TCGA dataset, we followed a quantitative methodology by experimenting with different algorithms and parameters. The final solution is a processing pipeline containing of these major components: data collection and preprocessing, dimensionality reduction using a DL based approach, clustering and then finally classifying the newly identified subtypes using another DL algorithm.

Out of the three unsupervised subtypes we clustered, we identified two subtypes of lung cancer that were different when compared to the existing subtypes in TCGA. The results of this study potentially open up a new way of profiling patients based on the new subtypes and extend research on enabling a data-driven approach to disease understanding, ultimately providing personalized treatment to patients. Moreover, the ANN classifier in this solution can help to identify which subtype of cancer the patient is suffering from. It performed with a high performance of 0.98 F1 score when the model was tested with unseen test data.

It is also crucial to acknowledge a few challenges and shortcomings of the study, as these provide opportunities for further investigation into this area. The TCGA dataset, comprised of total 1149 samples at the time when this study was done. Also, it lacks sufficient data for rare subtypes like LUAD_CN and LUAD_SN, underscoring the need for expanded datasets to deepen our understanding of these rare subtypes as well. Increasing data volume improves understanding and boosts clinical experts' confidence. DL algorithms also require large datasets for effective training, emphasizing the need for more data to ensure reliable model outcomes [7].

Further, a simple check was used to calculate the percentage of samples from existing subtypes that are present for each unsupervised subtype to consider them as new or unique subtypes when compared to existing subtypes present in TCGA. However, future advancement can explore a better approach of finding the dominant causal genes for each subtype to demonstrate the uniqueness of the newly identified subtypes with respect to the existing subtypes. Some researchers have used explainable AI techniques like SHAP to achieve this [14].

This research can be viewed as an effort made to contribute to the ongoing fight against cancer, specifically lung cancer. This research solution prototype can be extended to other cancer types as well to gain insights into them, and also paving the way for precision medicine tailored to address individual subtypes. Also, integrating the environmental and demographic data of the patients along with the gene expression data can enhance the solution's scope and applicability. Further optimization potential lies in exploring varied parameters, applying different loss functions for autoencoders, and alternative neural network architectures to further refine the performance. Ultimately, this work underscores AI's pivotal role in unsupervised cancer subtype identification. Since it is a data-driven approach, hence, sufficient confidence-building and a willingness to engage with the data-driven findings by the clinical experts will be necessary for its deployment in real-life applications like patient profiling and future work on producing medicines based on the causal genes of the subtypes.

References

1. Roser, M., Ritchie, H.: Cancer (2019). [Online]. At: https://ourworldindata.org/cancer
2. Chen, J.W., Dhahbi, J.: Lung adenocarcinoma and lung squamous cell carcinoma cancer classification, biomarker identification, and gene expression analysis using overlapping feature selection methods. Nat. Sci. Rep. **13323**(11), 1–15 (2021)
3. The Cancer Genome Atlas Program (TCGA). National Cancer Institute (2023). [Online]. At: https://www.cancer.gov/about-nci/organization/ccg/research/structural-genomics/tcga
4. Ostrander, E.A.: CENTRAL DOGMA. National Human Genome Research Institute (2023). [Online]. At: https://www.genome.gov/genetics-glossary/Central-Dogma
5. Kukurba, K.R., Montgomery, S.B.: RNA Sequencing and analysis. NCBI Cold Spring Harbor Protocols **2015**(11), 951–969 (2015)
6. Krzyszczyk, P., et al.: The growing role of precision and personalized medicine for cancer treatment. Technology. Pubmed Technol. Singap. World Sci. **6**(3), 79–100 (2019)
7. Sarker, I.H.: Deep learning: a comprehensive overview on techniques, taxonomy, applications and research directions. Pubmed SN Comput. Sci. **2**(420), 1–20 (2021)
8. Liu, Y., Wang, X.-D., Qiu, M., Zhao, H.: Machine learning for cancer subtype prediction with FSA method. In: Proceedings of the 4th International Conference in Smart Computing and Communication, pp. 387–397 (2019)
9. Amrane, M., Oukid, S., Gagaoua, I., Ensarİ, T.: Breast cancer classification using machine learning. IEEE Electr. Electron. Comput. Sci. Biomed. Eng. Meet. **2018**, 1–4 (2018)
10. Nidheesh, N., Nazeer, K.A.A., Ameer, P.M.: A Hierarchical clustering algorithm based on Silhouette Indexfor cancer subtype discovery from genomic data. Neural Comput. Appl. **32**, 11459–11476 (2020)

11. Wu, J., et al.: Unsupervised clustering of quantitative image phenotypes reveals breast cancer subtypes with distinct prognoses and molecular pathways. Pubmed Clin. Cancer Res. **23**(13), 3334–3342 (2018)
12. WookYang, J., Song, D.H., JungAn, H., Seo, S.B.: Classification of subtypes including LCNEC in lung cancer biopsy slides using convolutional neural network from scratch. Nature **12**(1830), 1–11 (2022)
13. Ali, A.M., et al.: A machine learning approach for the classification of kidney cancer subtypes using miRNA genome data. MDPI Appl. Sci. **8**(12), 1–14 (2018)
14. Park, K.H., Batbaatar, E., Piao, Y., Theera-Umpon, N., Ryu, K.H.: Deep learning feature extraction approach for hematopoietic cancer subtype classification. Int. J. Environ. Res. Public Health **18**(4), 1–24 (2021)
15. Muhammad, H., et al.: Unsupervised subtyping of cholangiocarcinoma using a deep clustering convolutional autoencoder. Med. Image Comput. Comput. Assist. Interv. **11764**, 604–612 (2019)
16. Kenallaki, M.: Convolutional Neural Networks and their Application in Cancer Diagnosis based on RNA-Sequencing (2022). [Online]. At: https://pergamos.lib.uoa.gr/uoa/dl/object/2976042/file.pdf
17. Rukhsar, L., et al.: Analyzing RNA-Seq gene expression data using deep learning approaches for cancer classification. MDPI Appl. Sci. **12**(4), 1–17 (2022)
18. Zhao, Y., et al.: TPM, FPKM, or normalized Counts? A comparative study of quantification measures for the analysis of RNA-seq data from the nci patient-derived models repository. J. Transl. Med. **19**(269), 1–41 (2021)
19. Pomaznoy, M., Sethi, A., Greenbaum, J., Peters, B.: Identifying inaccuracies in gene expression estimates from unstranded RNA-seq data. Sci. Rep. **9**(16342), 1–10 (2019)
20. Sadeghyan, S.: A New Robust Feature Selection Method Using Variance-Based Sensitivity Analysis (2018). [Online]. At: https://arxiv.org/pdf/1804.05092.pdf
21. Hu, C., Hou, X., Lu, Y.: Improving the architecture of an autoencoder for dimension reduction. In: Proceedings of 11th IEEE International Conference on Ubiquitous Intelligence & Computing, pp. 855–858 (2014)
22. Dogo, E.M., Afolabi, O.J., Twala, B.: On the relative impact of optimizers on convolutional neural networks with varying depth and width for image classification. Appl. Sci. **12**(11976), 1–36 (2022)
23. Bowyer, K.W., et al.: {SMOTE:} Synthetic Minority Over-sampling Technique. CORR arXiv. [Online]. At: http://arxiv.org/abs/1106.1813
24. KerasTuner, [Online]. Available: https://keras.io/keras_tuner/. Accessed 21 June 2024
25. Hexmoor, H., Chapter 6 - diffusion and contagion. In: Computational Network Science, pp. 45–64. Morgan Kaufmann (2014)
26. Jackson, B.R.: The dangers of false-positive and false-negative test results: false-positive results as a function of pretest probability. Pubmed Clin. Lab. Med. **28**(2), 1–19 (2008)
27. Kulkarni, A., Batarseh, F.A., Chong, D.: Chapter 5: foundations of data imbalance and solutions for a data democracy. In: Nexus of Artificial Intelligence, Software Development, and Knowledge Engineering (2021)

Loss Function Role in Processing Sequences with Heavy-Tailed Distributions

Mikhail Gritskikh[1], Artem Isakov[1,2], Natalia Gusarova[1], Dmitriy Dobrenko[1], Ivan Tomilov[1], and Aleksandra Vatian[1](✉)

[1] ITMO University, Saint-Petersburg 197101, Russian Federation
isakovao@itmo.ru, alexvatyan@gmail.com
[2] World-Class Research Center for Personalized Medicine, Almazov National Medical Research Centre, Saint-Petersburg 194156, Russia
https://en.itmo.ru/

Abstract. Predicting the behavior of sequences in presence of long-tailed distributions is an important task in many applied industries, especially in medicine. Transformers and LSTM networks are most often used as handlers for this purpose, but the main efforts are aimed at suppressing deviations from normality rather than at their informative use. The paper discusses an alternative approach - the possibility of solving an auto-regression problem under conditions of input distributions with heavy tails while preserving information about significant deviations. Namely, we theoretically and experimentally consider the impact of the choice of loss function for processing sequences with heavy-tailed distributions performed by transformers and LSTMs. As a loss function, we analyze the Mean Squared Error (MSE), Mean Absolute Error (L1 Loss), and Cauchy Loss functions. On synthetic and real (COVID-19) data we showed that Cauchy and MAE losses perform 40% better than MSE loss according to sMAPE metric.

1 Introduction

Predicting the behavior of sequences, including time-series data, is an important task in many applied industries, such as finance, education, medicine, etc. The choice of IT tools for solving this problem is largely determined by the statistical characteristics of the processed sequence, which, as practice shows, do not always obey the usual and convenient for processing Gaussian distribution. Works of recent years confirm that heavy-tails and chaotic behavior naturally appear in many ways in ML [1, 2]. As a tool for processing sequences of various types, including time-series data with complex statistical characteristics, sequence models, such as LSTM networks [3], are widely used in IT. In recent years, Transformers have become the de facto standard for this purpose [4]. Transformers-based models have emerged as powerful tools for various applications, particularly in handling time-series data, including for non-stationary time-series characterized by statistical properties, that may change over time (for example, [5]). At the same time, in existing works, the main efforts are aimed at suppressing deviations from normality rather than at their informative use.

As for the processing of medical data, here until recent years, the classical approach [6] dominated, in which the distribution of the data under study is considered as known

V. Julian et al. (Eds.): IDEAL 2024, LNCS 15346, pp. 361–374, 2025.
https://doi.org/10.1007/978-3-031-77731-8_33

(primarily normal), and large deviations from it are considered as outliers that should be removed from the sample. However, the challenges of recent years, such as the COVID-19 pandemic and the like, the strong diversification of patients according to the set of diseases they have, the fight against orphan diseases, etc., have inspired another approach to processing medical data, in which the distribution function is a priori specified only at the level of a family of distributions, all elements of the sample are considered as informative, and the task is to search for local dependencies, primarily predicting the behavior of the sequence over the next few steps. Thus, the information necessary to make an ad-hoc decision for a specific patient can be obtained based on his previous medical data. Transformers, as effective sequence models, seem to be a good ML solution for such cases. However, for them, the problem of prediction under conditions of input distributions with heavy tails while preserving information about significant deviations has practically not been considered.

This work is aimed at filling this gap. Namely, we consider the impact of the choice of loss function for processing sequences with heavy-tailed distributions performed by transformers and LSTMs. In brief, our contributions are as follows:

1. We conducted a theoretical assessment of the gradients of various loss functions during the transition from the normal distribution of the input sequence to the heavy-tailed distribution. To model a heavy-tailed input distribution, we use the Lévy alpha-stable distribution. As a loss function, we analyze the Mean Squared Error (MSE), Mean Absolute Error (L1 Loss), and Cauchy Loss functions. We have shown that self-attention mechanism can struggle with heavy-tailed distributions due to its propensity to overemphasize large outliers, resulting in less robust predictions, however, this problem is somewhat mitigated by the use of multihead attention. We have shown that when working in regression mode, the gradient error function for Cauchy decreases, for MAE it remains stable, and for MSE it increases, which affects the stability of gradient descent when training the network.
2. Using synthetic and real data, we experimentally studied the behavior of the prediction quality for various combinations of the distribution function of the input sequences and the transformer error function. The results obtained confirmed the validity of the theoretical estimates. Namely, we showed Cauchy and MAE losses per-form 40% better than MSE loss according to sMAPE metric on test set.

2 Related Works

Sequence Models in Processing Heavy-Tailed Distributions. The general theoretical foundations of the operation of the transformer and the self-attention mechanism are considered in the works of [7–9], but the influence of the input distribution is not affected in them. The flexibility of transformers with respect to input distribution is experimentally evaluated in [10] where they are shown to be good meta-output-predictors when adapting to unseen settings, non-i.i.d. noise, and time varying dynamics. In [3] ConvL-STM neural network is created to be appropriate for non-stationary time series data with non-Gaussian behavior.

LSTMs and transformers are used in modeling heavy-tailed sequences [3, 5, 11]. For example, in [11] transformers and LSTMs are used as tools for data-driven modeling of deviations in the generation frequency of the power supply system, which arises due to various sociogenic factors. The resulting sequences are processed by a dense network to predict the mean vectors and the covariance matrix. The negative log likelihood functions acts as the loss functions, and Gauss, Student-t and the Cauchy distribution are considered as target distribution functions. It has been experimentally shown that the best approximation to real data is achieved in the latter case (Cauchy distribution).

When processing heavy-tailed sequences, classification problems are considered primarily [12–17], but autoregression problems are practically not considered. Since processing heavy-tailed input leads to the complexity and slowdown of calculations, the main problem is the fight against deviations from normality. For this purpose, various technical solutions are used, including changes in the self-attention module, the multi-layer perceptron (MLP) module, and the entire transformer-based model [13, 14], as well as replacing the SGD optimization method with Adam [15, 16]. To combat the temporal heterogeneity of sequences, a specialized transformer model is proposed [17], which embeds the burstiness and memory coefficient into the self-attention module, enhancing the learning process with insights derived from the bursty patterns.

Sequence Models in Processing Heavy-Tailed Distributions in Medical Domain. The fact that parametric distributions to health care data often exhibit heavy tails is noted in a number of works [18–20]. In particular, it is shown that it is advisable to model the characteristics of the geographical empirical distributions for the numbers of both cumulative and daily confirmed COVID-19 cases and deaths at county, city, and state levels by heavy-tailed distribution [21]. Nevertheless, many works are of a purely ascertaining nature [22, 23], or fulfill fitting parameters of the sample to a given distribution [19, 21].

The explosion in popularity of transformers was also reflected in the domain of medical applications, as evidenced by reviews [24–26]. However, a detailed analysis shows that most of the works presented in them exploit traditional transformer classification tasks for processing language sequences, such as Clinical Relationship Extraction etc., adapted for the specifics of the domain (see, for example, [27, 28]). To this end, the vast majority of works use BEHRT with a quadratic error function as the basic transformer architecture [29]. As an example of more specific tasks, we can note the classification of abnormalities using the electroencephalogram [30], as well as the presence of stress and arrhythmia using the electrocardiogram [31].

However, there are works where LSTM is used as a sequential model. For example, in [32] the problem of cancer classification in conditions of incomplete data is solved, in [33]—the problem of multi-class classification of cardiovascular diseases.

In general, the analysis indicates a strong bias towards classification problems as opposed to regression problems. For example, the review [26] presents 21 works (for 2020–2022) related to the use of transformers in medical applications, while all of them solve the classification problem, and only in one [34] simultaneously solves the regression problem. The work [34] is aimed at predicting pregnancy complications. Dynamic changes in parameters (for example, the natural weight gain of a patient) are modeled here by a shift in the average value of the parameter distribution function, but not by a transition to a long-tailed distribution. In this work, a transformer with a conventional quadratic error function was used.

Thus, the analysis showed that, despite the obvious demand for an operational solution of regression problems for sequences with heavy tails, the problem considered in this work is not presented in the literature available to the authors, which once again confirms its relevance.

3 Methods and Materials

3.1 Theoretical Evaluation of Attention Mechanism for Heavy-Tailed Distributions

The self-attention mechanism, central to transformer models, computes the attention scores as follows:

$$\text{Attention}(Q, K, V) = \text{softmax}\left(\frac{QK^T}{\sqrt{d_k}}\right)V$$

where Q, K, and V are the query, key, and value matrices, and d_k is the dimensionality of the key vectors.

The softmax function is defined as:

$$\text{softmax}(z_i) = \frac{\exp(z_i)}{\sum_j \exp(z_j)}$$

For the scenario of heavy-tailed distributions, the values in QK^T can vary significantly due to the presence of large outliers. Let's denote a large outlier by e:

When $z_i = e$ and $z_j \approx 0$ for $j \neq i$, the softmax function outputs:

$$\text{softmax}(e) \approx \frac{\exp(e)}{\exp(e) + \sum_{j \neq i} \exp(0)} \approx \frac{\exp(e)}{\exp(e) + (n-1)} \approx 1$$

This results in the attention mechanism focusing almost entirely on the large outlier, ignoring other relevant information.

When multiple values in QK^T are large due to outliers, say $e_1, e_2, \ldots, e_k$ and others are small:

$$\text{softmax}(z_i) \approx \frac{\exp(z_i)}{\sum_{j=1}^{k} \exp(e_j) + \sum_{j=k+1}^{n} \exp(0)} \approx \frac{\exp(z_i)}{\sum_{j=1}^{k} \exp(e_j) + (n-k)}$$

In this case, the distribution of attention scores is heavily skewed towards the outliers, leading to unstable and unreliable predictions.

Given the above scenarios, the self-attention mechanism can struggle with heavy-tailed distributions due to its propensity to overemphasize large outliers, resulting in less robust predictions.

There is an approach that can partially mitigate the impact of heavy-tailed distributions in transformer models. Multi-head attention allows the model to attend to information from different representation subspaces at different positions. This mechanism helps to stabilize the model and makes it more robust to extreme values by distributing the attention process across multiple heads.

By projecting the input into multiple subspaces, each head processes a different representation of the input. This redundancy allows the model to mitigate the effect of extreme values that might dominate in a single representation space. If an extreme value negatively impacts one head, other heads can still provide stable outputs.

Each head in the multi-head attention mechanism has its own set of learned linear projections for queries Q, keys K, and values V:

$$Q_i = XW_i^Q, \quad K_i = XW_i^K, \quad V_i = XW_i^V$$

where W_i^Q, W_i^K and W_i^V are the projection matrices for the i-th head, and X is the input sequence matrix.

Different heads can focus on different parts of the sequence, capturing various aspects of the data. This diversity in focus ensures that the model does not overly rely on any single extreme value, as other heads can capture relevant context.

$$\text{Attention}(Q_i, K_i, V_i) = \text{softmax}\left(\frac{Q_i K_i^T}{\sqrt{d_k}}\right) V_i$$

The outputs of the individual attention heads are concatenated and projected to form the final output:

$$\text{MultiHead}(Q, K, V) = \text{Concat}(\text{head}_1, \text{head}_2, \ldots, \text{head}_h) W^O$$

where $\text{head}_i = \text{Attention}(Q_i, K_i, V_i)$ and W^O is the output projection matrix.

The concatenation and subsequent linear transformation effectively average the contributions of all heads. This averaging reduces the sensitivity to extreme values in any single head, as the final output is a combined representation.

$$\text{Output} = \frac{1}{h}\sum\nolimits_{i=1}^{h} W^O \text{head}_i$$

Even though multi-head attention can mitigate the impact of extreme values, the attention mechanism in transformers remains vulnerable by outliers.

3.2 Theoretical Evaluation of Regression Loss Functions

In this section, we evaluate different regression loss functions, particularly focusing on their behavior under large errors. Specifically, we analyze the Mean Squared Error (MSE), Mean Absolute Error (L1 Loss), and Cauchy Loss functions. Our theoretical evaluation emphasizes the suitability of the Cauchy Loss in handling large errors due to its heavy-tailed nature.

$$f(y|\hat{y}, \gamma) = \frac{1}{\pi\gamma\left[1 + \left(\frac{y-\hat{y}}{\gamma}\right)^2\right]}$$

Given a set of observations $\{y_i\}$ and a corresponding set of predicted values $\{\widehat{y_i}\}$, the negative log-likelihood for the Cauchy distribution is:

$$-\log L(y|\hat{y},\gamma) = -\sum\nolimits_{i=1}^{n} \log\left(\frac{1}{\pi\gamma\left[1+\left(\frac{y_i-\widehat{y_i}}{\gamma}\right)^2\right]}\right)$$

Simplifying this, we get:

$$-\log L(y|\hat{y},\gamma) = n\log(\pi\gamma) + \sum_{i=1}^{n} \log\left(1+\left(\frac{y_i-\widehat{y_i}}{\gamma}\right)^2\right)$$

Removing the constant terms (which do not affect the minimization process), the Cauchy loss can be represented as:

$$L_{Cauchy}(y,\hat{y}) = \sum\nolimits_{i=1}^{n} \log\left(1+\left(\frac{y_i-\widehat{y_i}}{\gamma}\right)^2\right)$$

Theoretical comparison of gradients of the loss functions also shows that in case of large errors Cauchy loss MAE are more suitable.

Given the following definition for MSE:

$$\mathrm{MSE} = \frac{1}{n}\sum\nolimits_{i=1}^{n} (y_i-\widehat{y_i})^2$$

The gradient for this loss:

$$\frac{\partial \mathrm{MSE}}{\partial \widehat{y_i}} = -\frac{2}{n}(y_i-\widehat{y_i})$$

Given the following definition for MAE:

$$\mathrm{L1\ Loss} = \frac{1}{n}\sum\nolimits_{i=1}^{n} |y_i-\widehat{y_i}|$$

The gradient for this loss:

$$\frac{\partial \mathrm{L1\ Loss}}{\partial \widehat{y_i}} = -\frac{1}{n}\mathrm{sign}(y_i-\widehat{y_i})$$

Given the following definition for Cauchy Loss:

$$\mathrm{Cauchy\ Loss} = \frac{1}{2}\log\left(1+\left(\frac{y_i-\widehat{y_i}}{\gamma}\right)^2\right)$$

The gradient for this loss:

$$\frac{\partial \mathrm{Cauchy\ Loss}}{\partial \widehat{y_i}} = -\frac{(y_i-\widehat{y_i})/\gamma^2}{1+\left(\frac{y_i-\widehat{y_i}}{\gamma}\right)^2}$$

Comparing MSE and Cauchy Loss let's consider a large error e from a heavy-tailed distribution:

- MSE: $\frac{\partial \text{MSE}}{\partial \hat{y}} \approx -2e$
- Cauchy Loss: $\frac{\partial \text{Cauchy Loss}}{\partial \hat{y}} \approx -\frac{e/\gamma^2}{(e/\gamma)^2} \approx -\frac{1}{e}$

For large , $\frac{1}{e} \ll e$, showing that the gradient of the Cauchy Loss decreases for large errors, while the gradient of MSE increases.

Comparing MAE and Cauchy Loss let's consider a large error e from a heavy-tailed distribution:

- MAE: $\frac{\partial \text{L1 Loss}}{\partial \hat{y}} = \text{sign(e)}$
- Cauchy Loss: $\frac{\partial \text{Cauchy Loss}}{\partial \hat{y}} \approx -\frac{e/\gamma^2}{(e/\gamma)^2} \approx -\frac{1}{e}$

For large , $\frac{1}{e} \ll 1$, showing that the gradient of the Cauchy Loss is even smaller than the constant gradient of MAE for large errors. Given that self-attention mechanism can focus more on extreme values, MAE and Cauchy loss functions seem to be the best fit for transformer model optimization in presence of heavy-tailed distributions.

3.3 Metrics

To assess the performance of the regression models and their effectiveness in predicting heavy-tailed time-series data, we employ the Symmetric Mean Absolute Percentage Error (sMAPE) as our primary evaluation metric. sMAPE is particularly suitable for time-series forecasting and regression tasks due to its scale-independent nature and balanced treatment of over- and under-forecasting errors.

The Symmetric Mean Absolute Percentage Error (sMAPE) is defined as:

$$\text{sMAPE} = \frac{100\backslash\%}{n} \sum\nolimits_{i=1}^{n} \frac{|y_i - \hat{y}_i|}{(|y_i| + |\hat{y}_i|)/2}$$

where y_i is the actual value, $\hat{y}_i$ is the predicted value, and n is the number of observations.

Taleb's Kappa metric is used to assess the degree of extremity in a distribution, particularly useful for identifying heavy-tailed behavior. The Kappa metric is defined as the ratio of the maximum to the mean of the data:

$$\kappa = \frac{\max(|y_i - \hat{y}_i|)}{\mu(|y_i - \hat{y}_i|)}$$

where $\max(|y_i - \hat{y}_i|)$ is the maximum absolute error and $\mu(|y_i - \hat{y}_i|)$ is the mean absolute error. A higher Kappa value indicates the presence of extreme values, characteristic of heavy-tailed distributions.

The Dispersion Scaling Plot is a graphical tool used to visualize the dispersion of errors in relation to the scale of the actual values. This plot helps in understanding whether the error dispersion remains consistent across different scales of data, which is particularly important for heavy-tailed distributions.

By employing sMAPE as our primary evaluation metric, we aim to achieve a comprehensive and balanced understanding of the models' predictive capabilities, particularly in the context of heavy-tailed time-series data. Taleb's Kappa metric and Dispersion scaling plot provide confirmation for heavy-tailed nature of our dataset.

3.4 Datasets

In this research we use both synthetic and real datasets to evaluate performance of models with different loss functions. The synthetic dataset is obtained from heavy tailed distribution with applied kernels.

The Lévy alpha-stable distribution, a generalization of the Gaussian distribution, is characterized by four parameters: α (stability parameter), β (skewness parameter), σ (scale parameter), and μ (location parameter). Its probability density function does not have a closed form, but it is defined via its characteristic function.

For $\alpha \neq 1$:

$$\phi(t) = \exp\left(i\mu t - |\sigma t|^{\alpha}\left(1 + i\beta\frac{t}{|t|}\tan\left(\frac{\pi\alpha}{2}\right)\right)\right)$$

For $\alpha = 1$:

$$\phi(t) = \exp\left(i\mu t - |\sigma t|\left(1 + i\beta\frac{2}{\pi}\ln|t|\right)\right)$$

To generate heavy-tailed time-series, we define the following parameters: L - length of the series; Initial alpha (α_{i}) - stability parameter at the beginning of the series; Final alpha (α_{fin}) - stability parameter at the end of the series; Initial fraction (fr_i) - fraction of the series with initial alpha; Transition fraction (fr_t) - fraction of the series transitioning from α_i to α_f. The sequence is divided into three parts: the initial period, transition period, and final period, with lengths:

$$L_i = \lfloor L \cdot fr_i \rfloor; \ \lfloor L_t = L \cdot fr_t \rfloor; \ L_f = L - L_i - L_t$$

The alpha values for these periods are:

$$\alpha_{\mathrm{initial}} = \alpha_i; \ \alpha_{\mathrm{transition}}(t) = \alpha_i + \frac{t}{L_t}(\alpha_{\mathrm{fin}} - \alpha_i); \ \alpha_{\mathrm{final}} = \alpha_{\mathrm{fin}}$$

The overall sequence is generated by concatenating these segments with their respective α values.

The smoothing is achieved through various kernels: Gaussian, Cosine, and a combined Cosine-Gaussian kernel. The Gaussian kernel G is defined as:

$$G(i, \sigma) = \frac{1}{\sqrt{2\psi^2}}\exp\left(-\frac{i^2}{2\sigma^2}\right)$$

where i and σ is the standard deviation. The kernel is then normalized:

$$G_{\mathrm{normalized}} = \frac{G}{\sum_{i=1}^{n} G(i, \sigma)}$$

The Cosine kernel C is defined as:

$$C(i, T) = \frac{1}{2}\left[1 + \cos\left(\frac{2\pi i}{T}\right)\right]$$

where i is the index, and T is the period. The kernel is normalized:

$$C_{\text{normalized}} = \frac{C}{\sum_{i=1}^{n} C(i, T)}$$

The combined Cosine-Gaussian kernel K is obtained by element-wise multiplication of the Gaussian and Cosine kernels:

$$K(i) = C(i, T) \cdot G(i, \sigma)$$

Normalization is applied to the combined kernel:

$$K_{\text{normalized}} = \frac{K}{\sum_{i=1}^{n} K(i)}$$

Combined Cosine-Gaussian Kernel merges the properties of both Gaussian and Cosine kernels, providing periodic smoothing with attenuation based on standard deviation. The smoothing of a time-series x using a kernel K is performed via convolution:

$$y(t) = \sum_{i=1}^{n} x(t - i)K(i)$$

where $y(t)$ is the smoothed time-series.

The researched approach was tested on a real-world dataset collected by the Our World in Data (OWID) organization, which tracks COVID-19 new case dynamics across different countries. The dataset provides a comprehensive collection of daily reported new cases of COVID-19 across multiple regions. This allows for a robust evaluation in predicting real-world, heavy-tailed time-series data, which is characterized by abrupt changes and prolonged periods of stability, like those generated by the Lévy stable distribution. The examples of generated and real-world time series illustrated on Fig. 1.

The data employed for both synthetic and real-world experiments have similar structure. These datasets cover the research needs to estimate different loss function on heavy tailed distributions.

3.5 Experiment Design

We conducted a series of experiments by generating datasets with three distinct stability periods and combining them with three different final stability parameters (α). The stability periods are defined as follows:

1. Short Stability Period - initial frac = (0.05, 0.1); transition frac = (0.05, 0.1)
2. Moderate Stability Period - initial frac = (0.3, 0.4); transition frac = (0.05, 0.1)
3. Long Stability Period - initial frac = (0.65, 0.75); transition frac = (0.05, 0.1)

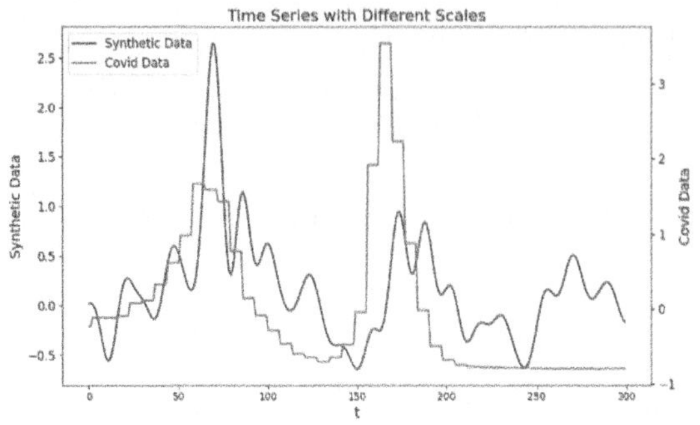

Fig. 1. Example of generated time series from heavy-tailed distribution.

These stability periods were combined with the following final stability parameters:

$$\alpha = 2; 1.9; 1.5$$

For each combination of stability period and final α, we applied two types of smoothing techniques - Gaussian Smoothing and Combined Gaussian-Cosine Smoothing.

Both LSTM and Transformer models were trained using the following loss functions on real-world data: Mean Squared Error (MSE), Mean Absolute Error (MAE) and Cauchy Loss.

The experiments were structured to analyze the performance under varying durations of stability periods and different final stability parameters (α), with two types of smoothing applied: Gaussian and combined Gaussian-Cosine. Additionally, both LSTM and Transformer models were trained using three different loss functions: MSE, MAE, and Cauchy Loss on real-world data.

4 Results and Discussion

4.1 Input Data Distribution Analysis

Kappa metric and Dispersion Scaling plot were used to illustrate heavy-tailed nature of synthetic and COVID-19 datasets. Figure 2 shows that both synthetic and real-world datasets have heavy-tailed nature.

4.2 Impact of Loss Function on Prediction Quality

The results of study of loss function impact on transformers and LSTM prediction quality are presented in Table 1. The table analysis shows that the Cauchy Loss and MAE loss functions perform similarly, at the same time these functions consistently outperform MSE loss function.

In Table 2 we present the results of experiment for real-world data. We see that transformer model performs better than LSTM in 1-step evaluation but underperforms in multi-step inference. Additionally, we see that Cauchy loss consistently outperforms other loss functions for transformer but underperforms for LSTM which indicates lower impact of heavy-tailed data on LSTM prediction.

The experimental results can be summarized as follows:

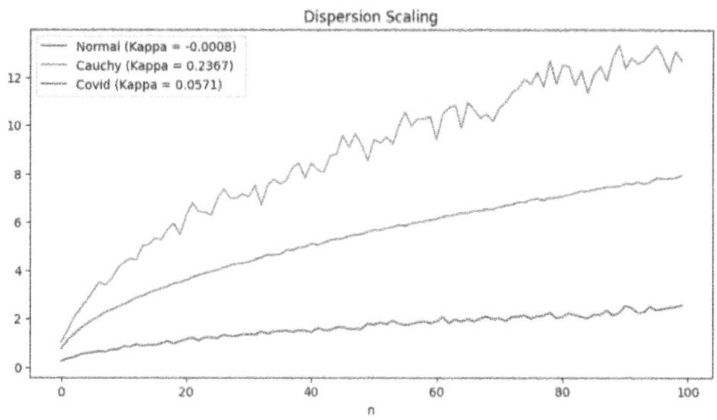

Fig. 2. Taleb's Kappa and Dispersion Scaling for analyzed dataset.

Table 1. Comparison of Transformer and LSTM model with different loss types on synthetic dataset.

Model type	Loss type	Eval sMAPE	Infer sMAPE
LSTM	Cauchy	127	127
LSTM	MAE	124	124
LSTM	MSE	143	146
Transformer	Cauchy	22	145
Transformer	MAE	21	152
Transformer	MSE	45	149

Table 2. Comparison of Transformer and LSTM model with different loss types on COVID-19 dataset.

Model type	Loss type	Eval sMAPE	Infer sMAPE
LSTM	Cauchy	117	117
LSTM	MAE	118	121
LSTM	MSE	88	92
Transformer	Cauchy	24	114
Transformer	MAE	36	134
Transformer	MSE	27	98

1. Transformer model performs well with 1-step prediction. However, in inference it shows significant deviation from the target time series comparing to LSTM.
2. The L1 loss function performs comparably to the Cauchy Loss, suggesting its potential as a viable alternative in some scenarios.

3. The MSE loss function underperforms compared to both Cauchy and L1 loss functions, indicating its limitations in handling heavy-tailed time-series data.

5 Conclusion

This research evaluates transformer models for predicting heavy-tailed time-series data and compares them to LSTM models. Our findings highlight that transformer models are vulnerable to outliers, but using MAE and Cauchy Loss effectively mitigates the impact of extreme values. These loss functions enhance robustness and ensure more accurate predictions.

Transformer models consistently outperform LSTM in 1-step prediction. This highlights the transformer's superior ability to capture long-range dependencies and complex data patterns.

We recommend using transformer models with MAE or Cauchy Loss in presence of heavy-tailed distributions, as they offer better resilience to extreme values. In conclusion, this combination effectively addresses the vulnerability to outliers and leverages the strengths of transformer models, making them a superior choice for predicting heavy-tailed time-series data.

Acknowledgments. This work was supported by Russian Science Foundation, Grant 23-11-00346.

Disclosure of Interests. Authors declare no competing interests.

References

1. Gurbuzbalaban M., et al.: Heavy Tails in ML: Structure, Stability, Dynamics. https://neurips.cc/virtual/2023/workshop/66530
2. Song, W., Chen, D., Zio, E.: Heavy tail and long-range dependence for skewed time series prediction based on a fractional Weibull process. Fractal Fract. **8**(1), 7 (2024). https://doi.org/10.3390/fractalfract8010007
3. Szarek, D., et al.: Non-Gaussian feature distribution forecasting based on ConvLSTM neural network and its application to robust machine condition prognosis. ACM Digit. Libr. **230**(1), 120588 (2023). https://doi.org/10.1016/j.eswa.2023.120588
4. Vaswani, A., et al.: Attention is all you need. In: 31st Conference on Neural Information Processing Systems (NIPS 2017), Long Beach, CA, USA
5. Mousavi, S.M., Ellsworth, W.L., Zhu, W., et al.: Earthquake transformer—an attentive deep-learning model for simultaneous earthquake detection and phase picking. Nat. Commun. **11**, 3952 (2020)
6. Machin, D., Campbell, M.J., Walters, S.J.: Medical Statistics: A Textbook for the Health Sciences, 5th edn., p. 448. Wiley-Blackwell (2021)
7. Geshkovski, B., et al.: A Mathematical Perspective on Transformers (2024). arXiv:2312.10794v3
8. Vuckovic, J., Baratin, A., Combes, R.T.: A Mathematical Theory of Attention. http://arxiv.org/abs/2007.02876v2
9. Sanford, C., Hsu, D., Telgarsky, M.: Representational Strengths and Limitations of Transformers (2023). arXiv:2306.02896v2

10. Balim, H., et al.: Can Transformers Learn Optimal Filtering for Unknown Systems? (2024). arXiv:2308.08536v3. https://doi.org/10.1080/15598608.2013.824823
11. Liu, B., Coblenz, M., Grothe, O.: Predicting grid frequency short-term dynamics with Gaussian processes and sequence modeling. In: E-Energy '24: The 15th ACM International Conference on Future and Sustainable Energy Systems, Singapore, Singapore (2024)
12. Zeng, F., et al.: Multivariate time series anomaly detection with adversarial transformer architecture in the Internet of Things. Futur. Gener. Comput. Syst. **144**, 244–255 (2023)
13. Chen, J., Kunstner, F., Schmidt, M.: Heavy-tailed noise does not explain the gap between SGD and Adam on transformers. In: OPT2021: 13th Annual Workshop on Optimization for Machine Learning, pp. 55
14. Du, J., et al.: Handling heavy-tailed input of transformer inference on GPUs. In: ICS '22: Proceedings of the 36th ACM International Conference on Supercomputing, no. 38, pp. 1–11 (2022). https://doi.org/10.1145/3524059.3532372
15. Kunstner, F., et al.: Heavy-Tailed Class Imbalance and Why Adam Outperforms Gradient Descent on Language Models. https://arxiv.org/html/2402.19449v1
16. Ahn, K., et al.: Linear Attention is (Maybe) All You Need (to Understand Transformer Optimization) (2024). arXiv:2310.01082v2
17. Lee, B., et al.: Burst and memory-aware transformer: capturing temporal heterogeneity. Front. Comput. Neurosci. **17**, 1292842 (2023). https://doi.org/10.3389/fncom.2023.1292842
18. Zhao, W., et al.: Type-I heavy tailed family with applications in medicine, engineering and insurance. PLoS ONE **15**(8), e0237462 (2020). https://doi.org/10.1371/journal.pone.0237462
19. Gardiner, J., et al.: Fitting heavy-tailed distributions to health care data by parametric and Bayesian methods. J. Stat. Theory Pract. **8**(4), 619 (2014)
20. Burton, C.: Heavy tailed distributions of effect sizes in systematic reviews of complex interventions. PLoS ONE (2012). https://doi.org/10.1371/journal.pone.0034222
21. Liu, P., Zheng, Y.: Heavy-tailed distributions of confirmed COVID-19 cases and deaths in spatiotemporal space. PLoS ONE (2023). https://doi.org/10.1371/journal.pone.0294445
22. Karlsson, M., et al.: Getting the right tail right: modeling tails of health expenditure distributions. In: ZEW - Centre for European Economic Research Discussion Paper No. 23–045, pp. 41 (2023)
23. DeWitt, M.E., et al.: The Challenges of Surveying Heavy-tail Distributions for Use in Infectious Disease Dynamics (2023). https://doi.org/10.1101/2023.07.05.23292248v1
24. Nerella, S.: et al.: Transformers in Healthcare: A Survey. https://arxiv.org/pdf/2307.00067
25. Denecke, K., May, A., Rivera-Romero, O.: Transformer models in healthcare: a survey and thematic analysis of potentials, shortcomings and risks. J. Med. Syst. **48**(1), 23 (2024)
26. Siebra, C.A., Kurpicz-Briki, M., Wac, K.: Transformers in health: a systematic review on architectures for longitudinal data analysis. Artif. Intell. Rev. **57**, 32 (2024). https://doi.org/10.1007/s10462-023-10677-z
27. Rasmy, L., Xiang, Y., Xie, Z., et al.: Med-BERT: pretrained contextualized embeddings on large-scale structured electronic health records for disease prediction. npj Digit. Med. **4**, 86 (2021). https://doi.org/10.1038/s41746-021-00455-y
28. Luo, J., et al.: Hitanet: hierarchical time-aware attention networks for risk prediction on electronic health records. In: Proceedings of the 26th ACM SIGKDD International Conference on Knowledge Discovery & Data Mining, pp. 647–656 (2020)
29. Li, Y., Rao, S., Solares, J.R.A., et al.: BEHRT: transformer for electronic health records. Sci. Rep. **10**, 7155 (2020). https://doi.org/10.1038/s41598-020-62922-y
30. Pu, X., et al.: EEGDnet: fusing non-local and local self-similarity for EEG signal denoising with transformer Comput. Biol. Med. Part A **151**, 106248 (2022)
31. Che, C., et al.: Constrained transformer network for ECG signal processing and arrhythmia classification. BMC Med. Inform. Decis. Mak. **21**(1), 184 (2021). https://doi.org/10.1186/s12911-021-01546-2

32. Wu, X., et al.: Long short-term memory model – a deep learning approach for medical data with irregularity in cancer predication with tumor markers. Comput. Biol. Med. **144**, 105362 (2022)
33. Guo, A., Beheshti, R., Khan, Y.M., et al.: Predicting cardiovascular health trajectories in time-series electronic health records with LSTM models. BMC Med. Inform. Decis. Mak. **21**, 5 (2021)
34. Ren H., et al.: RAPT: pre-training of time-aware transformer for learning robust healthcare representation. In: KDD '21: Proceedings of the 27th ACM SIGKDD Conference on Knowledge Discovery & Data Mining, pp. 3503–3511 (2021). https://doi.org/10.1145/3447548.3467069

Cooperative-Competitive Decision-Making in Resource Management: A Reinforcement Learning Perspective

Artem Isakov[1,2](✉), Danil Peregorodiev[1], Pavel Brunko[1], Ivan Tomilov[1], Natalia Gusarova[1], and Alexandra Vatian[1]

[1] ITMO University, Saint-Petersburg 197101, Russia
[2] World-Class Research Center for Personalized Medicine, Almazov National Medical Research Centre, Saint-Petersburg 194156, Russia
aoisakov@itmo.ru
https://itmo.ru/

Abstract. This study presents a multi-agent approach to modeling deci-sion-making processes for cooperative-competitive game scenarios generated by long-tailed distributions. We introduce a novel simulation environment that aims to replicate real-world resource allocation challenges, particularly in healthcare settings. The environment incorporates dynamic changes in agents' perceptions and utilizes heavy-tailed distributions to model unpredictable events. We adapt the Multi-Agent Deep Deterministic Policy Gradient algorithm to this environment, demonstrating its effectiveness in learning and strategy development. Our approach includes a role-based system and a two-component reward function that guides agents through the learning process. The results, evaluated using metrics such as cumulative reward, average episodic reward, and approximate Kullback-Leibler divergence, indicate successful learning and sufficent agent strategies. This research contributes to the field of multi-agent systems and resource management by providing a framework that can generalize learned policies to varying numbers of agents and resources, making it applicable to real-world scenarios with changing conditions.

Keywords: Multi-agent systems · Human-in-the-loop · Reinforcement learning · Resource allocation · Cooperative-competitive games · Heavy-tailed distribution

1 Introduction

The development of multi-agent systems is a promising direction for improving resource management processes. Multi-agent systems represent participants in the shared use of resources as intelligent agents that execute a learned policy

V. Julian et al. (Eds.): IDEAL 2024, LNCS 15346, pp. 375–386, 2025.
https://doi.org/10.1007/978-3-031-77731-8_34

of actions. In this case, the optimized loss function usually responds to several challenges at once [1], for example, reducing one's own costs while observing a certain condition of social well-being [2,3].

As multi-agent systems grow in complexity, with increasing numbers of agents and more intricate interactions, the relationship between individual agent decisions at the micro-level and the system's emergent properties at the macro-level becomes increasingly unpredictable. This complexity, coupled with the trend of transitioning from classical analysis methods based on Gaussian mixed distributions to analyzing heavy-tailed distributions that describe rare but significant events [4] necessitates new approaches to multi-agent system design. The presence of heavy-tailed distributions can also be utilized to foster explainability via analyzing crutial experience point [5].

Given these and challenges described in [6], we have developed a multi-agent environment that complies with the identified trends. We have also adapted a multi-agent reinforcement learning (MARL) algorithm namely Multi-Agent Deep Deterministic Policy Gradient (MADDPG) [7] to optimize the allocation of resources among agents engaged in cooperative-competitive game relations.

A specific feature of our multi-agent environment is the mutation of the agents internal state over the course of the simulation, which is not accounted for in the mentioned closely related works. Also, while we agree on the use of random events in the environment, we ensure that our changes to the environment are reasonable and not completely random to ensure the robustness and interpretability of the learned policies.

That said, we have developed a multi-agents system for resource management that incorporates artificial intelligence methodologies. The domain specificity of our work lies in healthcare resource allocation. To ensure that it yields adequate solutions across all simulation scenarios and complies with ethical safety policies for AI applications in healthcare, we discussed implementing a Human-in-the-Loop (HITL) validation process.

This study makes several contributions to the field of multi-agent systems and resource management:

1. In the Sect. 3.3 we present a flexible modeling environment that manages dynamic state spaces and evolving agent parameters.
2. In the Sect. 3.4 we propose a methodology for modeling divergent behavior of agents based on a sophisticated reward function with underlying heavy-tailed distributions for some factors.
3. In the Sect. 4 we apply the developed methodology to the real-world problem of operating room allocation.
4. Our experiments suggest including agents with rewards drawn from a distribution with heavy tails. On the one hand, it leads to oscillations in the learning process. On the other hand, this can be controlled by tuning the parameters of the chosen probability distribution.

2 Background and Related Works

There are a few early works discussing applications of MARL algorithms in dynamically changing environments. Bellingham et al. [8] and Alighanbari et al. [9] used Mixed-Integer Linear Programming (MILP) techniques and receding horizon task assignment (RHTA) methodology to deal with re-planning through rapid recalculation, which led to significant computational costs. More recent work [10] proposed a novel methodology that guaranteed agents would learn generalizable policies throughout the process of learning in randomly changing environments. For a comprehensive overview of various implementations and widespread challenges in applying reinforcement learning techniques to real-world problems, the reader is directed to [11] and healtcare domain systemtic review [12].

Despite the general enthusiasm that had to be mentioned that with the emergence of the Deep Q-Network architecture [13] and the spread of the idea of approximating actual values with neural networks [14], reinforcement learning has transitioned into the category of deep machine learning methods, thus inheriting the opacity of the causal relationships. A comprehensive review of explainable solutions in reinforcement learning is provided by [15], which outlines several approaches to enhance understanding of underlying processes. The most prevalent methods include policy summarization [16,17], expert collaboration [18,19], and visualization [20,21]. In our work, we briefly explored the application of role-based systems to extend learned policies to new agents. The same approach can also facilitate policy generalization. Additionally, our discussion of the HITL concept exemplifies a form of human expert collaboration.

While they are less complex and reliable methods as Deep Q-Networks, the choice of the reinforcement learning algorithm was made in favor of MADDPG because of its ability to decentralize execution process.

That said the Deep Q-Networks apply centralize learning and centralize execution achieving robust results in several tasks, though they are inherent limitations preventing wide adaptation to real world scenario [22].

In contrast, the MADDPG algorithm divides the processes into centralized learning and decentralized execution. Thus, with some refinement, it becomes possible to generalize the learned policies to an arbitrary number of concurrently running agents, which is often required in real-world applications.

For this purpose, we have implemented the classification of agents involved in clinic resource allocation processes into roles based on the health status of the patients they are responsible for. Such separation scales are also actively used by emergency services in affected regions to determine the order of medical care [23].

3 Methods and Materials

3.1 Multi-agent System with Human in the Loop

The developed multi-agent system mentioned as a part of a larger complex of technical means with a supervising human expert (see Fig. 1).

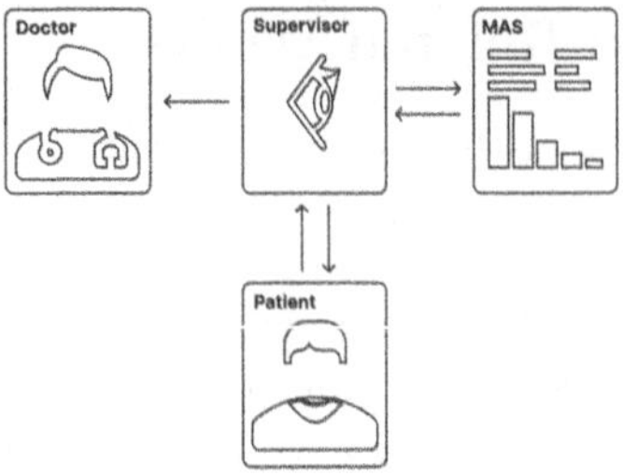

Fig. 1. Interaction of natural intelligence agents with artificial intelligence agents

As mentioned above, the key component of the multi-agent system is the MADDPG sub-redundant learning algorithm. The MADDPG algorithm is an extension of the Deep Deterministic Policy Gradient (DDPG) [24] algorithm designed for multi-agent environments. In MADDPG, each agent is trained using centralized training with decentralized execution, where the critic (value function) of each agent has access to the observations and actions of all agents, but the policy (actor) of each agent only has access to its own observations during execution. This allows agents to learn coordinated behaviors by considering the actions and observations of other agents during training, while still operating independently during execution. MADDPG effectively handles the non-stationarity and complexity of multi-agent interactions, enabling the development of sophisticated strategies in partialy observed cooperative-competitive settings. To optimize the MADDG parameters, we used the evolutionary hyperparameter optimization [25] provided by the Agile RL framework [26].

The multi-agent system is linked to a human expert who is given the opportunity to verify the results as well as make changes to them, in our case by mutual agreement with the patient. The information is then passed on, for example to the executive staff.

3.2 Simulation Scenario

The development of a multi-agent system commences [27] with the definition of a game scenario and the construction of simulation tools for cooperative-competitive behavior among two or more agents. PettingZoo [28] offers a suite of foundational classes and pre-configured gaming environments to expedite the production cycle, while allowing for the integration of arbitrary simulation logic.

Each custom PettingZoo environment adheres to one of two fundamental classes: AEC or ParallelEnv, reflecting the duality of gaming scenarios: turn-based games and simultaneous-action games. Irrespective of the chosen base class, PettingZoo facilitates the conversion between these two problem formulations through an advanced system of Python wrappers. Within the scope of the current research, we model cooperative-competitive game interactions with a parallel turn system in a partially observed environment [29].

To adapt the real-world resource management problem to the specified game environment framework, we define the state space S as a set of channels s with limited throughput capacity λ (see Fig. 2).

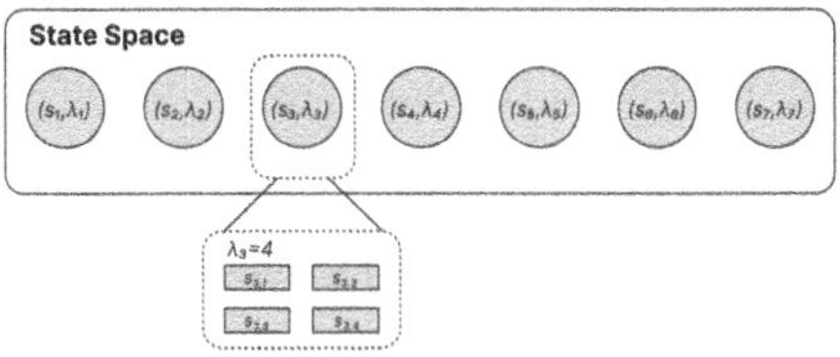

Fig. 2. State space

Thus, the game environment can be seamlessly adapted to address a range of applied queueing theory problems, whether it involves the distribution of vehicular traffic, parking spaces, ports on a virtual private server, or patients in a medical facility. To expand the range of possible game scenarios, we will introduce an additional time dimension. This requires linking each state with its corresponding temporal reference as follows (Fig. 3):

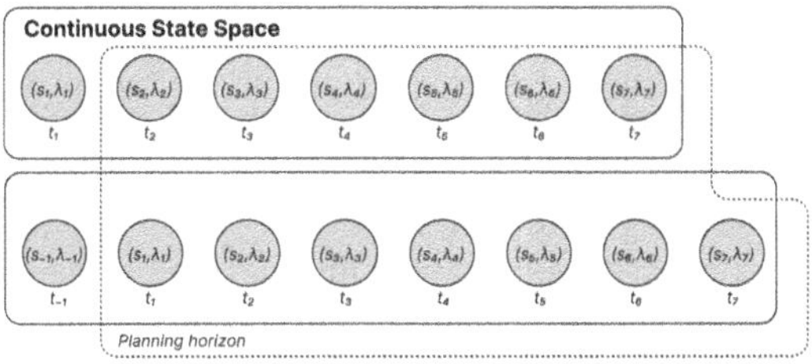

Fig. 3. Continuous state space with added time step

By incorporating a moving time window, we can model more sophisticated types of game scenarios.

The first and siplest game scenario involves utilizing a time window with step $t = 0$ to solve optimization problems where time acts as a variable influencing the intelligent system's policy, while the states themselves do not qualitatively change over time. For the second game scenario, we can employ a time window with a non-zero step t. For instance, when addressing daily flow planning, we shift the time window one step to the right daily, solving the same problem with qualitatively updated states. Assume we operate on a weekly schedule. Each subsequent day nominally becomes the first day of the planning horizon with a shift of $t = 1$, while qualitatively different from the previous one.

3.3 Simulation Environment

The game environment for the reinforcement learning algorithm is defined by the set of states $s \in S$, observations $o \in O$, and the actions $a \in A$ available to the agents and their corresponding rewards $r \in R$.

In the previous section, we gave an exhaustive description of the chosen game scenario. In brief, we define the space of available states $s_1\lambda_1, ..., s_n\lambda_n$ as a calendar sequence of days with limited throughput capacity, not exceeding a given planning horizon, usually one week (see Fig. 4). On demand, the planning horizon is shifted one day ahead, defining the mechanism of the intelligent system with dynamically time-varying simulation conditions.

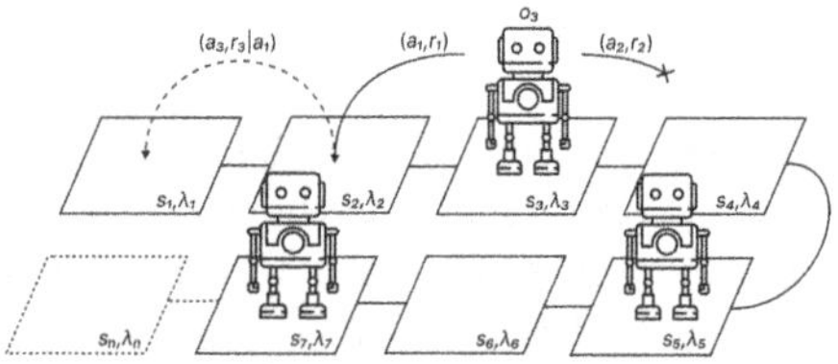

Fig. 4. Graphical representation of the game situation within the simulation framework

Within the developed game environment, each application is represented as an intelligent agent. Artificial intelligence agents are placed randomly in the environment. If necessary, the initial location of the agents can be set by following the priorities of the natural intelligence agent, i.e. the patient or administrator. In order to make the game contactable and to get more opportunities to analyze the peculiarities of cooperative-adversarial interaction between intelligent agents, it was decided to allow them to move one position to the left or right – closer or farther from the beginning of the planning horizon – per turn. In addition, the agents can take no action while maintaining their current position.

Intelligent agents inherit properties that reflect the patient's current health status: $c_i \in C$, the estimated complexity of the surgery, takes values from 0 to 1; $u_i \in U$, the estimated urgency to perform the surgery, takes values from 1 to 3; $p_i \in P$, the completeness of the collected information needed to perform the surgery, takes values from 0 to 1. In order to implement the role system required to generalize learned policies to a constantly changing number of patients, we conditionally divide the applications into twelve roles or unique combinations obtained by computing the Cartesian product of the $C \times U \times P$ sets.

An agent's set of internal properties changes over time in response to the agent's actions in the environment. Thus, if an agent moves its application to the second half of the planning horizon, there is a non-zero cumulative probability that its health state, described by parameters C and U, will deteriorate. The cumulative chance of a patient's condition worsening is called mutation. By placing a request for surgery in the second half of the planning horizon, for each extra day of delay, the agent receives an additional 2.5% to the probability that all or some of its parameters will mutate. Nevertheless, there is a positive change, the further away from the beginning of the planning horizon, the higher the probability that the completeness of the collected information P increases. Whether mutation will occur or not is determined at each step of the game simulation.

Each agent's observation space is intentionally limited to avoid overfitting. It includes a representation of its health state and current location in the planning horizon, as well as information about the number of submitted applications for the current day, three days before, and three days after. If necessary, the observation space can be expanded to include some biased representation of the number of applications for the entire planning horizon - the farther the date is from the agent's current location, the larger the magnitude of the random bias. In this way, the natural uncertainty of events over time can be modeled, incentivizing the agent to make decisions based on more reliable information. However, the degree of risk that agents can assume is part of the learned policy, and in some cases may differ significantly from conservative assumptions.

The task that artificial intelligence agents are underlying is to take the most favorable position in the scheduling horizon, while avoiding an overflow of requests for the selected day. The process terminates after a specified number of turns. In order to formally define the task at hand, we formed a two-component reward function that guides agents through the learning process.

3.4 Reward Function

The reward function in our environment quantifies the reward r_i for action a_i and is formulated as a combination of two key components - the base reward b_i and the scaling coefficient k_i for the i-th intelligent agent.

$$r_i = -\frac{b_i}{k_i} \sum_{n \in N} \frac{|s_n - \lambda_n|}{|a_i - n| + 1} \tag{1}$$

where b_i represents the base reward, k_i denotes the scaling factor reflecting the patient's physical condition, a_i - current alignment inside the planning horizon, s_n indicates the occupancy state of the environment and λ_n signifies the maximum number of requests that can be processed on the specified day $n \in N$ of the planning horizon.

The scaling factor modulates the magnitude of the base reward in accordance with changes induced by disease progression and is defined as:

$$k_i = argmax(1, ((c_i + (1 - p_i)) * u_i) \tag{2}$$

where p_i represents the completeness of provided information, c_i denotes the complexity of medical intervention, and u_i signifies the urgency of medical care delivery.

The base reward is a random variable. For most agents, it is chosen at initialization of the simulation environment from the Lévy alpha stable distribution with stability parameter α equal to 2. For a small subset of agents, the base reward is chosen with stability parameter α less than 2.

We propose selecting the stability parameter α within the interval (1, 2]. This is because at α=1, the Lévy stable distribution begins to assume the form of the heavy-tailed Cauchy distribution; at α=2, it becomes the normal Gaussian distribution.

During the selection of the optimal value, one can utilize the Optuna parameter optimization library [30] in conjunction with selected reinforcement learning algorithm. In this case, the optimisation will proceed until the selected metric reaches a threshold value set by the researcher or stops changing within a certain number of iterations.

Thus, we obtain an environment where the majority of agents are relatively equal in relation to each other, while some have an order of magnitude greater base reward for performing any actions. This way, we simulate the natural bias of the environment towards the agents operating within it. The nature of this bias can vary, although the source is singular - HITL, be it a VIP client, personal interest of an administrator, or a medical error leading to an underestimation of a patient's needs.

3.5 Metrics

To evaluate the performance, we utilized the following metrics: cumulative reward, average episodic reward, and approximate Kullback-Leibler (KL) divergence.

Cumulative reward represents the sum of all rewards obtained by the agents throughout the entire learning process. A consistently increasing cumulative reward indicates an improvement in the agents' strategies and their ability to achieve the goals set by the environment.

$$R_{cumulative} = \sum_{t=1}^{T} r_t \tag{3}$$

where r_t denotes the reward at step t, and T represents the number of steps in the i-th episode.

Average episodic reward is calculated as the mean reward obtained by the agents over a single episode (a period of interaction with the environment until the episode is completed). This metric allows for the assessment of the quality of the agents' strategies within the context of individual episodes.

$$R_{episodic} = 1/N \sum R_{cumulative} \tag{4}$$

where $R_{cumulative}$ represents the cumulative reward for the entire training process, and N denotes the number of training episodes.

Approximate KL divergence measures the difference between the probability distributions of the agents' actions. This metric is used to assess changes in the agents' strategies throughout the learning process and their ability to adapt to the environment. The KL divergence is approximated using the following formula:

$$KL(q,p) = \sum q(x) * log(\frac{q(x)}{p(x)}) \approx (ratio_t - 1) - log(ratio_t) \tag{5}$$

where $ratio_t$ is the ratio of the probability distributions $q(x)$ and $p(x)$, belonging to the new and old policies, respectively.

In summary, the three metrics collectively provide a comprehensive understanding of the learning process and agent performance. The increasing cumulative reward, rising average episodic reward, and stable, decreasing KL divergence values indicate successful learning and high-quality agent strategies. The statistical analysis confirms the significance of these findings, reinforcing the efficacy of the adapted MADDPG algorithm in our custom environment.

4 Results

We conducted a series of experiments with the adapted MADDPG algorithm in our custom environment. We used metrics described previously. The source code is available on Github: https://github.com/artemisak/rl-task-scheduler.

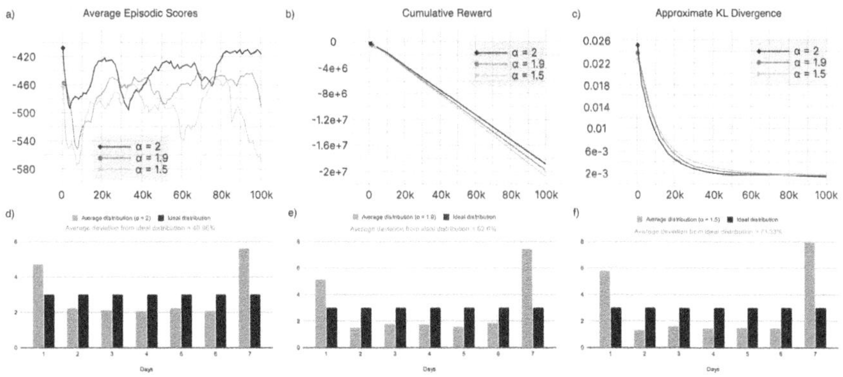

Fig. 5. Evaluating the performance of the learned policies

The results presented in Fig. 5 align with theoretical predictions. As α approaches 1 (corresponding to the Cauchy reward case), the average episodic scores (Fig. 5a) exhibit increased instability and generally lower values compared to the initial conditions. Conversely, when α equals 2, the average episodic scores demonstrate performance either comparable to or marginally exceeding the initial baseline. This suggests that the algorithm has successfully addressed the case, approximating the initial random distribution while incorporating adjustments based on the patient's current health status. As anticipated, the cumulative rewards (Fig. 5b) exhibit a decreasing trend over time. The approximate KL divergence (Fig. 5c) indicates that the algorithm's convergence rate is inversely proportional to α, with slower convergence observed for $\alpha < 2$. Figure 5d, 5e, and 5f present histograms depicting the mean planning occupancy across the planning horizon, derived from 1000 bootstrap resampled experiments.

5 Discussion

As was discussed previously in Background section there are numerous studies on the application of multi-agent reinforcement learning in industrial systems and healthcare.

In this study we have focused our efforts on adapting an existing and well-established contemporary multi-agent reinforcement learning method to an experimental simulation environment that we have developed.

To achieve this, we reformulated the problem statement, transitioning from single-agent reinforcement learning and genetic algorithms to multi-agent reinforcement learning, where each agent represents an end-user, engaging in cooperative-competitive interactions with other agents.

The environment is constructed to meet the conditions of several real-time optimization tasks. Its state space consists of a set of channels with limited capacity and is designed in such a way that it can work with a gradually shifting time window, reflecting the natural variability of the state space.

In our case, each new agent in the multi-agent system is assigned one of the twelve previously learned MADDPG policies after a preliminary assessment of the patient's condition, allowing it to work with an arbitrary number of agents.

We also explored various facets of uncertainty inherent in all real-world tasks. Firstly, by introducing mutations, we endowed the artificial intelligence agents with the ability to account for changes occurring in the natural intelligence agents whose interests they represent. Secondly, we examined the possibility of introducing agents into the environment towards which the environment is biased.

By decomposing the reward function into components, one of which is drawn from a random distribution, we analyzed the algorithm's convergence in the presence of agents with rewards sampled from heavy-tailed distributions. The use of alpha-stable Lévy distributions for modeling rewards introduces the necessary unpredictability, reflecting real-world scenarios in which certain agents may have disproportionate influence or priority. The incorporation of heavy-tailed distributions creates a more complex and realistic environment, testing the multi-agent system's resilience to biases.

The results of experiments with different α level values demonstrate that the α level should be chosen greater than 1, but less than 2. In this case, we get agents to whom the environment is strongly biased, although the reinforcement algorithm still converges to a general solution.

6 Conclusion

We developed a multi-agent reinforcement learning approach for cooperative-competitive decision-making in resource management with heavy-tailed distributions. Key contributions include a flexible simulation environment, modeling rare events, application to operating room allocation, and demonstrating generalizability. Experiments showed successful policy learning, though stability was affected by the Lévy stable distribution parameter α. The framework generalizes to varying agent and resource numbers, applicable to real-world scenarios. Future work could explore human-in-the-loop integration.

Acknowledgments. This work was supported by the Ministry of Science and Higher Education of the Russian Federation, Goszadanie (State Assignment) No. 2019-1339.

Disclosure of Interests. The authors declare that there are no conflicts of interest regarding the publication of this article. This research was conducted in the absence of any commercial or financial relationships that could be construed as a potential conflict of interest.

References

1. Lei, J., Tan, D., Ma, X., Wang, Y.: Reinforcement learning based multi-parameter joint optimization in dense multi-hop wireless networks. Ad Hoc Netw. **154**, 103357 (2024)
2. Nicolas, P.G., Paul-Antoine, B.: Deep hierarchical reinforcement learning to manage the trade-off between sustainability and profitability in common pool resources systems. In: 2021 International Joint Conference on Neural Networks (IJCNN), pp. 1–7. IEEE (2021)
3. Pretorius, A., et al.: A game-theoretic analysis of networked system control for common-pool resource management using multi-agent reinforcement learning. Adv. Neural Inf. Process. Syst. **33**, 9983–9994 (2020)
4. Alqahtani, M., Grafke, T.: Instantons for rare events in heavy-tailed distributions. J. Phys. A: Math. Theor. **54**(17), 175001 (2021)
5. Liu, X., Liu, S., An, B., Gao, Y., Yang, S., Li, W.: Effective interpretable policy distillation via critical experience point identification. IEEE Intell. Syst. **38**(5), 28–36 (2023)
6. Canese, L., et al.: Multi-agent reinforcement learning: a review of challenges and applications. Appl. Sci. **11**(11), 4948 (2021)
7. Lowe, R., Wu, Y.I., Tamar, A., Harb, J., Pieter Abbeel, O., Mordatch, I.: Multi-agent actor-critic for mixed cooperative-competitive environments. Adv. Neural Inf. Process. Syst. **30**, (2017)
8. Bellingham, J., Tillerson, M., Richards, A., How, J.: Multi-task allocation and path planning for cooperating UAVs. In: Cooperative Control: Models, Applications and Algorithms, pp. 23–41 (2003)
9. Alighanbari, M.: Task assignment algorithms for teams of UAVs in dynamic environments (Doctoral dissertation, Massachusetts Institute of Technology) (2004)
10. Qie, H., Shi, D., Shen, T., Xu, X., Li, Y., Wang, L.: Joint optimization of multi-UAV target assignment and path planning based on multi-agent reinforcement learning. IEEE Access **7**, 146264–146272 (2019)

11. Panzer, M., Bender, B.: Deep reinforcement learning in production systems: a systematic literature review. Int. J. Prod. Res. **60**(13), 4316–4341 (2022)
12. Al-Hamadani, M.N., Fadhel, M.A., Alzubaidi, L., Harangi, B.: Reinforcement learning algorithms and applications in healthcare and robotics: a comprehensive and systematic review. Sensors **24**(8), 2461 (2024)
13. Zhao, Y., Wang, Y., Tan, Y., Zhang, J., Yu, H.: Dynamic jobshop scheduling algorithm based on deep Q network. IEEE Access **9**, 122995–123011 (2021)
14. Adcock, B., Dexter, N.: The gap between theory and practice in function approximation with deep neural networks. SIAM J. Math. Data Sci. **3**(2), 624–655 (2021)
15. Wells, L., Bednarz, T.: Explainable AI and reinforcement learning–a systematic review of current approaches and trends. Front. Artif. Intell. **4**, 550030 (2021)
16. Amir, O., Doshi-Velez, F., Sarne, D.: Summarizing agent strategies. Auton. Agent. Multi Agent Syst. **33**, 628–644 (2019)
17. Lage, I., Lifschitz, D., Doshi-Velez, F., Amir, O.: Toward robust policy summarization. In: Autonomous Agents and Multi-agent Systems, pp. 2081 (2019)
18. Ehsan, U., Tambwekar, P., Chan, L., Harrison, B., Riedl, M.O.: Automated rationale generation: a technique for explainable AI and its effects on human perceptions. In: Proceedings of the 24th International Conference on Intelligent User Interfaces, pp. 263–274 (2019)
19. Tabrez, A., Hayes, B.: Improving human-robot interaction through explainable reinforcement learning. In: 2019 14th ACM/IEEE International Conference on Human-Robot Interaction (HRI), pp. 751–753. IEEE (2019)
20. Dethise, A., Canini, M., Kandula, S.: Cracking open the black box: what observations can tell us about reinforcement learning agents. In: Proceedings of the 2019 Workshop on Network Meets AI & ML, pp. 29-36 (2019)
21. Pan, X., Chen, X., Cai, Q., Canny, J., Yu, F.: Semantic predictive control for explainable and efficient policy learning. In: 2019 International Conference on Robotics and Automation (ICRA), pp. 3203–3209 IEEE (2019)
22. Jang, B., Kim, M., Harerimana, G., Kim, J.W.: Q-learning algorithms: a comprehensive classification and applications. IEEE Access **7**, 133653–133667 (2019)
23. Williams, S., Crouch, R.: Emergency department patient classification systems: a systematic review. Accid. Emerg. Nurs. **14**(3), 160–170 (2006)
24. Silver, D., Lever, G., Heess, N., Degris, T., Wierstra, D., Riedmiller, M.: Deterministic policy gradient algorithms. In: International Conference on Machine Learning, pp. 387–395 (2014)
25. Morales-Hernández, A., Van Nieuwenhuyse, I., Rojas Gonzalez, S.: A survey on multi-objective hyperparameter optimization algorithms for machine learning. Artif. Intell. Rev. **56**(8), 8043–8093 (2023)
26. Ustaran-Anderegg, N., Pratt, M.: AgileRL [Computer software]. https://github.com/AgileRL/AgileRL
27. Railsback, S.F., Lytinen, S.L., Jackson, S.K.: Agent-based simulation platforms: review and development recommendations. Simulation **82**(9), 609–623 (2006)
28. Terry, J., et al.: Pettingzoo: gym for multi-agent reinforcement learning. Adv. Neural Inf. Process. Syst. **34**, 15032–15043 (2021)
29. Zhang, K., Yang, Z., Başar, T.: Multi-agent reinforcement learning: A selective overview of theories and algorithms. In: Handbook of Reinforcement Learning and Control, pp. 321–384 (2021)
30. Akiba, T., Sano, S., Yanase, T., Ohta, T., Koyama, M.: Optuna: a next-generation hyperparameter optimization framework. In: Proceedings of the 25th ACM SIGKDD International Conference on Knowledge Discovery & Data Mining, pp. 2623–2631 (2019)

Improving Speech Emotion Recognition: Novel Aggregation Strategies for Self-supervised Features

Oscar Valls[1], Fran Pastor-Naranjo[1](✉), Rocío del Amor[1], Lucía Gómez-Zaragozá[1], Javier Marín-Morales[1], Mariano Alcañiz Raya[1], and Valery Naranjo[1,2]

[1] Instituto Universitario de Investigación en Tecnología Centrada en el Ser Humano (HUMAN-Tech), Universitat Politécnica de Valencia, 46022 Valencia, Spain
{frapasna,madeam2,vnaranjo}@upv.es

[2] Valencian Graduate School and Research Network for Artificial Intelligence (valgrAI), 46022 Valencia, Spain

Abstract. Speech Emotion Recognition (SER) is crucial for enhancing human-machine voice interactions, allowing systems to better interpret the speaker's emotional state and improve user experience. Self-supervised learning (SSL) models have significantly advanced speech recognition by generating speech representations from large unlabeled corpus, which can then be fine-tuned with smaller labeled datasets for downstream tasks such as voice command detection, automatic transcription, speaker identification, and SER. However, SSL models are typically optimized for general tasks rather than emotion recognition. This poses a challenge, as the limited labeled data in SER can hinder the generalization capabilities of these models, making the choice of SSL architecture and training strategy vital. The most common procedure in Speech Emotion Recognition is performing an average time pooling of the SSL features to train classification models, which often leads to a loss of temporal relationships in the data. In this study, we introduce two novel SSL feature aggregation methods that leverage attention mechanisms to better capture temporal dependencies in speech data. These methods significantly enhance the extraction of relevant information from SSL features, leading to improvements in classification accuracy. Our proposed approach outperforms the standard average time pooling method, achieving up to a 6.3% increase in weighted accuracy (WA) on the IEMOCAP database.

Keywords: Affective Computing · Speech Emotion Recognition · Transformer Models · Self-Supervised Speech Features · Attention-Based Temporal Aggregation

1 Introduction

Human speech communication allows individuals to express various kinds of information, not only verbal content expressed with words but also other non-

V. Julian et al. (Eds.): IDEAL 2024, LNCS 15346, pp. 387–397, 2025.
https://doi.org/10.1007/978-3-031-77731-8_35

verbal information that can be interpreted from speech and informs about the speaker's affective state and behavior. Speech Emotion Recognition (SER) is a field of research that focuses on analyzing speech signals to predict human affective states and emotions. Multiple methodologies that address this task can be found in literature, but deep learning techniques, and more specifically, semi-supervised transformer-based architectures have become the state-of-the-art technology in Speech Emotion Recognition.

Transformer-based architectures allowed big advancements in multiple speech-related tasks [10], such as automatic speech recognition [6], speaker identification[12], or keyword spotting [1]. Thanks to self-attention mechanisms, transformers are well suited for comprehending speech signals, as they analyze the entire speech utterance at once, making them capable of understanding complex dependencies between elements from a speech sequence, regardless of their distance. Transformers' main drawback is that they require large amounts of data to be trained[7]. Therefore, traditional supervised training techniques are ineffective for those audio and speech tasks where the amount of labeled data is limited, making transformer architectures unable to generalize and get outperformed by other architectures. Self-supervised learning (SSL) models have become state-of-the-art in speech-related tasks [9]. These models are pre-trained to generate audio representations using a large unlabeled speech corpus. These representations can be used to perform many speech-related tasks using pre-trained models as feature extractors or fine-tuned on a limited amount of labeled data to perform downstream tasks such as Speech Emotion Recognition.

Transformer models divide speech signals into time frames to apply self-attention between those time frames, composing the whole speech sequence. The transformer output is a sequence of latent representations from each time frame of the audio given as input. The most common procedure in Speech Emotion Recognition using transformer-based models pre-trained under an SSL paradigm is to perform average time pooling, where the features are aggregated from frame to utterance, and given as input to a classifier to predict emotions.

In speech signals, emotional content can be interpreted globally from the whole sentence, but significant emotional signals often reside in specific frames, such as variations in volume, pitch, voice modulations or word stress. This suggests that extracting emotional content at various levels - word, sentence or utterance - provides a more nuanced understanding of the speaker's affective state [14]. Effective aggregation of these frame-level features is therefore essential for capturing detailed emotional information. However, traditional time-averaged methods aggregate features by averaging frames, which can dilute or hide critical emotional information. To the best of the authors' knowledge, no previous study has focused on finding optimal methods to aggregate speech time-frame-level to utterance-level features. This work proposes two different time aggregation methods (EmoToken and EMOQ-Former) that enhance classification performance using aggregated frame-level representations extracted from self-supervised audio transformer encoders. They do it by selecting the most relevant time frames to be represented in the aggregated features using attention mechanisms. The key contributions of this paper are summarized as follows:

- Emotion Token (EmoToken), we propose a novel transformer-based aggregation model that generates an Emotion Token to represent the most relevant information extracted from frame-level representations using self-attention mechanisms.
- We introduce Emotion Querying Transformer (EMOQ-Former). It generates a new feature space, making use of self-attention and cross-attention mechanisms with learnable input queries that are trained to retrieve emotional content represented in the speech signal.
- We trained and validated both aggregation models with IEMOCAP database for speech emotion recognition with the most used audio self-supervised transformers: HuBERT, WavLM, and UniSpeech-SAT.
- We improved accuracy results on Speech Emotion Recognition using IEMOCAP database, achieving up to +6.3% enhancement in accuracy compared to average time pooling.

2 Methodology

2.1 Self-supervised Speech Encoders

The widespread adoption of the Self-supervised approach for generating multipurpose speech representations has increased the research community's contributions with the objective of developing better pre-training strategies and model architectures. These may deliver better encoders capable of representing speech information to solve various downstream tasks more efficiently. In this study, we focused on three most used models in the literature to validate our time aggregation methods:

- **HuBERT** [8]: It uses a two-step pre-training approach. The first step involves creating pseudo-labels for unlabeled audio data by clustering hidden features extracted from the audio. In the second step, the model is pre-trained to predict these pseudo-labels, treating it as a masked prediction task similar to BERT's masked language modeling.
- **WavLM** [3]: It is designed to improve upon previous speech models by focusing on handling noisy and real-world audio data more effectively. Introducing a self-supervised pre-training strategy that leverages large amounts of unlabeled audio data, including those with background noises and various acoustic environments. This approach allows WavLM to achieve superior performance in speech recognition tasks, especially in challenging audio conditions.
- **UniSpeech-SAT** [4]: Its architecture is built upon the transformer model, and it introduces a Speaker-Aware Training mechanism (SAT) that enables the model to better capture speaker characteristics along with speech content. This is achieved by incorporating speaker embeddings during the pretraining phase, which helps the model distinguish between different speakers and adapt to new speakers more effectively.

These feature extractors have three versions each, we find *HuBERT BASE, LARGE* and *XLARGE*, and WavLM and UniSpeech-SAT *BASE, BASE+* and *LARGE*. These versions differ from each other in model dimensions and the amount of training data used in the pre-training phase.

2.2 EmoToken

Derived from their original applications in NLP, self-attention mechanisms were domain-adapted and proven effective in audio and image classification tasks. The BERT [5] model introduced the class token mechanism to generate a trained embedding that represented the classes learned during training. This approach has been shown to outperform other architectures, such as CNN or RNN-based models in audio and image domains. However, this is only possible when there is enough training data due to transformers' inherent difficulties in generalizing with insufficient training samples.

Inspired by [13], we propose a new transformer-based time aggregation model for Speech Emotion Recognition. Figure 1 depicts the EmoToken transformer architecture. The frozen SSL feature extractor generates speech embeddings $\mathbf{S} \in \mathbb{R}^{n_s \times T_s \times d_s}$ where n_s is the batch size, T_s is the number of time steps, and d_s is the dimension of speech embedding. The class token $\mathbf{CLS} \in \mathbb{R}^{n_s \times 1 \times d_s}$, is appended to the embeddings and forwarded through trainable transformer layers. The representation of the learned class token is then used to feed the classifier and predict emotions. The class token learns to retrieve the most relevant information for the classification, using self-attention mechanisms that establish relations between the class token and all frame-level representations.

2.3 EMOQ-Former

The idea of the querying transformer was first introduced in BLIP-2 [11] foundation model to bridge the gap between an image encoder and LLM features, making use of learnable input parameters named Q-queries. We adapted it to train the Q-queries to retrieve emotional information in speech signals. Figure 2 depicts EMOQ-Former model architecture.

The method that EMOQ-Former uses to generate emotion-specific speech features is as follows:
Let $\mathbf{q} \in \mathbb{R}^{n_q \times d_q}$ represent the Q-queries, where n_q is the number of Q-queries, d_q is the dimension of Q-queries and $\mathbf{S} \in \mathbb{R}^{n_s \times T_s \times d_s}$ represent the speech embedding, where n_s is the batch size, T_s is the number of time steps, and d_s is the dimension of speech embedding. We first input the Q-queries $\mathbf{q} \in \mathbb{R}^{n_q \times d_q}$ into the self-attention mechanism:

$$A_{\text{self}} = \text{softmax}\left(\frac{(\mathbf{q}W_{q\text{self}})(\mathbf{q}W_{k\text{self}})^T}{\sqrt{d_k}}\right)\mathbf{q}W_{v\text{self}}$$

where $W_{q\text{self}} \in \mathbb{R}^{d_q \times d_k}$, $W_{k\text{self}} \in \mathbb{R}^{d_q \times d_k}$, and $W_{v\text{self}} \in \mathbb{R}^{d_q \times d_v}$ are the learnable weight matrices for queries, keys, and values in the self-attention mechanism,

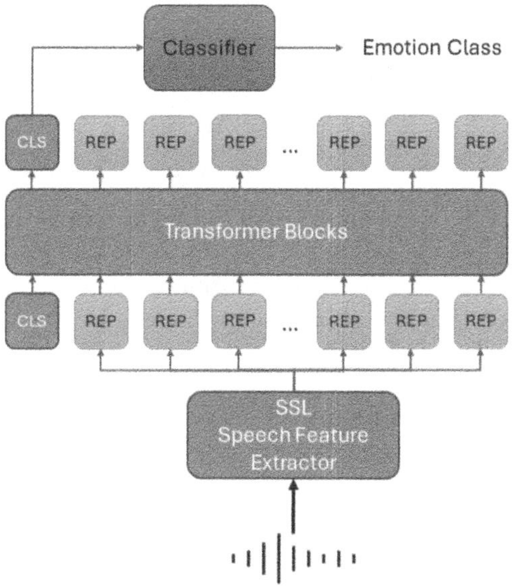

Fig. 1. EmoToken transformer architecture. The EmoToken is appended to the frame-level representations provided by the pre-trained model. The output representation of that token is then used to feed the classifier.

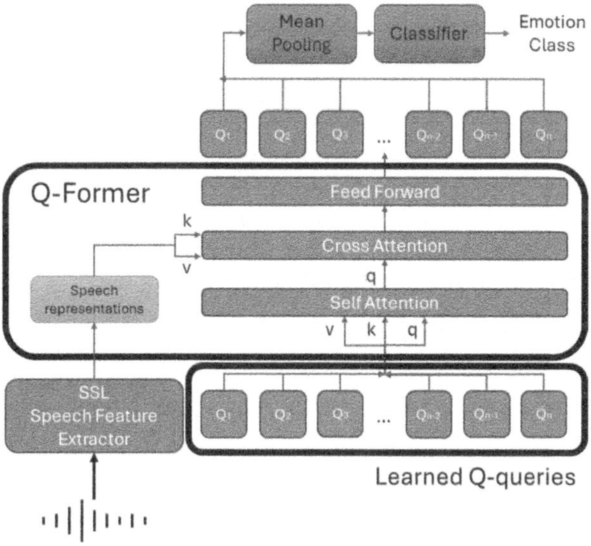

Fig. 2. EMOQ-Former architecture. The EMOQ-Former generates emotion-related feature space by combining learned queries with speech representations via cross-attention mechanisms. The output representations for the queries are averaged and fed into the classifier.

and d_k and d_v are the dimensions of keys and values. The output of the self-attention mechanism $A_{\text{self}} \in \mathbb{R}^{n_q \times d_v}$ are then used as queries for the cross-attention mechanism, while the speech embedding $\mathbf{S} \in \mathbb{R}^{n_s \times T_s \times d_s}$ serves as keys and values:

$$A_{\text{cross}} = \text{softmax}\left(\frac{A_{\text{self}} W_q (\mathbf{S} W_k)^T}{\sqrt{d_k}}\right) \mathbf{S} W_v$$

where $A_{\text{cross}} \in \mathbb{R}^{n_q \times n_q \times d_v}$ represents the cross-attention output, while $W_q \in \mathbb{R}^{d_v \times d_k}$, $W_k \in \mathbb{R}^{d_s \times d_k}$, and $W_v \in \mathbb{R}^{d_s \times d_v}$ are the learnable weight matrices for queries, keys, and values in the cross-attention mechanism.

This approach enables the attention mechanism to generate representations related to Q-queries within the speech embedding. The output of the EMOQ-Former denoted as the Q-Embedding $Q_e \in \mathbb{R}^{n_s \times n_q \times d_q}$, maintains a fixed length that is independent of the length of the input speech. This fixed-length representation leads to improved generalization performance across speech inputs of varying lengths.

3 Experimental Setup

3.1 IEMOCAP Database

We used the Interactive Emotional Dyadic Motion Capture (IEMOCAP) [2] database to perform our study and evaluate our methodology. IEMOCAP is the most widely used database in Speech Emotion Recognition. It consists of audiovisual recordings of dyadic interactions designed to elicit a wide range of emotional expressions. The database contains five sessions featuring two different actors (male and female) each, performing engaged in scripted and improvised scenarios, resulting in a diverse set of emotions, including happiness, sadness, anger, frustration, and neutrality. In our experiments, we reduced it to 4 classes: joy, anger, sadness, and neutral, including samples labeled as excitement in *joy* class. We performed a Speaker-Independent (SI) evaluation, where we saved session 5 for testing and used the other four sessions to perform 4-fold cross-validation for hyperparameter search with early stopping.

3.2 Training and Evaluation

After performing ablation studies to find the optimal model dimensions, we defined the following architecture dimensions for the models. EmoToken is composed of 8 transformer blocks with 2 attention heads whereas EMOQ-Former consists of 4 transformer blocks with 2 attention heads and 32 learned queries.

Ablation studies were performed using a single train-validation partition of IEMOCAP. We used sessions from 1 to 3 as the training subset, session 4 for validation, and HuBERT-BASE as a feature extractor. The number of trainable parameters for each aggregation model depends on the feature extractor embedding dimensions, but EMOQ-Former always has fewer trainable parameters than EmoToken transformer due to its reduced number of layers. A single linear head was used for multiclass classification using average time-pooling. The same linear head was appended to the EmoToken transformer and EMOQ-Former to classify after the aggregation.

We performed a hyper-parameter search with early stopping and chose the best training configuration based on the average of the best early-stopped validation accuracy from each fold. The optimal training hyper-parameters were searched using only the *BASE* versions from each SSL feature extractor and then applied directly to perform another cross-validation of the aggregation methods in bigger versions of each feature extractor to find the optimal number of epochs. Each aggregation method with each feature extractor was finally trained with its optimal hyper-parameters and number of epochs using the first 4 sessions of the IEMOCAP dataset and evaluated with session 5.

3.3 Evaluation Metrics

All results are reported in the two most spread Weighted Accuracy **(WA)** and Unweighted Accuracy **(UA)**. We used WA criteria in hyperparameter search and ablation experiments.

4 Results and Discussion

Table 1 shows test accuracy results, which provide an overview of EMOQ-Former **(QF)** and Emo-Token **(ET)** aggregations' effectiveness on speech emotion classification compared to classical average time pooling aggregation **(ATP)**.

Table 1. Test results on IEMOCAP with different aggregation methods: Average Time Pooling (ATP), EmoToken (ET), EMOQ-Former (QF)

Feature Extractors	WA (%)			UA (%)		
	ATP	*ET*	*QF*	*ATP*	*ET*	*QF*
HuBERT-BASE	61.2	61.0	**64.0**	62.8	65.0	**65.1**
HuBERT-LARGE	64.8	64.3	**67.9**	66.1	66.9	**68.5**
HuBERT-XLARGE	65.5	**68.1**	66.5	66.1	**70.0**	68.5
WavLM-BASE	55.3	58.4	**61.6**	57.2	58.4	**63.5**
WavLM-BASE+	59.2	60.3	**62.3**	60.3	61.2	**64.3**
WavLM-LARGE	62.4	**64.5**	60.8	63.0	**64.5**	62.7
UniSpeech-SAT-BASE	62.2	59.4	**62.5**	63.9	59.7	**64.7**
UniSpeech-SAT-BASE+	62.1	62.8	**63.8**	63.3	**65.3**	**65.3**
UniSpeech-SAT-LARGE	**56.5**	43.6	54.7	**58.8**	40.9	57.9

The best overall accuracy result was achieved by HuBERT *X-LARGE* features aggregated with EmoToken. In HuBERT and WavLM, EmoToken outperforms average time pooling and EMOQ-Former in their biggest version, achieving 68.1% and 64.5% of weighted accuracy respectively. EMOQ-Former stands as the best aggregator for lower-dimensioned features, as it performs better on smaller versions of all feature extractors. It gets better classification accuracies with medium and small-sized versions of all three feature extractors. LARGE or XLARGE versions of feature extractors are bigger in embedding dimensions. In those cases, EmoToken is a better technique for achieving better performance when aggregating higher-dimensional features, and EMOQ-Former works better when aggregating lightweight and less complex features. Compared to average time pooling, the highest performance increase (+6.3%) is produced by WavLM-BASE features aggregated with EMOQ-Former.

Looking at HuBERT-LARGE, WavLM-BASE, UniSpeech-SAT-BASE, and UniSpeech-SAT-BASE+, we find that the best-performing aggregation for these features' accuracy results are higher than the ATP features of their immediate bigger feature extractors. This enables using feature extractors with fewer parameters without compromising performance. HuBERT-LARGE and the aggregation model have 333M parameters, much less than HuBERT-XLARGE, which has 1B parameters; it also happens with WavLM-BASE+ and EMOQ-Former, which has 128M parameters, smaller than WavLM-LARGE (317M). This can lead to saving computational resources in Speech Recognition systems where the aggregation model can be appended to a lighter feature extractor to achieve high performance on Speech Emotion Recognition while self-supervised features remain the same and can be used for other purposes, also removing the need to allocate a specific fine-tuned model only for Speech Emotion Recognition.

Figure 3 depicts T-SNE visualizations of feature spaces generated with each aggregation method for each architecture's best-performing feature extractor version.

In general terms, both Emo-Token and EMOQ-Former provide better separability between classes than average time-pooled features. Comparing Emo-Token and EMOQ-Former, the first one generates visually better separable features than EMOQ-Former for three feature extractors. HuBERT-XLARGE and WavLM-LARGE features aggregated with EmoToken achieved better WA and UA, and those results are also translated into better separation between classes in T-SNE visualizations. This does not happen with UniSpeech-SAT features, where better separation can be spotted in T-SNE visualization but reports lower accuracy results.

In summary, attention-based methods are better for aggregating frame-level speech features. Nevertheless, there is no one-fits-all method, and the choice between Emotion Token, and EMOQ-Former should be based on the specific nature of the data and the features extracted. Using attention-based models to perform time-aggregation features is a better choice for designing Speech Emotion Recognition systems. Additionally, considerations such as computational efficiency and the nature of the task can guide the selection of an appropriate aggregation method, improving classification while saving computation time and resources.

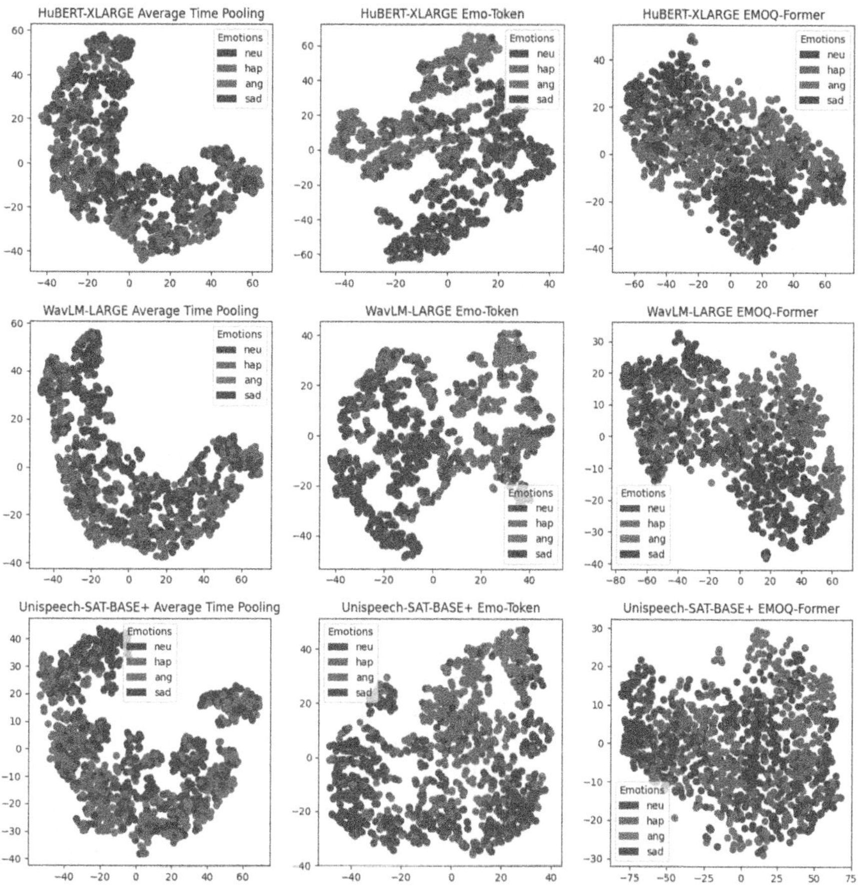

Fig. 3. T-SNE visualization of feature spaces generated by aggregation methods. Emotion classes are neu:Neutral, hap:Happiness, ang:Anger and sad:Sadness

5 Conclusion

In this paper, we proposed two transformer-based aggregation methods for self-supervised speech features as an alternative to average time pooling, which is adopted as the most common practice in speech emotion recognition. ATP's main disadvantage is the loss of local information, as the ATP strategy reduces the information to utterance-level, potentially masking relevant frame-level information. With Emo-Token and EMOQ-Former we performed time aggregation by applying attention mechanisms with different strategies. These methods generated optimal feature spaces for emotion recognition by retrieving relevant information related to emotion classes in the IEMOCAP database. We improved classification results compared to ATP in most of our experiments. This work sets a starting point for integrating transformer-based aggregation methods in Speech Emotion Recognition systems by also combining them with other common practices in this field, such as SSL model fine-tuning and cross-corpus training, with the objective of enhancing model performance while reducing the need to allocate bigger SSL models to get optimal features for Speech Emotion Recognition. Future work could focus on expanding these methods to other emotion recognition datasets and modalities, exploring their effectiveness in cross-corpus training for better generalization. Additionally, fine-tuning self-supervised models for emotion-specific tasks and investigating more efficient, real-time implementations are promising directions to enhance the robustness and scalability of transformer-based aggregation methods in Speech Emotion Recognition systems.

Acknowledgments. This work was supported by the Spanish Ministry of Science and Innovation for the DIPSY project (TED2021-131401B-C21), European Union's Horizon 2020 funded project "HELIOS" (No 825585), by the Universitat Politècnica de València (PAID-10-20), by the Generalitat Valenciana (ACIF/2021/187 and PROMETEO/2020/024) and by the Spanish Government (BEWORD PID2021-126061OB-C41).

References

1. Berg, A., O'Connor, M., Cruz, M.T.: Keyword Transformer: a Self-attention Model for Keyword Spotting (2021). arXiv preprint arXiv:2104.00769
2. Busso, C., et al.: IEMOCAP: interactive emotional dyadic motion capture database. Lang. Resour. Eval. **42**, 335–359 (2008)
3. Chen, S., et al.: Wavlm: large-scale self-supervised pre-training for full stack speech processing. IEEE J. Sel. Top. Signal Process. **16**(6), 1505–1518 (2022). https://doi.org/10.1109/jstsp.2022.3188113
4. Chen, S., et al.: Unispeech-sat: universal speech representation learning with speaker aware pre-training. In: ICASSP 2022 - 2022 IEEE International Conference on Acoustics, Speech and Signal Processing (ICASSP), pp. 6152–6156 (2022). https://doi.org/10.1109/ICASSP43922.2022.9747077
5. Devlin, J., Chang, M.W., Lee, K., Toutanova, K.: Bert: Pre-training of Deep Bidirectional Transformers for Language Understanding (2019)
6. Dong, L., Xu, S., Xu, B.: Speech-transformer: a no-recurrence sequence-to-sequence model for speech recognition. In: 2018 IEEE International Conference on Acoustics, Speech and Signal Processing (ICASSP), pp. 5884–5888 (2018). https://doi.org/10.1109/ICASSP.2018.8462506
7. Dosovitskiy, A., et al.: An Image is Worth 16 × 16 Words: Transformers for Image Recognition at Scale (2021)
8. Hsu, W.N., Bolte, B., Tsai, Y.H.H., Lakhotia, K., Salakhutdinov, R., Mohamed, A.: Hubert: Self-supervised Speech Representation Learning by Masked Prediction of Hidden Units (2021)
9. Jiang, D., et al.: A further study of unsupervised pretraining for transformer based speech recognition. In: ICASSP 2021 - 2021 IEEE International Conference on Acoustics, Speech and Signal Processing (ICASSP), pp. 6538–6542 (2021). https://doi.org/10.1109/ICASSP39728.2021.9414539
10. Latif, S., Zaidi, A., Cuayahuitl, H., Shamshad, F., Shoukat, M., Qadir, J.: Transformers in Speech Processing: A Survey (2023)
11. Li, J., Li, D., Savarese, S., Hoi, S.: Blip-2: Bootstrapping Language-Image Pre-training with Frozen Image Encoders and Large Language Models (2023)

12. Novoselov, S., Lavrentyeva, G., Avdeeva, A., Volokhov, V., Gusev, A.: Robust Speaker Recognition with Transformers Using wav2vec 2.0 (2022)
13. Shao, Z., et al.: Transmil: transformer based correlated multiple instance learning for whole slide image classification. Adv. Neural. Inf. Process. Syst. **34**, 2136–2147 (2021)
14. Wang, Y., et al.: Speech Swin-Transformer: Exploring a Hierarchical Transformer with Shifted Windows for Speech Emotion Recognition (2024)

Refining Multiple Instance Learning with Attention Regularization for Whole Slide Image Classification

Ilán Carretero[1(✉)], Pablo Meseguer[1,2], Rocío del Amor[1], and Valery Naranjo[1,2]

[1] Instituto Universitario de Investigación en Tecnología Centrada en el Ser Humano (HUMAN-Tech), Universitat Politécnica de Valencia, 46022 Valencia, Spain
{ilcarjuc,pabmees,madeam2,vnaranjo}@upv.es

[2] Valencian Graduate School and Research Network for Artificial Intelligence (valgrAI), 46022 Valencia, Spain

Abstract. Histopathological analysis of biopsy sections is crucial for the detection of cancer and the distinction between different tumor subtypes. To this end, pathologists identify certain key regions of the biopsy from which a diagnosis is derived. The classification of whole slide images (WSIs) can be addressed as a multiple instance learning (MIL) problem where only slide-level labels are available. In order to model the relevance scores of the different WSI patches, attention mechanisms are implemented within the MIL framework. However, excessive flexibility in the attention mechanisms for computing attention scores to patches may lead to nearly uniform attention distributions, potentially deteriorating the model's performance. In this paper, we introduce an unsupervised auxiliary loss function to recalibrate the attention mechanism enhancing emphasis on crucial patches and downscaling the influence of less relevant ones. The proposed MIL framework has been evaluated on invasive breast carcinoma (TCGA-BRCA) and renal cell carcinoma (TCGA-RCC) subtyping. The results obtained show that attention regularization not only improves the predictive capacity of the model but also significantly increases its interpretability by identifying regions and patterns of high diagnostic value.

Keywords: Hispathology imaging · Multiple Instance Learning · Attention mechanisms · Attention regularization · Explainable Deep Learning

1 Introduction

Histopathology involves studying and diagnosing diseases through microscopic examination of tissue samples extracted from patients. The histological analysis of these tissues plays a crucial role in clinical diagnosis. With the advent of digital pathology workflows that involve the scanning of biopsy slides at high

V. Julian et al. (Eds.): IDEAL 2024, LNCS 15346, pp. 398–409, 2025.
https://doi.org/10.1007/978-3-031-77731-8_36

resolution (whole slide imaging), Deep learning (DL) has significantly enhanced this field by providing diagnostic tools that often meet or exceed the performance of expert pathologists [3,13,18]. One major challenge in histopathology is the annotation of whole slide images (WSIs), which have ultra-high resolutions [8] (an average WSI is $40{,}000 \times 40{,}000$ pixels in size), making full annotation very time-consuming [17]. To overcome this limitation, weakly supervised learning (WSL) techniques rely on slide-level labels instead of detailed pixel-level annotations to address predictive tasks in histopathological imaging. This approach enables DL models to achieve high classification accuracy on WSIs and lead to scalable computational decision support systems, significantly improving diagnostic workflows in clinical practice [14].

Multiple Instance Learning (MIL) [12] is a WSL technique well-suited for histopathology, where WSIs are divided into smaller patches, treating the entire slide as a bag of these instances. The central assumption of MIL is that a bag is labeled based on the presence of at least one relevant instance: if any patch in a bag corresponds to the slide-level class assigned to the biopsy, the entire bag is labeled accordingly. MIL for WSIs prediction can be approached through instance-based or embedding-based methods. The instance-based approach derives the WSI label directly from individual tile predictions [1,7], whereas the embedding-based approach aggregates features from the tiles to form a higher-level bag representation [6,11]. Empirical evidence suggests that the embedding-based approach outperforms the instance-based method [15,19].

The success of MIL in histopathology depends both on feature extractors and attention methods that aggregate the patch-level feature representations. Feature extractors, such as convolutional neural networks (CNNs) [4,20], vision transformers [5,10] and the growing adoption of foundation models [2], are crucial for capturing detailed representations of patches. Attention mechanisms are essential for assigning varying levels of importance to different tiles, ensuring that the aggregate embedding, which represents the WSI, has a greater influence on the most significant ones [6,11]. While attention mechanisms help in adjusting the importance of each patch, it is well understood that a significant proportion of patches in biopsies are not relevant for slide-level classification [9]. Ignoring the necessity of imposing a priori conditions to recalibrate the patch weights can lead to a degradation in model performance and interpretability. In this way, irrelevant patches may be assigned undue importance, diluting the focus on critical regions and thereby impairing the model's ability to identify key pathological features accurately.

In this work, we introduce a novel auxiliary loss function designed to optimize the distribution of patch-level attention scores generated by the attention mechanisms. By incorporating this regularization term within the training regime of the multiple instance learning (MIL) framework, we achieve an overall improvement in model performance. This approach enforces a sharper differentiation among attention scores, resulting in distributions that are more concentrated and less uniformly spread. Consequently, the improved precision of the attention scores significantly enhances the models' interpretability and diagnostic utility.

2 Methodology

The overview of the WSIs classification pipeline can be seen in Fig. 1. It can be summarized into five main elements:

- A WSIs processing to automatically segment tissue regions and divide them into multiple squared lower-sized patches to fit the MIL paradigm.
- A foundation model encodes the information contained in each patch into a low-dimensional embedding with the most relevant information.
- An attention mechanism assigns a relevance score to each embedding, ensuring that each patch is weighted by its inferred attention score during aggregation.
- An auxiliary attention regularization loss function is operating on the attention scores. This penalty induces a redistribution of the patch importance assigned by the attention mechanism.
- A classifier layer using WSI-level embedding to perform the final prediction.

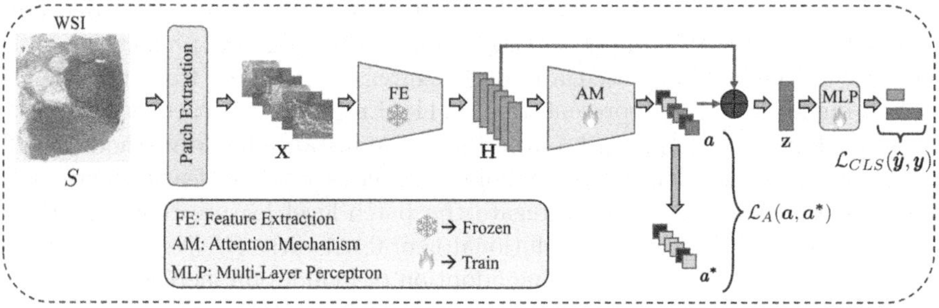

Fig. 1. Method overview. In this paper, we propose augmenting the MIL framework with an auxiliary attention regularization loss function designed to recalibrate attention scores. To do so, the WSI must first be preprocessed by splitting it into small patches. Then, using CONCH, feature representations of size 512 are extracted for each patch. These feature embeddings are used as input to the attention mechanism, from which the attention scores for each patch are extracted and used to weight them and create the WSI embedding. A classifier head is then used to generate the prediction. The attention regularization loss function complements the classification loss function, resulting in more focused and sparse attention scores.

2.1 WSI Preprocessing

The WSI preprocessing tool developed by [11] was used. First, the tissue region of each slide is automatically segmented and then accurately divided into non-overlapping patches. Therefore, we reformulate the representation of a whole slide image S to $\mathbf{X} = \{\mathbf{x_1}, \ldots, \mathbf{x_i}, \ldots, \mathbf{x_N}\}, \mathbf{x_i} \in \mathbb{R}^{n \times m}$ with X being the total number of tissue patches obtained by the WSI preprocessing toolbox, $n \times m$ the patch dimensions and N the number of patches.

2.2 Feature Extraction

The creation of representative embeddings of the patches from the feature extractor significantly affects the predictive capability of the model. Consequently, in order to extract highly informative representations, the contrastive learning from captions for histopathology model (CONCH) was used [10]. CONCH is a state-of-the-art visual-language foundation model tailored for computational pathology. The extensive amount of image-caption pairs used in CONCH pre-training allows the creation of highly representative embeddings. These reliable representations inherently lead to greater discrimination between different cancer subtypes. Consequently, simpler and lighter attention mechanisms can be used for the accurate identification of key patches, maintaining high performance while reducing computational demands. By applying CONCH on $\mathbf{X}$, the set of embeddings $\mathbf{H} = \{\mathbf{h}_1, \ldots, \mathbf{h}_i, \ldots, \mathbf{h}_N\}, \mathbf{h}_i \in \mathbb{R}^d$ is obtained, being d the embedding dimension.

2.3 Attention Module and Feature Aggregation

In this study, we utilized attention mechanisms such as Attention-Based Multiple Instance Learning (ABMIL) [6] and Clustering-constrained Attention Multiple Instance Learning Single attention Branch (CLAM-SB) [11]. ABMIL, introduced by Ilse et al. [6], incorporates a gated attention mechanism to develop a differentiable pooling operator capable of modeling nonlinear relationships. Analytically, it can be expressed as:

$$a_i = \frac{\exp\left(\mathbf{w}^\top \left(\tanh\left(\mathbf{V}\mathbf{h}_i\right) \odot \mathrm{sigm}\left(\mathbf{U}\mathbf{h}_i\right)\right)\right)}{\sum_{j=1}^{N} \exp\left(\mathbf{w}^\top \left(\tanh\left(\mathbf{V}\mathbf{h}_j\right) \odot \mathrm{sigm}\left(\mathbf{U}\mathbf{h}_j\right)\right)\right)}, \quad (1)$$

where $\mathbf{w} \in \mathbb{R}^{L\times 1}$, $\mathbf{U} \in \mathbb{R}^{L\times M}$ and $\mathbf{V} \in \mathbb{R}^{L\times M}$ are parameters, $tanh(\cdot)$ is the hyperbolic tangent, $sigm(\cdot)$ is the sigmoid function, $\odot$ is the element-wise multiplication and a_i is the i-th attention score.

CLAM-SB is based on the ABMIL framework by introducing a clustering-based approach [11]. CLAM-SB refines the attention mechanism by grouping similar instances into clusters and assigning attention weights to these clusters rather than individual instances. This adaptation allows for more robust handling of intra-bag variations and enhances the model's ability to focus on the most relevant regions within a WSI. The clustering process is integrated into the attention mechanism, enabling the model to adjust its focus based on the learned cluster representations dynamically.

Once the attention scores have been computed for each embedding, it is possible to obtain the slide-level embedding through:

$$\mathbf{z} = \sum_{i=1}^{N} a_i \mathbf{h}_i \quad (2)$$

notice that in both attention modules, it is fulfilled that $\sum_{i=1}^{N} a_i = 1$. This constraint is applied to make attention scores invariant w.r.t. the number of WSI patches.

2.4 Redistributing Attentions Scores: Attention Regularization Loss

Using attention methods, it is possible to distribute attention scores non-uniformly. However, without imposing constraints on the attention distribution, the resulting attention across the patches could become overly homogeneous. To enforce concentration of attention scores in an unsupervised fashion, it is first necessary to have a target distribution to compare with. The target distribution has been formulated as follows:

$$a^* = \begin{cases} \epsilon \text{ if } a \leq \mu \\ \psi \text{ if } a > \mu \end{cases} \tag{3}$$

where:

$$\epsilon = 1 \cdot 10^{-16}$$

$$\psi = \frac{1 - \epsilon \cdot \sum_{i=1}^{N} \mathbb{I}(a \leq \mu)}{\sum_{i=1}^{N} \mathbb{I}(a > \mu)}$$

$$\mu = \frac{\sum_{i=1}^{N} a_i}{N}$$

The formulation of the target attention distribution imposes that the attention mechanisms recalibrate the attention scores assigned to each patch, resulting in a downscaling of scores for less relevant patches and an amplification for more relevant ones. The generation of this target distribution is created dynamically from $\boldsymbol{a}$ obtained by the attention mechanism. The rationale for the dynamic part of the target attention is that in the training phase, the attention scores are adjusted, and their values are modified. Consequently, it is desired that this flexibility of the attention mechanisms remains unchanged. Also, the selection of μ as the threshold value for the difference between the assignment of attention scores is independent of the attention distribution and consequently robust because it only depends on the number of WSI patches.

Given the attention scores $\boldsymbol{a}$ and the target attention scores $\boldsymbol{a}^*$, it becomes feasible to utilize an auxiliary attention regularization loss function to enforce the distribution of $\boldsymbol{a}$ to closely match that of $\boldsymbol{a}^*$. The loss function can be approached from two different perspectives. If it is understood that one wants to reduce the distance between two vectors $\boldsymbol{a}$ and $\boldsymbol{a}^*$, one can use the $L1$ norm of the difference of both attention scores. In the scenario of considering $\boldsymbol{a}$ and $\boldsymbol{a}^*$ as two probability distributions[1], it is possible to use the Kullback-Leibler (KL) divergence. When using the KL divergence it is not possible to have zero-valued target attention scores, as it would lead to results outside the domain of

[1] Note that both $\boldsymbol{a}$ and $\boldsymbol{a}^*$ satisfy the properties of a probability density function.

the function. Hence, the residual value ϵ instead of zero. Thus the attention loss function can be described in the following ways:

$$\mathcal{L}_A = \begin{cases} \boldsymbol{a}^* \log\left(\frac{\boldsymbol{a}^*}{\boldsymbol{a}}\right) & \text{if } \boldsymbol{a}, \boldsymbol{a}^* \in \mathbb{P} \\ |\boldsymbol{a}^* - \boldsymbol{a}| & \text{if } \boldsymbol{a}, \boldsymbol{a}^* \in \mathbb{R}^n \end{cases} \tag{4}$$

where $\boldsymbol{a}, \boldsymbol{a}^* \in \mathbb{P}$ indicates that $\boldsymbol{a}, \boldsymbol{a}^*$ are considered probability distributions and $\boldsymbol{a}, \boldsymbol{a}^* \in \mathbb{R}^n$ indicates that $\boldsymbol{a}, \boldsymbol{a}^*$ considered are real vectors. For the sake of clarity, from now on we will refer to $\mathcal{L}_A = \boldsymbol{a}^* \log\left(\frac{\boldsymbol{a}^*}{\boldsymbol{a}}\right)$ as $\mathcal{L}_{KL}$ and $\mathcal{L}_A = |\boldsymbol{a}^* - \boldsymbol{a}|$ as $\mathcal{L}_{L1}$.

2.5 WSI-Level Prediction

Finally, a multi-layer perceptron (MLP) was used to obtain slide-level prediction from $\mathbf{z}$. Following [16], we initialized the weights of the MLP with the zero-shot prototypes, i.e. the L2-normalized text embedding of each class, obtained using the prompt ensemble approach proposed in [10] for each subtyping classification task. We followed this simple yet effective approach as it is demonstrated to surpass random-initialization linear probing.

For the classification task, the cross entropy was used as a loss function, which can be expressed as:

$$\mathcal{L}_{CLS} = -\sum_{i=1}^{N}\sum_{c=1}^{C} y_{i,c} \log(p_{i,c}) \tag{5}$$

where C is the number of classes, $y_{i,c}$ is a binary indicator if class label c is the correct classification for sample i and $p_{i,c}$ is the predicted probability that sample i belongs to class c. Consequently, the total loss of the model is:

$$\mathcal{L} = \mathcal{L}_{CLS} + \lambda \mathcal{L}_A \tag{6}$$

where λ is a hyperparameter to be optimized to control the impact of the attention loss function.

3 Experiments and Results

Datasets. Experimentation was carried out on two datasets from The Cancer Genomic Atlas (TCGA), publicly available at the National Cancer Institute (NIH) genomic data commons (https://portal.gdc.cancer.gov).

TCGA-BRCA (invasive breast carcinoma subtyping) comprises two different subtypes: Invasive Duct Carcinoma (IDC) and Invasive Lobular Carcinoma (ILC). The database contains 948 diagnostic WSIs, of which 757 are IDC and 191 are ILC. The patches were extracted at 20x magnification, and the mean number of patches extracted per slide is 9682.

TCGA-RCC (renal cell carcinoma subtyping) comprises three different subtypes: Kidney Renal Clear Cell Carcinoma (KIRC), Kidney Renal Papillary Cell Carcinoma (KIRP) and Kidney Chromophobe Renal Cell Carcinoma (KICH). The database contains 873 diagnostic WSIs, of which 107 are from KIRC, 270 from KIRP and 496 from KICH. The patches were extracted at 20x magnification, and the mean number of patches extracted per slide is 13208.

Experimental Setup. From each WSI, 256×256 patches were extracted non-overlapping, from which the embeddings were extracted using CONCH. In both datasets, 70% of the data was used for training and 30% for testing. Of the training dataset, 20% was left to validate and optimize hyperparameters. The models were trained with three different pseudo-random initializations to analyze the results obtained under different starting conditions. The accuracy (ACC), the area under the receiver operating characteristic curve (AUC), and the F1-score were utilized as evaluation metrics for all datasets. In the case of the TCGA-RCC dataset, which is multi-class, the average one-versus-rest AUC and the F1 macro average were used. The hyperparameter λ corresponding to the attention loss function $\mathcal{L}_A$ was obtained through a grid search on the validation set.

Implementation Details. In the training phase, the AdamW optimizer with a learning rate of $\eta = 1 \cdot 10^{-4}$ and weight decay of $1 \cdot 10^{-5}$ was used for both ABMIL and CLAM-SB. Being in a MIL context with WSIs with different number of patches, a batch of 1 was used. The models were trained during 20 epochs. All the experiments were conducted on the NVIDIA DGX A100 system. The embeddings extracted with CONCH and the employed partitions can be shared on demand.

3.1 Classification Performance

The performance of the attention regularization loss was evaluated for the ABMIL and CLAM-SB aggregators and the TCGA-BRCA and TGCA-RCC databases. The results obtained are presented in Table 1.

Adding the attention loss function to the MIL attention modules outperforms both datasets in terms of ACC, AUC and F1 concerning the ABMIL and CLAM-SB baseline models. Thus it can be seen that for both databases and all metrics, the best results (in bold) are always by applying one of the attention loss functions. In the case of the TCGA-BRCA dataset, the model with the best overall performance is CLAM-SB + KL with a relevant improvement over CLAM-SB in accuracy and F1 score of 3.52% and 4.08%, respectively. In contrast, for the TCGA-RCC dataset, the model with the best general performance is ABMIL + L1, with significant improvements in both accuracy and F1 score of 1.53% and 2.27%, respectively.

To demonstrate the effectiveness of the implemented attention loss function, it has also been evaluated to apply $\mathcal{L}_A$ as the squared difference between $\boldsymbol{a}$ and $\boldsymbol{a}^*$ ($\mathcal{L}_{L2}$). By applying the $L2$ norm of the differences, the penalty is reduced and

Table 1. Classification results in the test set for TCGA-BRCA and TCGA-RCC. Cases corresponding to the baseline ABMIL and CLAM-SB aggregators are shown in gray. The best MIL model and attention loss function for a given metric are indicated in bold. For each metric, the results belonging to the best sub-optimal MIL model are underlined. The sub-numbers indicate the difference between the baseline model and the models with the loss functions. The blue arrow indicates improvement while the red arrow indicates a decrease in performance. The terms KL, L1 and L2 are employed to refer to $\mathcal{L}_{KL}$, $\mathcal{L}_{L1}$ and $\mathcal{L}_{L2}$ respectively.

Method	TCGA-BRCA			TCGA-RCC		
	ACC(↑)	AUC(↑)	F1(↑)	ACC(↑)	AUC(↑)	F1(↑)
ABMIL	87.09	93.35	82.09	93.87	98.63	92.41
ABMIL + KL	89.08 (+1.99) ↑	93.59 (+0.24) ↑	84.28 (+2.19) ↑	95.02 (+1.15) ↑	98.71 (+0.08) ↑	94.03 (+1.62) ↑
ABMIL + L1	88.73 (+1.64) ↑	93.68 (+0.33) ↑	83.86 (+1.77) ↑	**95.40** (+1.53) ↑	**98.73** (+0.10) ↑	**94.68** (+2.27) ↑
ABMIL + L2	86.74 (−0.35) ↓	93.65 (+0.30) ↑	81.57 (−0.52) ↓	93.23 (−0.64) ↓	98.62 (−0.01) ↓	91.34 (−1.07) ↓
CLAM-SB	87.09	93.35	82.16	93.36	98.64	91.48
CLAM-SB + KL	**90.61** (+3.52) ↑	93.62 (+0.27) ↑	**86.24** (+4.08) ↑	93.10 (−0.26) ↓	98.23 (−0.41) ↓	90.60 (−0.88) ↓
CLAM-SB + L1	89.08 (+1.99) ↑	**93.73** (+0.38) ↑	84.28 (+2.12) ↑	94.38 (+1.02) ↑	98.57 (−0.07) ↓	92.87 (+1.39) ↑
CLAM-SB + L2	88.03 (+0.94) ↑	93.46 (+0.11) ↑	83.04 (+0.88) ↑	92.34 (−1.02) ↓	98.46 (−0.18) ↓	90.03 (−1.45) ↓

is not as significant concerning *KL* or *L1*. Regarding the quantitative results, it should be noted that in several of the metrics the predictive ability of the model when applying $\mathcal{L}_{L2}$ is lower than the predictive ability of the baseline model and therefore lower than applying $\mathcal{L}_{KL}$ and $\mathcal{L}_{L1}$. This demonstrates that the higher penalty generated by $\mathcal{L}_{KL}$ and $\mathcal{L}_{L1}$ on the dispersion in the attention distribution is necessary to improve the classification metrics consistently.

3.2 Analysis of Attention Scores

To evaluate the impact of incorporating the attention regularization loss function in the TCGA-BRCA and TCGA-RCC datasets, we plotted the attention scores of the WSIs. For each dataset, we calculated the mean attention scores below and above μ for the WSIs, comparing both the baseline models and the models enhanced with the attention regularization loss. The results of this comparison are presented in Fig. 2.

Similar insights can be drawn from both TCGA-BRCA and TCGA-RCC. When we analyze those attention scores $\leq \mu$, we observe that the attentions of the baseline model are more distributed, reaching values close to μ. However, when analyzing the models with the loss function $\mathcal{L}_A$, it can be seen how the attention values are more concentrated in values farther away from the mean. This means that when applying $\mathcal{L}_A$, the patches with attention values $\leq \mu$ become more residual and, therefore, have less influence on the WSI embedding $\boldsymbol{z}$.

When analyzing attention scores $> \mu$, the opposite effect occurs. Since $\mathcal{L}_A$ forces a greater concentration of attention scores in higher values, it is caused as a collateral effect by a greater dispersion in these attention values. Thus, it can be seen that in the baseline models (purple distributions), the attention scores are more concentrated near μ, while when applying $\mathcal{L}_A$ (green distributions), the attention values are farther away from μ.

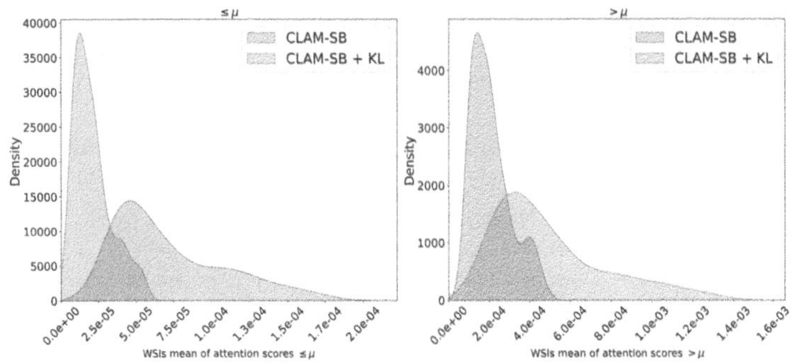

(a) TCGA-BRCA

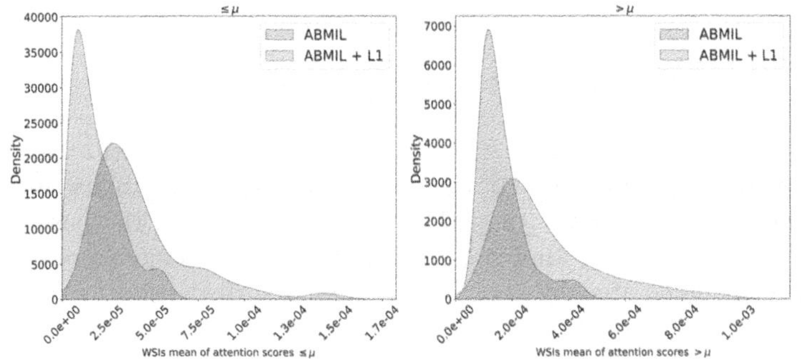

(b) TCGA-RCC

Fig. 2. Kernel density estimates of the distribution of attention scores for baseline models (purple) and models with the attention regularization loss function (green). The left column shows the mean attention scores for the WSIs $\leq \mu$. The right column shows the mean of the attention scores $> \mu$. (Color figure online)

3.3 Attention Visualization and Interpretability

In a more in-depth analysis, the attention maps generated for TCGA-BRCA and TCGA-RCC with the baseline models and the models with $\mathcal{L}_A$ are visualized. The qualitative results obtained are illustrated in Fig. 3.

In both datasets the effect of $\mathcal{L}_A$ on the attention maps is significant. As can be seen in both ABMIL and CLAM-SB, the differences between the attention scores of the different patches are smaller compared to adding $\mathcal{L}_A$ in both cases. This behavior results in a smaller difference between those patches that are considered relevant and those patches that are not to the attention maps of ABMIL + L1 (TCGA-RCC) and CLAM-SB + KL (TCGA-BRCA).

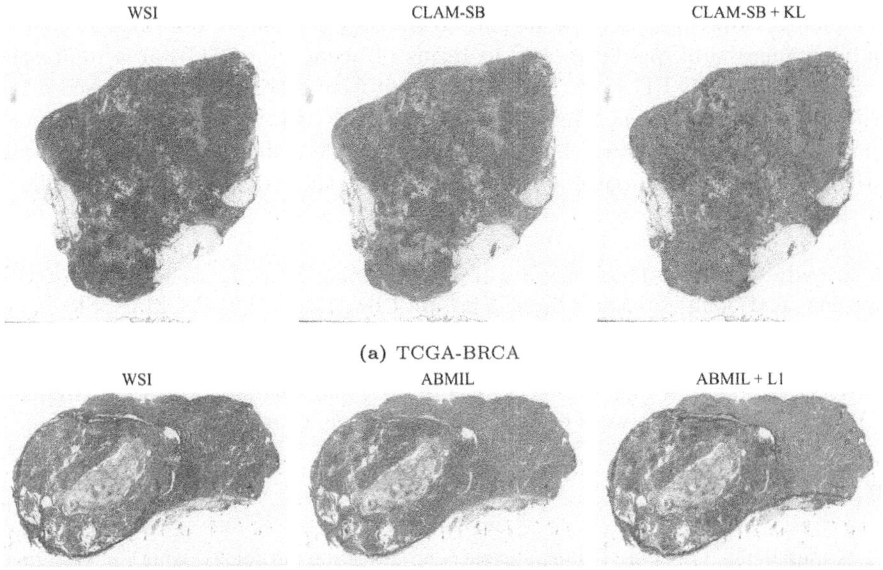

(b) TCGA-RCC

Fig. 3. Visual examples of the attention maps produced by different models for a TCGA-BRCA WSI and a TCGA-RCC WSI. The left column shows the original slide. The center column shows the attention map of the baseline models. The right column shows the attention maps when applying the attention loss function. The higher the intensity of the red color, the higher the attention score for that patch. (Color figure online)

When applying $\mathcal{L}_A$ the differences between patches of higher and lower relevance become larger. These differences lead to greater interpretability when using $\mathcal{L}_A$, since, as shown in Fig. 3, the model is forced to focus on certain patches. Specifically, the patches on which the model focuses are grouped into defined regions in both cases. This finding is of particular interest because the model is not only able to identify regions associated with the cancerous subtype but also highlights which regions are of greatest importance for that subtype in the WSI.

4 Conclusion

Histopathological image analysis leveraging deep learning techniques has demonstrated exceptional results in slide-level predictive tasks. While the performance of these models is critically important, the generation of interpretable results that expert pathologists can rely on for even more accurate diagnoses is equally necessary. In this paper, we quantitatively and qualitatively demonstrate that predictive performance does not necessarily have to be inversely related to explainability. By incorporating an unsupervised attention regularization loss function, we show that it is feasible to enhance model performance while simultaneously producing more interpretable attention maps. The results, obtained from two public cancer subtype databases, in terms of accuracy (ACC), area under the curve (AUC), and F1 score, as well as qualitative assessments, underscore the efficacy of our approach. These findings pave the way for introducing methods that generate sparser, and thus more interpretable, attention scores within the MIL paradigm in histopathology, while also boosting the predictive capabilities of the models.

Acknowledgments. This work has received funding from the Spanish Ministry of Economy and Competitiveness through project PID2022-140189OB-C21 (ASSIST) and from Generalitat Valenciana through COMTACTS2 (CIPROM/2022/20). The work of Rocío del Amor and Pablo Meseguer has been supported by the Spanish Ministry of Universities under an FPU Grant (FPU20/05263) and valgrAI - Valencian Graduate School and Research Network of Artificial Intelligence, respectively.

References

1. Campanella, G., et al.: Clinical-grade computational pathology using weakly supervised deep learning on whole slide images. Nat. Med. **25**(8), 1301–1309 (2019)
2. Chen, R.J., et al.: Towards a general-purpose foundation model for computational pathology. Nat. Med. **30**(3), 850–862 (2024)
3. Das, K., Conjeti, S., Roy, A.G., Chatterjee, J., Sheet, D.: Multiple instance learning of deep convolutional neural networks for breast histopathology whole slide classification. In: 2018 IEEE 15th International Symposium on Biomedical Imaging (ISBI 2018), pp. 578–581. IEEE (2018)
4. Hou, L., Samaras, D., Kurc, T.M., Gao, Y., Davis, J.E., Saltz, J.H.: Patch-based convolutional neural network for whole slide tissue image classification. In: Proceedings of the IEEE Conference on Computer Vision and Pattern Recognition, pp. 2424–2433 (2016)
5. Huang, Z., Bianchi, F., Yuksekgonul, M., Montine, T., Zou, J.: Leveraging medical twitter to build a visual–language foundation model for pathology AI. bioRxiv, pp. 2023–03 (2023)
6. Ilse, M., Tomczak, J., Welling, M.: Attention-based deep multiple instance learning. In: International Conference on Machine Learning, pp. 2127–2136. PMLR (2018)
7. Landini, G., Martinelli, G., Piccinini, F.: Colour deconvolution: stain unmixing in histological imaging. Bioinformatics **37**(10), 1485–1487 (2021)
8. Li, B., Li, Y., Eliceiri, K.W.: Dual-stream multiple instance learning network for whole slide image classification with self-supervised contrastive learning. In: Proceedings of the IEEE/CVF Conference on Computer Vision and Pattern Recognition, pp. 14318–14328 (2021)
9. Litjens, G., et al.: 1399 H&E-stained sentinel lymph node sections of breast cancer patients: the CAMELYON dataset. GigaScience **7**(6), giy065 (2018)
10. Lu, M.Y., et al.: A visual-language foundation model for computational pathology. Nat. Med. **30**(3), 863–874 (2024)
11. Lu, M.Y., Williamson, D.F., Chen, T.Y., Chen, R.J., Barbieri, M., Mahmood, F.: Data-efficient and weakly supervised computational pathology on whole-slide images. Nat. Biomed. Eng. **5**(6), 555–570 (2021)
12. Maron, O., Lozano-Pérez, T.: A framework for multiple-instance learning. In: Advances in Neural Information Processing Systems, vol. 10 (1997)

13. Melendez, J., et al.: A novel multiple-instance learning-based approach to computer-aided detection of tuberculosis on chest X-rays. IEEE Trans. Med. Imaging **34**(1), 179–192 (2014)
14. Morales, S., Engan, K., Naranjo, V.: Artificial intelligence in computational pathology-challenges and future directions. Digit. Signal Process. **119**, 103196 (2021)
15. Sharma, Y., Shrivastava, A., Ehsan, L., Moskaluk, C.A., Syed, S., Brown, D.: Cluster-to-conquer: a framework for end-to-end multi-instance learning for whole slide image classification. In: Medical Imaging with Deep Learning, pp. 682–698. PMLR (2021)
16. Silva-Rodríguez, J., Hajimiri, S., Ben Ayed, I., Dolz, J.: A closer look at the few-shot adaptation of large vision-language models. In: Proceedings of the IEEE/CVF Conference on Computer Vision and Pattern Recognition (CVPR), pp. 23681–23690 (2024)
17. Srinidhi, C.L., Ciga, O., Martel, A.L.: Deep neural network models for computational histopathology: a survey. Med. Image Anal. **67**, 101813 (2021)
18. Wang, X., et al.: Predicting gastric cancer outcome from resected lymph node histopathology images using deep learning. Nat. Commun. **12**(1), 1637 (2021)
19. Wang, X., Yan, Y., Tang, P., Bai, X., Liu, W.: Revisiting multiple instance neural networks. Pattern Recogn. **74**, 15–24 (2018)
20. Yan, Y., Wang, X., Guo, X., Fang, J., Liu, W., Huang, J.: Deep multi-instance learning with dynamic pooling. In: Asian Conference on Machine Learning, pp. 662–677. PMLR (2018)

Evaluating Performance and Trustworthiness of RAG Systems for Generating Administrative Text

Hugo Sánchez-Navalón[1(✉)], Carlos Monserrat[1], Dario Garigliotti[2], and Cèsar Ferri[1]

[1] Universitat Politècnica de València, Valencia, Spain
hsannav@upv.edu.es
[2] University of Bergen, Bergen, Norway

Abstract. As administrative language tends to be formal and exempt from double meanings or figurative expressions, it is a particular domain in which to explore the performance of Language Models. This paper presents a study on the feasibility of creating administrative texts-based RAG systems to serve as chatbots, analyzing the performance for this task of several Small and Large Language Models and defining ways of evaluating whether they hallucinate or not and whether they provide the user useful information or not. Conventional metrics depending on ground truth labels, such as cosine similarity or those from the ROUGE family, are explored, as well as new approaches to using other metrics not so popular in text evaluation, such as Euclidean and Manhattan distances. Moreover, all those objective metrics are compared with a subjective Likert scale to assess their performance at solving real users' problems and to find relations between subjective perceptions and objectively measured metrics for each of the RAG systems proposed. The results show that SLM models (such as NeuralChat) can perform as well as an LLM if RAG programming provides them with an appropriate context.

Keywords: RAG · LLM · SLM · Administrative Text · ROUGE · Cosine similarity · Question Answering · Chatbot

1 Introduction

This paper presents a study on the evaluation and efficacy of Retrieval Augmented Generative (RAG) models based on administrative texts. The objective is to assess the feasibility of building a chatbot capable of answering questions from different user profiles regarding various procedures related to the Public Administration of the Comunitat Valenciana, an autonomous community of more than 5 million inhabitants in the east of Spain. Generalitat Valenciana is the public entity that rules the region of Comunitat Valenciana.

V. Julian et al. (Eds.): IDEAL 2024, LNCS 15346, pp. 410–421, 2025.
https://doi.org/10.1007/978-3-031-77731-8_37

A formal register, exempt from double meanings or figurative expressions, characterizes the language of administrative texts. The questions typically asked by users are direct and relatively short. This suggests that building effective systems based on fewer parameters than state-of-the-art Large Language Models (LLMs) may be possible.

RAG systems are based on the original proposal by (Lewis et al., 2021) to help enhance Language Models responses with domain-specific information while skipping processes like fine-tuning, which may be costly. RAG model comprises two core components: a retriever and a generator. The retriever identifies relevant information from the external knowledge source based on the user's query or input. This retrieved information is then passed to the generator, a Language Model, which synthesizes a coherent and informative response incorporating the retrieved knowledge. This architecture avoids the high computational cost of building an entire Language Model over a specific context, as well as the considerable cost of fine-tuning already existing ones over a new context. Moreover, as they can easily identify the sources of their answers by listing the documents retrieved for each user's query, they provide a fairly high explainability for each of their generated texts.

Apart from that, the technical language that characterizes this context makes it difficult for some conventional users to understand some of the key concepts of certain documentation pieces. Language Models have shown good performance in rewriting text for different types of users to understand according to their needs, as educational-oriented research in Artificial Intelligence has shown [Gan et al., 2023]. Therefore, having well-performing RAG systems built over administrative texts could be a feasible solution to make them more understandable to different types of users and adaptable to their needs.

This paper is structured as follows. In Sect. 2, we summarise the relevant related works. The methodology employed in the paper is presented in Sect. 3. We include the results from an experimental evaluation of our method in Sect. 4. Finally, Sect. 5 concludes the paper with a recapitulation of our work and future extensions.

2 Related Work

The landscape of Retrieval-Augmented Generation (RAG) systems has been rapidly evolving since they were introduced back in 2020. In (Gan et al., 2024), the authors propose a comprehensive survey of RAG systems, including the importance of optimizing their retriever and generative subsystems and a bunch of evaluation approaches that involve experimenting with quality scores and robustness-related ones. They also present a hybrid approach of fine-tuning along with RAG construction to enhance its performance further.

A significant challenge in using Language Models (LMs) for generating administrative texts, which often contain objective information, is the risk of hallucinations. In (Zhang et al., 2023), the authors provide an overview of the hallucination issue and its types, focusing on fact-conflicting hallucination (generation of text that contradicts established knowledge). They show benchmark construction along with good-performing metrics to measure hallucination, involving human, model-based, and rule-based evaluation. They also offer ways to mitigate it during pre-training, fine-tuning, and inference.

While fine-tuning has been widely regarded as a primary method for improving model performance, (Ovadia et al., 2024) challenges this notion by showing how RAG, if correctly implemented, could potentially outperform it. As they uncover *forgetting* and *immemorization* as two of the main problems that lead to factual errors, they conclude that giving specific context for each generation seems a more reliable choice, even though fine-tuning remains a good option for many use-cases. Considering specifically objective information related to RAG content, as we expect from administrative texts, in (Lála et al., 2023), the authors discuss the feasibility of building a RAG based on technic-scientific knowledge. It turns out to be "more cost-effective than humans while retaining its accuracy on par with human researchers". This suggests that RAG construction may be a feasible approach for the context given in administrative texts (as proposed in the present paper) because they handle characteristics similar to those of scientific papers.

Moreover, considering the subfield of Legal Question Answering, in which our work is included, in (Abdallah et al., 2023), the authors provide an extensive explanation of the state of the art of these systems. They review the challenges we face (related to reliability, domain expertise, and data availability), the Machine Learning approaches usually given, and some surveys about their general performance. Transformer-based systems seem to be more effective than classic Machine Learning algorithms, which indicates that our approach based on Language Models is more suitable to catch up with state-of-the-art results than any other one. In addition, in (Garigliotti et al., 2024), the authors explore a way of evaluating RAG-driven self-supported Question Answering by creating a custom dataset including evaluation questions, ground truth answers and context relation, finding that LLMs such as GPT-4 are capable of reliably generating answers to most kinds of questions if proper context is provided.

3 Materials and Methods

This section details the methodological framework employed in our research, beginning with the specific approach to RAG implementation. As can be seen in Fig. 1, we first handle context obtention and then text generation through Language Models. Afterward, a benchmark consisting of labeled questions is fed into the RAG systems and evaluated through objective and subjective measures.

3.1 Approach

A Selenium bot was used to retrieve all downloadable PDF content in which all technical and administrative information is stored within the Generalitat Valenciana website. A Breadth-First-Search on its categories tree was done for that

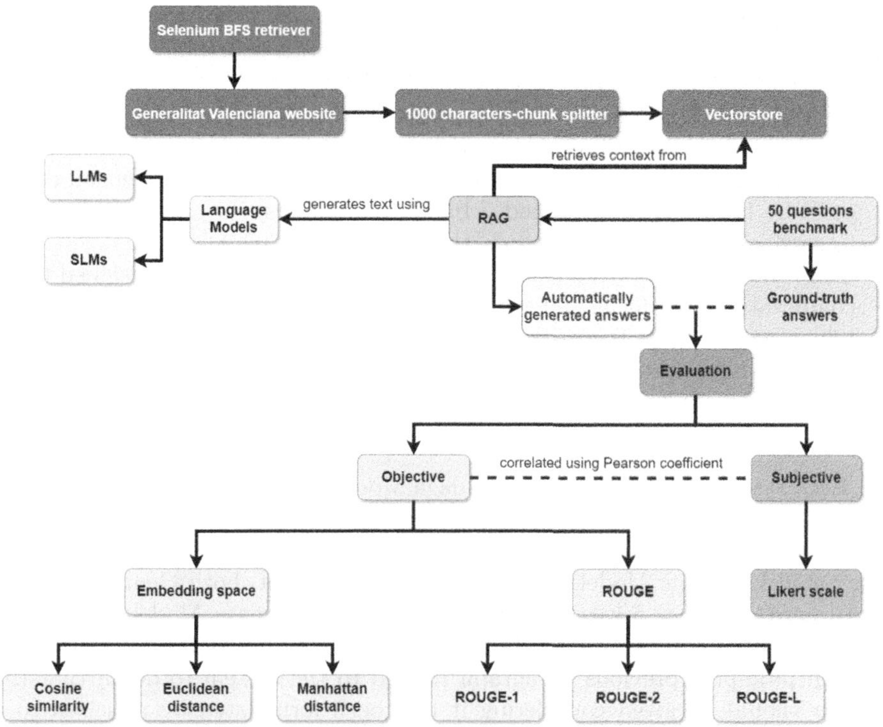

Fig. 1. Workflow of RAG building and evaluation process.

purpose. A total of 1,619 PDF documents were downloaded, which sum a size of 90.4 MB. Each of them is referred to a possible interaction between a citizen and the Valencian Regional Government for a particular purpose, and each is divided into up to four main parts, being "Basic Information", "Appliance", "Procedure", and "Resolution". Some examples of these interactions are tax returns, gambling self-prohibition, child adoption, and cultural events authorization.

We used dense retrieving for context obtention, a method for retrieving documents semantically similar to a query by encoding them into high-dimensional vectors and comparing their distances in the resulting vector space, as proposed in (Karpukhin et al., 2020). We first segmented the entire corpus of text into manageable chunks of 1,000 characters, ensuring a 100-character overlap between adjacent chunks. This overlap helps capture context that might be split across chunk boundaries. The embeddings were generated using *text-embedding-ada-002-v2* by OpenAI. The resulting vector representations were stored in a Chroma vector database. Chroma is a powerful and efficient vector database designed specifically for similarity search tasks, as detailed in (Huber et al., 2024).

Generator parts were built by using a set of Language Models (both Small Language Models and Large Language Models), including:

- **GPT Models.** 3.5, 4 and 4o. Widely known family of models developed by OpenAI that exhibit near-human level capabilities in some of the scenarios they have been tested on (OpenAI et al., 2024).
- **Phi 3 Mini.** A 3.8-billion parameters Language Model developed by Microsoft. It is defined as "instruction-tuned" (Abdin et al., 2004).
- **NeuralChat.** Based on the original Mistral, it is a fine-tuned version developed by Intel based on around 7 billion parameters. Its primary purpose is to serve as a base for building chatbots (Lv et al., 2023).
- **Gemma.** Family of open-source models developed by Google designed to be computationally efficient. Two versions have been used, involving 2 and 7 billion parameters (Team et al., 2024).
- **LLaMA 3.** It is a family of open-source Language Models developed by Meta AI aimed at multi-purpose tasks like creative writing or coding. The version used is based on 7 billion parameters (Touvron et al., 2023).

Each system used one of the Language Models and a cosine similarity-based retriever operating on the vector database. The k most similar passages to the user's query were selected to provide context to the language model in each case. For Small Language Models (SLMs: Phi 3, NeuralChat, both Gemma versions and LLaMA 3), $k = 5$ was used, while for Large Language Models (LLMs: all GPT variants), $k = 20$. Augmentation was addressed by adding to the context of each prompt a previous paragraph related to the system to detail that the answer should be given considering it had been formulated by a citizen to a particular Public Administration seeking an objective and truthful answer.

GPT LLMs were accessed through the OpenAI API. SLMs were run on a private Ollama server allocated in a computer with an AMD Ryzen 7 5800H CPU, a Nvidia Geforce GTX 1650 with 4 GB DDR4 GPU, and 16 GB of DDR4 RAM. Communication with all of them was handled through LangChain's Python library.

3.2 Evaluation

Evaluation questions were generated with the help of the GPT-4o model. Under the premise that each potential technical question would relate to one of the available administrative procedures, a random sample of one hundred procedures was used to generate one potential user question per procedure, using the prompt *"Generate a question which can be answered with the following text. It has to be a possible question that a citizen could do in order to get information related with the text."*, and adding then the whole text of the selected document. Subsequently, fifty questions were manually selected, prioritizing those most likely to be posed by potential users while ensuring a relative diversity in the topics addressed (procedures, requirements, subsidy programs, selection criteria for specific allocations, etc.). These fifty questions were then manually answered, drawing upon information about the corresponding procedure to establish a ground truth for each. The sum of questions and ground truth labels is what we have used as test collection. As an example, one generated question was *"What is the discount percentage on the monthly housing quota for large or single-parent families of general category and special category?"*, and its ground truth answer was determined to be *"The discount on the monthly housing fee will be 15% for large family or single-parent family title of general category, and 40% for large family or single-parent family title of special category"*.

Generated questions were fed into all RAG systems to get automatically generated answers. After that, they were manually evaluated by human subjective comparison to ground truth answers, according to the following previously defined Likert scale:

1. The response either explicitly states that the answer is unknown or is entirely incorrect and contains significant factual inaccuracies (hallucinations).
2. The response may be tangentially related to the question but contains major hallucinations that render it useless or potentially misleading.
3. The response contains some correct elements but fails to address the most relevant aspects of the question, severely limiting its usefulness to the user.
4. The response is correct and likely helpful but lacks important information.
5. The response is correct and provides an appropriate level of detail. There may be minor discrepancies compared to the established ground truth, but these are negligible and do not alter the overall meaning or usefulness of the response.

Concerning the objective evaluation of texts and assessing similarities between them, one of the core concepts in the literature has been representing words or documents as dense vectors (embeddings) in a high-dimensional space and later measuring their similarity using cosine similarity. (Mikolov et al., 2013) demonstrated the effectiveness of cosine similarity in capturing semantic relationships between words. However, its efficacy has recently been questioned, and some research led by Netflix has shown its limitations, proposing evaluating with other metrics so that similarities are not arbitrarily drawn, as detailed in (Steck et al., 2024).

Embeddings of each ground truth and each synthetically generated answer were calculated and stored. Considering these recent debates about the arbitrariness or not of cosine similarity, it was computed along with Euclidean and Manhattan distances for each of the ground truth-generated answer pairs. The reasoning is that while cosine similarity accounts for both vectors laying in the same direction in the vectorial space defined by the embedding model, Euclidean and Manhattan account for those representations being geographically close. As administrative texts are considered fairly exempt from double meanings and other semantic issues normally addressed by cosine similarity, metrics based on spatial distance could be more meaningful and more effective at reflecting details likely to be hidden by cosine similarity in this context. To make them more meaningful in interpretation, each of the calculated values for Euclidean and Manhattan distances has been subtracted from 1 to get metrics where the higher the value, the greater the precision and fidelity to the ground truth.

Apart from that, metrics from ROUGE (Recall-Oriented Understudy for Gisting Evaluation) family have been used. They are a widely used set of metrics for evaluating the quality of automatic summaries and machine translations by comparing them to human-generated reference texts, as proposed in (Lin, 2004). The metrics are based on the overlap of n-grams (contiguous sequences of n words) between the system-generated text and the reference text.

Several variations of ROUGE exist, each focusing on different aspects of the text, of which it is worth mentioning:

- **ROUGE-N.** Measures the overlap of n-grams between the system and reference texts. Common values for N are 1 (unigrams) and 2 (bigrams).
- **ROUGE-L.** Assesses the longest common subsequence (LCS) between the system and reference texts, which can capture sentence-level structures better than ROUGE-N.

ROUGE scores range from 0 to 1, with higher scores indicating a stronger similarity between the system-generated text and the reference text. While ROUGE has been criticized for its limitations in capturing meaning and coherence, as in (Ganesan, 2018), it remains a popular and practical tool for evaluating text generation systems due to its simplicity and computational efficiency. Therefore, for each synthetically generated answer, ROUGE-1, ROUGE-2, and ROUGE-L have also been computed.

Finally, the correlation of each calculated metric to the corresponding Likert labels has been computed individually for each of the RAGs built as a way to assess which of them seem more likely to reflect response accuracy.

4 Results

As Fig. 2 shows, Large Language Models tested tend to show a slightly higher average in Likert subjective punctuation than Small Language Models, with GPT-4o showing the highest recorded average. However, there are no significant differences at α of 0.95 with the punctuation obtained by RAGs involving NeuralChat and Phi 3. GPT-4o only shows significant differences with both versions of Gemma and LLaMA 3, while GPT-3.5 and GPT-4 only differ significantly from the 2 billion parameters version of Gemma. As can be seen, NeuralChat almost catches up with the performance of GPT-4 in the experiments run.

Cosine similarity in this context seems to show little improvement to Euclidean and Manhattan-based similarities in most cases, with Manhattan outperforming it in the case of Phi 3 evaluation, as Fig. 3 shows, even though all three metrics show poor performance in that example given. It is worth mentioning that cosine, Euclidean, and Manhattan-defined metrics are highly sensitive to how each generative model rearranges information when expressing its answer, as from the conclusions which can be drawn from (Steck et al., 2024). It can be asserted that, in the case of GPT-4-based generation, all three metrics seem to be good predictors of the reliability of the answer in this specific context. Additionally, it can be seen that these metrics are also moderately informative in the

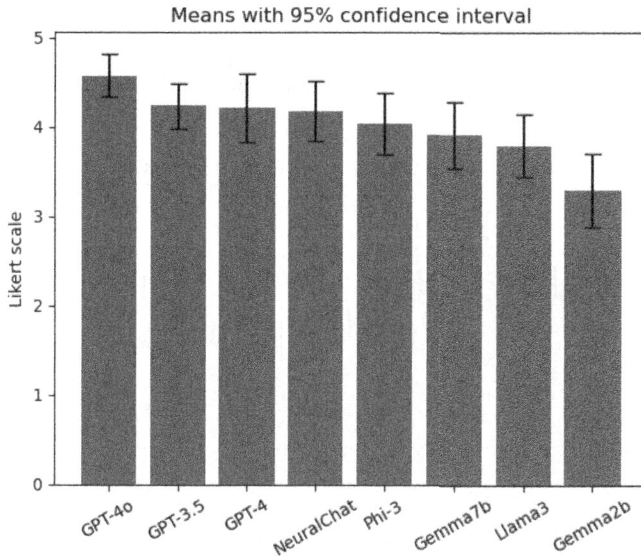

Fig. 2. Average of subjective Likert scores of the answers generated by each of the RAG systems. Shown with a confidence interval with $\alpha = 0.95$.

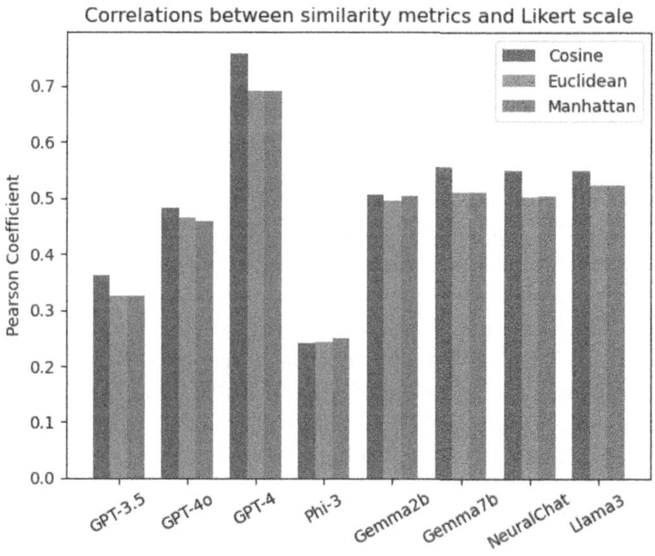

Fig. 3. Pearson correlation coefficient between subjective Likert punctuation and each of the embedding-related metrics for each RAG system.

case of SLMs with the exception of Phi 3. The fact that their lower complexity makes them more propense to replicate pieces of the context retrieved seems to leverage the association between the metrics measured in the embedding space and the loyalty to the actual expected answer more than it shows in GPT-3.5. Overall, as all observed correlations are strictly positive, it can be ensured they have certain mutual information in common with human subjective labels.

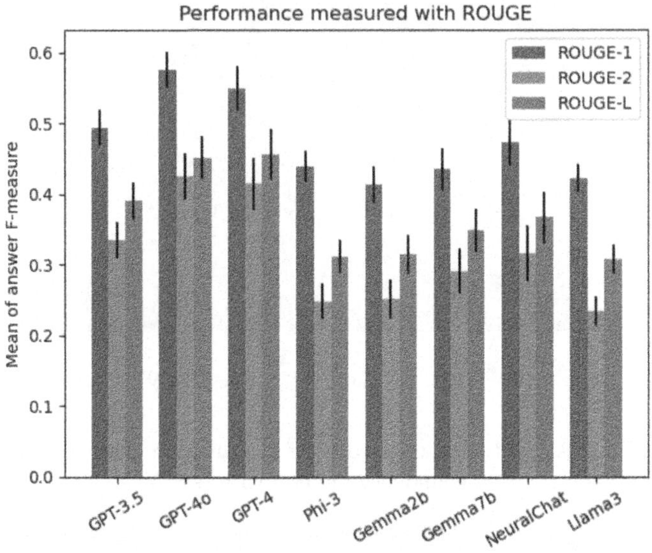

Fig. 4. Average of answer F-measure for each of the RAG systems built. Shown with a confidence interval with $\alpha = 0.95$ for each RAG system.

In the case of ROUGE metrics, a slight but clear outperform can be seen in GPT-4o and GPT-4 generated answers, which are statistically different from every other observed metric at an α of 0.95, as Fig. 4 shows. NeuralChat and GPT-3.5 show near-equivalent performance with no significant differences, and the rest of the SLMs seem to show similar metrics, which are slightly worse than those from GPT-3.5 when they appear as significantly different.

When measuring correspondence between ROUGE metrics and human subjective labels, we see that the highest correspondence belongs to the simplest model employed, Gemma in its 2 billion parameters version, as Fig. 5 shows. Again, the lowest correspondence appears in Phi 3. Apart from the finding that ROUGE-2 shows the highest correlation with subjective labeling in the simplest models (Phi 3 and Gemma with 2 billion parameters), followed by ROUGE-L and then ROUGE-1, and noting that GPT-3.5 and GPT-4o exhibit relatively lower ROUGE-1 scores compared to the other two metrics, no other consistent patterns emerge across the models when comparing these metrics.

Considering the observed relationships, it is evident that ROUGE metrics offer a quantitative insight that correlates appreciably with subjective evaluations captured through Likert scales. This correlation highlights the potential of ROUGE metrics to serve as a reliable metric for assessing semantic accuracy in

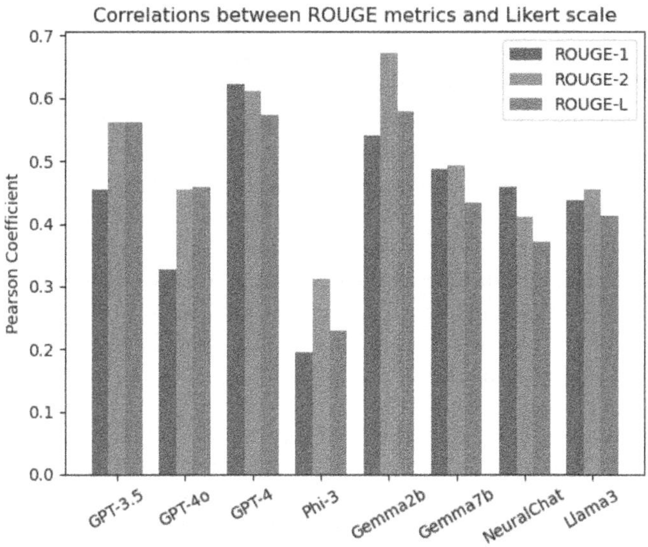

Fig. 5. Pearson correlation coefficient between subjective Likert punctuation and each of the ROUGE metrics measured for each RAG system.

the responses generated by SLMs. Consequently, including ROUGE-based evaluations could enhance the robustness of the assessment framework for SLMs, providing an objective measure that aligns well with human judgment. This integration could also facilitate more nuanced and precise model performance evaluations, driving further advancements in developing semantically coherent language models.

5 Conclusions

While a RAG built with GPT-4o appears to be the best option for ensuring optimal semantic accuracy in an administrative context, its size and high computational cost suggest that exploring alternatives like NeuralChat might be a viable solution. NeuralChat, which can operate on conditional hardware and demonstrates comparable performance, could make the system more affordable and accessible to the general public.

This article has also analysed the different existing metrics for evaluating RAG models. Concerning this analysis, it has been observed that the cosine-based distance and, above all, the ROUGE-2 metric present a high correlation with the evaluations of the responses made by humans following the Likert scale.

Finally, the objective evaluation of NeuralChat's semantic agreement with expected answers through the embedding space metrics mentioned (including cosine similarity, Euclidean distance, Manhattan distance, and ROUGE metrics) opens new possibilities for model performance enhancement. Related to that, one future work could involve training machine learning models to predict the likelihood of the chatbot's answers being correct. When these models predict an answer as potentially incorrect, an alternative approach could be employed using a GPT-4o-based RAG. This secondary system would handle a smaller subset of questions, thereby ensuring a more accurate and efficient resolution of queries that NeuralChat fails to address satisfactorily. This hybrid approach could leverage models to improve response accuracy and reliability.

The fact that NeuralChat shows better results than other SLMs in terms of subjective labeling suggests that the fact that it has been fine-tuned to work in chatbot contexts might enhance its performance compared to other models with a similar number of parameters. Therefore, a feasible way of further improving it could be to fine-tune over NeuralChat in this specific administrative context by adopting the hybrid approach proposed by (Gao et al., 2024).

Acknowledgments. This work was funded by the Norwegian Research Council grant 329745 Machine Teaching for Explainable AI, CIPROM/2022/6 (FASS-LOW) funded by Generalitat Valenciana, the EC H2020-EU grant agreement No. 952215 (TAILOR), and Spanish grant PID2021-122830OB-C42 (SFERA) funded by MCIN/AEI/10.13039/501100011033 and "ERDF A way of making Europe". Authors thank the Cátedra de Inteligencia Artificial aplicada a la Administración Pública of Universitat Politècnica de València (UPV).

References

Abdallah, A., Piryani, B., Jatowt, A.: Exploring the state of the art in legal QA systems. J. Big Data **10**(1) (2023)

Abdin, M., et al.: Phi-3 technical report: a highly capable language model locally on your phone (2024)

Gan, W., Qi, Z., Wu, J., Lin, J. C.-W.: Large language models in education: vision and opportunities (2023)

Ganesan, K.: Rouge 2.0: updated and improved measures for evaluation of summarization tasks (2018)

Gao, Y., et al.: Retrieval-augmented generation for large language models: a survey (2024)

Garigliotti, D., Johansen, B., Kallestad, J.V., Cho, S.-E., Ferri, C.: EquinorQA: large language models for question answering over proprietary data. In: ECAI 2024 - 27th European Conference on Artificial Intelligence - Including 13th Conference on Prestigious Applications of Intelligent Systems (PAIS 2024). IOS Press (2024)

Huber, J. et al.: Chroma docs—docs.trychroma.com (2024). https://docs.trychroma.com/. Accessed 11 June 2024

Karpukhin, V., et al.: Dense passage retrieval for open-domain question answering (2020)

Lewis, P., et al.: Retrieval-augmented generation for knowledge-intensive NLP tasks (2021)

Lin, C.-Y.: ROUGE: a package for automatic evaluation of summaries, p. 10 (2004)

Lv, L., et al.: NeuralChat: a customizable chatbot framework (2023)

Lála, J., O'Donoghue, O., Shtedritski, A., Cox, S., Rodriques, S.G., White, A.D.: PaperQA: retrieval-augmented generative agent for scientific research (2023)

Mikolov, T., Chen, K., Corrado, G., Dean, J.: Efficient estimation of word representations in vector space (2013)

OpenAI, et al.: GPT-4 technical report (2024)

Ovadia, O., Brief, M., Mishaeli, M., Elisha, O.: Fine-tuning or retrieval? Comparing knowledge injection in LLMs (2024)

Steck, H., Ekanadham, C., Kallus, N.: Is cosine-similarity of embeddings really about similarity? In: Companion Proceedings of the ACM on Web Conference 2024, WWW 2024. ACM (2024)

Team, G., et al.: Gemma: open models based on Gemini research and technology (2024)

Touvron, H., et al.: LLaMA: Open and efficient foundation language models (2023)

Zhang, Y., et al.: Siren's song in the AI ocean: a survey on hallucination in large language models (2023)

Blueprint of Tomorrow: Contrasting Off-Line and On-Line Drawing Tasks for Alzheimer's Disease Screening

Nina Hosseini-Kivanani[1(✉)], Elena Salobrar-García[2,3], Lorena Elvira-Hurtado[3,4], Mario Salas[4], Christoph Schommer[1], and Luis A. Leiva[1]

[1] University of Luxembourg, Esch-sur-Alzette, Luxembourg
{nina.hosseinikivanani,christoph.schommer,luis.leiva}@uni.lu
[2] Ramon Castroviejo Institute of Ophthalmologic Research, Madrid, Spain
elenasalobrar@med.ucm.es
[3] Universidad Complutense of Madrid, Madrid, Spain
marelvir@ucm.es
[4] Memory Unit, Geriatrics Service, Hospital Clínico San Carlos, Madrid, Spain
mario.salas@salud.madrid.org

Abstract. Alzheimer's disease (AD) is the leading cause of dementia. Although there is currently no cure for AD, early detection of cognitive decline can help clinicians mitigate its impact. Recently, Machine Learning (ML) approaches have been developed to automatically analyze handwriting and hand-drawing tasks to support the early diagnosis of AD. In this paper, we study pentagon and clock drawing tests using both off-line (scanned image pixels) and on-line (discrete point sequences) data as input to several ML models (i.e., DensNet, ResNet, EfficientNet, RNN, LSTM, and GRU). Our study is the first to determine the most effective modality (on-line vs. off-line) and drawing tasks to distinguish healthy controls from AD patients (binary classification) as well as two stages of AD severity (multi-class classification). Our results suggest that, contrary to other domains, the off-line modality outperforms the on-line one, sometimes by a large margin: 90% vs. 60% accuracy in binary classification and 53% vs. 82% accuracy in multi-class classification. This suggests that, for drawing tasks and small-scale datasets, image-based representations may be more effective in predicting AD than those relying on more complex data representations.

Keywords: Alzheimer disease · Off-line handwriting · On-line handwriting · Deep learning · Data augmentation · Classification

1 Introduction

Neurodegenerative disorders are among the primary causes of disability worldwide, marked by an irreversible loss of neurons that culminates in progressive neurological decline, manifesting as motor and cognitive impairments.

V. Julian et al. (Eds.): IDEAL 2024, LNCS 15346, pp. 422–433, 2025.
https://doi.org/10.1007/978-3-031-77731-8_38

Alzheimer's disease (AD) and Parkinson's disease (PD) are particularly notable for their widespread prevalence, which affects approximately 50 million and 10 million individuals, respectively. AD, in particular, is closely linked with cognitive deficits, affecting memory, attention, language comprehension, and spatial awareness [32]. The global demographic trend toward an older population underscores the critical need for early AD detection and intervention as the most prevalent form of dementia. However, the present diagnostic landscape reveals a concerning trend, with an estimated 75% of dementia cases worldwide going undiagnosed and rates of early-stage detection being considerably lower [10]. Enhancing screening methodologies in accessible settings, particularly in primary healthcare, emerges as a strategic response to improve diagnostic rates [8]. Research indicates that primary healthcare practitioners face substantial challenges in the early detection of dementia and in the timely referral of patients to specialized care [10]. Thus, the development and implementation of accessible and efficient screening tools for use in primary healthcare or by individuals at home are crucial steps toward closing the diagnostic gap, potentially leading to improved detection rates of AD.

Handwriting and hand-drawing tasks[1] entail the coordination of fine motor movements and cognitive processes, making them popular as a psychometric tool to evaluate and diagnose AD [35], leveraging the correlation between declining drawing abilities and the onset of AD. Deterioration in handwriting skills, characterized by inconsistencies in size, spacing, and letter formation, indicates a progression of the disease [7]. Recent studies (e.g., [17,23]) have shown the potential of handwriting-related tasks to reveal specific cognitive deficits indicative of AD. Researchers have explored various automated methods, including drawing tasks [17], neuroimaging [31], and gait assessments [11], to capture cognitive impairments across multiple domains. However, the current need for healthcare professionals' reliance on manual analysis highlights a significant bottleneck. This puts forward the importance of developing automated tools to make the AD screening process easier, quicker, and more affordable, particularly in non-specialist settings.

Our contributions are straightforward yet significant. Firstly, we gathered handwriting data from both AD patients and healthy individuals. The data includes two types of drawings—pentagons and clocks—captured *simultaneously* in two formats: off-line, as scanned images, and on-line, as sequences of discrete points. This dual-method collection allows us to: (i) Compare how different hand-drawing tasks perform in classification tests (ii) Assess various machine learning models to see which best classifies the data (iii) Examine the differences in using static images versus dynamic, point-by-point data in model performance.

Secondly, our experiments focus on distinguishing AD (mild AD and moderate AD) patients from healthy controls using state-of-the-art neural network models. We used pre-trained convolutional neural networks (CNNs) for analyz-

[1] We consider 'handwriting' and 'hand-drawing' synonymous because both tasks involve the same neurophysiological and peripheral processes involved in motor control.

ing the off-line data and recurrent neural networks (RNNs) for the on-line data. Additionally, we used data augmentation techniques to enhance the models' ability to generalize. Interestingly, our findings reveal that the off-line modality consistently outperformed the on-line one, achieving higher accuracy in both binary (90% versus 60%) and multi-class (53% versus 82%) classification tasks. This suggests that simpler, static image-based approaches may be more effective for tasks like drawing analysis in AD research than those relying on more complex, temporal data, at least when working with small-scale datasets.

2 Related Work

Technological advancements in ML and computer vision have significantly enhanced the efficiency and objectivity of remote patient monitoring systems by providing real-time data for improved care beyond conventional healthcare settings [6]. Handwriting analysis has been shown to be effective in detecting cognitive decline and changes in motor skills in AD, thus serving as an effective diagnostic tool [19]. However, despite these advances, there is no research aimed at comparing model performance using *both* off-line and on-line data from *identical* patient/healthy cohorts [5,30].

Recent studies have used CNNs (e.g., [3,18,19]) and RNNs (e.g., [2,6,18]) for early detection of AD, showing that Deep Learning (DL) models can significantly enhance the accuracy of diagnosing AD in its early stages. We can find notable works that studied each modality separately (cf. off-line [3,15,17] and on-line [6,21,23]), suggesting that on-line data are preferred over off-line data, given that on-line handwriting provides a feature-rich representation, including, e.g., temporal and spatial sequences of discrete points that are not available in the off-line representation. Collectively, these studies highlight the potential of ML technologies not only to revolutionize AD diagnosis but also to enable more personalized and timely therapeutic interventions, ultimately improving patient outcomes. Most of these studies have relied on some form of data augmentation, given the limited number of samples in clinical datasets. In this regard, Dao et al. [6] used Generative Adversarial Networks (GANs) as an alternative to data augmentation, and trained AD classifiers with RNNs that achieved 89% accuracy. However, GANs require a significant amount of data to begin with, which is often not available in most cases.

Finally, we should mention relevant studies that have compared different drawing symbols for AD screening, such as clocks drawngs [1] using CNN models, which achieved an AUC score of 81%. By combining clock drawing with age and education using logistic regression, their model improved to 91%. Pentagon drawings [26] reached an accuracy of 93% using GoogLeNet for binary classification, distinguishing between correct and incorrect pentagon drawings from patients only. Another study focused on letters [6], and obtained high accuracy by using DL models for detecting and classifying early-stage of ADs patients based on on-line handwriting loop patterns. In sum, it remains unclear which is the most adequate input modality for AD screening and also what the most adequate drawing symbols are to achieve competitive performance.

By systematically addressing this gap in the research literature, our study paves the way for a more holistic understanding of AD classification models, opening promising directions to more accurate AD screening approaches in the future. For example, [23] reported that combining multiple drawing tasks improves detection accuracy by capturing different cognitive impairments, achieving a classification accuracy of 75.2%.

3 Materials and Methods

3.1 Participants

Thirty-three individuals were recruited from the Memory Unit of the Hospital Clinico San Carlos (HCSC) in Madrid between January 2023 and January 2024. The group consisted of 22 patients and 11 healthy controls (HCs), all aged between 70 and 89. Participants were asked to both clocks and pentagons, which are well-established symbols in cognitive assessment tasks [9]. All participants had normal vision and hearing. They underwent a neuropsychological assessment of their drawing tasks in a clinical setting to minimize distractions and reduce background noise. Each participant was individually assessed, beginning with an informed consent form. Cognitive status was evaluated using the Mini-Mental State Examination (MMSE) [39]. Patients with AD were classified into mild AD and moderate AD, based on guidelines from the National Institute of Neurological and Communicative Disorders and Stroke (NINCDS), the Alzheimer's Disease and Related Disorders Association (ADRDA) workgroup [28], and the Diagnostic and Statistical Manual of Mental Disorders, Fifth Edition (DSM-5) [13]. Statistical analysis (ANOVA test, after verification of normality and homoscedasticity) confirmed that there were no significant age differences between the healthy, mild AD, and moderate AD groups ($F(2, 55) = 2.04, p > .139$) (Table 1).

Table 1. Demographics and the Number of Drawing Tasks of this Study.

Drawing Task	Num. of drawing	HC	Mild AD	Moderate AD
Pentagon	33	11	8	14
Clock	33	11	8	14
Total	**66**	**22**	**16**	**28**
Gender (F & M)		8F, 3M	5F, 3M	12F, 4M
Age (Mean ± SD)		82.64 ± 2.46	76.5 ± 5.75	78.94 ± 4.78
MMSE (Mean ± SD)		29.9 ± 0.83	25.33 ± 1.21	22.37 ± 3.58

3.2 Drawing Tasks and Preprocessing

Participants were instructed to draw the pentagons and clocks using a Repaper tablet (size: 10.9-inch)[2] with a blank sheet attached and a regular pen that had an accelerometer connected to the Repaper app via Bluetooth for data capturing (see Fig. 1). This setup was designed to provide a familiar pen-and-paper experience to participants while being able to capture on-line and off-line data simultaneously. The participants were asked to draw each symbol from memory. We collected 66 drawings in total.

The on-line data (discrete point sequences) were stored as SVG files (as per the Repaper app) and then converted to JSON format, comprising sequences of {x,y,t} points. The off-line data (image pixels) were stored as PDF files (scanned with the HP Color LaserJet Pro scanner) and then converted to PNG images and resized to square size (224 × 224 px) as this is standard for CNN models. We applied the canny edge detector to enhance the quality of the scanned images.

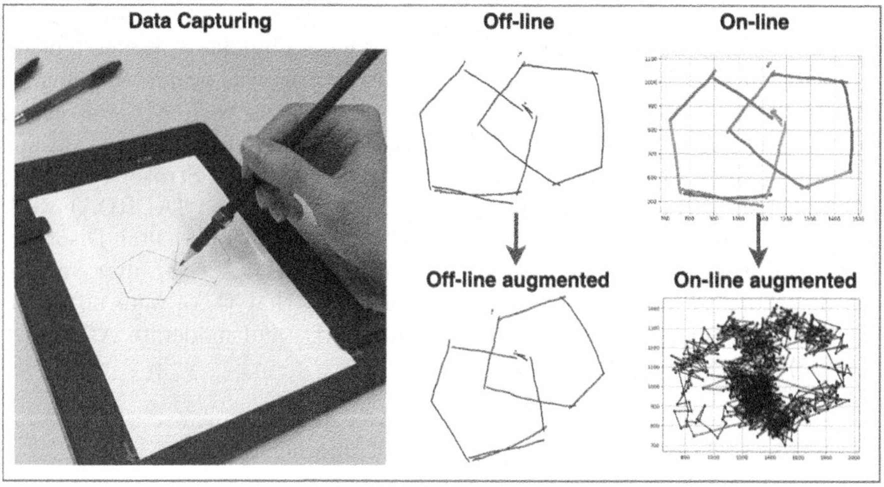

Fig. 1. Example of pentagon drawing on a tablet (left) and drawing samples in off-line and on-line version (right), before and after applying data augmentation.

3.3 Data Augmentation

We created synthetic samples to make the models more robust and generalizable [12]. For off-line data, we applied the usual geometric transformations, where suitable.[3] For on-line data, jittering, scaling, and warping have been pro-

[2] https://www.iskn.co/eu.

[3] For example, pentagons can be flipped in horizontal or vertical axes, whereas clocks cannot be flipped because it would destroy their semantics.

posed [22]; however, more recently, Maslych et al. [27] found that an "All Variability Chain" (AVC) of transformations (gaussian, frame-skip, spatial, perspective, rotate, scale) provided a significant boost in classification performance with RNNs, achieving state-of-the-art accuracy in gesture recognition. Therefore, we adopt the AVC approach to augment our on-line data; see Fig. 1. After data augmentation, we concluded to a dataset consisting of 300 images (off-line data) and 300 point sequences (on-line data), where all groups were balanced up to 100 observations; e.g., 92 variations of the 8 pentagons from the Mild AD group were created; see Fig. 2.

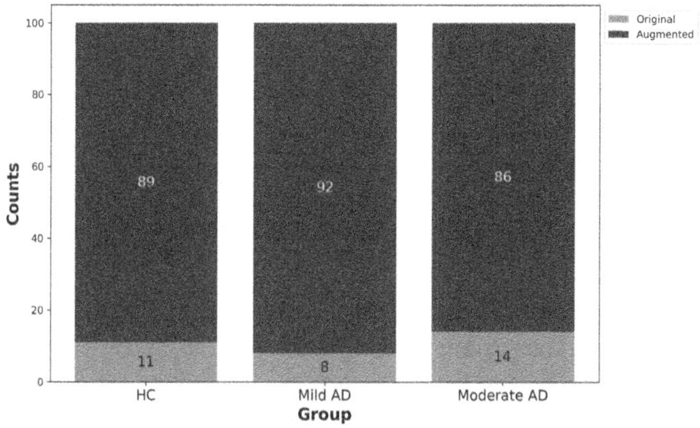

Fig. 2. Class distribution of Pentagon drawings before and after data augmentation.

To evaluate the quality of the augmented data, we used the Structural Similarity Index Measure (SSIM) [41] for off-line data and Dynamic Time Warping (DTW) [34] for on-line data. We use SSIM to compare augmented data against original images to ensure that key structural details are preserved even as variations are introduced. This balance is critical since, while the augmented data are inherently different, maintaining structural similarity ensures that the variations remain realistic and relevant for training robust models. By using SSIM, we can confirm that the augmentation process does not distort the data to the extent that it loses its representative characteristics. In this sense, structural similarity is beneficial, as it ensures that the augmented data faithfully represents the original data. SSIM values range from 0.7 to 0.8 ($M = 0.75$, $SD = 0.03$), whereas DTW values range between 123 and 5678 ($M = 2000$, $SD = 850$), indicating that augmented images are not near-duplicates of the original data but rather new images that eventually should help improving model performance.

3.4 Models

Convolutional Neural Nets. We use three state-of-the-art pre-trained CNNs for analyzing the off-line data: ResNet50 [14], DenseNet121 [20], and EfficientNet [37]. ResNet and DenseNet use residual connections, which are instrumental to train very deep models. While ResNet performs an element-wise addition to pass the output to the next layer, DenseNet connects all layers directly to each other through concatenation. However, EfficientNet uses a uniform compound scaling technique that achieves the same performance as state-of-the-art CNNs but with much better efficiency. These CNNs were trained on the large ImageNet dataset, and we fine-tuned them to our AD dataset by transfer learning [42].

Recurrent Neural Nets. Since transfer learning for on-line data is not currently possible, as there are no public pre-trained models available, we train three RNNs from scratch: Vanilla RNN, LSTM [16], and GRU [4]. LSTM is an improvement over vanilla RNNs by adding long-term memory, making them ideal for complex sequences. GRU is a simplification of LSTM while retaining the same performance, making them ideal for cases where computational efficiency is crucial. These RNN models include a hidden layer of 100 units with hyperbolic tangent activation and 0.1 dropout, followed by a softmax output layer. We experimented with other combinations of layers and different hidden units, but we did not observe improvements with regard to this configuration.

Training and Evaluation. All CNNs and RNNs are trained with the popular Adam optimizer, with a learning rate of 0.001 and decay rates $\beta_1 = \beta_2 = 0.99$. The loss function is categorical cross-entropy, consistent with our binary and multi-class classification tasks. All models use a batch size of 32 (images or sequences) and use up to 50 epochs for training with early stopping (patience of 10 epochs) to prevent overfitting. We train each model on 80% of the data and test on the remaining 20% of both on-line and off-line data. We then used stratified 5-fold cross-validation on the training set only, ensuring that each fold was representative of the whole by maintaining approximately the same percentage of samples of each class as in the training subset. We computed the classification accuracy (Acc) and Area Under the ROC curve (AUC) to assess model performance.

4 Results and Discussion

Figure 3 summarizes the results of our experiments, depicting the differences between CNN classifiers (top row) for off-line data (ResNet, DenseNet, and EfficientNet) and RNN classifiers (bottom row) for on-line data (vanilla RNN, LSTM, and GRU). These results are instrumental for assessing the efficacy of binary (distinguishing AD patients from HC) and multi-class classification (distinguishing HC, mild AD, and moderate AD) tasks in the context of AD screening using hand-writing tasks (i.e., pentagon and clock).

In binary classification of pentagon drawings, the before-augmentation scenario showed modest performance across models. After-augmentation, however, there was an increase in classifier accuracy, particularly with EfficientNet, which improved substantially, from 60% to 90%. A similar trend was observed in the classification of clock drawings, where the after-augmentation results underscored the effective impact of our data augmentation strategies on model performance.

For multi-class classification, the differential impact of data augmentation was again significant. EfficientNet was the best performer, especially for pentagon drawings, confirming the model's ability to differentiate between various stages of AD severity. The trend of improved after-augmentation accuracy was consistent across other models, although the degree of improvement varied.

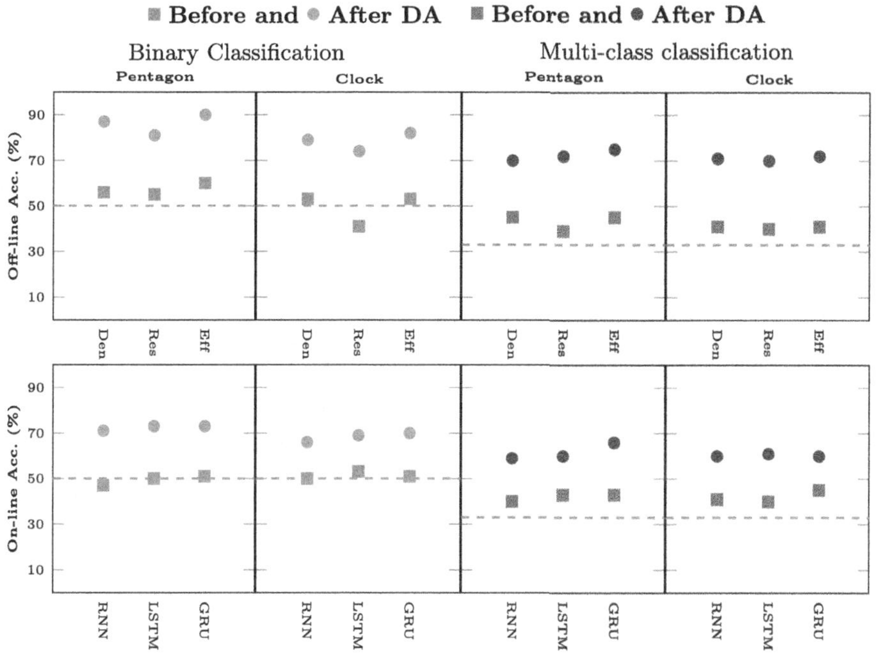

Fig. 3. Experiment results. CNN classifiers for off-line data (Res: ResNet, Den: DenseNet, Eff: EfficientNet) are depicted in the top row, whereas RNN classifiers for on-line data (vanilla RNN, LSTM, GRU) are depicted in the bottom row. The dashed lines represent the performance of a random classifier, illustrating the empirical lower bound.

Our results indicate that off-line data, when enhanced through strategic data augmentation, provides a more stable and consistent basis for AD classification compared to on-line data. This stability can be attributed to the static nature of off-line data, which, unlike on-line data, is less affected by the variabilities introduced by the temporal and dynamic components of on-line drawing. In non-clinical domains, researchers have shown that on-line data is preferred over off-line data (e.g., [24,38]), given the rich patterns and movement dynamics involved [25]. In our experiments, however, we observed that pre-trained CNNs outperformed RNNs trained from scratch due to the size and variability in online data that our RNNs were not able to capture as effectively as the CNNs. This variability also seems to affect tasks differently; for example, more cognitively demanding drawing tasks allowed for an easier distinction of AD patient handwriting from HCs; see e.g., Fig. 3.

Our study builds upon previous research that focused on one drawing type for AD screening [3,19]. However, our findings suggest that both pentagon and clock drawings are suitable for AD screening, with EfficientNet achieving the highest performance in binary classification (90% accuracy, 92% AUC) and in multi-class classification settings (75% accuracy, 79% AUC) for using pentagons, followed by clocks (Binary classification: 82% accuracy and 79% AUC, multi-class classification: 70% accuracy and 75% AUC), highlighting the effectiveness of these simple tasks. Our results also indicate that the performance gap before and after data augmentation differs across tasks, with larger differences in off-line data. Previous results by Maslych et al. [27] reported improved performance of an RNN model in several handwriting tasks using AVC for data augmentation, although all the tasks were focused on gesture recognition and a specific dataset. In any case, it seems clear that data augmentation is necessary for both off-line and on-line drawing tasks.

Without data augmentation, most models behaved like a random classifier; see the dashed lines in Fig. 3. After data augmentation, however, we observed a significant improvement in model performance across all tasks for both binary and multi-class classification scenarios. These improvements were more apparent for CNN models, which somehow disagrees with previous findings in AD screening that reported similar performance for RNNs [2,29].

Interestingly, Souillard-Mandar et al. [36] found that the digital Clock Drawing Test had superior diagnostic performance compared to traditional paper-and-pencil methods for differentiating healthy individuals from cognitive impairment subjects (only binary classification), using traditional ML models without any data augmentation. Previous studies have demonstrated the effectiveness of various augmentation strategies in similar contexts for off-line data [17,33]. In our work, we designed and optimized our augmentation strategy to ensure its suitability for our specific datasets. This involved iterative testing and refinement of different augmentation techniques, such as geometric transformations.

5 Limitations and Future Work

The main limitation of our study is the small sample size of the original dataset, which had to be augmented with suitable variations in order to fine-tune the CNN models and, more importantly, to train the RNN models from scratch. However, we should remark that dealing with small sample sizes is a well-known and pervasive issue among clinical studies [21,40]. Recruiting participants with AD is very challenging due to strict criteria and ethical concerns. Despite our best efforts, it took us one year to recruit 33 suitable participants. On the other hand, future work should consider different data augmentation approaches for on-line trajectories, since we have observed that the AVC method [27] is suboptimal for AD screening. Additionally, an interesting direction for future research could be assessing whether simpler, custom-built CNN models can achieve comparable or superior results with reduced complexity. Ultimately, despite these shortcomings, our findings show promise and could lead to practical clinical applications.

6 Conclusion

We have analyzed the impact of off-line vs. on-line handwriting data for AD screening using pentagon and clock drawings with suitable data augmentation techniques. We trained several CNNs and RNNs for binary and multi-class classification settings. Our results show that data augmentation is always beneficial and that pentagons have better discriminative power than clocks. Our results also show that CNNs outperform RNNs in all settings, contradicting what was previously known in non-clinical work.

Our observed improvements in performance suggest that while our current strategy enhances model robustness, further optimizations could indeed yield even better results, a possibility that we aim to explore in future work. We acknowledge that there are numerous opportunities for additional refinement, and future work will continue to explore and optimize these techniques to maximize their efficacy in enhancing model robustness and accuracy. Our code and models are available upon reasonable request.

Acknowlegments. Work supported by the UCM research group (Grupo de investigacion básica en Ciencias de la Visión del IIORC, UCM-GR17-920105), the Horizon 2020 FET program of the European Union (grant CHIST-ERA-20-BCI-001), and the European Innovation Council Pathfinder program (grant 101071147).

References

1. Amini, S., et al.: An artificial intelligence-assisted method for dementia detection using images from the clock drawing test. J. Alzheimers Dis. **83**(2) (2021)
2. Bensalah, A., Parziale, A., De Gregorio, G., Marcelli, A., Fornés, A., Lladós, J.: I can't believe it's not better: in-air movement for Alzheimer handwriting synthetic generation. In: Parziale, A., Diaz, M., Melo, F. (eds.) IGS 2023. LNCS, vol. 14285, pp. 136–148. Springer, Cham (2023). https://doi.org/10.1007/978-3-031-45461-5_10
3. Chen, S., Stromer, D., Alabdalrahim, H.A., Schwab, S., Weih, M., Maier, A.: Automatic dementia screening and scoring by applying deep learning on clock-drawing tests. Sci. Rep. **10**(1) (2020)
4. Cho, K., Merrienboer, B.V., Gulcehre, C., Bougares, F., Schwenk, H., Bengio, Y.: Learning phrase representations using RNN encoder-decoder for statistical machine translation. In: Proceedings of the EMNLP (2014)
5. Cilia, N.D., De Gregorio, G., De Stefano, C., Fontanella, F., Marcelli, A., Parziale, A.: Diagnosing Alzheimer's disease from on-line handwriting: a novel dataset and performance benchmarking. J. Eng. Appl. Artif. Intell. **111** (2022)

6. Dao, Q., El-Yacoubi, M.A., Rigaud, A.S.: Detection of Alzheimer disease on online handwriting using 1D convolutional neural network. IEEE Access **11** (2023)
7. Delazer, M., Zamarian, L., Djamshidian, A.: Handwriting in Alzheimer's disease. J. Alzheimer's Dis. **82**(2) (2021)
8. Eichler, T., et al.: Rates of formal diagnosis of dementia in primary care: the effect of screening. Alzheimers Dement. Diagn. Assess. Dis. Monit. **1**(1) (2015)
9. Folstein, M.F., Folstein, S.E., McHugh, P.R.: "Mini-mental state": a practical method for grading the cognitive state of patients for the clinician. J. Psychiatr. Res. **12**(3) (1975)
10. Gauthier, S., Rosa-Neto, P., Morais, J.A., Webster, C.: World Alzheimer report 2021: journey through the diagnosis of dementia. Alzheimers Dis. Int. **2022** (2021)
11. Ghoraani, B., Boettcher, L.N., Hssayeni, M.D., Rosenfeld, A., Tolea, M.I., Galvin, J.E.: Detection of mild cognitive impairment and Alzheimer's disease using dual-task gait assessments and machine learning. Biomed. Signal Process. Control **64** (2021)
12. Goodfellow, I., Bengio, Y., Courville, A.: Deep Learning. MIT Press (2016)
13. Guha, M.: Diagnostic and statistical manual of mental disorders: DSM-5 (5th edition). Reference Rev. **28**(3) (2014)
14. He, K., Zhang, X., Ren, S., Sun, J.: Deep residual learning for image recognition. In: Proceedings of the CVPR (2016)
15. Higaki, Y.: Clock-drawing test and cube-copying test to quickly screen dementia: in combination with the mini-mental state examination scores. Intern. Med. (2023)
16. Hochreiter, S., Schmidhuber, J.: Long short-term memory. Neural Comput. **9**(8) (1997)
17. Hosseini-Kivanani, N., et al.: Ink of insight: data augmentation for dementia screening through deep learning. In: Proceedings of the ICMHI (2024)
18. Hosseini-Kivanani, N., Salobrar-García, E., Elvira-Hurtado, L., López-Cuenca, M., Schommer, C., Leiva, L.A.: Predicting Alzheimer's disease and mild cognitive impairment with off-line and on-line house drawing tests. In: Proceedings of the e-Science. IEEE (2024)
19. Hosseini-Kivanani, N., Schommer, C., Leiva, L.A.: The magic number: impact of sample size for dementia screening using transfer learning and data augmentation of clock drawing test images. In: Proceedings of the Healthcom (2023)
20. Huang, G., Liu, Z., van der Maaten, L., Weinberger, K.Q.: Densely connected convolutional networks. In: Proceedings of the CVPR (2017)
21. Impedovo, D., Pirlo, G.: Dynamic handwriting analysis for the assessment of neurodegenerative diseases: a pattern recognition perspective. IEEE Rev. Biomed. Eng. **12** (2018)
22. Iwana, B.K., Uchida, S.: An empirical survey of data augmentation for time series classification with neural networks. PLOS One **16**(7) (2021)
23. Kobayashi, M., Yamada, Y., Shinkawa, K., Nemoto, M., Nemoto, K., Arai, T.: Automated early detection of Alzheimer's disease by capturing impairments in multiple cognitive domains with multiple drawing tasks. J. Alzheimers Dis. **88**(3) (2022)
24. Leiva, L.A., Alabau, V., Romero, V., Toselli, A.H., Vidal, E.: Context-aware gestures for mixed-initiative text editing UIs. Interact. Comput. **27**(6) (2015)
25. Leiva, L.A., Diaz, M., Ferrer, M.A., Plamondon, R.: Human or machine? It is not what you write, but how you write it. In: Proceedings of the ICPR (2021)
26. Maruta, J., et al.: Deep convolutional neural networks for automated scoring of pentagon copying test results. Sci. Rep. **12**(1) (2022)

27. Maslych, M., Taranta, E.M., Aldilati, M., Laviola, J.J.: Effective 2D stroke-based gesture augmentation for RNNs. In: Proceedings of the CHI (2023)
28. McKhann, G., Drachman, D., Folstein, M., Katzman, R., Price, D., Stadlan, E.M.: Clinical diagnosis of Alzheimer's disease: report of the NINCDS-ADRDA work group under the auspices of department of health and human services task force on Alzheimer's disease. Neurology **34**(7) (1984)
29. Mwamsojo, N., et al.: Reservoir computing for early stage Alzheimer's disease detection. IEEE Access **10** (2022)
30. Müller, S., et al.: Increased diagnostic accuracy of digital vs. conventional clock drawing test for discrimination of patients in the early course of Alzheimer's disease from cognitively healthy individuals. Front. Aging Neurosci. **11** (2017)
31. Odusami, M., Maskeliūnas, R., Damaševičius, R.: An intelligent system for early recognition of Alzheimer's disease using neuroimaging. Sensors **22** (2022)
32. Perry, R.J., Hodges, J.R.: Attention and executive deficits in Alzheimer's disease: a critical review. Brain **122**(3) (1999)
33. Raksasat, R., et al.: Attentive pairwise interaction network for AI-assisted clock drawing test assessment of early visuospatial deficits (2023)
34. Senin, P.: Dynamic time warping algorithm review. Inf. Comput. Sci. Dep. Univ. Hawaii **855**(1-23) (2008)
35. Smith, A.D.: On the use of drawing tasks in neuropsychological assessment. Neuropsychology **23**(2) (2009)
36. Souillard-Mandar, W., et al.: Learning classification models of cognitive conditions from subtle behaviors in the digital clock drawing test. Mach. Learn. **102** (2015)
37. Tan, M., Le, Q.: EfficientNet: rethinking model scaling for convolutional neural networks. In: Proceedings of the ICML (2019)
38. Tappert, C.C., Mosley, P.H.: Recent advances in pen computing. Technical report 166, Pace University (2001)
39. Tombaugh, T., McIntyre, N.J.: The mini-mental state examination: a comprehensive review. J. Am. Geriatr. Soc. **40**(9) (1992)
40. Vessio, G.: Dynamic handwriting analysis for neurodegenerative disease assessment: a literary review. Appl. Sci. **9**(21) (2019)
41. Wang, Z., Bovik, A.C., Sheikh, H.R., Simoncelli, E.P.: Image quality assessment: from error visibility to structural similarity. IEEE Trans. Image Process. **13**(4) (2004)
42. Weiss, K., Khoshgoftaar, T.M., Wang, D.: A survey of transfer learning. J. Big Data **3**(1) (2016)

Digital Mental Health Apps: Key Features and User Engagement for Better Wellness

Cristiana Rocha[1], Diogo Martinho[1](✉), Luis Conceição[1], Paulo Novais[2], and Goreti Marreiros[1]

[1] Institute of Engineering, Polytechnic of Porto, Porto, Portugal
{1191610,diepm,msc,mgt}@isep.ipp.pt
[2] ALGORITMI Centre, University of Minho, Guimarães, Portugal
pjon@di.uminho.pt

Abstract. Given the global impact of mental disorders and the increasing prevalence of Major Depressive Disorder (MDD) as one of the main causes of disability, it is imperative to invest in solutions for self-management of mental health. Thus, the evolution of Digital Mental Health reveals a significant impact, with major investment in technologies that drive the development of projects that aim to innovate the benefits and results of this area. As part of the development of the RM4Health project, this work evaluated the role of web, desktop and/or mobile applications in the self-management of mental health, highlighting existing strengths and gaps. The study covered 205 applications for the management of mental health, detailing the panorama of current Behavioral Management practices, Persuasive Systems Design (PSD), Behavioral Economics (BE) principles and Security and Privacy strategies. The results obtained in this survey allow the definition of recommendations for the development of more effective, engaging, accessible and safe solutions, to promote autonomy in the management of the mental health condition. A high diversity of resources was observed, however areas of underutilization were identified, especially in BE approaches. It is recommended to combine Artificial Intelligence, Smart Coach and Gamification strategies with robust security and privacy measures. Furthermore, the development of comprehensive and intuitive solutions that prioritize key behavioral management functionalities, PSD resources and BE principles that involve the user in their own health management, regardless of whether they have been diagnosed with a mental disorder, were suggested. This study could guide future research and development in order to improve the quality of the user experience and drive advances in the field of Digital Mental Health.

Keywords: Major Depressive Disorder · Digital Mental Health · Artificial Intelligence

1 Introduction

Mental health is a global public health issue [1]. A Lancet Psychiatry study [2] analyzing 12 disorders in 204 countries between 1990–2019 showed a 48% rise in diagnoses, from 654.8 million to 970.1 million. COVID-19 also caused a 28% rise in major depressive disorder (MDD), adding 53 million cases [3]. MDD is especially concerning, with

V. Julian et al. (Eds.): IDEAL 2024, LNCS 15346, pp. 434–446, 2025.
https://doi.org/10.1007/978-3-031-77731-8_39

15% attempting suicide [4]. In the WHO European region, over 150 million people were diagnosed with mental health issues last year [5]. Digital Mental Health (DMH) technologies offer new treatment opportunities, but high dropout rates due to insufficient personalization and support hinder their success [6]. In 2023, apps for anxiety and depression management made up 29.3% of the mental health app market, valued at $6.2 billion, and projected to grow to $17.5 billion by 2030 at a 15.2% CAGR [7]. To improve engagement, technologies like AI [8], gamification [9, 10], and Smart Coaching are promising [11, 12]. AI personalizes user experiences, predicting risks and meeting individual needs [13–15]. Gamification boosts engagement through rewards and challenges [16, 17]. Smart Coaches offer tailored support, improving adherence [11, 18]. These strategies enhance well-being, reduce relapse, and improve life quality [19]. Consequently, this analysis focuses on strategies used by DMH applications to enhance user adherence, engagement, and interaction. Key behavioral management strategies, including sleep quality, nutrition, physical activity, medication adherence, and self-care, are essential for improving mental health. Good sleep reduces insomnia [20], balanced diets boost mood and energy [21], while regular medication minimizes relapse risks [22]. Activities like relaxation [23], social interaction [24, 25], and hobbies [26] help reduce stress and promote happiness. Professional assistance [27] and practicing gratitude and self-compassion [28] build emotional resilience and self-acceptance. Data security and privacy issues will also be considered [29]. Persuasive Systems Design (PSD) [30] and Behavioral Economics (BE) also play crucial roles, promoting user engagement, social connections, and trustworthiness [31]. Ensuring data privacy and security through multifactor authentication and encryption is essential for protecting user data [32]. Together, AI, Gamification, and Smart Coaching can significantly improve mental health outcomes by enhancing app effectiveness and user experience.

2 Problem Statement and Objective

Despite the growth of DMH solutions, many apps face adoption and effectiveness issues due to poor interfaces, lack of behavioral tools, weak persuasive design, and over-reliance on professional intervention. This limits user autonomy and reduces self-management [33]. Drawing from the RM4Health project [34], this article reviews gaps in mental health apps, focusing on MDD management. It highlights strategies to boost engagement, covering behavioral management, PSD, BE principles, and data security. The study aims to improve user autonomy and app interaction by answering three questions: "What are the behavioral management resources and strategies in mental health applications?", "What are the PSD resources and BE principles in mental health applications and what strategies are used to increase user engagement and adherence?" and "What strategies are adopted to manage data security and privacy in mental health applications?". A review of apps on major platforms provides insights and recommendations for more effective and secure mental health apps.

3 Metodology

This section outlines the methodology process applied, starting with the selection of specific keywords to identify mental health management solutions across various application stores. After defining eligibility criteria and identifying relevant resources, the solutions targeted for analysis were selected and categorized.

3.1 Information Sources and Research Strategy

The study reviewed mobile, web, and desktop solutions focused on managing depressive disorders available on Google Play, the App Store, and the Microsoft Store from January 2024 to April 2024. Keywords used in the search included: "Depression", "Depressive Disorder", "Major Depressive Disorder", "Clinical Depression", "Mental Disorder", "Mood Disorder", "Psychiatric Disorder", "Mental Health", "Emotional Well-being", "Mental Wellness", "Psychological Health" or "Mental Illness". Translations of these terms into Portuguese were also considered.

3.2 Eligibility Criteria

The analysis included solutions for managing various disorders due to the limited availability of personalized MDD options. Eligible solutions were available on Google Play, the App Store, or the Microsoft Store, in web, mobile, or desktop formats, and in English or Portuguese. Both free and paid apps were considered, excluding those not updated since before January 1, 2010, or unrelated to mental health, such as wallpaper apps. Each app had to offer at least one depression management feature, such as mood tracking or therapeutic techniques. Apps for other disorders were included due to the scarcity of MDD-specific options.

3.3 Selection and Categorization

Application selection was conducted by the authors using a blind methodology, with three rounds of decision-making to ensure consistency and resolve discrepancies. Initially, 506 solutions were identified, mainly from Google Play and the App Store. After removing duplicates (applications listed in both stores), 327 applications remained. Of these, 205 were deemed eligible based on the following criteria: (1) resources for managing depression, such as online therapy; (2) availability in English and/or Portuguese; (3) presence on Google Play, the App Store, or the Microsoft Store; (4) web, mobile, or desktop format; and (5) last updated after January 1, 2010. Data summarization occurred in three phases: 1) Categorization of eligible applications; 2) Distribution of behavioral management resources among eligible applications; and 3) Analysis and distribution of behavioral management resources, PSD resources, BE principles, and security and privacy strategies among the twenty most relevant applications.

4 Results and Discussion

The results cover the distribution and categorization of eligible applications, and the analysis of Behavioral Management Resources, PSD resources, BE principles, and Security and Privacy strategies.

4.1 Categorization of Eligible Applications

The analysis of mental health applications provided valuable market insights. Most applications (80%) fell under the "Health and Fitness" category, followed by "Medical Care" (11.22%) and "Lifestyle" (6.83%), highlighting a clear market trend. However, concerns arise regarding the diversity and effectiveness of mental health approaches. Notable categories included "Mood Tracker" (20.38%), "Self-Help" (18.81%), and "Alternative Methods" (18.65%). In contrast, there was a noticeable gap in categories related to gamification and social interaction, such as "Game", "Chatbot", and "Support Community". Intervention focus was balanced between depression (28.08%) and anxiety/stress (27.60%), reflecting the interconnected nature of these mental disorders. The use of multimedia formats was diverse, with text (31.80%) being the most common, followed by graphics (23.56%) and audio (17.46%), enhancing accessibility and user experience. Personalization for depression was present in 53.66% of applications, but only 0.98% offered features specifically for MDD, indicating a limited range of tailored solutions. The temporal analysis showed an increase in launches in 2018 and 2020, coinciding with heightened demand due to the COVID-19 pandemic's impact on mental health. The subsequent decline in 2023 and 2024 suggests a potential market shift or stagnation. User feedback was generally positive (51.71%), with a notable number of reviews (34.15%) and downloads (16.59%), indicating widespread adoption and recognition. Regarding pricing, most applications offered free access with optional purchases (73.17%), followed by free acquisition and, to a lesser extent, paid models, reflecting diverse pricing strategies to cater to different market segments.

4.2 Analysis and Distribution of Behavioral Management Resources Among Eligible Applications

A detailed examination of behavioral management resources revealed a broad and varied distribution. "Behavioral Change" (15.24%) was the most common resource, followed by "Mood and Symptom Control" (13.90%) and "Relaxation Techniques" (11.05%). These approaches support the adoption of healthy habits, coping techniques, emotional self-management, and health monitoring, significantly influencing thought patterns and behavior and empowering users to manage their health autonomously. Less frequently used resources included "Self-Care and Self-Compassion" (7.04%), "Negative Behavioral Patterns" (6.95%), "Practice of Positive Affirmations" (6.77%), and "Practice of Gratitude" (4.46%) which are crucial for addressing dysfunctional thinking and fostering a constructive internal narrative. The underutilization of resources for "Medication" (1.43%), "Healthy Eating" (4.19%), "Hobbies and Leisure Activities" (4.28%), "Professional Help" (5.44%), "Regular Physical Exercise" (5.79%), "Socialization and Interaction" (6.60%), and "Sleep Quality" (6.86%) highlights significant opportunities for investment, given their proven benefits for promoting a healthy lifestyle and improving mental health outcomes. Figure 1 illustrates the of Behavioral Management Resources among eligible applications.

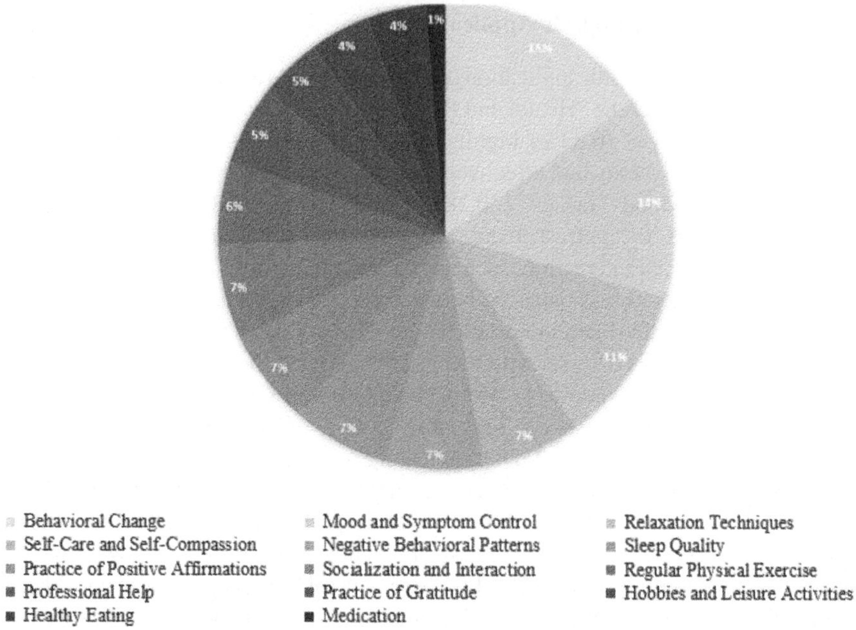

Fig. 1. Distribution of Behavioral Management Resources of the 205 eligible applications.

4.3 Analysis and Distribution of Behavioral Management Resources, PSD Resources, BE Principles and Security and Privacy Strategies Among the Top Twenty Applications

Main Selected Applications. Despite identifying 205 applications, the study focused on a detailed analysis of the most recent and popular applications capable of supporting management and depressive disorders, ensuring the study of the most current solutions innovative and impactful, specifically those updated on or after January 1, 2022. These selected applications were required to include personalized depression management resources, have an average rating between 4 and 5 stars, and possess at least 1000 reviews and over 100000 downloads. As a result, the following applications were noted: "Calm"; "Shmoody"; "Youper"; "Amaha"; "What's Up?"; "Woebot"; "MoodTools"; "MindDoc"; "Mindshine"; "Anxiety Relief - Head On"; "Being: self therapy & CBT ai"; "Wysa"; "7 Cups"; "Conversa Comigo"; "MindDiary: Diário de humor"; "Now&Me – Therap, Self-Care"; "Tochi"; "Ada"; "Hiwell" and "Metamorfosis Diário Emocional".

Analysis and Distribution of Behavioral Management Resources. The analysis revealed a varied allocation of behavioral resources. "Behavioral Change" and "Mood and/or Symptoms" (11.39%) were most prevalent, indicating a focus on altering thought patterns and monitoring symptoms. "Relaxation Techniques" (10.76%) highlighted the importance of stress management and emotional balance. However, resources like "Healthy Eating" (4.43%), "Regular Physical Exercise" (5.06%), "Socialization and

Interaction" (6.96%), and "Positive Affirmations" (7.59%) are underutilized, suggesting areas for increased investment. The absence of "Medicines" indicates a gap specially in cases requiring pharmacological intervention, pointing to a need for integrating these resources into digital depression treatments. Figure 2 illustrates the of Behavioral Management Resources among the top twenty applications.

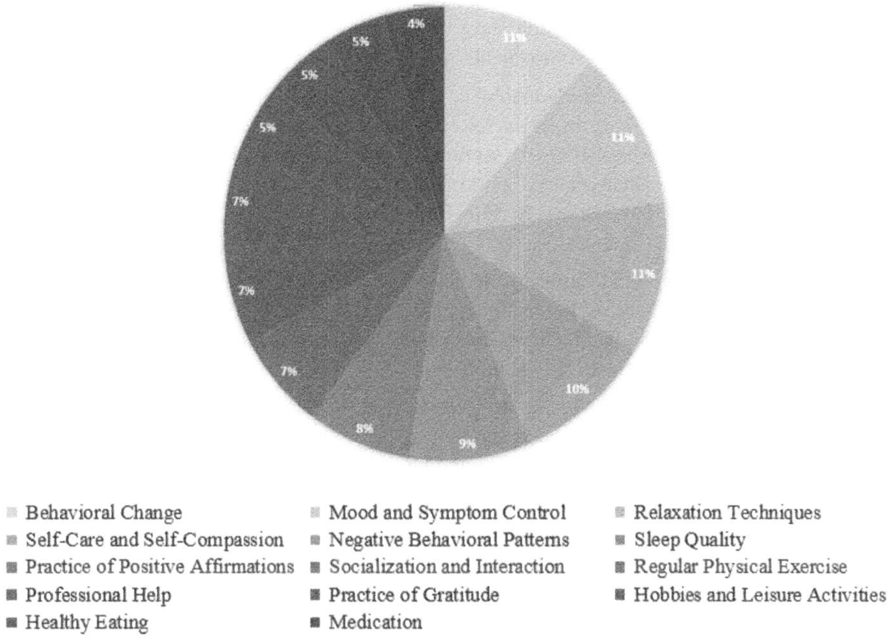

Fig. 2. Distribution of Behavioral Management Resources of the 20 eligible applications.

Analysis and Distribution of PSD Resources. The distribution showed a variety of functionalities, including Credibility Support, Dialogue Support, and Primary Task Support. However, 35% of applications lacked Social Support resources, which limits comprehensive user support and suggests a need for development in this area.

Primary Task Support. All applications offered "Test" and "Reduction" features, with most providing "Tailoring," "Personalization," "Self-Monitoring," and "Tunneling" (90%). There is a noted scarcity of "Simulation" resources, which offer practical, visual representations of therapeutic techniques and exercises. Incorporating simulation features could enhance user experience by providing real-time, animated feedback, making it easier for users to adopt and benefit from recommended techniques. Table 1 summarizes the predominant strategies in each Primary Task Support approach.

Dialogue Support. The analysis revealed a widespread use of "Predilection," with notable emphasis on "Suggestion" and "Similarity," which foster personal connection and empathy. "Social Role" was featured in nearly half of the applications, but "Reminders," "Rewards," and "Praise" were less common. This suggests opportunities for

Table 1. Primary Task Support approaches.

Principle	Most Referenced Approaches	Description
Personalization (90%)	Recommendations, Interactions, Personalized Support; Activity Suggestions based on user; Emotion and Mood Customization; Personalized Health Assessments	Personalized support based on individual needs and preferences for tailored emotional well-being
Reduction (90%)	Guided Meditations; Mindfulness Exercises; Short Mood-Boosting Actions; Anonymous Support; Focused Psychotherapy; Mini-Therapies; Chatbot	Methods to reduce anxiety, improve mood, and manage stress with simplified, user-specific tasks
Rehearsal (90%)	Mindfulness Masterclasses; Guided Relaxation; Coping Skills Practice; Grounding Game; Daily Reflections	Exercises and simulations to develop coping, mindfulness, and reflection skills for everyday situations
Self-monitoring (90%)	Daily Mood Monitoring; Anxiety and Depression Tracking; Progress and Learning Tracking; Therapy Session Monitoring	Tracking and visualizing emotional and behavioral progress to promote mental well-being
Tailoring (90%)	Adaptable Resources; Intervention Filtering; Online Therapy; Personalized Interactions; Meditation and Mindfulness Selection; Specialist Advice	Tailored approaches addressing specific concerns like anxiety or depression with individualized support
Tunneling (90%)	Guided Meditation Series; Initial Check-up; Targeted Interventions; Sequential Courses; Mood Logging; Structured Exercises; Personalized Guidance	Sequential programs guiding users through relaxation, self-assessment, and emotional growth in staged content
Simulation (10%)	Explanatory Animations for Therapeutic Practices; Personalized Guidance with Real-Time Feedback; Adaptability to Individual User Needs	Interactive, visual experiences to facilitate learning and practicing therapeutic techniques, enhancing emotional skills

increased investment in these areas to enhance motivation and support users' therapeutic practices. Table 2 summarizes the predominant strategies.

Table 2. Dialogue Support approaches.

Principle	Most Referenced Approaches	Description
Liking (100%)	Soft Colors & Minimalist Interface. Relaxing design elements to foster a calming experience	Design Elements. Welcoming and calming visuals using soft colors and an intuitive interface
Suggestion (85%)	Personalized Recommendations. Tailored exercises, courses, and feedback based on user needs	User-Centric Recommendations. Personalized activities and feedback to enhance mental health
Similarity (70%)	Emotional Connection. Use of relatable language and visual examples for better user engagement	Engaging Communication. Understandable language and relatable examples to connect with users
Social Role (55%)	Social & Virtual Support. Avatars and community features for interactive and supportive experiences	Interactive Support. Community features and avatars for a supportive, engaging experience
Reminders (20%)	Personalized Reminders. Custom reminders for medication, wellness, and journaling	Activity Reminders. Custom notifications for maintaining mental health routines
Rewards (15%)	Incentives & Rewards. Progress scoring and symbolic rewards for achieving mental health goals	Progress Recognition. Rewards and incentives to motivate and acknowledge user progress
Praise (10%)	Motivational Messages. Congratulatory and encouraging messages for user achievements	Encouragement. Messages to celebrate and support user milestones and achievements

Social Support. These resources showed uneven distribution. "Social Facilitation" and "Cooperation" were common, while "Social Learning," "Recognition," and "Normative Influence" were less frequent. The absence of "Social Comparison" and "Competition" likely stems from concerns about stress and emotional pressure, although careful implementation of these resources could encourage individual progress and goal setting without fostering unhealthy competition. Table 3 summarizes the key strategies.

Credibility Support. The distribution emphasized "Reliability," with significant focus on "Superficial Credibility" and "Expertise," indicating a strong effort to convey trustworthiness and competence. However, "Authority" and "Third Party Endorsements" were less prevalent, suggesting areas for improvement to enhance perceived credibility. Greater validation from recognized sources and external endorsements could strengthen user acceptance and adherence. These findings are detailed in Table 4.

Analysis and Distribution of BE Principles. The analysis showed significant underutilization of BE principles, with only 10% of applications using "Fresh Start Effect"

Table 3. Social Support approaches.

Principle	Most Referenced Approaches	Description
Social Facilitation (60%)	Anonymous Support Community. Platform for social interaction and mutual support, tracking and observing peer activities	Supportive Environment. Facilitates mutual support with activity tracking and social interaction
Cooperation (50%)	Cooperative Group Activities. Encouragement of group exercises, therapy groups, and professional contact for collaborative mental health goals	Cooperative Goals. Promotes group collaboration for achieving mental health objectives
Recognition (35%)	Supportive Environment. Community messages, Q&A, user testimonials, and chatbot-shared experiences for support and recognition	Community Engagement. Provides support through messages, Q&A, testimonials, and shared experiences
Social Learning (35%)	Educational Resources. Shared experiences, therapist-selected resources, chatrooms, forums, and expert guidance for learning and emotional well-being	Learning Opportunities. Offers educational resources and professional guidance for effective mental health strategies
Normative Influence (20%)	Inspirational Stories. Sharing success stories, inspirational chatbots, and professional testimonials to motivate users	Motivational Support. Inspires users with success stories and testimonials
Social Comparison (0%)	Lack of Social Comparison. Absence of social comparison features, focusing on privacy and emotional well-being	Privacy Focus. Avoids social comparison to maintain privacy and emotional safety
Competition (0%)	Lack of Competitive Features. Absence of competitive elements to avoid negative comparisons and respect sensitivity	Ethical Approach. Avoids competitive features to prevent negative comparisons and respect user sensitivity

and 5% incorporating "Pre-Commitment Pledges" and "Loss Aversion". No applications employed "Financial Incentives". Implementing strategies like "Loss Aversion" and "Fresh Start Effect" could improve user commitment and adherence to mental health goals, highlighting the need for better investment in BE practices to influence user behavior positively.

Table 4. Credibility Support approaches.

Principle	Most Referenced Approaches	Description
Trustworthiness (100%)	Scientific and Professional Credibility. Research-based techniques, expert development, compliance with standards, positive feedback, and data security	Scientific and Professional Credibility. Ensures transparency and reliability through scientific evidence and expert involvement
Superficial Credibility (95%)	User Experience and Transparency. Minimal ads, positive testimonials, simple interface, free trials, and affordable pricing	User Experience and Transparency. Builds trust with minimal ads, intuitive design, and positive user feedback
Expertise (90%)	Expert Team and Practices. Specialized therapeutic approaches, content from licensed professionals, and access to certified experts	Expert Team and Practices. Showcases reliability through specialized and professional content development
Verifiability (70%)	Verification and Transparency. Research references, detailed website, study participation, and clear terms of use	Verification and Transparency. Allows users to verify information with research references and direct access
Real-world Feel (65%)	Interactive Support: Direct feedback, chatbots, virtual therapists, community support, and expert access	Interactive Support. Provides real-time support and interaction with professionals
Authority (45%)	Professional Recognition. Contributions and recognition from mental health specialists, licensed professionals, and reputable sources	Professional Recognition. Highlights credibility through involvement and recognition by qualified professionals and institutions

Analysis and Distribution of Security and Privacy Strategies. The review revealed strong commitment to "Privacy Policy", "Data Collection," and "Data Encryption", ensuring user data protection. However, "Data Sharing" received less emphasis, reflecting a cautious approach to third-party information sharing. Areas such as "Data Removal", "Data Retention Time", "Security Updates and Fixes", "Consent", and "Privacy Settings" were moderately addressed, indicating room for improvement. Clear and accessible information about data protection could enhance user trust and clarity.

5 Conclusion and Future Work

The analysis of DMH applications highlights a strong focus on behavioral change techniques, mood and symptom monitoring, and therapy and relaxation practices. However, there is notable underutilization in medication management, professional help, and promoting healthy habits, such as balanced eating, leisure activities, physical exercise, sleep hygiene, and socialization. To enhance user experience, personalized guidance and real-time feedback through animations and simulations are recommended. Additionally, incorporating reminders, rewards, and praise can boost motivation and progress in mental health journeys. Cautious exploration of social comparison and competition may also encourage individual progress. Reinforcing application credibility through validation from recognized sources and third-party endorsements can improve acceptance and effectiveness. Investments in underutilized BE principles are suggested to increase adherence and motivation. Despite robust security and privacy policies, there is room for improvement in clarity and accessibility regarding data protection. Limitations of this study include potential sample representativeness issues, the exclusion of emerging innovations, and the limitations of a single-perspective evaluation. Future development should focus on integrating established and promising approaches to create more effective, engaging, and secure solutions that promote user autonomy and adherence. This study aims to stimulate crucial discussions and advance technology for mental health, offering hope and opportunity for millions of people.

Acknowledgments. This research work was developed under the project RM4Health (ITEA-2021-21022-RM4Health), funded by the European Regional Development Fund (ERDF) within the project number COMPETE2030-FEDER-00391100, and funded by National Funds through the Portuguese FCT—Fundação para a Ciência e a Tecnologia under the R&D Units Project Scope, UIDB/00760/2020 (https://doi.org/10.54499/UIDB/00760/2020).

References

1. Depressive disorder (depression)
2. GBD 2019 Mental Disorders Collaborators: Global, regional, and national burden of 12 mental disorders in 204 countries and territories, 1990–2019: a systematic analysis for the global burden of disease study 2019. Lancet Psychiatry **9**, 137 (2022)
3. New Global Burden of Disease analyses show depression and anxiety among the top causes of health loss worldwide, and a significant increase due to the COVID-19 pandemic | Institute for Health Metrics and Evaluation
4. Artificial intelligence in mental health research: new WHO study on applications and challenges
5. COVID-19 pandemic triggers 25% increase in prevalence of anxiety and depression worldwide
6. Litke, S.G., et al.: Mobile technologies for supporting mental health in youths: scoping review of effectiveness, limitations, and inclusivity. JMIR Ment. Health. **10**, e46949 (2023)
7. Mental Health Apps Market Size And Share Report (2030)
8. Olawade, D.B., Wada, O.Z., Odetayo, A., David-Olawade, A.C., Asaolu, F., Eberhardt, J.: Enhancing mental health with artificial intelligence: current trends and future prospects. J. Med. Surg. Public Health **3**, 100099 (2024)

9. Cheng, C., Ebrahimi, O.V.: Gamification: a novel approach to mental health promotion. Curr. Psychiatry Rep. **25**, 577–586 (2023)
10. Martinho, D., Carneiro, J., Corchado, J.M., Marreiros, G.: A systematic review of gamification techniques applied to elderly care. Artif. Intell. Rev. **53**, 4863–4901 (2020)
11. Hornstein, S., Zantvoort, K., Lueken, U., Funk, B., Hilbert, K.: Personalization strategies in digital mental health interventions: a systematic review and conceptual framework for depressive symptoms. Front. Digit. Health. **5**, 1170002 (2023)
12. Conceição, L., Rodrigues, V., Meira, J., Marreiros, G., Novais, P.: Supporting argumentation dialogues in group decision support systems: an approach based on dynamic clustering. Appl. Sci. **12**, 10893 (2022)
13. Shimada, K.: The role of artificial intelligence in mental health: a review. Sci. Insights **43**, 1119–1127 (2023)
14. Minerva, F., Giubilini, A.: Is AI the future of mental healthcare? Topoi **42**, 809 (2023)
15. Souza, J., et al.: A remote monitoring platform for the management of lower limb vascular diseases. Stud. Health Technol. Inform. **302**, 1013–1014 (2023)
16. Aschentrup, L., Steimer, P.A., Dadaczynski, K., Mc Call, T., Fischer, F., Wrona, K.J.: Effectiveness of gamified digital interventions in mental health prevention and health promotion among adults: a scoping review. BMC Public Health **24**, 1–18 (2024)
17. Martinho, D., Crista, V., Carneiro, J., Corchado, J.M., Marreiros, G.: An intelligent coaching prototype for elderly care (2022)
18. Martinho, D., et al.: A hybrid model to classify patients with chronic obstructive respiratory diseases. J. Med. Syst. **45** (2021)
19. Column - Leveraging Gamification and AI: The Future of Mental Health Care - MedTech Intelligence
20. Iwaya, L.H., Babar, M.A., Rashid, A., Wijayarathna, C.: On the privacy of mental health apps: an empirical investigation and its implications for app development. Empir. Softw. Eng. **28** (2023)
21. Scott, A.J., Webb, T.L., Martyn-St James, M., Rowse, G., Weich, S.: Improving sleep quality leads to better mental health: a meta-analysis of randomised controlled trials. Sleep Med. Rev. **60** (2021)
22. Grajek, M., et al.: Nutrition and mental health: a review of current knowledge about the impact of diet on mental health. Front. Nutr. **9** (2022)
23. McMillan, S.S., Stewart, V., Wheeler, A.J., Kelly, F., Stapleton, H.: Medication management in the context of mental illness: an exploratory study of young people living in Australia. BMC Public Health **20** (2020)
24. Hamdani, S.U., et al.: Effectiveness of relaxation techniques 'as an active ingredient of psychological interventions' to reduce distress, anxiety and depression in adolescents: a systematic review and meta-analysis. Int. J. Ment. Health Syst. **16** (2022)
25. Wickramaratne, P.J., et al.: Social connectedness as a determinant of mental health: a scoping review. PLoS One **17** (2022)
26. Kuczynski, A.M., Halvorson, M.A., Slater, L.R., Kanter, J.W.: The effect of social interaction quantity and quality on depressed mood and loneliness: a daily diary study. J. Soc. Pers. Relat. **39**, 734–756 (2022)
27. Fancourt, D., Aughterson, H., Finn, S., Walker, E., Steptoe, A.: How leisure activities affect health: a narrative review and multi-level theoretical framework of mechanisms of action. Lancet Psychiatry **8**, 329 (2021)
28. Stanyon, M., et al.: Impact of involvement in mental health professional education on patient educators: a qualitative systematic review. BMJ Open. **14** (2024)
29. Komase, Y., et al.: Effects of gratitude intervention on mental health and well-being among workers: a systematic review. J. Occup. Health **63**, 12290 (2021)

30. Oinas-Kukkonen, H., Harjumaa, M.: Persuasive systems design: key issues, process model, and system features. Commun. Assoc. Inf. Syst. **24**, 28 (2009)
31. Altman, M.: Behavioral economics. J. Exp. Anal. Behav. **42**, 435–452 (1984)
32. Malik, A.S., Acharya, S., Humane, S.: Exploring the impact of security technologies on mental health: a comprehensive review. Cureus **16** (2024)
33. Wu, A., Scult, M.A., Barnes, E.D., Betancourt, J.A., Falk, A., Gunning, F.M.: Smartphone apps for depression and anxiety: a systematic review and meta-analysis of techniques to increase engagement. NPJ Digit. Med. **4** (2021)
34. ITEA 4 · Project · 21022 RM4Health

Automatic PDF Document Classification with Machine Learning

Sócrates Llácer Luna[1], Dario Garigliotti[2], Fernando Martínez Plumed[1](✉), and César Ferri Ramírez[1]

[1] Universitat Politècnica de València, València, Spain
fmartinez@dsic.upv.es
[2] University of Bergen, Bergen, Norway

Abstract. Universitat Politècnica de València (UPV) faces challenges in managing its Alfresco document repository, which contains 600,000 PDF files, of which only 100,000 are correctly categorised. Manual classification is laborious and error-prone, hindering information retrieval and advanced search capabilities. This project presents an automated pipeline that integrates optical character recognition (OCR) and machine learning to efficiently classify documents. Our approach distinguishes between scanned and digital documents, accurately extracts text and categorises it into 51 predefined categories using models such as BERT and RF. By improving document organisation and accessibility, this work optimises UPV's document management and paves the way for advanced search technologies and real-time classification systems.

Keywords: Document Classification · OCR · Machine Learning · Alfresco Repository

1 Introduction

In the digital age, document digitisation and efficient access to information are crucial for the operation and development of modern institutions. The Universitat Politècnica de València (UPV), a public university in Spain, is currently facing a challenge in managing its Alfresco document repository, which contains approximately 600,000 PDF files. In particular, only 100,000 of these documents are correctly categorised, creating an imbalance that hinders efficient information retrieval and the implementation of advanced search and analysis systems.

Manual categorisation of such a large number of documents is not only time-consuming and labour-intensive, but also prone to error, further complicating information management. As a result, there is an urgent need for an automated solution that can handle the diversity and complexity of the documents in the repository while ensuring accurate classification.

Another significant challenge is the imbalance of labelled examples, with many classes having very few correctly categorised instances (see Fig. 1). This imbalance complicates the training of machine learning models, as classes with

V. Julian et al. (Eds.): IDEAL 2024, LNCS 15346, pp. 447–459, 2025.
https://doi.org/10.1007/978-3-031-77731-8_40

fewer examples may be underrepresented, leading to biased or inaccurate classification results.

This project aims to address these challenges by introducing an automated pipeline that uses optical character recognition (OCR) and machine learning algorithms for document classification. The proposed solution is designed to distinguish between scanned and digital documents, accurately extract text and use this text to feed classification models. Our approach aims to improve the organisation and accessibility of documents within the Alfresco repository, thus optimising document management at UPV and facilitating the integration of new search and analysis technologies.

The main contributions of this work include:

- Development of an automated document classification pipeline tailored to 51 predefined categories, improving the granularity and accuracy of document organisation.
- Implementing techniques to differentiate between scanned and digital documents, ensuring appropriate handling of different document types.
- Testing of different text vectorisation methods, including traditional term frequency inverse document frequency (TF-IDF) and advanced transformer-based embeddings.
- Comprehensive evaluation of multiple classification models, including BERT and Random Forest, to determine the most effective approach to document categorisation.

2 Background

Optical Character Recognition (OCR) technology has revolutionised the conversion of various types of documents into machine-readable text. Early OCR systems relied on pattern matching and feature recognition [7]. More recent developments have incorporated machine learning techniques to improve accuracy and versatility. One of the most popular OCR engines today is Tesseract, originally developed by Hewlett-Packard and now maintained by Google [24]. Tesseract offers multi-language support and seamless integration with image processing libraries such as OpenCV [5].

PyTesseract, a Python wrapper for Tesseract, has become a widely used tool in the OCR community due to its ease of use and high accuracy. It allows for pre-processing steps such as noise reduction, contrast adjustment and rotation, which are essential for improving the quality of OCR results [18]. Applications of OCR range from digitising printed archives to automating data entry processes and improving text search capabilities within large document repositories [25].

Document classification [4] is a well-researched area in machine learning and natural language processing (NLP). Current methods range from traditional algorithms to deep learning models. Traditional approaches include algorithms such as Naive Bayes [19], Support Vector Machines (SVM) [15], and k-Nearest Neighbours (k-NN) [1]. Other supervised learning algorithms, including decision trees [23], random forests [6], and gradient boosting techniques such as XGBoost [9], rely on labelled data for training. These models have been used extensively in various text classification tasks and have been shown to be effective in different scenarios. They usually rely on textual features extracted using techniques such as Term Frequency-Inverse Document Frequency (TF-IDF) [10].

In contrast, recent advances in NLP have mainly focused on classifying short texts, such as sentences or paragraphs, using Transformer-based models such as BERT [11,17]. However, these models struggle with long documents due to the computational overhead associated with the self-attention mechanism in standard transformers [27]. To overcome this, various methods have been proposed, such as truncating documents to fit within the token boundary [2,8] or using alternative architectures such as the longformer, which uses sparse attention to efficiently handle long sequences [3,29], the latter often showing superior performance [16].

3 Empirical Methodology

The empirical methodology for our document classification system involves several distinct stages, from data acquisition to model evaluation. The whole process attempts to ensure a seamless and efficient document classification pipeline, aimed at handling the diverse and voluminous nature of the documents in the Alfresco repository at UPV.

Data. The dataset used in this study was taken from the Alfresco document repository at the Universitat Politècnica de València (UPV). The repository contains approximately 600,000 PDF files, of which only a fraction —approximately 100,000— are correctly categorised into predefined categories. The labelled subset are unevenly distributed across 51 pre-defined categories (see Fig. 1). This imbalance posed additional challenges for model training, particularly in ensuring that under-represented categories were accurately classified.

The documents in the repository represent a wide variety of types and content. These include, but are not limited to: *academic records* such as transcripts, certificates and degrees; *administrative records*, including financial records, correspondence and official regulations; *research papers and reports*, consisting of theses, dissertations and project reports; and *miscellaneous documents*, comprising minutes of meetings, memoranda and other miscellaneous forms.

Prepossessing. Once the documents were downloaded using the Alfresco API [21], the next task was to determine whether each file was a scanned image or a digitally created document. This distinction is important because the method of text extraction differs significantly between the two types. For scanned documents, we used OCR using *PyTesseract* [22] to convert images into machine-readable text. The OCR process also included the application of enhancements such as contrast adjustment, rotation correction and DPI adjustment using the *OpenCV* library [20] to ensure high quality text extraction.

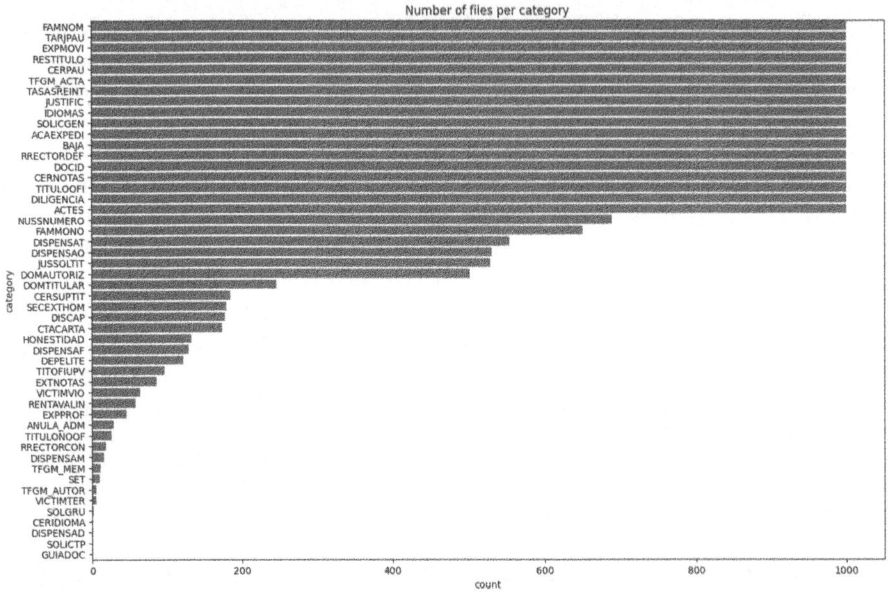

Fig. 1. Class distribution of the labeled documents, highlighting the imbalance across different categories.

Once the text was extracted, it underwent a series of preprocessing steps to prepare it for classification. These steps included (1) the removal of special characters and numbers that did not contribute to the semantic understanding of the texts were removed; (2) lemmatisation and stemming to reduce words to their root forms, ensuring that different inflections of a word were treated similarly; and (3) the elimination of stop words (e.g., 'and', 'the', etc.) that have no significant meaning were removed. We explored different models for lemmatization, including the SpaCy model *'es_core_news_sm'*[1] and the more advanced *'es_dep_news_trf'*[2] model, which leverages transformer-based techniques to achieve higher accuracy.

After preprocessing, the text data was vectorised. We used two different models for this: the *Term Frequency-Inverse Document Frequency (TF-IDF)*, a traditional text vectorization method that transforms documents into numerical features that reflect the importance of words within the corpus; and the Spacy model *'es_dep_news_trf'*, an advanced model that provides vector representations of the texts using transformer-based techniques. As we show in our comparative analysis, TF-IDF proved to be more robust, providing better performance for this specific classification task compared to the more advanced Spacy model.

[1] An spanish pipeline optimised for CPU that can be found in https://spacy.io/models/es.

[2] An Spanish transformer pipeline that can be found in https://huggingface.co/spacy/es_dep_news_trf.

Document Classification Algorithms. We explored several algorithms to find the most effective approach. We started by testing traditional models that have been shown to perform well in the related literature such as *Support Vector Machines* (SVM) [26] and *Random Forest* (RF) [6]. Both models provided a strong basis for classification performance. The *SVM* model was particularly useful in dealing with the complexity of our diverse document set, due to its effectiveness in high-dimensional spaces. However, the RF algorithm showed better performance. Known for its effectiveness with large datasets and its ability to provide robust predictions by averaging the results of multiple decision trees, also reducing overfitting.

To improve classification accuracy, we also used BERT-based language models. The choice of BERT over Longformer [3] for this study is justified by several key considerations: BERT's proven effectiveness and extensive support in the NLP community make it a reliable choice for categorising the document segments typical of this task [2]. In addition, BERT offers computational feasibility and robust performance in capturing contextual embeddings, even for moderately long sequences [12], which is well suited to the needs of the project. Its pre-trained models and ease of integration through libraries such as Huggingface's Transformers [28] facilitate implementation, making BERT a pragmatic and effective choice given the constraints and goals of the study.

Training and Evaluation. Given the size of the labelled dataset, which consists of approximately 100,000 instances, we decided to use split-validation rather than cross-validation to ensure computational efficiency and faster processing while still maintaining a reliable estimate of model performance. The models were thus trained and evaluated on a labelled subset of the data (80%). We used *precision* (which measures the proportion of true positives among the predicted positives, indicating how many of the predicted categories were correctly identified.), *recall* (which measures the proportion of true positives among the actual positives, reflecting the ability of the model to identify all relevant instances), and *F1-score* (which is the harmonic mean of precision and recall, providing a balance between the two metrics to evaluate the overall performance of the model) as the primary metrics for performance evaluation [13]. The use of a validation set (20% of the labelled subset of the data) ensured the generalisability of the models to unseen data for classification into the 51 predefined categories. The best performing models will be used to categorise the remaining unlabelled documents (500K) in the model deployment phase (which is outside the scope of this work).

BERT training also involved tokenisation (we used *bert-base-uncased*[3]), input sequence truncation and padding, followed by fine-tuning the pre-trained transformer model for our specific task (we used *BertForSequenceClassification*[4]). This computationally intensive process was accelerated using GPU resources.

[3] https://huggingface.co/google-bert/bert-base-uncased.

[4] https://huggingface.co/docs/transformers/model_doc/bert.

4 Results

Here we present the performance results and detailed analysis of the classification models used in the automated document classification pipeline. The analysis includes overall model performance metrics and a more granular examination of specific document classes.

4.1 Model Performance Overview

For our experiments, we focused primarily on traditional vectorisation methods, in particular TF-IDF, due to its superior performance (i.e., better category separation) as observed in our initial evaluations (that will be explained in the following subsections). Figure 2 shows the comparison of precision, recall and F1 scores for the SVM, RF and BERT models on the validation set. The RF model consistently outperformed the others on all metrics —accuracy (0.841), F1 score (0.838), precision (0.847) and recall (0.841). This indicates RF's superior ability to correctly classify documents while minimising false positives and negatives. In contrast, BERT and SVM had comparable but slightly lower performance scores.

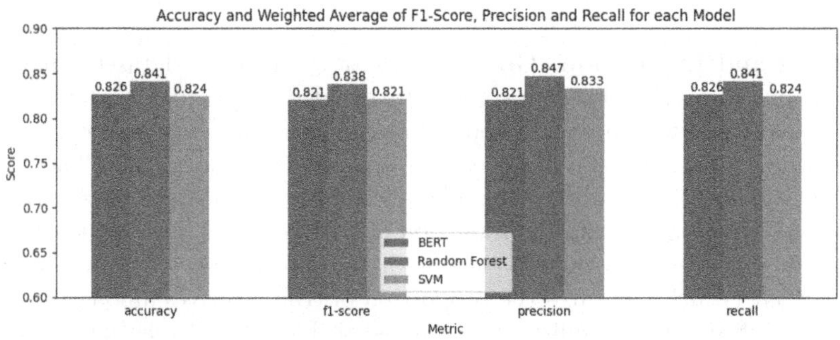

Fig. 2. Comparison of evaluations of the 3 models experimented.

4.2 Class-Wise Analysis

To provide a more comprehensive understanding, we discuss the performance of the models on specific document classes (i.e., TITOFIUPV, BAJA, ACTES, DOCID, CERNOTAS, TARJPAU, DOMTITULAR, ACAEXPEDI) that are representative of common document types within the dataset, and represent the types of documents in which UPV management is most interested. These classes also serve as benchmarks to illustrate the effectiveness of the model across different levels of content and structural complexity. Table 1 shows the summary of the results for this classes. The F1-scores, precision, and recall for these highlighted classes across the models are shown in Fig. 3.

Table 1. Class-wise Performance Comparison of Models

Class	Description	Model Performance
TITOFIUPV	Short documents containing the collection record of the official title and the corresponding degree title	BERT showed the highest precision and recall, demonstrating its ability to capture contextual information effectively, even in short documents with limited content
BAJA	Documents often containing emails and other hand-crafted documents with high template similarity	RF achieved the best recall, accurately identifying this particular class. BERT excelled in precision, using its understanding of context to distinguish these documents from others
ACTES	Minutes of various events, showing common structural elements	RF slightly outperformed the other models, reflecting its strength in dealing with structured documents. BERT and SVM also performed well, indicating the general effectiveness of all models for this class
DOCID	Scanned identification documents with limited text	BERT significantly outperformed SVM and RF in accuracy, benefiting from its deep understanding of limited textual context
CERNOTAS	Documents with similar structure but different content, representing different course marks	All models struggled, especially with accuracy. The diversity of content within similar structures was a challenge, with BERT performing worse than SVM and RF
TARJPAU	Documents containing university entrance exam grades, often bilingual (Spanish and Valencian)	Similar to CERNOTAS, but with less variety of content. BERT faced challenges due to the bilingual nature, but still performed comparatively well
DOMTITULAR	Bank account certificates, both scanned and digitised	BERT achieved higher recall than other models, aided by the presence of contextual expressions that helped to identify this class despite the limited information
ACAEXPEDI	Diverse documents containing different types of content, often relevant to other categories	Models showed moderate performance, reflecting the difficulty of classifying documents with mixed or overlapping content

The performance comparison between BERT, RF and SVM over different document classes reveals interesting insights. BERT excels at capturing contextual information for short documents such as those in the TITOFIUPV class, achieving the highest precision and recall. It also achieves remarkable precision for the BAJA class, but is outperformed by RF in terms of recall, highlighting RF's ability to accurately delineate complex classes. For the ACTES class, which is characterised by a common structure but different content, RF slightly outperforms BERT and SVM, highlighting its ability to deal with such scenarios.

However, BERT shows a significant improvement in accuracy for the DOCID class, illustrating its superior language understanding capabilities. Challenges for BERT arise with the CERNOTAS and TARJPAU classes. Despite its general strength in context understanding, BERT struggles with these classes due to their similar structures and varied content, as well as the bilingual challenges in TARJPAU. In the DOMTITULAR class, BERT's higher recall indicates his ability to grasp the context effectively, even with limited information. Conversely, the diverse nature of the ACAEXPEDI class is a challenge for all models, but RF and SVM show better performance in distinguishing document structure. BERT's difficulty here highlights its limitations with contextual distortion when faced with content diversity.

Overall, RF proved to be particularly effective in scenarios with common document structures or where high recall is crucial. BERT showed its strength in accuracy and context understanding, especially for documents with limited information. However, BERT struggled with diverse content or bilingual scenarios, highlighting its limitations in certain classification contexts. Therefore, for the task of automated document classification that we address, the RF model is recommended due to its higher overall reliability and accuracy.

4.3 Visualisation of Vectorisation

To further assess the effectiveness of our text vectorisation methods, we visualised document embeddings with t-SNE (t-distributed Stochastic Neighbour Embedding) [14] (which offers a visual representation of the high-dimensional data in two dimensions). These visualisations illustrate how well different document categories are separated in the vector space, providing insight into the clustering quality of our vectorisation techniques.

In this regard, Fig. 4 (top) shows the t-SNE visualisation of document embeddings using the SpaCy model *'es_dep_news_trf'*. Despite the use of advanced transformer-based techniques, the SpaCy model showed some category overlap, indicating room for improved differentiation. Conversely, Fig. 4 (bottom) shows the t-SNE visualisation for the TF-IDF method. This traditional vectorisation technique showed better category separation, consistent with its higher classification performance, suggesting that the concrete problem and specific dataset of this study benefited from this simpler yet effective approach.

Overall, TF-IDF proved to be a more effective vectorisation method for this classification task compared to the advanced SpaCy model. By optimising the vectorisation approach, we improved the overall performance and robustness of our classification models, in particular the RF model, which handled the document diversity in the dataset exceptionally well. We have seen that, in general, a careful selection of both the classification model and the vectorisation technique is crucial for achieving optimal results in automated document classification tasks.

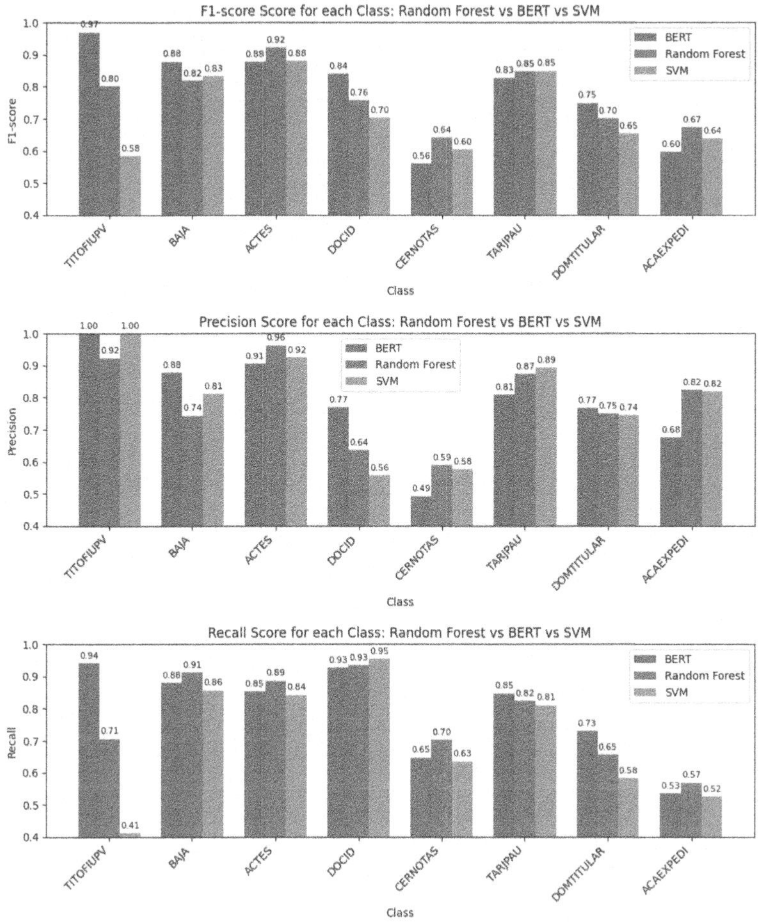

Fig. 3. Performance metrics for highlighted classes across different models.

4.4 Interpretation of Results

The results of this study highlight the effectiveness of our automated classification pipeline in addressing the challenges faced by the UPV in managing its extensive document repository. Among the models evaluated, RF demonstrated superior performance in accurately classifying documents into predefined categories, in line with our goal of improving document organisation and accessibility. The RF model's ability to handle the complex task of multi-class classification was particularly noteworthy, as it skilfully used TF-IDF vectorisation to manage the semantic inconsistencies inherent in the dataset due to OCR inaccuracies. While BERT's deep learning capabilities allowed it to capture the nuanced complexities in the text, its performance was not as consistently robust across the diverse document set as RF.

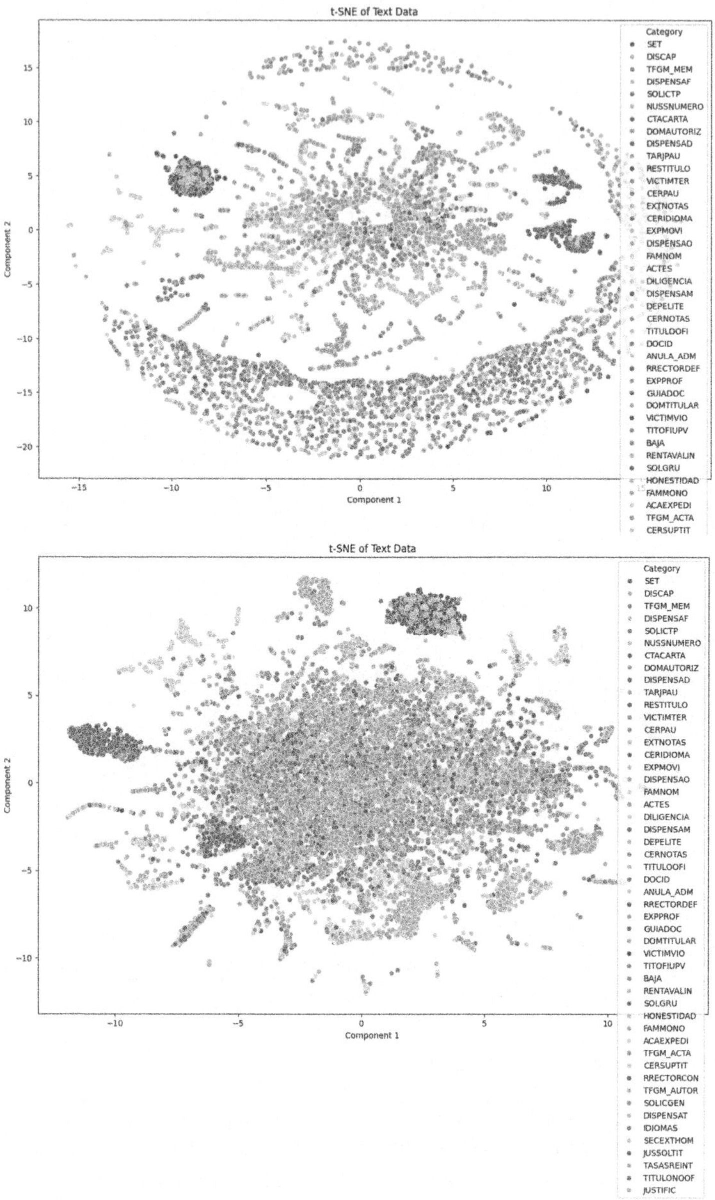

Fig. 4. t-SNE visualization of the embeddings, highlighting the similarity across different categories using TF-IDF (top) and the SpaCy model *'es_dep_news_trf'* (bottom).

Despite the promising results, there were several limitations to this study. Firstly, the variety of document types, ranging from highly structured forms to unstructured text, posed challenges that could not be fully addressed by a single approach. The quality of the documents and the presence of multiple languages further complicated the OCR process, adversely affecting the accuracy of text extraction and consequently the classification performance. Inconsistent OCR text extraction, particularly in documents with complex layouts or noisy backgrounds, often led to misclassifications.

5 Conclusions and Future Work

This study demonstrates the effectiveness of our automated classification pipeline in managing the large and diverse document repository at the UPV. RF outperformed BERT and SVM in all evaluated metrics (accuracy, F1 score, precision and recall) in most of the classes and overall, making it particularly suitable for the diverse document types present in UPV's Alfresco repository. While BERT showed strength in handling contextually rich documents, it was less effective across broader categories compared to RF. In addition, the traditional TF-IDF vectorisation method provided better category separation, which significantly improved the overall performance of the model. The ability of RF to effectively use TF-IDF embeddings demonstrated its robustness in dealing with data inconsistencies due to OCR inaccuracies.

Future work should focus on several specific actions to build on the current study. First, an in-depth analysis of document classes using unsupervised machine learning techniques, such as clustering, will help to identify natural groupings within the data and potentially reveal new, meaningful categories. Secondly, implementing a hierarchical cascade model can break down the classification task into smaller, more manageable components, which can improve overall classification accuracy. Finally, increasing the volume of accurately labelled samples will create a more balanced and comprehensive training dataset, reducing bias towards underrepresented classes and improving the model's ability to generalise.

By addressing these areas, future iterations of the document classification pipeline can achieve greater accuracy, reliability and efficiency, significantly enhancing the performance of UPV's document management system.

Acknowledgments. This work was funded by the Norwegian Research Council grant 329745 Machine Teaching for Explainable AI, CIPROM/2022/6 (FASSLOW) funded by Generalitat Valenciana, the EC H2020-EU grant agreement No. 952215 (TAILOR), and Spanish grant PID2021-122830OB-C42 (SFERA) funded by MCIN/AEI/10.13039/501100011033 and "ERDF A way of making Europe". Authors thank the Cátedra de Inteligencia Artificial aplicada a la Administración Pblica of Universitat Politècnica de València (UPV).

References

1. Aas, K., Eikvil, L.: Text categorisation: a survey (1999)
2. Adhikari, A., Ram, A., Tang, R., Lin, J.: DocBERT: BERT for document classification. arXiv:1904.08398 (2019)

3. Beltagy, I., Peters, M.E., Cohan, A.: LongFormer: the long-document transformer. arXiv:2004.05150 (2020)
4. Borko, H., Bernick, M.: Automatic document classification. J. ACM (JACM) **10**(2), 151–162 (1963)
5. Bradski, G.: The OpenCV library. Dr. Dobb's J.: Softw. Tools Prof. Program. **25**(11), 120–123 (2000)
6. Breiman, L.: Random forests. Mach. Learn. **45**(1), 5–32 (2001)
7. Casey, R.G., Lecolinet, E.: A survey of methods and strategies in character segmentation. IEEE Trans. Pattern Anal. Mach. Intell. **18**(7), 690–706 (1996)
8. Chalkidis, I., Fergadiotis, M., Kotitsas, S., Malakasiotis, P., Aletras, N., Androutsopoulos, I.: An empirical study on large-scale multi-label text classification including few and zero-shot labels. arXiv:2010.01653 (2020)
9. Chen, T., Guestrin, C.: XGBoost: a scalable tree boosting system. In: Proceedings of the 22nd ACM SIGKDD International Conference on Knowledge Discovery and Data Mining, pp. 785–794 (2016)
10. Christian, H., Agus, M.P., Suhartono, D.: Single document automatic text summarization using term frequency-inverse document frequency (TF-IDF). ComTech: Comput. Math. Eng. Appl. **7**(4), 285–294 (2016)
11. Devlin, J., Chang, M.W., Lee, K., Toutanova, K.: BERT: pre-training of deep bidirectional transformers for language understanding. arXiv:1810.04805 (2018)
12. Ding, M., Zhou, C., Yang, H., Tang, J.: CogLTX: applying BERT to long texts. Adv. Neural. Inf. Process. Syst. **33**, 12792–12804 (2020)
13. Ferri, C., Hernández-Orallo, J., Modroiu, R.: An experimental comparison of performance measures for classification. Pattern Recogn. Lett. **30**(1), 27–38 (2009)
14. Hinton, G.E., Roweis, S.: Stochastic neighbor embedding. Adv. Neural. Inf. Process. Syst. **15** (2002)
15. Joachims, T.: Text categorization with support vector machines: learning with many relevant features. In: Nédellec, C., Rouveirol, C. (eds.) ECML 1998. LNCS, vol. 1398, pp. 137–142. Springer, Heidelberg (1998). https://doi.org/10.1007/BFb0026683
16. Johnson, A.E., et al.: MIMIC-III, a freely accessible critical care database. Sci. Data **3**(1), 1–9 (2016)
17. Liu, Y., et al.: RoBERTa: a robustly optimized BERT pretraining approach. arXiv:1907.11692 (2019)
18. Malathi, T., Selvamuthukumaran, D., Chandar, C.D., Niranjan, V., Swashthika, A.: An experimental performance analysis on robotics process automation (RPA) with open source OCR engines: Microsoft OCR and google tesseract OCR. In: IOP Conference Series: Materials Science and Engineering, vol. 1059, p. 012004 (2021)
19. McCallum, A., Nigam, K., et al.: A comparison of event models for naive bayes text classification. In: AAAI-98 Workshop on Learning for Text Categorization, Madison, WI, vol. 752, pp. 41–48 (1998)
20. OpenCV Contributors: OpenCV Documentation. https://docs.opencv.org/4.x/. Accessed 17 July 2024
21. Potts, J.: Alfresco Developer Guide. Packt Publishing Ltd. (2008)
22. PyTesseract Contributors: Pytesseract: A python wrapper for Google's tesseract-OCR engine (2024). https://pypi.org/project/pytesseract/. Accessed 08 July 2024
23. Quinlan, J.R.: Induction of decision trees. Mach. Learn. **1**(1), 81–106 (1986)
24. Smith, R.: An overview of the tesseract OCR engine. In: Ninth International Conference on Document Analysis and Recognition (ICDAR 2007), vol. 2, pp. 629–633. IEEE (2007)

25. Srivastava, S., Verma, A., Sharma, S.: Optical character recognition techniques: a review. In: 2022 IEEE International Students' Conference on Electrical, Electronics and Computer Science (SCEECS), pp. 1–6. IEEE (2022)
26. Steinwart, I., Christmann, A.: Support Vector Machines. Springer, Heidelberg (2008). https://doi.org/10.1007/978-0-387-77242-4
27. Vaswani, A., et al.: Attention is all you need. Adv. Neural. Inf. Process. Syst. 5998–6008 (2017)
28. Wolf, T., et al.: HuggingFace's transformers: state-of-the-art natural language processing. arXiv:1910.03771 (2019)
29. Zaheer, M., et al.: Big bird: transformers for longer sequences. Adv. Neural. Inf. Process. Syst. **33**, 17283–17297 (2020)

Contributions on Mixtures of Polynomials for Hybrid Bayesian Networks

Juan Carlos Luengo[1], Darío Ramos-López[1,2], Rafael Rumí[1,2](✉), and Ana D. Maldonado[1,2]

[1] Department of Mathematics, University of Almería, Almería, Spain
rrumi@ual.es

[2] Center for Development and Transfer of Mathematical Research to Industry, University of Almería, Almería, Spain

Abstract. Bayesian networks based on mixtures of polynomials (MoPs) are used to model multivariate hybrid or continuous probability distributions. MoPs try to provide a flexible yet simple model to deal with probabilistic reasoning, approximating complex or empirical probability distributions. We review the main models based on MoPs in the literature, introducing the mixtures of polynomials with tails (tMoPs) and derive analytically their main statistical properties: (cumulative) distribution function, moments, variance and moment-generating function. Also, we present an algorithm for learning a tMOPs density from data. This learning algorithm is tested with several standard probability distributions. In these experiments, the tMOPs models yield satisfactory results in comparison to other MoPs available alternatives. Finally, we sketch the process of building a meta-model that can directly provide a tMOPs expression for a specific density among a given family. This idea is tested with the well-known Beta distribution (two shape parameters), giving promising results.

Keywords: hybrid Bayesian networks · mixtures of polynomials · density estimation

1 Introduction

Bayesian networks (BNs) are flexible and powerful probabilistic graphical models useful in probabilistic reasoning under uncertainty and decision-making problems [4,11]. BNs have been employed in a large number of fields and real applications in many different domains [1–3,9]. BNs are more explainable and understandable models than other state-of-the-art techniques.

The most interesting aspects of BNs are revealed when there are both continuous and discrete variables (hybrid networks). In this scenario, it is crucial to have a uniform and coherent representation of the probability distributions used, so that by operating them, a distribution in the same family is obtained, otherwise the standard inference algorithms cannot be applied.

V. Julian et al. (Eds.): IDEAL 2024, LNCS 15346, pp. 460–471, 2025.
https://doi.org/10.1007/978-3-031-77731-8_41

In addition to using discrete BNs (with prior discretization of the necessary variables) or the well known conditional linear Gaussian (CLG) model (which is not truly closed, as the posterior distributions are mixtures of Gaussians), few alternatives exist that satisfy this requirement. One possibility is the use of Mixtures of Truncated Exponentials (MTEs; [6,10]). Other option are BNs based on mixtures of polynomials [15], in which we focus on this paper. MTEs and MoPs schemes are sometimes unified in the literature under the name Mixture of Truncated Basis Functions (MoTBFs; [5,13]). The main advantage of MoPs over MTEs are the interpretability, since polynomials are simple to understand, and the operational ease, since multiply and integrate polynomials is trivially closed. However, this methodology has only been developed partially, still most of the well-known BN techniques (learning, inference for example) have not been adapted to the MOPs case, and results about MoPs are scattered. We try to integrate these results in this paper, including the new tMOPs model, compare them and relate the MOPs distribution to some other known distributions, as well as present a basic theoretical study about the MOPs distribution. We will focus only in the study on learning procedure of the distribution, leaving apart their usage within BN. However, as it is known, in BNs live together marginal (univariate) distributions together with conditional distributions. The main problem in hybrid BNs lies in learning accurately the marginal distribution, since for the conditional ones, only the domain of the parent variables is divided, and then a marginal density learned for the non-conditioning variable.

2 Bayesian Networks Based on Mixtures of Polynomials

BNs based on MoPs [8,14,15] are used to model multivariate continuous probability distributions. Like MTEs [10], MoPs try to provide a flexible yet simple model to deal with probabilistic reasoning and approximate complex or empirical probability distributions.

In a nutshell, MoPs are defined as piecewise functions that are polynomials in each subdomain. In this section, we review the work that has been performed in this topic, and then build on them by presenting an approach that we will refer to as the tMOPs model.

2.1 Previous Works

MoPs were originally introduced in [15] with the next definition:

Definition 1 (Mixtures of polynomials, MoP). *A one-dimensional real function $f : \mathbb{R} \to \mathbb{R}$ is said to be a mixture of polynomials (MoP) if it is a non-negative piecewise function of the form*

$$f(x) = \begin{cases} a_{0i} + a_{1i}x + a_{2i}x^2 + \ldots + a_{ni}x^n & \textit{for } x \in A_i, \quad i = 1, \ldots, k \\ 0 & \textit{otherwise} \end{cases} \tag{1}$$

where $A_1, \ldots, A_k$ (subdomains) are disjoint intervals in $\mathbb{R}$ that do not depend on x, and $a_{0i}, a_{1i}, a_{2i} \ldots a_{ni}$ (coefficients) are real constants for all i.

Such a function is said to be a k-piece, n-degree MOPs function or potential. In addition, if it integrates up to one, it is said to be a MOPs density.

The main motivation for working with models based on MoPs is the ease of working with them both analytically and computationally. Furthermore, they are closed under addition, differentiation, multiplication, marginalization and integration. Additionally, learning the parameters is easier than in MTEs, as there are many ways to obtain them in the literature (Taylor expansion, splines, regression...).

However, if the amount k of subdomains or the degree n are large, these general MoPs may become too complex models and prone to overfitting.

Shenoy-West MOPs Approximation. Shenoy and West present in [16] a method based on the Taylor series expansion, so it requires the analytical expression of the density function to be approximated. The non-negativity of the Taylor approximation over the entire domain is not guaranteed. In case it presents some roots, two possibilities are provided: either increase the degree of the polynomial, or increase the number of subintervals, thereby increasing considerably not only the accuracy but also the complexity of the fit.

B-Splines MOPs Approximation. In [8], the authors present a new method of learning a MOPs based on uniform B-splines. The domain Ω is splited in L_x+1 equi-spaced subintervals with knots $\{a_0, ..., a_{L_x}\}$ to define $M_X = L_X + r_X - 1$ B-splines with order r_X

MoTBF Model. MoTBFs, introduced in [12] is a general model based on the abstract notion of real-valued basis functions $\psi(\cdot)$, so that an MoTBF can be defined as a function $g_k : \Omega \rightarrow \mathbb{R}_0^+$ that:

$$g_k = \sum_{i=0}^{k} a_i \psi_i(x)$$

where $a_i \in \mathbb{R}$. It can be noted that particular cases of this type of functions are MoPs and MTEs.

To fit a MoTBF, teh corresponding CDF is fitted, minimizing the difference between the empirical distributions and the MoTBF CDF (see details in [12]).

3 tMOPs Model

This method is a particular case of the original MOPs definition made in [15] where only a maximum of three pieces are taken: a polynomial function in the central one and two constant tails in the other two subintervals (that appear only if they are necessary).

Definition 2. *Mixtures of polynomials with tails function (tMoP) A one-dimensional function $f : \mathbb{R} \to \mathbb{R}$ is said to be a tailed mixture of polynomials function if it is a piecewise function of the form*

$$f(x) = \begin{cases} b_1 & \text{for } x \in [l_0, l_1], \\ \sum_{i=0}^{n} a_i x^i & \text{for } x \in [l_1, l_2], \\ b_2 & \text{for } x \in [l_2, l_3], \\ 0 & \text{otherwise} \end{cases} \tag{2}$$

where $l_0 \leq l_1 < l_2 \leq l_3$ and the coefficients $a_0, a_1, a_2 \ldots a_n$ are constants. The values b_1 and b_2 are called the MOPs tails and they are relevant only if $l_0 < l_1$ or $l_2 < l_3$ (i.e., when $[l_0, l_1]$, $[l_2, l_3]$ or both are proper intervals).

When we try to fit a density using a polynomial it is common that there are areas where it is negative, especially those where the probability is close to 0. Achieving non-negativity in a polynomial fit is not an easy task since it generally causes a loss of accuracy, especially in the areas near the extremes of the domain. To avoid this mismatch, the polynomial fit is restricted to the area where it is positive (this process is explained in detail in Sect. 3.2) using a 0-degree polynomial (a constant) to fit the tails. In that case we can get the advantages similar to discretization in the tails, and a better fit in the center of the distribution.

3.1 tMOPs Properties

We will study the main features of a tMOPs defined on the interval $\Omega = [\alpha_1, \alpha_2]$ whose central polynomial has roots in β_1 and β_2 so that $\beta_1 < \beta_2$. Then, according to Definition 2, $l_0 = \alpha_1, l_1 = \beta_1, l_2 = \beta_2$ and $l_3 = \alpha_2$.

Cumulative Density Function. In this case we have to consider the three pieces in which a tMOPs is split. We will note as F_c for the central part where the polynomial is defined and F_l and F_r for the left and right tails:

$$\begin{aligned} F_l(x) &= \int_{\alpha_1}^{x} b_1 \, dt = b_1 t]_{\alpha_1}^{x} = b_1(x - \alpha_1) \\ F_c(x) &= \int_{\alpha_1}^{\beta_1} b_1 \, dt + \int_{\beta_1}^{x} \sum_{i=0}^{n} a_i t^i \, dt = b_1 t]_{\alpha_1}^{\beta_1} + \sum_{i=0}^{n} \frac{a_i t^{i+1}}{i+1}\Bigg]_{\beta_1}^{x} = b_1(\beta_1 - \alpha_1) + \sum_{i=0}^{n} \frac{a_i(x^{i+1} - \beta_1^{i+1})}{i+1} \\ F_r(x) &= \int_{\alpha_1}^{\beta_1} b_1 \, dt + \int_{\beta_1}^{\beta_2} \sum_{i=0}^{n} a_i t^i \, dt + \int_{\beta_2}^{x} b_2 \, dt = b_1 t]_{\alpha_1}^{\beta_1} + \sum_{i=0}^{n} \frac{a_i t^{i+1}}{i+1}\Bigg]_{\beta_1}^{\beta_2} + b_2 t]_{\beta_2}^{x} \\ &= b_1(\beta_1 - \alpha_1) + \sum_{i=0}^{n} \frac{a_i(\beta_2^{i+1} - \beta_1^{i+1})}{i+1} + b_2(x - \beta_2) \end{aligned} \tag{3}$$

So the cumulative density function for a tMOPs is:

$$F(x) = \begin{cases} b_1(x-\alpha_1) & \text{for } x \in [\alpha_1, \beta_1], \\ b_1(\beta_1-\alpha_1) + \sum_{i=0}^{n} \frac{a_i(x^{i+1}-\beta_1^{i+1})}{i+1} & \text{for } x \in [\beta_1, \beta_2], \\ b_1(\beta_1-\alpha_1) + \sum_{i=0}^{n} \frac{a_i(\beta_2^{i+1}-\beta_1^{i+1})}{i+1} + b_2(x-\beta_2) & \text{for } x \in [\beta_2, \alpha_2], \end{cases} \tag{4}$$

Moments

$$\begin{aligned} E[X^m] &= \int_{\Omega} x^m \cdot f(x)dx \\ &= \frac{b_1(\beta_1^{m+1}-\alpha_1^{m+1})}{m+1} + \sum_{i=0}^{n} \frac{a_i(\beta_2^{i+m+1}-\beta_1^{i+m+1})}{i+m+1} + \frac{b_2(\alpha_2^{m+1}-\beta_2^{m+1})}{m+1} \end{aligned} \tag{5}$$

Variance

$$\begin{aligned} \sigma^2 = E[X^2] - E[X]^2 =& \frac{b_1(\beta_1^3-\alpha_1^3)}{3} + \sum_{i=0}^{n} \frac{a_i(\beta_2^{i+3}-\beta_1^{i+3})}{i+3} + \frac{b_2(\alpha_2^3-\beta_2^3)}{3} \\ & - \left(\frac{b_1(\beta_1^2-\alpha_1^2)}{2} + \sum_{i=0}^{n} \frac{a_i(\beta_2^{i+2}-\beta_1^{i+2})}{i+2} + \frac{b_2(\alpha_2^2-\beta_2^2)}{2}\right)^2 \end{aligned} \tag{6}$$

Moment Generating Function. The successive differentiation of a polynomial $p(x) = \sum_{i=0}^{n} a_i x^i$ are:

$$p^{(k)}(x) = \sum_{i=k}^{n} \frac{i!}{(i-k)!} a_i x^{i-k} \tag{7}$$

Integrating by parts, the moments generating function is:

$$\begin{aligned} M_x(t) = E[e^{tx}] &= \int_{\Omega} e^{tx} \cdot f(x)dx \\ &= \frac{b_1(e^{t\beta_1}-e^{t\alpha_1})}{t} + \sum_{i=0}^{n} (-1)^i \frac{e^{t\beta_2}p^{(i)}(\beta_2) - e^{t\beta_1}p^{(i)}(\beta_1)}{t^{i+1}} + \frac{b_2(e^{t\alpha_2}-e^{t\beta_2})}{t} \end{aligned} \tag{8}$$

3.2 Learning tMoPs Parameters

Considering a sample of size n of X in its domain $\Omega_X = [X_{min}, X_{max}]$, we can note $\{X_{min} = X_1 \leq X_2 \leq \ldots \leq X_{max} = X_n\}$, our goal is to estimate a univariate tMOPs density as defined in 2.

We will split this process in two steps, firstly we will calculate the set of kernel points and secondly we will calculate the parameters for the polynomial fit and the tails, if necessary.

Obtaining Balanced Kernel Points. We start estimating an empirical density from the sample through the Kernel Smoothing technique [18]. We consider a set of kernel points $\{(x, y)\}$, with y being the kernel density value at the point x in the domain of X. Initially, we take 500 equispaced points in Ω_X as follows.

In the way the kernel points are obtained, the density in the areas near the extremes is lower. In order to avoid this distortion, the sample points are mirrored to the right taking as symmetry point X_{max} (likewise to the left) forming a sample of size $3n$ as follows: $\{X_n^{(l)}, \ldots, X_1^{(l)}, X_1, \ldots, X_n, X_n^{(r)}, \ldots, X_1^{(r)}\}$. From that sample we compute the kernel points (x, y) which are subsequently restricted to the domain Ω_X having kernel points (x', y') such that $x' \in \Omega_X$.

The next step is to match the kernel point density to the distribution of the sample along Ω_X. The set of kernel points is equispaced; however, it is common in samples to have areas with higher kernel point density and areas with lower kernel point density. This is corrected by splitting the domain into 10 disjoint subintervals $I_1, \ldots, I_{10}$ and calculating the proportion of points from the original sample in each of these intervals as follows:

$$p_i = \frac{\#\{X_j : \quad X_j \in I_i\}}{n}$$

Since the kernel points are equispaced, in each interval I_i there will be 10% of the points; however, in those intervals where $p_i < 0.1$ we randomly eliminate kernel points until their proportion is p_i.

Building a tMOPs Density from a Polynomial Fit. Now a polynomial fit of the above kernel points must be performed using least squares regression.

The process begins by establishing a maximum degree for the polynomial, which we will call n and fitting a polynomial $p(x)$ of that degree. Several scenarios are possible in such a case.

If the polynomial has no roots within the domain, then $p(x) > 0$, $\forall x \in \Omega_X$ and the tMOPs will consist of a single truncated polynomial, without the need to set tails, which will eventually be normalized to ensure that the tMOPs is a density.

If there are roots, it is necessary to select the interval in which the polynomial will be defined, leaving the rest as one or two tails. Let $R = \{X_{min} = r_0, r_1, \ldots, r_k, X_{max} = r_{k+1}\}$ be the the set of roots in Ω_X adding the two endpoints and $R_i = [r_i, r_{i+1}]$ each of the subregions.

Among the regions R_i where $p(x) > 0$ we calculate the one with the highest density in terms of kernel points, R_A, and also the one that covers a larger range of the domain, R_B. If they are different, two tMoPs, $f_A(x)$ and $f_B(x)$, are created with their central parts in R_A and R_B respectively and in the side regions, if any, one or two tails are created dividing the proportion of kernel points by the length of the interval. Among the two possible models, the one that minimizes the mean square error with respect to the kernel points is selected.

Then the degree is reduced and the above scheme is followed. This process continues as the BIC [17] increases. The process concludes by normalizing the polynomial along with the tail or tails so that it becomes a density function.

4 Approximation to Standard Univariate PDF's Using MoPs

MoPs can be used to fit any probability distribution, but in order to test their goodness of fit with the most common distributions, experiments have been carried out with the six families detailed below, varying their parameters in order to cover a wider range of possibilities. It should be noted that in those distributions where the domain is infinite, it has been truncated so that the probability in it is higher than 0.99.

The distributions used are as follows:

- Exponential of parameter 1 defined on the interval $[0, 6]$
- Standard normal $N(0, 1)$ on the interval $[-3, 3]$
- Gamma with three parameter combinations: $\Gamma(6, 1)$ on the interval $[0, 15]$, $\Gamma(8, 1)$ on the interval $[0, 17]$ and $\Gamma(11, 1)$ on the interval $[0, 20]$
- Beta with three parameter combinations: $\beta(2, 2)$, $\beta(2.7, 1.3)$ and $\beta(1.3, 2.7)$
- Lognormal with three parameter combinations: $LN(0, 0.25)$ on the interval $[0, 4]$, $LN(0, 0.5)$ on the interval $[0, 6]$ and $LN(0, 1)$ on the interval $[0, 11]$
- Bimodal as a linear combination of two normal distributions: $0.25 \cdot N(0, 1) + 0.75 \cdot N(6, 2)$ on the interval $[-3, 12]$

Among the four methods for learning a univariate distribution described in Sect. 2.1, the Shenoy method, the MOTBF and tMoPs have been performed. As for the method based on B-splines, it has been theoretically developed in [12] as explained in the previous paragraphs, however, it was not possible to access an implemented version with which to perform the corresponding experiments.

Shenoy's method [16] has been implemented using R software, but with certain restrictions detailed below. The Taylor expansion series, which we will call $t(x)$, on the interval Ω_X has been computed with respect to its midpoint. Also the maximum degree has been $n = 7$, yet the non-negativity of the polynomial in that region is not guaranteed, therefore, in case $t(x) < 0$ for some $x \in \Omega$, the degree has been reduced. If that situation is repeated, ultimately a uniform distribution has been employed; i.e. $n = 0$.

The number of intervals into which the domain has been subdivided is also not a parameter established in advance. To standardize the experiments, 1, 2, 3 or 4 equispaced intervals have been carried out.

Considering that the number of parameters should be similar to have a fair comparison with the rest of the models, as the number of intervals increases, experiments have been performed with a lower degree. That is the reason why with 1 interval the degree is 7, with 2 intervals we have tested degrees 7 and 4, with 3 intervals, 7 and 3 and with 4 intervals, 7 and 2.

Unlike Shenoy's method, in MoTBF and tMOPs it is not necessary to know the empirical distribution, but only a sample. In these cases, random samples of size 50, 100 and 1000 have been taken. With each sample size the experiment was repeated 10 times and, as in the Taylor polynomials, the maximum degree allowed was 7.

4.1 Results

Since the empirical distribution is available in all experiments, the Kullback-Leibler divergence (KL) has been used to measure the goodness of each fit. Considering a density function $f(x)$ and its MOPs approximation $\bar{f}(x)$, it is defined as:

$$D_{KL}(f(x), \bar{f}(x)) = \int_{\Omega} f(x)\, log \frac{f(x)}{\bar{f}(x)} dx$$

As described above, different experiments have been carried out in each method. In Shenoy's method there are 7 types varying the number of intervals and the degree of Taylor expansion. As expected, the higher the degree and the greater the number of intervals, the better the fit is. In all the distributions tested, the model with 4 intervals and degree 7 is the one that minimizes the KL divergence.

The MoTBF and tMOPs models have three variations and, like the previous one, the fit is better as the sample is larger. In fact, using Friedman's test, the null hypothesis of equality is rejected. The results obtained with the best option of each method, i.e. the method developed by Shenoy with 4 intervals and degree 7 and the MoTBF and tMOPs methods with samples of size 1000, are detailed in Table 1.

It can be seen that the results using the Taylor series expansion are considerably better than the other two methods, especially in the exponential, normal and beta distribution with parameters $(2, 2)$. These results are not surprising since to carry out this method it is needed to know the empirical distribution instead of a sample. In this case it can, and indeed it does, make use of a great deal of extra information with respect to the other two methods.

Using again the Friedman's test, the hypothesis of equality between these three methods is rejected, offering Shenoy's method significantly better results, while the hypothesis of equality between the two methods based on samples is not rejected, having both similar results.

In the first plot of Fig. 1 we see the results of the fit for a gamma distribution of parameters $(6, 1)$ in the interval $[0, 15]$. The left-hand side shows how the tMOPs model creates a small tail avoiding the mismatches produced by the MoTBF polynomial fit at the extremes of the interval.

In the second plot of Fig. 1, representing a beta distribution of parameters $(1.3, 2.7)$, it can be clearly seen how something similar happens in the right part; in fact those probabilities close to 0 have caused that the Shenoy method in that interval has to be fitted with a uniform, as if it were a tail. In contrast, the tail created by the tMOPs is considerably smaller.

Table 1. Kullback-Leibler divergence for the fit of known distributions using MoPs models

PDF	$E(1)$	$N(0,1)$	$\Gamma(6,1)$	$\Gamma(8,1)$	$\Gamma(11,1)$	$\beta(2,2)$
Shenoy	1.739e−15	2.283e−10	1.957e−04	3.029e−04	6.454e−05	0
MoTBF	9.359e−04	5.230e−03	7.631e−03	7.571e−03	8.535e−03	1.736e−03
tMOPs	5.057e−03	8.155e−03	5.388e−03	6.121e−03	6.806e−04	5.313e−03
PDF	$\beta(2.7,1.3)$	$\beta(1.3,2.7)$	$LN(1,0.25)$	$LN(1,0.5)$	$LN(1,1)$	$BiMod$
Shenoy	1.304e−02	1.304e−03	6.545e−02	1.332e−01	5.057e−02	2.633e−03
MoTBF	4.692e−03	6.040e−03	1.542e−01	1.088e−01	4.146e−02	2.093e−02
tMOPs	6.024e−03	5.216e−03	6.969e−01	1.117e−01	4.635e−02	2.297e−02

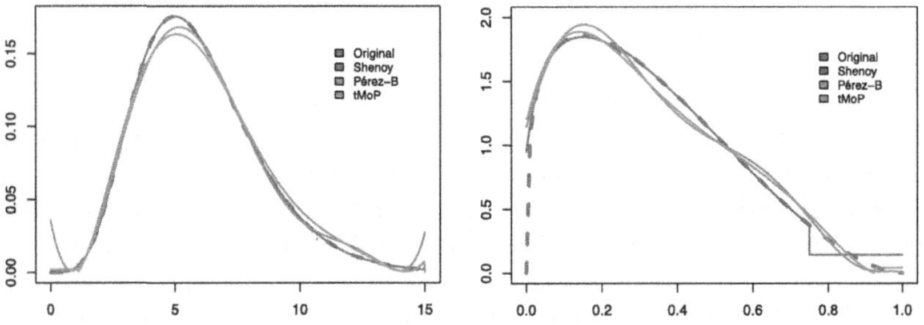

Fig. 1. Fit of a $\Gamma(6,1)$ and a $\beta(1.3,2.7)$ density using the alternative MOPs models mentioned in Sect. 4.

5 tMOPs Meta-model for the Beta Distribution

In order to avoid the computational complexity of fitting new distributions, a new method has been implemented to learn a Beta distribution of any parameter from a finite set of previously fitted Beta distributions.

Thanks to the ease with which MoPs can be combined, it is also possible to combine two tMoPs efficiently without the need to learn a new model from a sample.

As an initiation we have restricted the Beta distribution to the parameter set $\alpha \in [0,1]$ and $\beta \in [0,1]$.

This PDF has been chosen mainly for two reasons: one is that the variation in the parameters causes significant changes in the shape of the density and the other is that the domain of definition of the distribution is always the interval $[0,1]$, so it is not necessary to truncate the sample.

The previous step is to learn a set of Beta distributions within that parameter interval. In this case, a grid has been created with equispaced parameters at distance 0.1; i.e., $\alpha_i = 0.1i, i = 0, ..., 10$ and $\beta_i = 0.1i, i = 0, ..., 10$, making a total of 121 distributions.

It is noteworthy that the tMoPs are easily saved in three vectors: one with the coefficients of the polynomial, another with the roots and a third one with the probabilities of the tails.

When we want to fit a distribution $B(\alpha', \beta')$ we search the four points between which it is located and, using the Euclidean distance between the parameters we select the two closest ones, which can be denoted by $B(\alpha_n, \beta_n)$ and (α_m, β_m) whose distances are d_n and d_m.

The new polynomial will be a weighted mean of the polynomials of the previous distributions taking the weights inversely proportional to d_n and d_m. Regarding the tails, if any, the same procedure is performed.

Lets suppose that the roots of the two learned distributions are $r_{n_1} < r_{n_2}$ and $r_{m_1} < r_{m_2}$, in that case the roots of the new model will be:

$$(\max\{r_{n_1}, r_{m_1}\}, \min\{r_{n_2}, r_{m_2}\})$$

So the non-negativity of the new polynomial is guaranteed. In the case one distribution has zero or one root, the maximum or minimum will be used instead.

As the final step, the tMOPs thus obtained is normalized.

An example of this method to estimate the $\beta(0.31, 0.92)$ can be seen in Fig. 2. It shows that the difference between the tMOPs calculated from the previous database and the tMOPs calculated using the exact parameters are very similar, whereas the computational cost has been reduced considerably.

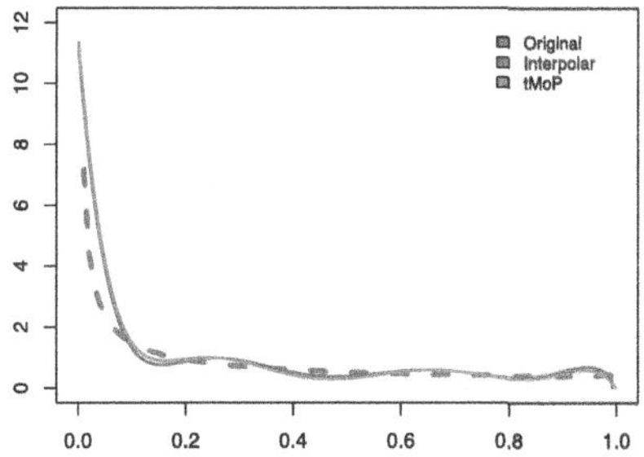

Fig. 2. Fit of a $\beta(0.31, 0.92)$ PDF using a grid of tMoPs equispaced in the domain of both parameters.

6 Conclusions and Future Work

In this work we have analyzed hybrid Bayesian networks based on mixtures of polynomials (MoPs). We reviewed the various MOPs alternatives in the literature, highlighting their main features. Then we focused on the MoPs with tails (tMoPs) and increased their theoretical and applied knowledge in different ways.

First, several analytical properties of the tMoPs have been derived. These expressions allow direct computation of important statistics and related functions for a tMoP, given its parameters. Despite their usefulness and importance, they have not been reported anywhere else in the literature, to the best of our knowledge.

In addition, we give the main ideas of an algorithm to learn the tMOPs coefficients from data. Most of these ideas were originally introduced in [7] with a focus on classification and regression problems using a random sample, but some of them have been improved or extended since. On the other hand, we have tested this learning algorithm with several standard probability distributions (Exponential, Normal, Gamma, Beta, log-Normal and bimodal). In the experiments, the tMOPs models yielded satisfactory results in comparison to other MoPs available alternatives. We plan to go deeper into this problem, especially to include conditional density functions in a general hybrid scheme.

Finally, we sketch the process of building a meta-model that can directly provide a tMOPs expression for a specific density among a given family. This idea is tested with the well-known Beta distribution (two shape parameters), giving promising results. Shortly, we intend to extend this idea to build general meta-models than can transform any usual density function into a tMoP.

Globally, all these results show that tMOPs models are a valuable tool for probabilistic graphical models with hybrid variables. Also, general knowledge about MoPs and tMoPs has been increased by providing new results and insights for them.

Acknowledgments. This work is supported by the research projects PID2021-124472NB-I00 (D. Ramos-López) and PID2022-139293NB-C31 (D. Ramos-López and R. Rumí) funded by MCIN/AEI/10.13039/501100011033 "ERDF A way of making Europe" funds. Both D. Ramos-López and R. Rumí are supported also by the Center for Development and Transfer of Mathematical Research to Industry CDTIME (University of Almería), and by the Junta de Andalucía research groups FQM-229 and FQM-224 respectively.

Disclosure of Interests. The authors have no competing interests to declare that are relevant to the content of this article

References

1. Cabañas, R., et al.: Financial data analysis with PGMs using AMIDST. In: IEEE 16th International Conference on Data Mining (ICDM 2016) Workshops, pp. 1284–1287 (2016). https://doi.org/10.1109/ICDMW.2016.0185
2. Cai, B., Huang, L., Xie, M.: Bayesian networks in fault diagnosis. IEEE Trans. Industr. Inf. **13**(5), 2227–2240 (2017). https://doi.org/10.1109/TII.2017.2695583
3. Castro-Luna, G.M., Martínez-Finkelshtein, A., Ramos-López, D.: Robust keratoconus detection with Bayesian network classifier for Placido-based corneal indices. Cont. Lens Anterior Eye **43**(4), 366–372 (2020). https://doi.org/10.1016/j.clae.2019.12.006

4. Korb, K., A. Nicholson, E.: Bayesian Artificial Intelligence. CRC Press/Chapman & Hall Book (2010)
5. Langseth, H., Nielsen, T.D., Pérez-Bernabé, I., Salmerón, A.: Learning mixtures of truncated basis functions from data. Int. J. Approximate Reasoning **55**, 940–956 (2014). https://doi.org/10.1016/j.ijar.2013.09.012
6. Langseth, H., Nielsen, T.D., Rumí, R., Salmerón, A.: Mixtures of truncated basis functions. Int. J. Approximate Reasoning **53**, 212–227 (2012). https://doi.org/10.1016/j.ijar.2011.10.004
7. Luengo, J.C., Rumí, R.: Naive Bayes classifier with mixtured of polynomials. In: Proceedings of the International Conference on Pattern Recognition Applications and Methods (ICPRAM-2015), pp. 14–24 (2015)
8. López-Cruz, P.L., Nielsen, T.D., Bielza, C., Larrañaga, P.: Learning mixtures of polynomials of conditional densities from data. In: Bielza, C., et al. (eds.) CAEPIA 2013. LNCS (LNAI), vol. 8109, pp. 363–372. Springer, Heidelberg (2013). https://doi.org/10.1007/978-3-642-40643-0_37
9. Maldonado, A.D., Valdivielso, A., Rescia, A., Aguilera, P.A.: Probabilistic graphical models for species richness prediction: are current protected areas effective to face climate emergency. Glob. Ecol. Conserv. **23**, e01162 (2020)
10. Moral, S., Rumi, R., Salmerón, A.: Mixtures of truncated exponentials in hybrid Bayesian networks. In: Benferhat, S., Besnard, P. (eds.) ECSQARU 2001. LNCS (LNAI), vol. 2143, pp. 156–167. Springer, Heidelberg (2001). https://doi.org/10.1007/3-540-44652-4_15
11. Pearl, J.: Probabilistic Reasoning in Intelligent Systems: Networks of Plausible Inference. Morgan-Kaufmann, San Mateo (1988)
12. Pérez-Bernabé, I., Fernández, A., Rumí, R., Salmerón, A.: Parameter learning in hybrid Bayesian networks using prior knowledge. Data Min. Knowl. Disc. **30**, 576–604 (2016). https://doi.org/10.1007/s10618-015-0429-7
13. Pérez-Bernabé, I., Maldonado, A., Salmerón, A., Nielsen, T.: MoTBFs: an R package for learning hybrid Bayesian networks using mixtures of truncated basis functions. R J. **12**(2), 321–341 (2021). https://doi.org/10.32614/RJ-2021-019 https://doi.org/10.32614/RJ-2021-019
14. Shenoy, P.P.: Two issues in using mixtures of polynomials for inference in hybrid Bayesian networks. Int. J. Approximate Reasoning **53**, 847–866 (2012). https://doi.org/10.1016/j.ijar.2012.01.008
15. Shenoy, P.P., West, J.C.: Mixtures of polynomials in hybrid Bayesian networks with deterministic variables. In: Proceedings of the 8th Workshop on Uncertainty Processing (WUPES 2009), pp. 202–212 (2009)
16. Shenoy, P.P., West, J.C.: Inference in hybrid Bayesian networks using mixtures of polynomials. Int. J. Approximate Reasoning **52**, 641–657 (2011). https://doi.org/10.1016/j.ijar.2010.09.003
17. Stone, M.: Comments on model selection criteria of Akaike and Schwartz. J. Roy. Stat. Soc. B **41**, 276–278 (1979)
18. Wand, M., Jones, M.: Kernel Smoothing. Chapman & Hall, Boca Raton (1995)

Age-Unbiased Facial Emotion Recognition with Regularizing Self-Attention Value Vector

Jaeil Park and Sung-Bae Cho(✉)

Deptartment of Computer Science, Yonsei University, Seoul 03722, Korea
{wodlf603,sbcho}@yonsei.ac.kr

Abstract. Facial emotion recognition (FER) is a challenging task that involves analyzing emotions based on attributes entangled with the face in an image. The FER model inherently possesses age bias because age-related features such as wrinkles and skin texture can create confusion in understanding the emotions of the elderly and decrease accuracy. This paper proposes a FER method that mitigates age bias by applying adversarial regularization to the value vector of self-attention, which is related to age attributes, to minimize information loss. The proposed method focuses on removing biases without affecting the landmarks that make up facial emotions, as key and query vectors in the attention are associated with facial landmarks while value vector is associated with textural features of the image. An age discriminator interrupts age information in value vectors and restores information through the emotion classifier to minimize excessive loss. Experimental results with two FER datasets, RAF-DB and FACES, show that the proposed method yields superior accuracy while reducing bias compared to the *state-of-the-art* methods.

Keywords: Facial emotion recognition · bias mitigation · self-attention regularization · adversarial autoencoder

1 Introduction

Facial emotion recognition (FER) is a challenging task that identifies emotions from facial images, playing a significant role in various domains such as mental care analysis, identification of driver fatigue in business operations, and enhancement of human-computer interaction systems. Since it is difficult to interpret entangled complex information in facial images, various approaches based on neural networks that emulate human cognitive processes have been investigated [1]. Recently, extracting features of images by transformer with self-attention has achieved remarkable accuracy. In addition, as FER is intricately related to the detection and analysis of facial landmarks, various models that utilize facial landmark show high performance [2].

However, current neural network models often result in biases in recognition because of imbalanced data [3]. Particularly, age biases in emotion recognition have been studied due to the challenges associated with accurately interpreting emotional expressions in older individuals [4]. The elderly is more frequently misclassified as sad, a phenomenon

V. Julian et al. (Eds.): IDEAL 2024, LNCS 15346, pp. 472–480, 2025.
https://doi.org/10.1007/978-3-031-77731-8_42

potentially attributed to facial features related to age, such as wrinkles and skin texture [5]. This bias greatly affects fair emotion recognition. To mitigate this bias, several researchers explored bias mitigation strategies involving modifications to either the FER data or the models themselves [3]. However, they are associated with accuracy loss, attributable to the issues such as the loss of key features constructing facial landmarks. Moreover, most of them are hard to apply to various FER models.

To overcome these problems, we propose a FER method that mitigates the age bias using an adversarial regularization to the value vector component of attention related to age attributes. While key and query vectors in the attention are associated with facial landmarks and value vector is associated with textural features of the image [6, 7], our method focuses on removing bias without impacting the landmarks that manifest facial emotions. An autoencoder with an age discriminator disrupts the age information and restores information through an emotion classifier to minimize excessive loss. Experiments with two FER datasets, RAF-DB and FACES, will demonstrate that the proposed method maintains accuracy while mitigating biases compared to the *state-of-the-art* methods.

Table 1. Related works for facial emotion recognition and its bias mitigation methods.

Type	Method	Description
FER	Transformer	Vision transformer (ViT) with attention pooling [8]
		Separation of handling landmark detection and image processing [2]
Bias Mitigation for FER	Data Alteration	Aging details removal using generated mask of face images [9]
	Data Constraint	Age group constrained FER for specific age group [9]
	Adversarial Learning	Adversarial learning with various concatenated features (gender, race, age) [3]
		Adversarial learning with various disentangled features (gender, race, age) [3]

2 Related Works

2.1 FER Models and Their Bias

Recent developments in FER have been driven by sophisticated machine learning and computer vision techniques. A key to this progress lies in benchmark datasets such as real-world affective faces database (RAF-DB) [10] and FACES [11] which have become foundational in training and evaluating FER models. These datasets offer a diverse range of emotional states, enabling the development of robust algorithms. We use datasets with age labels. A significant leap in FER research is the adoption of deep learning, particularly transformer, as shown in Table 1.

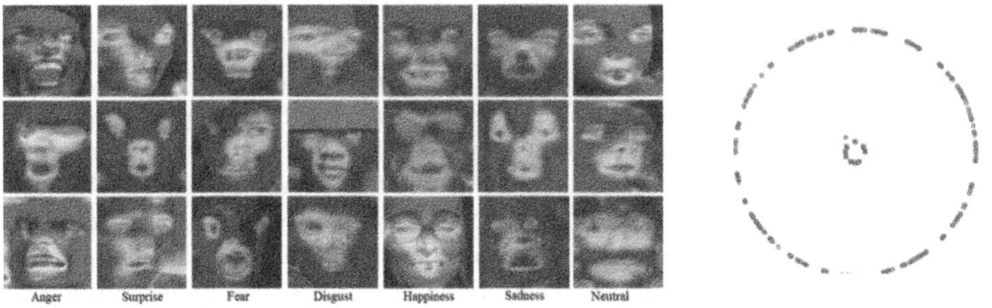

Fig. 1. Association between attention and attributes in the FER model. Left (a): Photo extracted with the attention map of a FER. It can be observed that there is a deep association with landmarks. Right (b): tSNE results of the value vector and the age attribute.

However, bias and fairness in FER emerge as critical concerns. Gou et al. highlighted biases in age, gender, and ethnicity recognition, emphasizing the need for unbiased algorithms and diverse training datasets [9]. This aspect of research underlines the importance of ethical considerations in algorithm development.

2.2 Bias Mitigation Methods for FER Models

In the field of FER, various methods have been explored for bias mitigation, focusing on data or model modification, as shown in Table 1. However, these bias mitigation methods often require complex preprocessing that is challenging to apply to general models. Moreover, they are directly affected by information loss due to generation processes or adversarial learning, leading to a paradoxical decrease in accuracy. This presents a significant challenge in effectively removing bias while preserving essential information for accurate recognition.

3 Methodology

3.1 Value Vector of Self-Attention and Age Bias

This section describes a brief overview of the self-attention mechanism in transformer, and its relationship with an age bias. In the context of images, for example, the vision transformer (ViT) applies this approach by segmenting images into patches and transforming these sequences into other sequences, demonstrating strong performance with self-supervised learning. The notations for computational process are as follows: The size of the hidden state vector is denoted as d_{model}, and the number of heads in multi head self-attention is represented as h. The dimensions of the query, key, and value vectors are $d_q = d_k = d_v = \frac{d_{model}}{h}$.

At each attention head, encoded word vectors are transformed into three matrices: a query matrix $Q \in R^{(T \times d_q)}$, a key matrix $K \in R^{(T \times d_k)}$, and a value matrix $V \in R^{(T \times d_v)}$.

This transformation occurs h times, resulting in $h \times 3$ matrices. These are used to calculate the attention-weighted combination of value vectors as follows:

$$A(Q, K) = Softmax\left(\frac{QK^T}{\sqrt{d_{model}}}\right), Attention(Q, K, V) = A(Q, K)V \quad (1)$$

$$MultiHeadAttention = Concat(H_1, \ldots, H_h)W^O \quad (2)$$

where $H_i = Attention\left(QW_i^Q, KW_i^K, VW_i^V\right)$, W_i^Q, W_i^K, and W_i^V are the linear transformation matrices at the i-th head and T is transpose of matrix. A is called an attention map which represents relations of each vector as a weight of similarity [12].

Figure 1(a) illustrates the attention map in FER. The attention map, representing relations of image patches, exhibits characteristics associated with landmarks. In contrast, various research reported that value vectors represent local information [6]. Additionally, we found a high correlation between facial age information and local information in images [13]. Figure 1(b) shows the relationship between the distribution of the value vector and age, confirming the correlation between the actual age and the value vector. Consequently, we aim to mitigate age bias in the value vector V by exploring the relationship between V and facial age information.

3.2 Bias Mitigation by Value Regularization

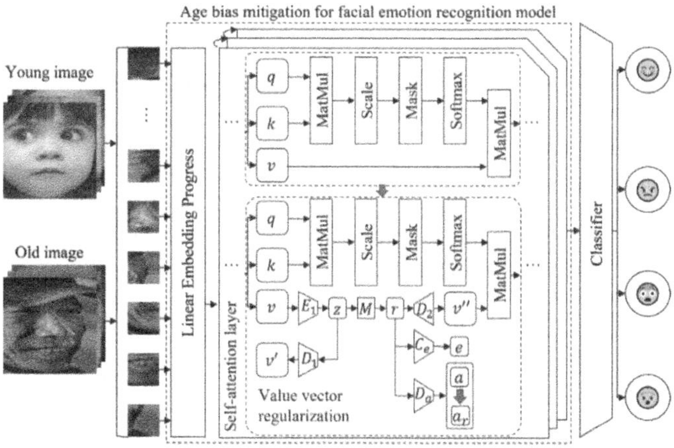

Fig. 2. The overall architecture of the proposed model.

Figure 2 illustrates the architecture of a facial emotion recognition model that mitigates bias using the value vector of self-attention. The model processes image patches as inputs, executing a transformer. Within this architecture, the value vector of the self-attention is regularized through an autoencoder using multi-layered encoder and decoder.

Denoting the function that conducts this process as f_{AE}, all the self-attention with the proposed autoencoder is represented in Eq. (3).

$$Attention(Q, K, V) = Softmax\left(\frac{QK^T}{\sqrt{d_{model}}}\right)f_{AE}(V) \tag{3}$$

The autoencoder f_{AE} aims to generate a fair representation by focusing on the latent vector regularization process in adversarial autoencoders. Drawing on the concept of using latent vectors to converge and regularize data distributions, the proposed method seeks to adjust the data distributions of value vectors with age features, adopting a disentangled representation approach [14, 15]. Suppose that $V = \{v_1, \ldots, v_t\}$. Using an encoder E_1 and a decoder D_1, the proposed method creates a latent vector z that contains all the information in embedded tokens v and can be decoded into $v\prime$ that contains the same information as the original data.

To separate the learning direction of age information removal and original information retrieval, we generate a mutated latent vector r that passes through the layer M and inherits the information from the vector [16]. Ensuring that the mutated latent vector retains the information about the original vector as much as possible, the vectors are also decoded with a decoder D_2 and compared with v''. Mutated latent vectors are classified by age a by an age discriminator D_a, thereby the discriminator learns the distribution of two age groups of latent vectors. Moreover, we employ an emotion classifier C_e to prevent excessive information loss. C_e learns the emotional attributes of the latent vector e. With this structure, function f_{AE} can be written as Eq. (4):

$$\begin{gathered} f_{AE}(v) = v'' = (D_2 \circ M \circ E_1)(v) \\ \text{where } (D_1 \circ E_1)(v) \sim v,\ v'' \sim v, \\ (D_a \circ M \circ E_1)(v) \sim a,\ (C_e \circ M \circ E_1)(v) \sim e \end{gathered} \tag{4}$$

Subsequently, the discriminator D_a is fixed, and the age label of the mutated latent vector a is randomized as a_r. As z and r are regularized while D_a maintains to discriminate the age label of them appropriately, z and r are learned not to be distinguished by D_a. By iteration of learning the discriminator and randomizing the label, z and r are learned to be disentangled for an age feature while maintaining the emotional attributes with C_e.

4 Experiments

4.1 Experimental Setting

In this paper, we employ the real-world affective faces database (RAF-DB) [10] and FACES [11] for both training and evaluation. These datasets are utilized specifically for the basic emotion subset, classifying images into seven expressions (surprise, fear, disgust, happiness, anger, sadness, and neutral) and bifurcating age groups into young (up to 69 years old for RAF-DB, young / middle for FACES) and old (over 70 years old for RAF-DB, old for FACES) [4]. Age bias starts to appear from the 40s from a cognitive perspective [11], but there is a change in trend around the age of 70, leading to a significant accuracy drop [17]. Therefore, we target primarily those over 70 years old, as they are highly relevant to the age group where the problem is most pronounced. Notably, while these datasets provide labels for age attributes, they are not curated with a focus on group balance. This results in an imbalanced distribution of samples across expressions and age groups in both training and testing splits, contributing to age biases in the models as indicated by Xu et al. [3].

Fairness evaluation is conducted using the equal opportunity and equal odds metrics as proposed by Hardt et al. [18]. Equal opportunity is assessed through the absolute difference in the true positive rate between the two groups as shown in Eq. (5). Equal odds additionally compute the absolute difference in the false positive rate between the groups as shown in Eq. (6).

$$EOP = \Sigma_{a \in A} |TPR_{a=1} - TPR_{a=0}| \tag{5}$$

$$EOD = \Sigma_{a \in A} \frac{|TPR_{a=1} - TPR_{a=0}| + |FPR_{a=1} - FPR_{a=0}|}{2} \tag{6}$$

4.2 Bias Mitigation on FER

Table 2 compares the overall performance and Table 3 shows its example of debiasing result from bias data as "sadness" on old faces. The proposed method shows the most significant improvement in fairness metrics compared to other methods, and the proposed method enhances the accuracy significantly. When the proposed method is applied to APViT, bias is mitigated properly. The methods of mitigating bias through aware and unaware approaches are found to significantly decrease accuracy due to a lack of information restoration capability.

Table 2. Experimental results with two datasets (EOP: equal opportunity, EOD: equal odds).

Model	RAF-DB			FACES		
	Acc	EOP	EOD	Acc	EOP	EOD
APViT [8]	0.920	0.261	0.178	0.800	0.361	0.194
APViT [8] + Aware [7]	0.913	0.251	0.153	0.798	0.350	0.183
APViT [8] + Unaware [7]	0.876	0.243	0.150	0.776	0.342	0.181
POSTERv2 [2]	0.922	0.323	0.169	0.819	0.347	0.184
POSTERv2 [2] + Aware [7]	0.917	0.312	0.156	0.812	0.325	0.165
POSTERv2 [2] + Unaware [7]	0.896	0.319	0.149	0.797	0.292	0.164
APViT [8] + Ours	0.920	0.260	0.139	0.808	0.331	0.171
POSTERv2 [2] + Ours	0.929	0.261	0.143	0.826	0.286	0.153

Table 3. Per-method prediction on old faces misclassified as 'Sadness'

Target label		Neutral	Neutral	Neutral	Neutral
Predicted label	Baseline	Sadness	Sadness	Sadness	Sadness
	Aware	Disgust	Sadness	Sadness	Neutral
	Unaware	Neutral	Neutral	Sadness	Disgust
	Ours	Neutral	Neutral	Neutral	Sadness

4.3 Ablation Study

Table 4 summarizes the ablation study across the three categories. First, we investigate bias mitigation in attention by regularizing the query, key, value, or all of them. While regularizing the query or key reduces accuracy, regularizing the value successfully mitigates age bias and improves accuracy. Regularizing all three yields intermediate results. Second, altering the dimension of the latent vector shows that both accuracy and bias mitigation improve as the dimension increases, with 64 identified as the optimal dimension beyond which no significant improvements are observed. Finally, removing specific modules from the autoencoder reveals that eliminating the age discriminator results in the similar performance to the baseline, while removing the emotion classifier further reduces bias but causes a notable drop in accuracy.

Table 4. Experimental result on ablation study.

Category	Type	Accuracy	Equal opportunity	Equal odd
Baseline		0.9219	0.3227	0.1694
Mitigation target	Query	0.8431	0.3823	0.1911
	Key	0.8751	0.3743	0.1837
	Value	0.9291	0.2605	0.1432
	QKV	0.9121	0.2401	0.1613
Latent vector dimension	32	0.9207	0.2549	0.1674
	64	0.9291	0.2605	0.1432
	128	0.9290	0.2591	0.1481
	256	0.9292	0.2611	0.1402
Absence of module	D_a	0.9117	0.3241	0.1702
	C_e	0.9107	0.2603	0.1427

5 Concluding Remarks

In this paper, we propose a facial emotion recognition model that mitigates age bias within it with regularizing value vector of self-attention. We argue that the value vector is associated with facial bias and propose a method to address this by creating a fair representation using an autoencoder targeting the value vector. Bias is mitigated through a discriminator, and to minimize information loss, adversarial learning is conducted using both a decoder and a classifier.

The effectiveness of the proposed method is verified in reducing age bias by training and testing it on two FER datasets that include age labels. The experimental results demonstrate that it not only enhances the fairness of the model's output but also maintains high accuracy, outperforming other existing bias mitigation techniques and FER models. Additionally, with confusion matrices, we confirm that the proposed mitigated model maintains the overall accuracy of predictions when compared to a baseline model. In the future, we will attempt to mitigate bias in other image recognition domains in addition to facial emotion recognition.

Acknowledgment. This work was supported by Yonsei Fellow Program funded by Lee Youn Jae, IITP grant funded by the Korean government (MSIT) (No. RS-2020-II201361, Artificial Intelligence Graduate School Program (Yonsei University); No. RS-2022-II220113, Developing a Sustainable Collaborative Multi-modal Lifelong Learning Framework), and Air Force Defense Research Sciences Program funded by Air Force Office of Scientific Research.

References

1. Canal, F.Z., et al.: A survey on facial emotion recognition techniques: a state-of-the-art literature review. Inf. Sci. **582**, 593–617 (2022)
2. Mao, J., Xu, R., Yin, X., Chang, Y., Nie, B., Huang, A.: POSTER++: a simpler and stronger facial expression recognition network. arXiv preprint arXiv:2301.12149 (2023)
3. Xu, T., White, J., Kalkan, S., Gunes, H.: Investigating bias and fairness in facial expression recognition. In: Bartoli, A., Fusiello, A. (eds.) ECCV 2020, vol. 12540, pp. 506–523. Springer, Cham (2020). https://doi.org/10.1007/978-3-030-65414-6_35
4. Kim, E., Bryant, D.A., Srikanth, D., Howard, A.: Age bias in emotion detection: An analysis of facial emotion recognition performance on young, middle-aged, and older adults. In: AAAI/ACM Conference on AI, Ethics, and Society, pp. 638–644 (2021)
5. Ganel, T., Sofer, C., Goodale, M.A.: Biases in human perception of facial age are present and more exaggerated in current AI technology. Sci. Rep. **12**(1), 22519 (2022)
6. Tewel, Y., Gal, R., Chechik, G., Atzmon, Y.: Key-locked rank one editing for text-to-image personalization. In: ACM SIGGRAPH 2023 Confernce, pp. 1–11 (2023)
7. Wang, H., Sanchez, V., Li, C.-T.: Improving face-based age estimation with attention-based dynamic patch fusion. IEEE Trans. on Image Processing **31**, 1084–1096 (2022)
8. Li, S., Deng, W., Du, J.: Reliable crowdsourcing and deep locality-preserving learning for expression recognition in the wild. In: IEEE Conference on Computer Vision and Pattern Recognition, pp. 2852–2861 (2017)
9. Ebner, N.C., Riediger, M., Lindenberger, U.: FACES—A database of facial expressions in young, middle-aged, and older women and men: Development and validation. Behav. Res. Methods **42**, 351–362 (2010)

10. Guo, G., Guo, R., Li, X.: Facial expression recognition influenced by human aging. IEEE Trans. on Affect. Comput. **4**(3), 291–298 (2013)
11. Xue, F., Wang, Q., Tan, Z., Ma, Z., Guo, G.: Vision transformer with attentive pooling for robust facial expression recognition. IEEE Trans. on Affective Computing **14**(4), 3244–3256 (2022)
12. Khan, S., Naseer, M., Hayat, M., Zamir, S.W., Khan, F.S., Shah, M.: Transformers in vision: a survey. ACM Comput. Surv. (CSUR) **54**(10s), 1–41 (2022)
13. Li, Z., Gong, D., Li, X., Tao, D.: Aging face recognition: a hierarchical learning model based on local patterns selection. IEEE Trans. Image Process. **25**(5), 2146–2154 (2016)
14. Kim, J.-Y., Cho, S.-B.: An information theoretic approach to reducing algorithmic bias for machine learning. Neurocomputing **500**, 26–38 (2022)
15. Moon, H.-J., Bu, S.-J., Cho, S.-B.: Learning disentangled representation of residential power demand peak via convolutional-recurrent triplet network. In: International Conference on Data Mining Workshops, pp. 757–761. IEEE (2020)
16. Park, J., Cho, S.-B.: Adversarial discriminator to mitigate gender bias in abusive language detection. In: European Conference on Artificial Intelligence, pp. 1851–1858 (2023)
17. Ruffman, T., Henry, J.D., Livingstone, V., Phillips, L.H.: A meta-analytic review of emotion recognition and aging: implications for neuropsychological models of aging. Neurosci. Biobehav. Rev. **32**(4), 863–881 (2008)
18. Hardt, M., Price, E., Srebro, N.: Equality of opportunity in supervised learning. In: Advances in Neural Information Processing Systems, vol. 29 (2016)

Assessing the Impact of Temporal Data Aggregation on the Reliability of Predictive Machine Learning Models

Ayah Barhrhouj(✉), Bouchra Ananou, and Mustapha Ouladsine

Aix-Marseille University, University of Toulon, CNRS, LIS (UMR 7020), Marseille, France
{ayah.barhrhouj,bouchra.ananou,mustapha.ouladsine}@univ-amu.fr

Abstract. In time series analysis, data aggregation is an essential preprocessing step that consolidates data points over specified time intervals, simplifying the data structure and reducing noise. This process is vital for enhancing the manageability of large, complex datasets, commonly encountered in various domains such as energy consumption forecasting. However, the choice of the most appropriate temporal aggregation (TA) frequency can significantly impact the *performance* and the *reliability* of predictive models. Different TA frequencies may perform better in different scenarios, and there is a lack of evidence on what indicators may determine the superiority of each one. In this study, we design and execute an empirical experiment framework to first explore the performance of various machine learning (ML) models using different TA frequencies on 04 multivariate time series datasets of ship fuel consumption. We further investigate the reliability of those models by studying the marginal contribution of the time series features in each aggregation period. Using the Random Forest Regressor, XGBoost Regressor, LightGBM, and Extra Trees predictive models along with the Shapley Additive Explanations to analyze feature contributions, we found that time series aggregation consistently produces accurate results regardless of the TA period. However, the models may exhibit significant biases, highlighting the need to examine their inner workings to select the appropriate TA period.

Keywords: Machine Learning · Explainable AI · Time-series · Temporal aggregation

1 Introduction

Time series analysis stands as a robust statistical method employed to analyze data points collected or recorded at specific time intervals. Its fundamental objective is to understand the underlying structure and function that produced the observations, facilitating the prediction of future values based on past observations [1]. Time series analysis has been successfully applied across a multitude of

V. Julian et al. (Eds.): IDEAL 2024, LNCS 15346, pp. 481–492, 2025.
https://doi.org/10.1007/978-3-031-77731-8_43

domains, including economics, finance, environmental science, and engineering [5,6]. For instance, in economics, time series models are used to predict market trends and economic indicators, help forecast stock prices and assess risks [5]. Environmental scientists utilize time series analysis to monitor and predict changes in climate patterns and pollutant levels [6]. Notably, in fields such as energy consumption forecasting, time series analysis plays a pivotal role. Accurate predictions in energy consumption are vital for efficient energy management, supply-demand equilibrium, and cost reduction [9].

In the specific context of predicting ship fuel consumption, time series analysis often entails working with aggregated data, despite the availability of finer temporal resolutions [14]. Researchers opt for aggregation to reduce noise, reflect operational processes, or mitigate the missing values impact. Aggregation necessitates careful consideration, including the choice of temporal reference (e.g., calendar or operational schedule) and the temporal unit for analysis (e.g., minutes, hours). Decisions regarding the definition of these units vary based on the temporal reference and study characteristics, such as geographic location or weather conditions. Additionally, local engine patterns, such as consumed energy type (e.g., fuel or gas), can impact data collection and analysis, thus influencing model results [13].

In time series analysis, data aggregation is an essential preprocessing step that consolidates data points over specified time intervals, simplifying the data structure and reducing noise [3]. This process is particularly vital for enhancing the manageability of large, complex datasets, which are commonly encountered in various domains. Nonetheless, the choice of the most appropriate temporal aggregation frequency is not trivial and can significantly impact the performance and reliability of predictive models [11]. Temporal aggregation involves averaging or summing data points within specified time intervals, such as minutes, hours, days, or weeks. Different TA frequencies can lead to different levels of data granularity, which in turn can influence the model's ability to capture underlying patterns and trends [12].

Despite its significance, the optimal choice of TA frequency remains an open question. Different frequencies may yield varying performance in different scenarios, and there is a lack of comprehensive evidence on the indicators that determine the superiority of one frequency over another [2]. This gap in the literature highlights the need for a systematic investigation into the effects of TA frequency not only on model performance but also on its reliability. In this study, we aim to address this gap by designing and executing an empirical experiment framework to explore the impact of different TA frequencies on the models accuracy and reliability using 04 multivariate time series datasets of ship fuel consumption. Ship fuel consumption data is a real world case for this analysis due to its complexity and the presence of various influencing factors such as weather conditions, ship speed, and engine patterns. By examining this specific application, we can derive insights that may be applicable to other domains with similar characteristics. The findings from this study provide valuable insights into the effects of TA frequency on model performance, offering practical guidelines for selecting the appropriate TA period and ultimately enhancing the predictive power and reliability of ML models in time series analysis.

2 Methodology

2.1 Data Sources

The data used in this study was collected from an international maritime company. Initially, the data was stored in multiple files encompassing a wide range of heterogeneous information, including GPS, navigation, engine, propulsion, flowmeters, and meteorological measurements. These files were integrated into a platform, where for each ship, the files were merged into a single comprehensive database. This process ensured the availability of all data for each sensor or piece of equipment. An overview of each ship data used in this study is provided in Table 1.

Table 1. Overview of the studied time series datasets.

Ship	Start Date	End Date	Dataset dimensionality
SHIP 1	Jan 26, 2024	May 29, 2024	98.812 instances; 133 features
SHIP 2	Aug 02, 2023	May 24, 2024	124.949 instances; 168 features
SHIP 3	Oct 28, 2022	Jul 31, 2023	117.024 instances; 64 features
SHIP 4	Mar 14, 2022	May 04, 2023	20.492 instances; 96 features

2.2 Temporal Data Aggregation

We focused our analysis on datasets spanning different time frames for each ship. Each dataset covered specific voyage periods for different ships as detailed in Table 1, with varying levels of granularity ranging from each second to one minute. To analyze ship fuel consumption patterns comprehensively, we aggregated the high-frequency data into 10-min, 20-min, 30-min, 40-min, 50-min, and 60-min intervals, using the average method. This averaging process aimed to smooth out fluctuations, mitigate missing data and provide a clearer representation of fuel consumption trends over time. Each aggregation interval allowed us to capture different levels of granularity in the data, from more granular (10 min) to less granular (60 min), reflecting varying operational insights and computational requirements. Subsequently, we developed separate predictive models for each temporal data aggregation frequency, enabling a comprehensive assessment of ship fuel consumption patterns across various TA frequencies.

2.3 Predictive Analysis

Figure 1 provides an overview of the proposed framework for the performance and the reliability analysis of predictive ship fuel consumption models for different temporal data aggregation frequencies. The framework is mainly composed of different phases:

Data Preparation. The data was collected from various files, recorded at intervals of a few seconds or a minute. To facilitate analysis, the data was initially aggregated every 1 min. This aggregation step helped to reduce the data complexity and smooth out short-term fluctuations. Following aggregation, the raw data was then preprocessed. This includes normalization and feature extraction, to transform it into a suitable form for modeling. Data cleaning techniques were applied to handle missing values, remove outliers, and correct inconsistencies.

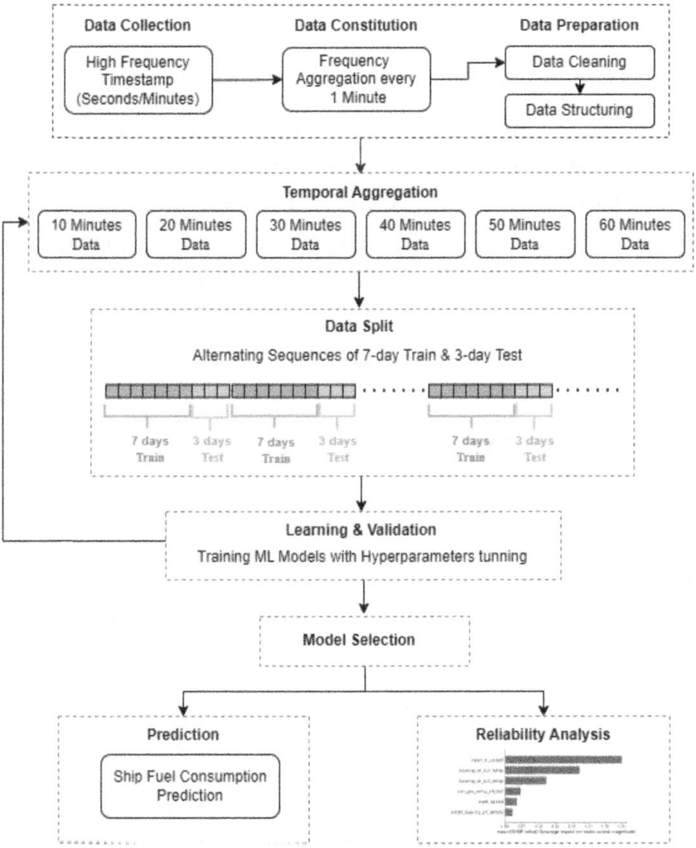

Fig. 1. Performance & reliability analysis framework of predictive ship fuel consumption models for different temporal data aggregation frequencies.

Machine Learning Pipeline. For each considered temporal aggregation period, the data was divided into alternating sequences of 7 days for training and 3 days for testing. This approach allowed for consistent and comprehensive assessment of the model's performance, ensuring that the training and testing datasets were temporally distinct yet representative of the overall data distribution [10]. This method helps to capture temporal patterns and dependencies in the ship fuel consumption data, providing a realistic evaluation of the model's predictive capabilities. For the prediction, we rely on the Random Forest Regressor, XGBoost Regressor, LightGBM, and Extra Trees ML models.

Grid-search exploration of hyperparameters with cross-validation was performed for each predictive model. Models were evaluated based on their results on the test set in terms of the coefficient of determination (R2), Root Mean Square Error (RMSE), and Mean Absolute Error (MAE). The model with the best prediction performance was selected for the reliability analysis.

2.4 Reliability Analysis

To explore the reliability of ship fuel consumption predictions made by the best-performing model, we rely mainly on the Shapley Additive Explanations (SHAP) method [7]. SHAP assigns a score to each feature in a given data point, quantifying its contribution to the model's prediction. These scores, based on Shapley values from cooperative game theory, allocate the value of a collective effort to individual participants. By interpreting SHAP values, we gain a deeper understanding of how individual features influence predictions, thereby enabling the audit of the machine learning models reliability [4]. The exact Shapley values φ_j for feature j are computed as follows in the Eq. 1:

$$\varphi_j = \sum_{S \subseteq X \setminus j} \frac{|S|!(|X| - |S| - 1)!}{|X|!} [f(X_j) - f_S(X_j)] \tag{1}$$

where X represents the set of input features, $S \subseteq X \setminus j$ denotes a subset of features excluding feature j, $f(X_j)$ represents the model prediction for the input sample X_j, and $f_S(X_j)$ is the model prediction for the input sample X_j with the features in S set to their expected values.

Applying SHAP to fuel consumption predictions allows us to dissect the contributions of each ship feature to the energy consumption. This analysis highlights which features are most influential and their interactions in affecting the output [8]. Through these analyses, we will be able to verify if the predictive model consistently relies on the correct attributes (features) to predict fuel consumption, regardless of the data aggregation period.

3 Results and Discussion

This study marks the first attempt, to our knowledge, to explore the impact of temporal data aggregation on time-series analysis efficiency and reliability within the maritime domain, specifically focusing on ship fuel consumption prediction.

3.1 Predictive Performance Analysis

We investigate how aggregating ship operational data, encompassing parameters such as sensor data, engine performance, and navigation information, into different temporal intervals influences predictive modeling accuracy. We explore the effects of aggregating data at different intervals of 10, 20, 30, 40, 50, and 60 min on the performance of predictive models. State of the art ML techniques are employed to predict fuel consumption. Performance metrics, including R2, RMSE, and MAE are used to assess model accuracy across datasets from four different ships. Performance evaluation results for each ship dataset are presented in Tables 2, 3, 4, and 5.

Table 2. Evaluation of model performance in predicting fuel consumption of the SHIP 1 for different TA frequencies.

TA Frequency	Random Forest			Extra Trees			LightGBM			XGBoost		
	R2	RMSE	MAE	R2	RMSE	MAE	R2	RMSE	MAE	R2	RMSE	MAE
10-min	0.876	535.489	414.966	0.882	521.026	395.258	0.923	421.519	318.968	0.918	435.171	319.871
20-min	0.889	511.710	401.590	0.896	495.763	382.204	0.929	409.630	315.626	0.920	433.026	332.666
30-min	0.888	518.039	418.215	0.899	493.722	390.172	**0.937**	**387.900**	**304.115**	0.929	411.877	321.080
40-min	0.873	550.849	443.792	0.882	530.613	419.900	0.927	416.206	304.628	0.925	422.435	333.395
50-min	0.878	535.153	435.318	0.891	505.660	406.413	0.931	401.174	309.361	0.928	411.542	329.770
60-min	0.878	538.083	429.576	0.890	509.707	399.061	0.929	409.166	312.992	0.929	410.270	314.920

Table 3. Evaluation of model performance in predicting fuel consumption of the SHIP 2 for different TA frequencies.

TA Frequency	Random Forest			Extra Trees			LightGBM			XGBoost		
	R2	RMSE	MAE	R2	RMSE	MAE	R2	RMSE	MAE	R2	RMSE	MAE
10-min	0.889	488.430	387.083	0.907	445.730	354.016	0.949	329.163	**236.761**	0.949	329.786	244.599
20-min	0.884	585.661	429.161	0.901	539.501	396.018	0.952	377.447	273.883	0.945	402.662	298.794
30-min	0.881	520.359	403.572	0.899	479.337	358.697	0.946	348.653	268.047	0.943	358.308	277.468
40-min	0.878	500.445	401.129	0.898	458.351	364.665	0.950	320.557	239.484	0.944	337.760	257.675
50-min	0.891	501.675	392.539	0.893	497.841	388.438	**0.957**	**315.900**	242.671	0.949	342.773	261.586
60-min	0.886	496.904	373.862	0.900	465.364	349.271	0.943	350.001	280.833	0.931	385.866	307.869

For the SHIP 1 (Table 2), the range of R2 was 0.873–0.889 for Random Forest, 0.882–0.899 for Extra Trees, 0.923–0.937 for LightGBM, and 0.918–0.929 for XGBoost. The LightGBM model outperformed the other models across all TA frequencies. At the 30-min aggregation interval, LightGBM achieved the highest R2 value of 0.937, with RMSE and MAE values of 387.900 and 304.115, respectively. This ensures that the 30-min TA interval is optimal for this dataset, providing the best balance between prediction accuracy and error metrics. XGBoost

Table 4. Evaluation of model performance in predicting fuel consumption of the SHIP 3 for different TA frequencies.

TA Frequency	Random Forest			Extra Trees			LightGBM			XGBoost		
	R2	RMSE	MAE	R2	RMSE	MAE	R2	RMSE	MAE	R2	RMSE	MAE
10-min	0.839	545.516	375.484	0.881	468.023	317.240	**0.937**	**340.648**	**171.743**	0.920	384.687	213.431
20-min	0.817	585.491	368.960	0.892	449.996	313.051	0.935	348.643	204.673	0.908	414.754	255.788
30-min	0.821	683.644	0516.420	0.891	533.948	438.801	0.927	434.874	231.314	0.910	483.470	295.337
40-min	0.820	629.542	441.457	0.886	500.572	359.665	0.922	412.911	236.382	0.914	435.016	266.868
50-min	0.816	647.328	478.064	0.862	561.102	397.303	0.922	420.956	232.917	0.907	459.441	275.836
60-min	0.819	697.595	517.053	0.884	556.959	441.525	0.926	445.564	213.778	0.912	486.270	284.522

Table 5. Evaluation of model performance in predicting fuel consumption of the SHIP 4 for different TA frequencies.

TA Frequency	Random Forest			Extra Trees			LightGBM			XGBoost		
	R2	RMSE	MAE	R2	RMSE	MAE	R2	RMSE	MAE	R2	RMSE	MAE
10-min	0.886	556.103	428.310	0.897	528.721	401.336	0.935	**417.807**	**182.445**	0.913	485.697	251.408
20-min	0.881	665.340	476.491	0.896	621.574	448.237	0.937	481.299	279.593	0.918	550.890	365.462
30-min	0.882	691.917	523.178	0.901	635.952	484.294	**0.949**	452.512	273.977	0.926	549.071	381.443
40-min	0.871	693.332	542.347	0.894	629.672	497.281	0.946	446.622	281.718	0.924	530.027	374.014
50-min	0.878	660.031	503.368	0.898	603.944	466.885	0.945	443.728	245.434	0.922	527.085	359.997
60-min	0.877	652.831	500.434	0.899	591.895	453.426	0.942	446.317	244.438	0.926	505.893	330.156

also performed well, particularly at the 30-min TA frequency, while RF showed the lowest overall performance evaluation.

For the SHIP 2 (Table 3), the range of R2 was 0.881–0.891 for Random Forest, 0.893–0.907 for Extra Trees, 0.943–0.957 for LightGBM, and 0.931–0.949 for XGBoost. LightGBM showed superior performance, particularly at the 50-min TA frequency, where it achieved an R2 of 0.957 and an RMSE of 315.900. Notably, the best MAE is observed at the 10-min TA frequency. The best TA frequency was preferentially selected with the highest coefficient of determination R2. Therefore, the 50-min aggregation frequency is deemed optimal. The RF model showed the lowest performance, especially at 30-min frequency.

For the SHIP 3 (Table 4), the range of R2 was 0.816–0.839 for Random Forest, 0.862–0.892 for Extra Trees, 0.922–0.937 for LightGBM, and 0.907–0.920 for XGBoost. LightGBM demonstrated robust performance across all TA frequencies, with the best results at the 10-min TA frequency, achieving an R2 of 0.937, RMSE of 340.648, and MAE of 171.743. This highlights the model's capability to handle more granular data effectively. XGBoost also showed strong performance at the 10-min TA frequency.

For the SHIP 4 (Table 5), the range of R2 was 0.871–0.886 for Random Forest, 0.894–0.901 for Extra Trees, 0.935–0.949 for LightGBM, and 0.913-0.926 for XGBoost. LightGBM achieved the highest R2 of 0.949 at the 30-min TA frequency, with an RMSE of 452.512, and MAE of 273.977. Interestingly, the best RMSE and MAE were observed at the 10-min TA frequency. However, the 30-min TA frequency is considered optimal based on the highest R2 score.

Overall, LightGBM predictive model emerged as the best-performing model for predicting ship fuel consumption across different TA frequencies and ships. It consistently achieved high R2 values and low error metrics. It is worth-noting that the optimal TA frequency varies from dataset to others with a progressive increase and decrease of predictive accuracy with greater temporal data aggregation periods. These findings emphasize the importance of an in-depth investigation of the inner working of the models for selecting appropriate TA frequencies to achieve accurate and reliable predictions.

3.2 Reliability of Predictive Models

As discussed in Sect. 2.4, we calculated the marginal contributions of the different attributes to the fuel consumption using the shapley additive explanations. Figures 2, 3, 4, and 5 illustrate the SHAP feature importance plots based on the LightGBM model for predicting ship fuel consumption at different temporal aggregation frequencies.

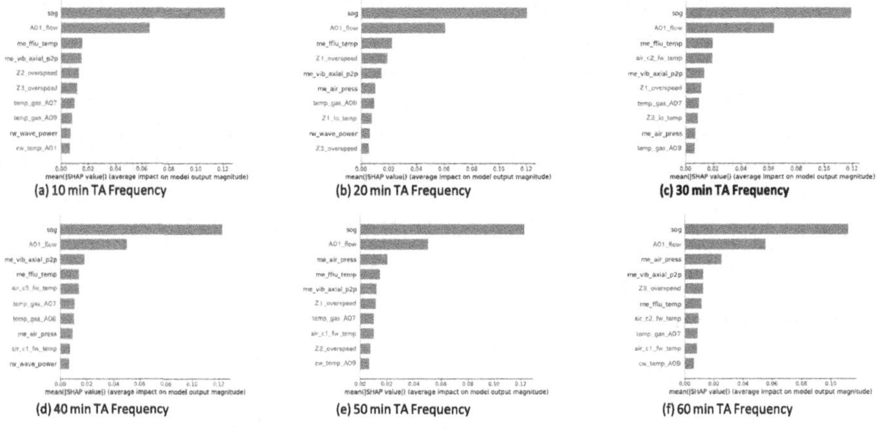

Fig. 2. Feature importance for the SHIP 1.

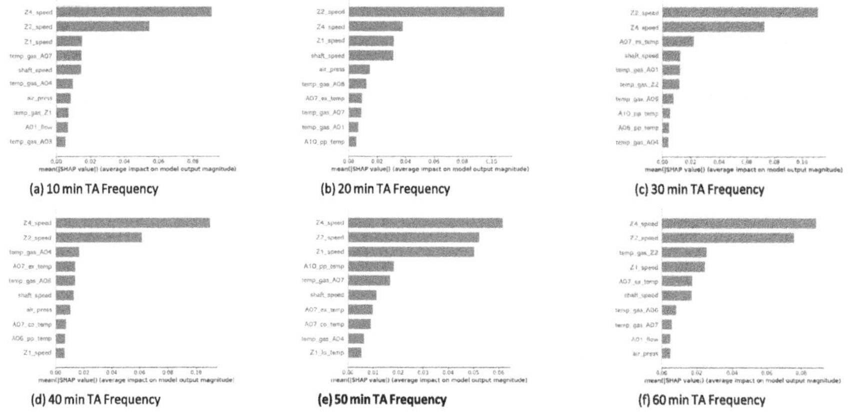

Fig. 3. SHAP feature importance for the SHIP 2.

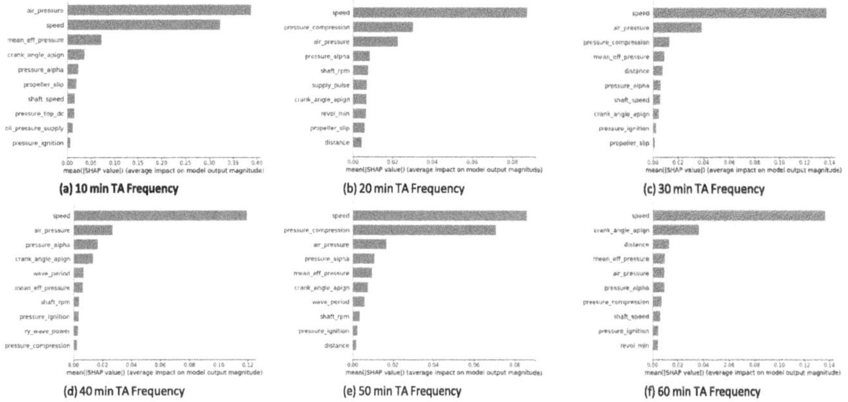

Fig. 4. SHAP feature importance for the SHIP 3.

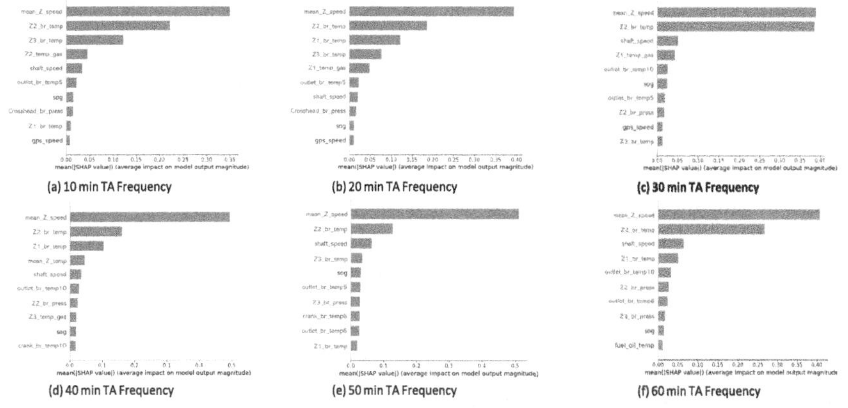

Fig. 5. SHAP feature importance for the SHIP 4.

For the SHIP 1 dataset (Fig. 2), across all TA frequencies, the feature "sog" (speed over ground) consistently emerges as the most significant predictor of fuel consumption. Its high mean SHAP values indicates its substantial contribution to the model's output. This highlights the critical role of speed in determining fuel usage, aligning with maritime operational knowledge. Another consistently important feature is "A01_flow" (flow of equipment A01), which ranks highly in all TA frequencies. This suggests that monitoring and managing the flow of this equipment is crucial for accurate fuel consumption predictions, as it closely relates to system performance and overall efficiency. In the 10-min TA frequency, features like "me_ffiu_temp" (foc fuel temperature) and "me_vib_axial_p2p" (axial vibration) show significant impact. In the 20-min TA frequency, "Z1_overspeed" (overspeed of equipment Z1) and "me_air_press" (air pressure) become more pronounced. In the 30-min TA frequency, which is selected as the optimal aggregation frequency, "air_c2_fw_temp" (air input temperature at equipment c2) gains importance, along with the previously significant features. This indicates that temperature-related parameters play a more critical role in the model's predictions. In the 40-min TA frequency, "A01_flow" and "me_vib_axial_p2p" maintain their high significance, similar to the 10-min TA frequency. However, "air_c3_fw_temp" also becomes relevant, highlighting the continued importance of certain temperature-related parameters. In the 50-min TA frequency, "me_air_press" and "me_ffiu_temp" are again prominent. In the 60-min TA frequency, the importance of features like "me_air_press" and "me_ffiu_temp" remains significant.

For the SHIP 2 (Fig. 3), there are 3 equipment Z[1], which almost always operate at the same speed. In the 10-min TA frequency data, the speeds of equipment Z ("Z1_speed", "Z2_speed", and "Z4_speed") consistently emerge as the most significant predictor of fuel consumption, however, with different impacts which contradicts the maritime operational knowledge. Besides the speeds of equipment Z, "temp_gas_A07" (exhaust gas temperature from equipment A07) and "shaft_speed" (shaft speed) show significant impact. In the 20-min TA frequency, the prominence of equipment Z speeds remains, with additional importance given to "shaft_speed" and "air_press" (air pressure). In the 30-min TA frequency, "A07_ex_temp" (temperature at equipment A07) becomes more pronounced. However, the speed of equipment Z1 does not appear among the 10 most important attributes for fuel consumption. The same behavior is observed in the predictive model based on the 40-min TA frequency data. At the 50-min TA frequency, which is selected as the optimal aggregation frequency (along with the 40-min TA frequency that showed similar R2 score but biased learning), the speeds of equipment Z ("Z1_speed", "Z2_speed", and "Z4_speed") consistently emerge as the most significant predictors of fuel consumption with the same marginal contribution which is confirmed by the maritime operational knowledge.

[1] Note that these 3 Z equipments play the same role and have the same impact on the fuel consumption.

For the SHIP 3 (Fig. 4), across all TA frequencies, the feature "speed" consistently emerges as the most significant predictor of fuel consumption. In the 10 and 20-min TA frequency, which are selected as the optimal aggregation frequencies for the predictive models with an R2 of 0.937 and 0.935 respectively, we observed that the models rely on completely different attributes. For example, the "air_pressure" attributes exhibits a marginal contribution of 0.38 in the 10-min TA, and an average of only 0.03 in the other TA frequencies. The same observation applies to the "speed" attribute and other attributes (e.g., "distance", "pressure_compression") that appear and disappear from one TA frequency to another.

For the SHIP 4 (Fig. 5), the predictive models show a nearly identical predictive R2 score of ~0.94 and across all TA frequencies. In these models, the feature "mean_Z_speed" (mean speed of equipment Z) consistently emerges as the most significant predictor of fuel consumption. Similarly, for the SHIP 2, SHIP 4 contains equipment Z that plays the same role and therefore the same impact on the fuel consumption. However, we observe that the "Z_br_temp" (bearing oil temperature of equipment Z) exhibit different marginal impacts on fuel consumption across the different TA frequencies. This observation contradicts established maritime operational knowledge. The model with a 30-min TA frequency demonstrates the highest R2 score but shows a clear bias. Conversely, the model with a 20-min TA frequency, ranked fifth with an R2 score of 0.937, is more reliable.

The feature importance analyses for the 04 datasets across different TA frequencies reveal consistent key predictors of fuel consumption. This analysis allowed us to identify the predictors that most significantly impact the fuel consumption. More specifically and interestingly, the study demonstrated that accuracy is no longer the sole criterion for identifying the most suitable TA frequency. Instead, it is crucial to delve deeper into the model inner working to verify that it relies on the correct predictors, and therefore identify the adequate TA frequency.

4 Conclusion

This study underscores the critical importance of temporal aggregation frequency in time series analysis, particularly in the context of predicting ship fuel consumption. Through an empirical experiment framework utilizing 04 multivariate time series datasets, we investigated the performance and the reliability of different ML models using various TA frequencies. Our findings reveal that the choice of TA frequency significantly influences the model performance. We observed that regardless of the TA period, time series aggregation consistently produced accurate results, underscoring its utility in simplifying data structure and reducing noise. However, our analysis also revealed that models may exhibit critical biases in particular TA frequencies, highlighting the necessity of deeper investigation into their inner workings to select the appropriate TA period. The findings from this study have practical implications for decision-makers in the maritime industry and beyond, offering guidelines for selecting the optimal frequency for enhanced predictive model accuracy and reliability. Moving forward, future research could explore additional ML algorithms and validation datasets.

References

1. Box, G.E.P., Jenkins, G.M., Reinsel, G.C., Ljung, G.M.: Time Series Analysis. Wiley Series in Probability and Statistics, 5th edn. Nashville, TN (2015)
2. Dakos, V., et al.: Methods for detecting early warnings of critical transitions in time series illustrated using simulated ecological data. PLoS ONE **7**(7) (2012). https://doi.org/10.1371/journal.pone.0041010
3. García, S., Luengo, J., Herrera, F.: Data Preprocessing in Data Mining. Springer, Cham (2015). https://doi.org/10.1007/978-3-319-10247-4
4. Garouani, M., Bouneffa, M.: Unlocking the black box: towards interactive explainable automated machine learning. In: Quaresma, P., Camacho, D., Yin, H., Gonçalves, T., Julian, V., Tallón-Ballesteros, A.J. (eds.) IDEAL 2023. LNCS, vol. 14404, pp. 458–469. Springer, Cham (2023). https://doi.org/10.1007/978-3-031-48232-8_42
5. Idrees, S.M., Alam, M.A., Agarwal, P.: A prediction approach for stock market volatility based on time series data. IEEE Access **7**, 17287–17298 (2019). https://doi.org/10.1109/access.2019.2895252
6. Lam, H.C., Li, A.M., Chan, E.Y., Goggins, W.B.: The short-term association between asthma hospitalisations, ambient temperature, other meteorological factors and air pollutants in Hong Kong: a time-series study. Thorax **71**(12), 1097–1109 (2016). https://doi.org/10.1136/thoraxjnl-2015-208054
7. Lundberg, S.M., Lee, S.I.: A unified approach to interpreting model predictions. In: Advances in Neural Information Processing Systems, vol. 30 (2017). https://doi.org/10.48550/ARXIV.1705.07874
8. Molnar, C.: Interpretable machine learning. Lulu.com, Morrisville, NC (2020). https://christophm.github.io/interpretable-ml-book/
9. Petrucci, A., Barone, G., Buonomano, A., Athienitis, A.: Modelling of a multi-stage energy management control routine for energy demand forecasting of smart communities using a recurrent neural network. Energy Convers. Manag. **268**, 115995 (2022). https://doi.org/10.1016/j.enconman.2022.115995
10. Prakash, R., Desai, J., Piplani, R.: An optimal data-splitting algorithm for aircraft sequencing on a single runway. Ann. Oper. Res. **309**(2), 587–610 (2021). https://doi.org/10.1007/s10479-021-04351-2
11. Rostami-Tabar, B., Babai, M.Z., Ali, M., Boylan, J.E.: The impact of temporal aggregation on supply chains with arma(1, 1) demand processes. Eur. J Oper. Res. 920–932 (2019). https://doi.org/10.1016/j.ejor.2018.09.010
12. Rostami-Tabar, B., Goltsos, T.E., Wang, S.: Forecasting for lead-time period by temporal aggregation: whether to combine and how. Comput. Ind. **145**, 103803 (2023). https://doi.org/10.1016/j.compind.2022.103803
13. Seithe, G.J., Bonou, A., Giannopoulos, D., Georgopoulou, C.A., Founti, M.: Maritime transport in a life cycle perspective: how fuels, vessel types, and operational profiles influence energy demand and greenhouse gas emissions. Energies **13**(11), 2739 (2020). https://doi.org/10.3390/en13112739
14. Zhou, Y., et al.: Predicting ship fuel consumption using a combination of metocean and on-board data. Ocean Eng. **285**, 115509 (2023). https://doi.org/10.1016/j.oceaneng.2023.115509

Topic Modeling in Telegram Channels During the Russia-Ukraine Conflict

Arsenii Tretiakov, Sergio D'Antonio-Maceiras, and Alejandro Martín(✉)

Department of Computer System Engineering, Universidad Politécnica de Madrid, Calle de Alan Turing, 28031 Madrid, Spain
a.tretiakov@alumnos.upm.es, {sergio.dantonio,alejandro.martin}@upm.es

Abstract. Telegram has become a preferred platform for far-right activism, conspiracy theories, political propaganda, and misinformation, each with its own target audience. This study investigates the application of multilanguage machine learning techniques for extracting topics from political content on Telegram channels. This research improves the understanding of political information and narratives in different languages on Telegram over time, contributing to the study of digital communication and information warfare within the content of the Telegram channels sphere. Through the exploitation of an extensive dataset on the subject, this work contributes to the state-of-the-art of topic analysis by providing information on prevailing discourses, including the escalating of armed conflicts between Russian and Ukraine, and political tensions in Europe and the USA.

Keywords: Telegram · topic modelling · BERTopic · social networks · propaganda · far-fight activism

1 Introduction

Telegram is currently the fastest-growing social media platform, featuring a wide array of channels on various topics, including political discourse. It is popular not only in post-USSR countries, but also in the Middle East, Europe, and the USA, with more than 800 million monthly active users worldwide and significant numbers from India, Russia, Indonesia, the USA, and Brazil[1]. According to studies on far-right activity on Telegram [5,24], these types of channel can rank from 5,000 to 300,000 subscribers or more, while some channels have more than 1M followers. Among them, Telegram has become one of the preferred platforms for European and American far-right activism, conspiracy theories, anti-immigrant, anti-feminism discourses, and political extremism among others. [10]. Such groups of channels form an "echo chamber" [16], which can be defined as the far-right core community that has been identified, where continuous reciprocal interactions reinforce shared political biases [7]. In modern information warfare, different governments or political movements can strategically

[1] https://worldpopulationreview.com/country-rankings/telegram-users-by-country.

V. Julian et al. (Eds.): IDEAL 2024, LNCS 15346, pp. 493–504, 2025.
https://doi.org/10.1007/978-3-031-77731-8_44

use Telegram to disseminate information to domestic and international audiences who are less engaged with traditional media sources [25]. This platform, especially through its political and news channels, has become a central hub for sharing targeted information, managed by individual or multiple administrators, and is distinguished by its high viewership and citation rate [23]. Telegram has notably evolved into a key medium for broadcasting updates on political events and actions in Ukraine, Russia, Europe, and the USA.

The messenger remains understudied in computer science, with fewer studies on its content and dynamics compared to Twitter or Facebook. Current research faces limitations such as language specificity or focus on particular social-political phenomena, restricting comparative analysis. Furthermore, researchers often use a single approach rather than multiple methodologies [11]. The relevance of the study lies in analysing the diversity of Telegram channels to examine how political content and propaganda evolve across different languages, affiliations, and time periods. Using a fine-tuned BERTopic model and data analysis methods, it assesses the impact of the platform on public opinion, social polarization, its role in modern information warfare, and political mobilisation, addressing Telegram's understudied nature in digital communication research [22].

The paper is structured as follows: in Related Work, the current state of research in the topic modeling on Telegram and social networks is introduced; section Methodology details the methodology employed in this research; the description of the pre-trained fine-tuned BERTopic model and data collection methods are outlined; in Results, the performance of the proposed method is presented, offering a comprehensive analysis of the data and its implications; Conclusions discusses findings within the broader field, drawing connections to existing literature and providing recommendations for further study.

2 Related Work

Although, as we have mentioned, Telegram is a less studied social network compared to others, several works can be found that apply topic modelling to understand the complexity and discourse structure inside this social platform.

Simon et al. [28] applied topic modelling of 362,837 documents to examine the main communities within the Dutch Telegramsphere. The topics were extracted from the Telegram dataset using BERTopic to provide word clusters and OpenAI GPT-4 for natural language interpretations; the dataset was composed by 7,711,975 messages from 10,633 channels by 335,088 unique users (29.9% bots and 70.1% humans) [20] about COVID disinformation. Zehring & Domahidi [31] applied Structural Topic Modelling to 2,497,679 Telegram messages from German anti-COVID and far-right movements, the study found that the Querdenken movement on Telegram is deeply linked to far-right, QAnon, and alternative media channels, mainly discussing coronavirus conspiracy theories, right-wing populism, and QAnon-related topics.

The Italian and English networks were built and key communities revealed, with their narratives characterised through topic modelling using Anchored Correlation Explanation; the dataset includes 1,980 chats with a total of 8.4 million messages [1]. Sprenkamp et al. [29] developed R2G ("Refugees to Government"), an application using state-of-the-art topic modelling with BERTopic to identify Ukrainian refugee needs from 170,915 messages in 33 Telegram groups to capture text relationships in Ukrainian, Russian, German, and English. Morrell [18] mapped discourse of Russian extremist communities on Telegram, which were identified using LDA and comparing Louvian and Girvan-Newman detection methods, based on 2,731 filtered messages from 19 far-right channels.

BERTopic's performance with various embedding models in preprocessed text was compared [17], revealing its ability to generate meaningful topics from a dataset of 3,286 Serbian tweets about COVID-19 vaccine hesitancy, containing 15 broad topics. Entity and central centroid-based incremental clustering method was empirically shown to outperform existing approaches [21], achieving robust results in processing 500,000 messages on the Twitter Events2012 dataset[2]. Ustyianovych et al. [30] collected text data from VKontakte and performed topic modelling using MLflow with Splunk algorithms. Finally, Russian information narratives from pro-Russian websites were analysed across 5.37 million Reddit comments and 1056 topics identified using the clustering and keyword extraction by BERTopic [9].

The predominance of BERTopic is not surprising. Since its launch in 2022 it has been widely used for its ability to process complex large-scale multilingual textual data in an unsupervised context, for its flexibility and ability to maintain topic coherence over time [26], and has proved to give better results compared with LDA, NMF, and top2vec [4]. Therefore, in this work, the library, with some differential specifications is the one used.

3 Methodology

3.1 Telegram Data

As mentioned before, Telegram channels may be private or public, allowing only administrators to publish posts. Private chats, groups, and channels can interact through message forwarding, displaying the original author [12] and reactions. Reactions to the posts are various emoji that the channel's admin allows users to leave under the post. Telegram API was used to collect publications and their metadata from channels. The collection process generated JSON files containing various media data types, which were stored in MongoDB.

In order to find channels for the dataset, a diverse range of open sources were utilised. This includes references to primary media outlets within each country, government institutions, and political organisations and figures. Channel ratings from platforms TgStat[3] and Tlgrm[4], partner channel lists, and forwards from other sources were leveraged. Academic articles on this topic also served as sources [15].

[2] https://figshare.com/articles/dataset/Twitter_event_datasets_2012-2016_/5100460/2.

[3] https://tgstat.ru/.

[4] https://tlgrm.ru/channels.

In the computational collection and analysis of Telegram data, it is important to note potential biases from the deletion of messages and entire channels [3]. The dataset consists of 1108 distinct channels in Russian (53.29% of the total dataset), English (16.45%), Ukrainian (10.02%), Spanish (8.88%), German (4.45%), Italian (2.71%), Portuguese (2.30%), and French (1.90%). The corpus of messages covers from 2015 until 1st July 2023. After preprocessing, it comprises 15,994,793 messages, reduced from 18,786,143. Further detailing the dataset's composition, channels were systematically classified and labeled based on their political orientation and biases (Table 1). Russian-language channels were categorized as pro-Kremlin, anti-Kremlin (opposition), pro-Ukrainian, and neutral. The neutral category includes media and news outlets without discernible political bias. In English, Spanish, French, and other languages, channels are categorized as far-right, pro-Kremlin, and traditional media outlets.

Table 1. Number of messages in the Telegram dataset by languages and political views.

	Far-Right	Pro-Kremlin	Pro-Ukrainian and Anti-Kremlin	Traditional Mass Media and Neutral
English	1,835,023	492,387	27,766	244,488
French	148,742	157,992	-	-
German	510,730	220,459	-	1,265
Italian	122,417	284,277	-	-
Portuguese	285,978	77,114	12,049	2,347
Russian	-	5,782,860	2,070,127	657,779
Spanish	526,024	132,047	20,495	779,778
Ukrainian	-	3,349	1,599,300	-

3.2 Topic Modelling

Following best practices, incremental learning ("Online Topic Modelling") was used to continuously adapt to new data [6]. This approach minimizes memory usage, integrates new data without full retraining, and dynamically identifies new topics[5]. The framework (Fig. 1) includes the BERT Embedding model, Incremental PCA (IPCA) as Dimension Reducer, MiniBatchKMeans as Clustering Engine, Vectorizer Model, and KeyBERT representation model.

Preprocessing Data. The multilingual corpus of Telegram messages poses challenges in data preparation due to the variety of languages, document sizes, and content diversity. Initially, all documents were normalized and cleaned to

[5] https://maartengr.github.io/BERTopic/getting_started/online/online.html.

remove emojis, URLs, and irrelevant languages, e.g. Arabic or Hebrew. Documents were lemmatized to handle the grammatical complexities of Russian and Ukrainian, ensuring the model does not mistake similar words with different endings as distinct. Preprocessing also involved removing empty messages, duplicates, and messages with 5 words and fewer.

Model Application. The model initiates the process by extracting embeddings using pre-trained models, such as Sentence-Transformers, then reduces dimensional to enhance clustering. The process concludes by organising embeddings into clusters that represent potential topics. The "paraphrase-multilingual-MiniLM-L12-v2" embedding model is selected for its proficiency in generating semantically wide representations across more than 50 languages (including Ukrainian). This model maps sentences to a 384-dimensional dense vector space and it has 117M parameters.

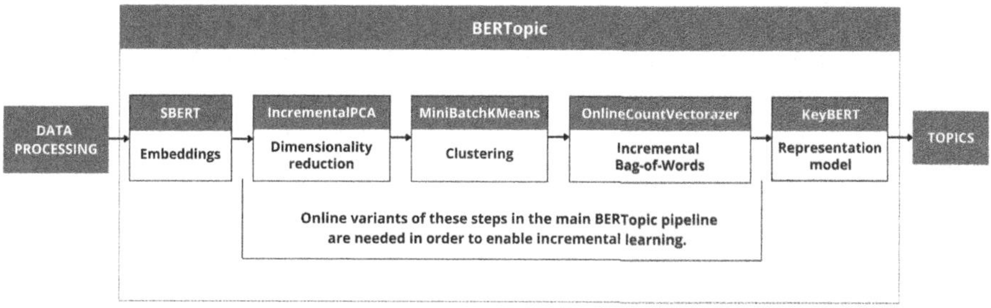

Fig. 1. Workflow and details of our BERTopic architecture.

OnlineCountVectorizer pre-processes documents by transforming text into a token count matrix. Filters, including a minimum document term frequency of 5 and a maximum document frequency of 50%, focus on relevant terms. A decay factor of 0.01 adjusts term significance based on occurrence rate. Stop words from the SpaCy library eliminate less informative terms. The IPCA reduces embeddings to 50 principal components, capturing significant data variance while being memory efficient for handling 15.9M messages and avoiding topic overlap. IPCA maintains continuity of principal component values over sequential updates, ensuring data consistency [14]. The MiniBatchKMeans algorithm, optimised for large datasets through stochastic gradient descent, was used for clustering. The elbow method determined 150 as the optimal number of clusters by identifying a point where the decrease in within-cluster sum of squared errors slows down significantly [27]. Using a fixed zero random state ensures replicability[6]. While MiniBatchKMeans might slightly underperform in cluster tightness, it offers significant speed and scalability advantages [19]. In an online

[6] https://scikit-learn.org/stable/modules/generated/sklearn.cluster.MiniBatchKMeans.html.

topic modeling, document chunks were processed iteratively, updating the topic model to maintain relevance and accuracy. A class-based TF-IDF procedure maintains topic coherence over time, allowing the model to adapt to new data in dynamic settings, and after training, the fine-tuned KeyBERTInspired representation model was employed with the CountVectorizer to filter out stop words [8].

Model Evaluation. Extracted topics were evaluated based on keyword frequency, Topic Coherence, and Topic Diversity. The Coherence Model, using the 'c_v' measure from the Gensim library, assesses semantic similarity between high-scoring words to gauge topic quality. Coherence scores were aggregated for an average across Telegram channels. The diversity score, calculated as the ratio of unique words to total words across topics, provides insight into the model's ability to generate meaningful and distinct topics.

4 Results

Results show a Coherence Score of 0.646, indicating a medium level of semantic similarity between topics, suggesting that the defined topics are meaningful and the words within each topic likely share a common context. 150 thematic clusters were obtained, which were received in different languages, predominantly Russian and Ukrainian, the results were translated with Deepl translator library. Additionally, the Topic Diversity score of 0.485 reflects a moderate level of uniqueness across the topics. Moreover, as the number of clusters decreases, the topic diversity metric, which indicates the variety of topics in the trained model, also diminishes. The 12 most representative topics represent 20.02% of all dataset (Fig. 2), it demonstrates the most frequent terms in largest topics, indicating weight of each term in the topic. These topics are repetitive and concern the politics of Ukraine and Russia, the details of the armed conflict. Table 2 shows examples of Telegram's messages with information about views and reactions.

Figure 3 shows the most popular topics and their frequency ratio by political view for each language. Pro-Kremlin topics are predominantly discussed in Russian sources and they are also prevalent in German and French sources, indicating cross-linguistic influence. Topic 2, one of the most frequent, mostly appears in Russian Pro-Kremlin sources. Neutral and Russian opposition topic 64 ('russia', 'ukraine', 'moscow', 'kazakhstan', 'belarus', 'detention') mentions political arrests in CIS countries (Kazakhstan and Belarus). Pro-Ukraine topics in Russian and Ukrainian show the most frequent topic 89. Pro-Kremlin topic 139 highlights news of missile or drone attacks on Ukrainian and Russian cities ('moscow', 'kharkiv', 'ukraine', 'road', 'report', 'vsu'[7], 'shelling'), focusing on specific cities and events.

Topics in Far-Right and Traditional Mass Media (Traditional) demonstrate a diverse range of interests. Topic 18 ('fbi', 'trump', 'president', 'indictment')

[7] Ukrainian Army.

Table 2. Examples of Telegram messages and their statistics

Topic	Message	Views	Reactions
89	I have awarded the title of Hero of Ukraine to Oleksandr Matsievskyi. A man who will be known by all Ukrainians for his bravery and for his "Glory to Ukraine!" - Zelensky.	1186431	31244
2	After yesterday the Russian Defence Ministry officially confirmed the information of the head of the PMC "Wagner" about the capture of Bakhmut, and the Kremlin reported that Vladimir Putin congratulated the PMC and the Russian Armed Forces on this event, Ukraine does not recognise the loss of the city.	13899	811
122	Russia's president, Vladimir Putin, was the fourth highest rated by the same Americans surveyed, a remarkable fact given the Western media campaign against the Eurasian country in the wake of the conflict in Ukraine.	10389	725
74	The military administration told how many missiles were launched at Zaporizhzhia: one of the missiles got stuck in the roof - inclusion.	6576	135
13	Putin's "cook" denies Kremlin's claims about Russia's war with NATO in Ukraine - ISW. He noted that Russia is fighting "exclusively with Ukrainians" equipped with NATO-provided equipment and some "Russophobic" mercenaries who voluntarily support Ukraine, but not with NATO itself.	403663	10875
110	The composition of Russian assault units is based on an analysis of Russian military doctrine.	60864	1373
64	A court in Kazakhstan has arrested Dmitry Kozak, a 19-year-old Russian anarchist detained at Moscow's request.	19410	251
104	Putin answered the question that a lot of people have been asking for the last year: Why didn't Russia start the SMO back in 2014, when the Ukrainian army was weak.	5940	231
103	"Of course, you can build abstruse schemes, as the intellectual Shepherd does. But life is banal: Russia is a post-Soviet country. The collapse of the Soviet Union had a delayed result. Take into account that the Ukrainian and Russian parties ruled the USSR and competed with each other. What Yeltsin avoided, Putin could not avoid."	72083	1117
31	Taliban bring armoured vehicles to the border with Iran. Local sources say that tanks, light armoured vehicles and large detachments of the Taliban movement are concentrated in the Eslam-Kale border crossing area on the border between Iran and Afghanistan.	4809	124
58	Putin is working in the Kremlin today despite yesterday's drone attack. There are no changes in his schedule, Peskov said. The press secretary said that the president remains calm, collected and clear in his assessments and commands against the background of the drone attack. No special address by Putin to Russians is planned.	217152	1870
116	The plane carrying Chinese President Xi Jinping has landed at Moscow's Vnukovo airport. The Chinese leader will stay in the capital for three days: he is expected to have several meetings with Russian President Vladimir Putin and members of the government.	8120	279

is prevalent in Far-Right English sources and topic about elections is in Traditional Mass Media. Vaccine discussions topic during the COVID-19 pandemic is relevant for Far-Right German and French channels. Topics 108 ('republican', 'democrat', 'political', 'election', 'voter') and 68 ('tweet', 'clip', 'protest',

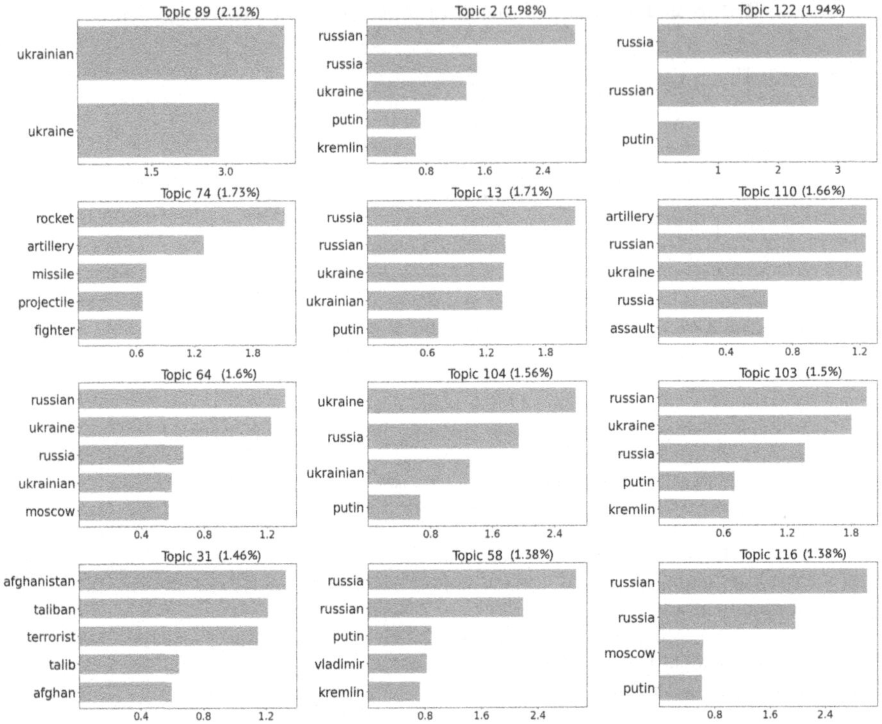

Fig. 2. Word Scores of the Most Frequent Topics and Their Percentage of the Entire Dataset.

'patriotfront'[8], 'social', 'post', 'peace', 'activist', 'trump') are primarily found in Italian and Portuguese sources within the Far-Right, indicating regional interests in right-wing politics.

The Fig. 4 changes in the popularity of topics and their growth in scope (frequency of mentions in documents) over time by quarter. It demonstrates a significant rise in discussions related to Russia and Ukraine, particularly peaking in early 2022, which corresponds with the onset of the conflict. Topic 13 appears in Q1 2020, and topic 89 culminates in significant spikes in Q1 2022 and Q2 2022. Similarly, topics 74 and 110 associated with 'missile', 'shell', 'artillery', 'fighter', 'ammunition' surged in Q2 2022, reflecting the escalation of military actions in Ukraine. Discussions on 'coronavirus' and related terms dominated earlier periods, particularly in 2020, indicating the global impact of the pandemic. Moreover, the figure highlights trump' in Q1 2021, which was relevant because of the growth of the QAnon movement [13]. Other topics, such as 'inflation' and 'economy', influenced by economic repercussions of the pandemic and geopolitical tensions.

[8] Patriot Front is an American white supremacist and neo-fascist hate group.

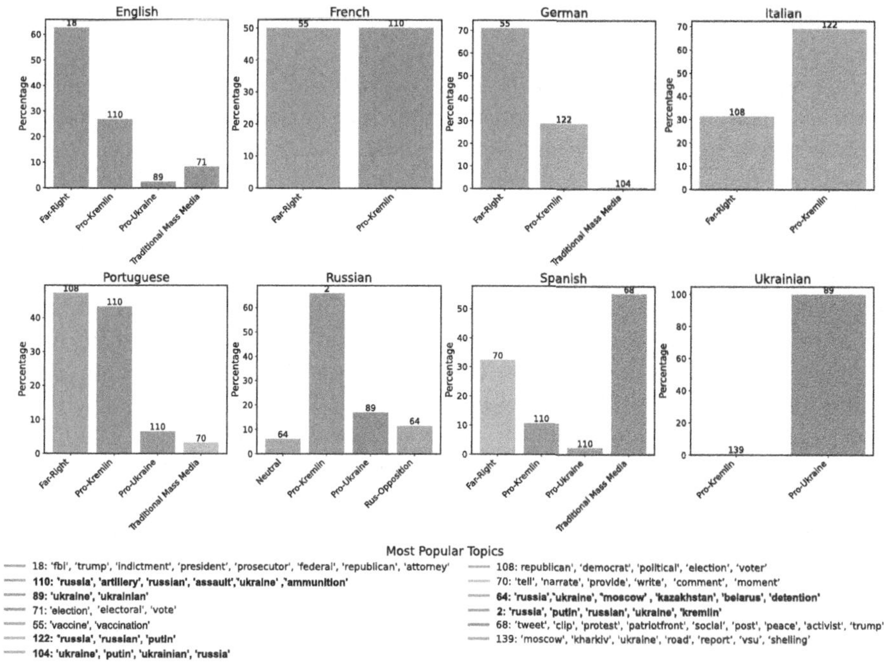

Fig. 3. The most popular topics on languages and political views.

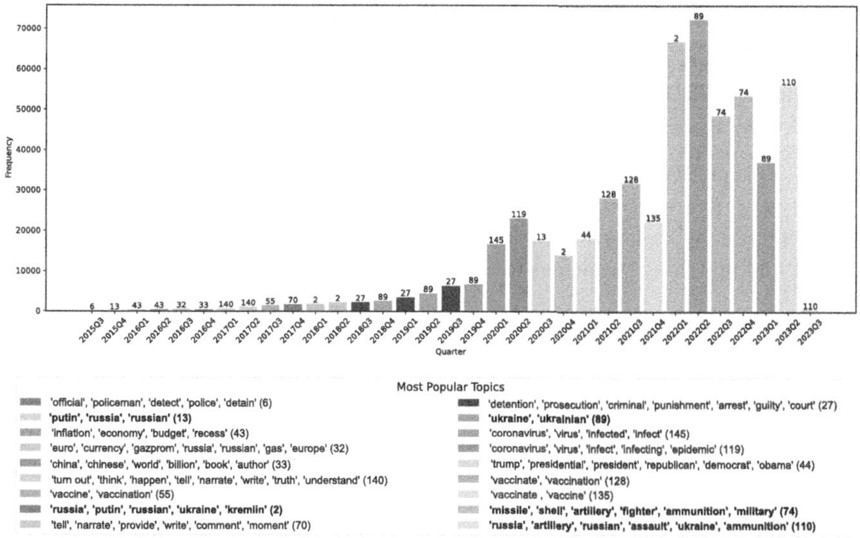

Fig. 4. Most Popular Topic by Quarter.

5 Conclusions

In this paper, BERTopic was utilized to analyze and categorize topics in Telegram channels, providing insights into the distribution and characteristics of content across various topics and time periods. The BERTopic analysis revealed a medium level of semantic similarity and a moderate level of uniqueness across 150 identified topics, encompassing key themes such as the Russian-Ukrainian political and armed conflict and the confrontation between Trump and Biden. The analysis demonstrated how the essence of posts in Telegram channels varied. This comprehensive approach allowed to obtain of the frequency and popularity of topics. Despite these promising results, limitations were noted, including the over-representation of Russian language content and the focus on specific topics like the armed conflict between Russia and Ukraine. One possible consequence of this is overlapping clusters, that can also be a consequence of applying the K-means method [2]. On the other side, the occurrence of similar topics in BERTopic models is due to the high granularity which allows the model to capture subtle nuances and variations within the data, enhancing the ability to explore and understand the information more comprehensively.

Future research should aim to extend the language segments of Telegram channels, include additional metadata for a more thorough data analysis, and application of multi-language pre-trained classification machine learning models, to detect and measure toxicity in political content on Telegram channels. Further validation with larger and more diverse datasets will be essential to generalize these findings and enhance the robustness of the analysis.

Acknowledgments. This work has been funded by Grant PLEC2021-007681 (XAI-DisInfodemics) and PID2020-117263GB-100 (FightDIS) funded by MCIN/AEI/10.13039/501100011033 and by "ERDF A way of making Europe", by the "European Union" or by the "European Union NextGenerationEU/PRTR", by grant PCI2022-134990-2 (MARTINI) of the CHISTERA IV Cofund 2021 program, funded by MCIN/AEI/10.13039/ 501100011033 and by the "European Union NextGenerationEU/PRTR", by Calouste Gulbenkian Foundation, under the project MuseAI - Detecting and matching suspicious claims with AI.

References

1. Alvisi, L., Tardelli, S., Tesconi, M.: Unraveling the Italian and English telegram conspiracy spheres through message forwarding (2024)
2. An, Y., Oh, H., Lee, J.: Marketing insights from reviews using topic modeling with BERTopic and deep clustering network. Appl. Sci. **13**(16), 9443 (2023)
3. Buehling, K.: Message deletion on telegram: affected data types and implications for computational analysis. Commun. Methods Measur. 1–23 (2023)
4. Egger, R., Yu, J.: A topic modeling comparison between LDA, NMF, Top2Vec, and BERTopic to demystify twitter posts. Front. Sociol. **7**, 886498 (2022)
5. Gallagher, A., O'Connor, C.: Layers of lies: a first look at Irish far-right activity on telegram. Inst. Strategic Dialogue **12** (2021)

6. Gómez-Pedrero, J.A., Estrada, J.C., Alonso, J., Quiroga, J.A., Vargas, J.: Incremental PCA algorithm for fringe pattern demodulation. Opt. Express **30**(8), 12278–12293 (2022)
7. Grindrod, P., Bovet, A.: Organization and evolution of the UK far-right network on telegram. Appl. Netw. Sci. **7** (2022)
8. Grootendorst, M.: BERTopic: neural topic modeling with a class-based TF-IDF procedure. arXiv preprint arXiv:2203.05794 (2022)
9. Hanley, H.W., Kumar, D., Durumeric, Z.: Happenstance: utilizing semantic search to track Russian state media narratives about the Russo-Ukrainian war on reddit. In: Proceedings of the International AAAI Conference on Web and Social Media, vol. 17, pp. 327–338 (2023)
10. Hoseini, M., Melo, P., Benevenuto, F., Feldmann, A., Zannettou, S.: On the globalization of the QAnon conspiracy theory through telegram. In: Proceedings of the 15th ACM Web Science Conference, pp. 75–85 (2023)
11. Khaund, T., Hussain, M.N., Shaik, M., Agarwal, N.: Telegram: data collection, opportunities and challenges. In: Lossio-Ventura, J.A., Valverde-Rebaza, J.C., Díaz, E., Alatrista-Salas, H. (eds.) SIMBig 2020. CCIS, vol. 1410, pp. 513–526. Springer, Cham (2021). https://doi.org/10.1007/978-3-030-76228-5_37
12. La Morgia, M., Mei, A., Mongardini, A.M.: TGDataset: a collection of over one hundred thousand telegram channels. arXiv preprint arXiv:2303.05345 (2023)
13. Lee, C.S., Merizalde, J., Colautti, J.D., An, J., Kwak, H.: Storm the capitol: linking offline political speech and online twitter extra-representational participation on QAnon and the January 6 insurrection. Front. Sociol. **7**, 876070 (2022)
14. Lippi, V., Ceccarelli, G.: Incremental principal component analysis exact implementation and continuity corrections. arXiv preprint arXiv:1901.07922 (2019)
15. Macdonald, S., et al.: The European far-right online: an exploratory twitter outlink analysis of German & French far-right online ecosystems (2022)
16. McLuhan, M., Gordon, W.: Understanding Media: The Extensions of Man. Gingko Press (2003). https://books.google.es/books?id=m7poAAAAIAAJ
17. Medvecki, D., Bašaragin, B., Ljajić, A., Milošević, N.: Multilingual transformer and BERTopic for short text topic modeling: the case of Serbian. In: Trajanovic, M., Filipovic, N., Zdravkovic, M. (eds.) ICIST 2023. LNNS, vol. 872, pp. 161–173. Springer, Cham (2024). https://doi.org/10.1007/978-3-031-50755-7_16
18. Morrell, S.: Mapping extremist discourse communities on telegram: the case of the Russian imperial movement and its affiliates. APSA Preprints (2023)
19. Newling, J., Fleuret, F.: Nested mini-batch k-means. In: Advances in Neural Information Processing Systems, vol. 29 (2016)
20. Ng, L.H.X., Kloo, I., Clark, S., Carley, K.M.: An exploratory analysis of Covid bot vs human disinformation dissemination stemming from the disinformation dozen on telegram. J. Comput. Soc. Sci. 1–26 (2024)
21. Nguyen, S., Ngo, B., Vo, C., Cao, T.: Hot topic detection on Twitter data streams with incremental clustering using named entities and central centroids. In: 2019 IEEE-RIVF International Conference on Computing and Communication Technologies (RIVF), pp. 1–6. IEEE (2019)
22. Olsson Gardell, E.K., Wagnsson, C., Wallenius, C.: The evolving security landscape: citizens' perceptions of feminism as an emerging security threat. Eur. J. Secur. Res. **7**(1), 67–86 (2022)
23. Poldi, F., Alaphilippe, A., Jonusaite, R., Sessa, M.G.: Disinformation on telegram: research and content moderation policies (2022)

24. Puyosa, I., Ponce de Léon, E.: Understanding telegram's ecosystem of far-right channels in the US. Atlantic Council-Digital Forensic Research Lab (2022). https://tinyurl.com/p5rt2buu
25. Salikov, A., et al.: Telegram as a means of political communication and its use by Russia's ruling elite. Politologija **3**(95), 83–110 (2019)
26. Schäfer, K., Choi, J.E., Vogel, I., Steinebach, M.: Unveiling the potential of BERTopic for multilingual fake news analysis – use case: Covid-19 (2024). https://arxiv.org/abs/2407.08417
27. Shi, C., Wei, B., Wei, S., Wang, W., Liu, H., Liu, J.: A quantitative discriminant method of elbow point for the optimal number of clusters in clustering algorithm. EURASIP J. Wirel. Commun. Netw. **2021**, 1–16 (2021)
28. Simon, M., Welbers, K., Kroon, A.C., Trilling, D.: Linked in the dark: a network approach to understanding information flows within the Dutch Telegramsphere. Inf. Commun. Soc. **26**(15), 3054–3078 (2023)
29. Sprenkamp, K., Zavolokina, L., Angst, M., Dolata, M.: Data-driven governance in crises: topic modelling for the identification of refugee needs. In: Proceedings of the 24th Annual International Conference on Digital Government Research, DGO 2023, pp. 1–11. Association for Computing Machinery, New York (2023). https://doi.org/10.1145/3598469.3598470
30. Ustyianovych, T., Kasianchuk, N., Falfushynska, H., Fedushko, S., Siemens, E.: Dynamic topic modelling of online discussions on the Russian war in Ukraine. In: Proceedings of International Conference on Applied Innovation in IT, vol. 11, 2, pp. 81–89. Anhalt University of Applied Sciences (2023)
31. Zehring, M., Domahidi, E.: German corona protest mobilizers on Telegram and their relations to the far right: a network and topic analysis. Soc. Media+ Soc. **9**(1), 20563051231155106 (2023)

Structural and Semantic Data Layers in Time Series Analyses

Alexander Graß[1,2(✉)], Christian Beecks[1,3], and Stefan Decker[1,2]

[1] Fraunhofer Institute for Applied Information Technology FIT, Sankt Augustin, Germany
{alexander.gras,christian.beecks,stefan.decker}@fit.fraunhofer.de
[2] RWTH Aachen University, Aachen, Germany
[3] University of Hagen, Hagen, Germany

Abstract. Understanding the inherent complexity of temporal data is crucial for effective time series analytics. One dimension of complexity is the level of structural depth at which analysis methods operate. These levels range from entire time series collections down to individual sequences of reduced dimensionality and length. Complementary to this type of complexity, is the quantity and expressiveness of knowledge associated with time series data, including labels and other features that provide valuable information. Both, the structural as well as the semantic layer, define the suitability and effectiveness of different analysis methods. In this paper we introduce a conceptual framework to support the automated selection of analytical time series approaches. To this end, we specify a context-free grammar to describe hierarchies and compositions of time series data, while also defining different classes of semantic information, resulting in a data-specific classification of time series analysis methods. Along with a demonstration via concrete examples, we provide a discussion on challenges, opportunities and future work associated with the proposed approach.

Keywords: Time Series · Data Analytics · Knowledge Discovery

1 Motivation

Time series analysis is a common task across various domains [1,3,4], but a generation of valuable insights from time series data is particularly demanding for companies without in-house experts. While intelligent solutions exist, to assist users with the selection and optimization of methods for predefined analysis tasks [6], automated and task-agnostic knowledge discovery remains a challenge [11]. One important reason is the dependency between analysis methods and different hierarchical data levels, ranging from multiple time series collections down to individual subsequences of reduced dimensionality. Thus, a suitable method for univariate time series analysis may not necessarily be feasible for collections of multivariate time series data and vice versa. The structural complexity of the

V. Julian et al. (Eds.): IDEAL 2024, LNCS 15346, pp. 505–511, 2025.
https://doi.org/10.1007/978-3-031-77731-8_45

data implies the compatibility and relevance of different analytical approaches. As time series data can be further enriched with contextual information, including domain-specific facts and prior knowledge such as labels and features [2], this structural complexity is additionally amplified at a semantic level. With the availability of prior knowledge, the methodical preference might vary, given that the inclusion of knowledge can substantially contribute to an improvement in accuracy or explainability [9]. Although approaches for a systematic classification of analysis algorithms exist, they define solutions for predefined analysis tasks [8,10]. Nonetheless, in many scenarios it is not immediately evident what conclusions can be drawn from existing data settings. In this paper we introduce a conceptual framework to enable a mapping of analysis methods to time series specifications. The goal is to find data-compatible analysis methods in the scope of knowledge discovery. In Sect. 2 we therefore present a context-free grammar to define hierarchies and compositions of time series data and further discuss the additional impact of semantic information on algorithm and model selection. In Sect. 3 we demonstrate the applicability of our presented approach, before in Sect. 4 and 5 we conclude with a discussion on opportunities, challenges and future work.

2 Data Layers in Time Series Analyses

Before presenting a framework for a data-specific selection of analytical approaches, the next two subsections will discuss solutions to describe the complexity of both, the structural and semantic data layers of time series data. Although further layers can be defined, including algorithmic properties or technical criteria, we identified them as not data-specific or as derivable information. The reason is that further influential factors, such as the amount of data or the quantity of labeled data instances, can be subsumed from the structural and the semantic information.

2.1 Structural Layer

The structural layer defines all hierarchies and corresponding compositions of time series data. Independent of semantic context, this layer explains the data complexity resulting from combinations of different time series granularities. Since real-world scenarios can involve any composition of available time series data, in Listing 1.1 we introduce a context-free grammar to generate specific combinations of time series data across different hierarchical levels. These hierarchical levels range from sets of time series collections (STSC) and individual time series collections (TSC) down to single time series (TS), including multivariate time series (MTS) and univariate time series (UTS). A recursive relationship to derive sliced subsequences with respect to index and dimensionality is implicitly defined by the grammar. At the base level, each time series is specified by a tuple of associated information including data information (D), label information (L) and prior knowledge (K).

```
1 STSC → stsc(STSC, TSC) | stsc(TSC)
2 TSC → tsc(TSC, TS) | tsc(TS)
3 TS → MTS | UTS
4 MTS → mts(TS, TS) | mts(UTS) | mts(D, L, K)
5 UTS → uts(UTS, UTS) | uts(D, L, K)
```

Listing 1.1. Context-free grammar to represent different hierarchical levels and compositions of time series data. Lowercase symbols allow for path tracking. Terminal symbols are D, L and K representing semantic information.

The following is a line-by-line description of the grammar:

1 A set of time series collections consists of one or more time series collections.
2 A time series collection consists of one or more time series.
3 A time series is either multivariate or univariate.
4 A multivariate time series can either be derived to a set of multivariate and/or univariate subsequences, to a single univariate time series or it's concrete data information.
5 A univariate time series can be derived to a set of univariate subsequences or it's concrete data information.

An additional illustration is provided in Fig. 1. As the tuple (D, L, K) is defined as semantic information, representing an adjacent layer of data complexity, it is discussed in the following subsection.

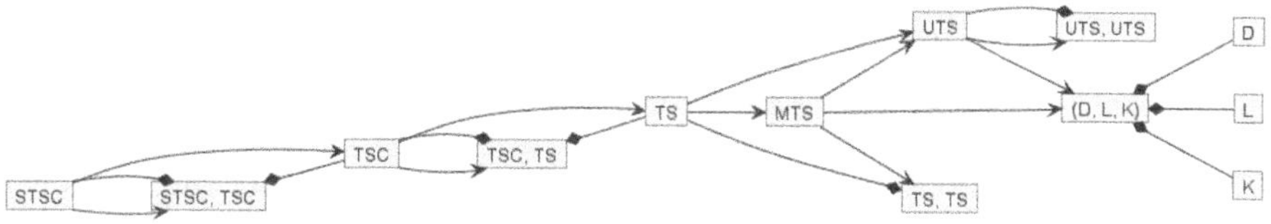

Fig. 1. Object diagram representing the grammar from Listing 1.1 as visual description of time series hierarchies. Arrows represent individual rules, while compositions are defined by diamond symbols.

2.2 Semantic Layer

Time series data exhibit temporal dependencies causing variations of characteristic elements such as seasonality, trend and noise. This heterogeneity complicates the process of identifying appropriate analytical methods, as the effectiveness of different approaches can vary based on the unique characteristics of the time series to be analyzed. Hence, as prior knowledge describing these characteristics might be available in advance, multiple analytical approaches exist to make use of it [9]. Further influential factors to determine a potential use of particular analytical methods encompass label and data information. Although data information could be considered as structural details, they reflect semantic insights by means of attribute names and raw data information. Table 1 provides a summary of semantic classes and their use. To achieve automated algorithm and model selection, decomposed time series entities represented by their semantic information must also be evaluated according to algorithmic requirements.

Table 1. Summary of semantic information.

Class	Description	Use
Data	Information associated with raw data like index, dimensions and data types	Alignment of method-specific requirements and particularities
Labels	Information representing the ground truth	(Semi-)Supervised learning
Knowledge	Prior data-related information and/or domain-specific facts	Informed data analysis and machine learning

Validation of Algorithmic Requirements

Let $T_{i,S}$ represent the semantic information of all classes $S = \{D, L, K\}$ for the i-th time series sequence derived from the structural layer. Equivalently $T_{i,D}$, $T_{i,L}$ and $T_{i,K}$ correspond to all attributes of classes D, L and K.

A validation of algorithm-specific requirements can be defined as a logical expression V of distinct conditions C_k.

$$V = C_1 * C_2 * ... * C_m$$

with $*$ being a placeholder for the logical operations $\{\wedge, \vee, \neg\}$.

Example 1. Conditions for a query-based similarity search using the Euclidean distance between labeled time series entities could be defined as:

- $C_1 := \forall i, j, \text{ if } i \neq j : \text{len}(T_{i,D}[\text{index}]) = \text{len}(T_{j,D}[\text{index}])$
- $C_2 := \forall i, j, \text{ if } i \neq j : T_{i,D}[\text{dtype}] = T_{j,D}[\text{dtype}]$
- $C_3 := \forall i : T_{i,L} \neq \emptyset$

With $C_1 \wedge C_2 \wedge C3$ stating that all time series have to be of the same length (C_1) and data type (C_2) and only correspond to labeled data (C_3).

3 Schematic Selection of Time Series Analysis Methods

To demonstrate the applicability of the conceptual framework, we continue with an arbitrary scenario based on Example 1 from the previous section. Figure 2 shows a setting consisting of two distinct univariate time series collections. To keep the example concise we dispense the semantic class of data information and reduce the amount of semantic details. We assume that for an ongoing knowledge discovery process, we validate the applicability of different data analysis methods at derived levels of structural time series granularity. We further assume that in a

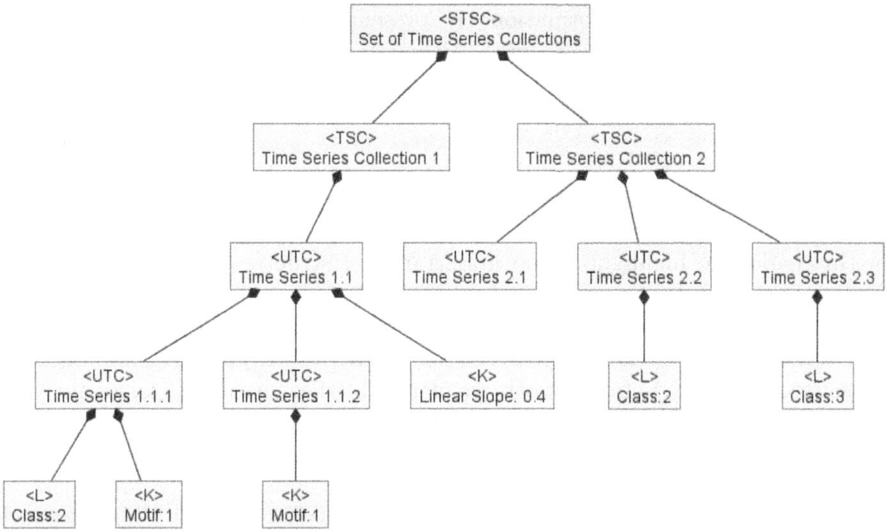

Fig. 2. Simplified visualization of a concrete data setting.

current iteration at level $\{TS, CTS\}$, we implicitly evaluate the question, whether the motif identified in time series 1.1 - represented by the subsequences 1.1.1 and 1.1.2 - can also be identified in collection 2. Although the approach defined in Example 1 is identified as one solution to this problem, condition C_3 of the semantic validation only allows for labeled data and we ignore time series 1.1.2 and time series 2.1. Additionally given a hypothetical difference in time series length, a validation of condition C_1 for the Euclidean distance fails. Provided that Dynamic Time Warping [7] is another available solution to query-based similarity search, which can be applied as a length-invariant distance measure, we finally compare the motif from time series 1.1.1 against all instances from time series collection 2 via Dynamic Time Warping.

While this is only one example for an automated selection of analytical methods, the next section will discuss further opportunities and related challenges.

4 Opportunities and Challenges

The here presented approach can be regarded a conceptual solution to support manual and automated knowledge discovery processes. By enabling a mapping from the space of time series compositions to potentially suitable analysis procedures, this approach can serve as a preliminary stage for AutoML methods. The automatic derivation of structural subsets and the semantic validation of methodical requirements, allows for a reduction of methods prior to a metric-based selection via evaluated analyses. Complementary to a potential utilization for individual analysis tasks, it could further enable a validation of entire pipelines [5]. As a storage of analytical findings from conducted tasks leads to an enrichment of semantic information, the presented solution is able to reuse this information for subsequent analyses. Although semantic information, including derived knowledge, can be arbitrarily complex, a more detailed classification would improve the process of validation and increase interpretability.

5 Conclusion and Future Work

In this paper we highlighted the importance of structural and semantic time series information for an automated selection of analytical methods. Apart from conceptual contributions, such as a context-free grammar to derive hierarchies of time series data, we demonstrated the applicability of the presented approach via a concrete example. As future work, we plan to further investigate on an ontology for a detailed classification of semantic information and efficient procedures to incorporate the presented framework in knowledge discovery processes.

Acknowledgment. This publication has received funding from the BMEL research and innovation programme under grant agreement No 28DK124E20 - KINLI. This paper reflects only the authors views and the commission is not responsible for any use that may be made of the information it contains.

References

1. Arul, M., Kareem, A.: Applications of shapelet transform to time series classification of earthquake, wind and wave data. Eng. Struct. **228**, 111564 (2021)
2. Christ, M., Braun, N., Neuffer, J., Kempa-Liehr, A.W.: Time series feature extraction on basis of scalable hypothesis tests (tsfresh-a python package). Neurocomputing **307**, 72–77 (2018)
3. Elgar, F.J., Stefaniak, A., Wohl, M.J.: The trouble with trust: time-series analysis of social capital, income inequality, and Covid-19 deaths in 84 countries. Soc. Sci. Med. **263**, 113365 (2020)
4. Elsaraiti, M., Ali, G., Musbah, H., Merabet, A., Little, T.: Time series analysis of electricity consumption forecasting using Arima model. In: 2021 IEEE Green Technologies Conference (GreenTech), pp. 259–262. IEEE (2021)
5. Graß, A., Beecks, C., Chala, S.A., Lange, C., Decker, S.: A knowledge graph for query-induced analyses of hierarchically structured time series information. In: Abelló, A., et al. (eds.) ADBIS 2023. CCIS, vol. 1850, pp. 174–184. Springer, Cham (2023). https://doi.org/10.1007/978-3-031-42941-5_16
6. Karmaker, S.K., Hassan, M.M., Smith, M.J., Xu, L., Zhai, C., Veeramachaneni, K.: AutoML to date and beyond: challenges and opportunities. ACM Comput. Surv. (CSUR) **54**(8), 1–36 (2021)
7. Müller, M.: Dynamic time warping. In: Information Retrieval for Music and Motion, pp. 69–84 (2007)
8. Shyam, R., Singh, R.: A taxonomy of machine learning techniques. J. Advancements Robot. **8**(3), 18–25 (2021)
9. Von Rueden, L., et al.: Informed machine learning-a taxonomy and survey of integrating prior knowledge into learning systems. IEEE Trans. Knowl. Data Eng. **35**(1), 614–633 (2021)

10. Ying, Y., Duan, J., Wang, C., Wang, Y., Huang, C., Xu, B.: Automated model selection for time-series anomaly detection. arXiv preprint arXiv:2009.04395 (2020)
11. Zhai, Y., Lv, Z., Zhao, J., Wang, W.: Knowledge discovery and variable scale evaluation for long series data. Artif. Intell. Rev. **56**(4), 3157–3180 (2023)

[illegible]

Author Index

V. Julian et al. (Eds.): IDEAL 2024, LNCS 15346, pp. 513–516, 2025.
https://doi.org/10.1007/978-3-031-77731-8

SPRINGER NATURE

GPSR Compliance

The European Union's (EU) General Product Safety Regulation (GPSR) is a set of rules that requires consumer products to be safe and our obligations to ensure this.

If you have any concerns about our products, you can contact us on ProductSafety@springernature.com

In case Publisher is established outside the EU, the EU authorized representative is:

Springer Nature Customer Service Center GmbH
Europaplatz 3
69115 Heidelberg, Germany

The manufacturer's authorised representative in the EU is Springer Nature Customer Service Centre GmbH, Europaplatz 3, 69115 Heidelberg, Germany. If you have any concerns regarding our products, please contact ProductSafety@springernature.com

Printed and bound by CPI Group (UK) Ltd, Croydon, CR0 4YY
30/09/2026
02225566-0019